Recent Titles in This Series

155 **William B. Jacob, Tsit-Yuen Lam, and Robert O. Robson, Editors,** Recent advances in real algebraic geometry and quadratic forms, 1994

154 **Michael Eastwood, Joseph Wolf, and Roger Zierau, Editors,** The Penrose transform and analytic cohomology in representation theory, 1993

153 **Richard S. Elman, Murray M. Schacher, and V. S. Varadarajan, Editors,** Linear algebraic groups and their representations, 1993

152 **Christopher K. McCord, Editor,** Nielsen theory and dynamical systems, 1993

151 **Matatyahu Rubin,** The reconstruction of trees from their automorphism groups, 1993

150 **Carl-Friedrich Bödigheimer and Richard M. Hain, Editors,** Mapping class groups and moduli spaces of Riemann surfaces, 1993

149 **Harry Cohn, Editor,** Doeblin and modern probability, 1993

148 **Jeffrey Fox and Peter Haskell, Editors,** Index theory and operator algebras, 1993

147 **Neil Robertson and Paul Seymour, Editors,** Graph structure theory, 1993

146 **Martin C. Tangora, Editor,** Algebraic topology, 1993

145 **Jeffrey Adams, Rebecca Herb, Stephen Kudla, Jian-Shu Li, Ron Lipsman, and Jonathan Rosenberg, Editors,** Representation theory of groups and algebras, 1993

144 **Bor-Luh Lin and William B. Johnson, Editors,** Banach spaces, 1993

143 **Marvin Knopp and Mark Sheingorn, Editors,** A tribute to Emil Grosswald: Number theory and related analysis, 1993

142 **Chung-Chun Yang and Sheng Gong, Editors,** Several complex variables in China, 1993

141 **A. Y. Cheer and C. P. van Dam, Editors,** Fluid dynamics in biology, 1993

140 **Eric L. Grinberg, Editor,** Geometric analysis, 1992

139 **Vinay Deodhar, Editor,** Kazhdan-Lusztig theory and related topics, 1992

138 **Donald St. P. Richards, Editor,** Hypergeometric functions on domains of positivity, Jack polynomials, and applications, 1992

137 **Alexander Nagel and Edgar Lee Stout, Editors,** The Madison symposium on complex analysis, 1992

136 **Ron Donagi, Editor,** Curves, Jacobians, and Abelian varieties, 1992

135 **Peter Walters, Editor,** Symbolic dynamics and its applications, 1992

134 **Murray Gerstenhaber and Jim Stasheff, Editors,** Deformation theory and quantum groups with applications to mathematical physics, 1992

133 **Alan Adolphson, Steven Sperber, and Marvin Tretkoff, Editors,** p-Adic methods in number theory and algebraic geometry, 1992

132 **Mark Gotay, Jerrold Marsden, and Vincent Moncrief, Editors,** Mathematical aspects of classical field theory, 1992

131 **L. A. Bokut′, Yu. L. Ershov, and A. I. Kostrikin, Editors,** Proceedings of the International Conference on Algebra Dedicated to the Memory of A. I. Mal′cev, Parts 1, 2, and 3, 1992

130 **L. Fuchs, K. R. Goodearl, J. T. Stafford, and C. Vinsonhaler, Editors,** Abelian groups and noncommutative rings, 1992

129 **John R. Graef and Jack K. Hale, Editors,** Oscillation and dynamics in delay equations, 1992

128 **Ridgley Lange and Shengwang Wang,** New approaches in spectral decomposition, 1992

127 **Vladimir Oliker and Andrejs Treibergs, Editors,** Geometry and nonlinear partial differential equations, 1992

126 **R. Keith Dennis, Claudio Pedrini, and Michael R. Stein, Editors,** Algebraic K-theory, commutative algebra, and algebraic geometry, 1992

125 **F. Thomas Bruss, Thomas S. Ferguson, and Stephen M. Samuels, Editors,** Strategies for sequential search and selection in real time, 1992

(*Continued in the back of this publication*)

Dedicated to Professor Donald Dubois

CONTEMPORARY MATHEMATICS

155

Recent Advances in Real Algebraic Geometry and Quadratic Forms

Proceedings of the RAGSQUAD Year,
Berkeley, 1990–1991

William B. Jacob
Tsit-Yuen Lam
Robert O. Robson
Editors

American Mathematical Society
Providence, Rhode Island

The papers in this volume grew out of a year long program in Real Algebraic Geometry and Quadratic Forms held at the University of California at Berkeley, during the academic year 1990–91 and a special session on the same topics held at the joint winter mathematics meeting in San Francisco during January of 1991. The special year came to be known as "RAGSQUAD". Funding was provided by NSF and NSA/MSP grants DMS-9003386 and MDA90-H-1015.

1991 *Mathematics Subject Classification*. Primary 00B25, 11Exx, 14Pxx.

Library of Congress Cataloging-in-Publication Data

Recent advances in real algebraic geometry and quadratic forms: proceedings of the RAGSQUAD year, Berkeley, 1990–1991/William B. Jacob, Tsit-Yuen Lam, Robert O. Robson, editors.

p. cm. — (Contemporary mathematics, ISSN 0271-4132; 155)

Papers resulting from a course held at the University of California at Berkeley, 1990–91 and the AMS Special Session on Real Algebraic Geometry and Quadratic Forms held in San Francisco, Jan. 1991, both comprising the special year dubbed RAGSQUAD.

ISBN 0-8218-5154-3 (acid-free)

1. Geometry, Algebraic. 2. Forms, Quadratic. I. Jacob, Bill. II. Lam, T. Y. (Tsit-Yuen), 1942– . III. Robson, Robert O., 1954– . IV. AMS Special Session on Real Algebraic Geometry and Quadratic Forms (1991: San Francisco, Calif.) V. Series: Contemporary mathematics (American Mathematical Society); v. 155.

QA564.R43 1993 93–6377

512′.74—dc20 CIP

♾ The paper used in this book is acid-free and falls within the guidelines
established to ensure permanence and durability.
♻ Printed on recycled paper.
All articles in this volume were printed from copy prepared by the authors.
Some articles were typeset using AMS-TeX or AMS-LaTeX,
the American Mathematical Society's TeX macro systems.

10 9 8 7 6 5 4 3 2 1 99 98 97 96 95 94

Contents

Recent Advances in Real Algebraic Geometry and Quadratic Forms: proceedings of the RAGSQUAD year, Berkeley, 1990-91

BILL JACOB, T. Y. LAM, ROBBY ROBSON, *EDITORS*

Preface

The papers in this volume grew out of a year long program in **Real Algebraic Geometry and Quadratic Forms** held at the University of California at Berkeley, during the academic year 1990–91 and a special session on the same topics held at the joint winter mathematics meetings in San Francisco during January of 1991. The program at Berkeley was organized by T. Y. Lam and Robby Robson, while the AMS special session was organized by Bill Jacob. The special year came to be known as "RAGSQUAD".

A list of RAGSQUAD participants appears below.

This volume of papers is dedicated to Professor Donald Dubois. The lead article *D. W. Dubois and the Pioneer Days of Real Algebraic Geometry*, contributed by Carlos Andradas and Tomás Recio, touches on the recent history of conferences, meetings, and special years in Real Algebraic Geometry and Quadratic Forms.

There is a long list of persons and organizations we wish to thank for material, scientific, and moral support. The University of California provided visiting positions, "University professorships", and academic exchanges for a number of participants. The Mathematical Sciences Research Institute provided research professorships, and both institutions provided support in many other ways as well. The AMS sponsored the special session noted above, and funding was provided by federal grants[1]. Many foreign participants received support from their own countries and institutions, and of course there were numerous graduate

---

1991 *Mathematics Subject Classification.* 00B25 11Exx 14Pxx.

[1]NSF and NSA/MSP grants DMS-9003386 and MDA90-H-1015.

students and office workers without whose efforts the program would not have run.

This volume contains scientific works resulting from the spirit of international collaboration which characterized the RAGSQUAD year. In more human terms, RAGSQUAD became a community built upon a foundation of intense mathematical friendship, not to mention some wonderful international pot luck dinners and even a RAGSQUAD concert. We hope that both the high quality of the mathematics and the RAGSQUAD spirit are reflected in this volume.

RAGSQUAD Participants

Longer Term Participants

S. Akbulut	M. E. Alonso	C. Andradas
R. Berr	L. Bröcker	R. Brown
G. Brumfiel	M. Buchner	J.-L. Colliot-Thélène
M. J. de la Puente	C. Delzell	M. Dickmann
B. Fein	R. Huber	M. Knebusch
M. Krüskemper	K. H. Leung	T. Y. Lam
D. Leep	J. Madden	M. Marshall
J. Merzel	R. Parimala	A. Pfister
A. Prestel	R. Robson	J. Ruiz
W. Scharlau	C. Scheiderer	R. Sridharan
G. Stengle	N. Schwartz	T. Smith
A. Wadsworth	C. Yeomans	

Additional AMS Special Session Participants and short–term visitors

R. Aravire	R. Baeza	S. Barton
R. Brown	T. Craven	R. Elman
R. Fitzgerald	L. González-Vega	M. Hirsch
J. Hsia	W. Jacob	L. Legorreta
C. Mulcahy	I. Pays	R. Perlis
V. Powers	T. Récio	M.-F. Roy
D. Shapiro	J. Shick	J.-P. Tignol
L. van den Dries	R. Ware	J. Yucas

Student Participants

M. J. Gonzalez-Lopéz	D. Hoffmann	Y.-S. Hwang
N. Kan	A. Klute	D. Tao

Contemporary Mathematics
Volume **155**, 1994

D. W. Dubois and The Pioneer Days of Real Algebraic Geometry

TOMÁS RECIO AND CARLOS ANDRADAS

Introduction

The Special Session on Real Algebraic Geometry and Quadratic Forms, organized by William B. Jacob in the 16-19 January, 1991; 863rd AMS meeting in San Francisco was, remarkably, exactly (except for a few days) 10 years later, in the same city and under the auspices of the same society, that the first formal session devoted to Real Algebraic Geometry was held. But not everything was exactly the same: most of us were ten years older but not necessarily wiser to cope with the great developments that had taken place, during the ten year period, on the subject. Some new faces had entered into scene but, on the other hand, some significative people were missing, like Gus Efroymson (whose early death occured in August 1983), or Prof. D. W. Dubois, the organizer of the first meeting, who was not in San Francisco this time, having retired from mathematical activities in 1985. The joint publication of the proceedings of the 1991 Special Session and the conferences of the Special Year (1990-1991) devoted to Real Algebraic Geometry and Quadratic Forms by the Mathematics Department of the University of California at Berkeley, offers the opportunity that many colleagues were awaiting to honour Don Dubois. The following lines are intended to be both a dedication and a biographical sketch of his life and mathematical activity. They are mainly based on our fragile personal memory and therefore we apologize in advance for the involuntary mistakes and omissions .

1991 *Mathematics Subject Classification.* 01A70,14P99.
This paper is submitted in final form and no version of it will be submitted for publication elsewhere.

The encounter

I (Tomás Recio) met Don Dubois, accidentally, on a rainy winter day in Madrid (February 15, 1979); a day that we both will never forget. The story of this casual meeting shows how the profession of mathematician, quite far from the benefits of public recognition and understanding of our work, helps sometimes to discover new friends at unexpected places. I was working in Madrid at the Spanish Higher Council for Scientific Research (C.S.I.C), in the Institute of Mathematics. At that moment all I knew about him was little more than the existence of some mathematician called D.W. Dubois. Indeed, having written my dissertation on semianalytic geometry I was familiar with Risler's formulation of the Real analytic and algebraic Nullstellensätze as it appears in Risler's[1] "Un theoreme des zeros en geometrie algebrique et analytique reelles" in which the name of Dubois is listed in the references. Also, in 1977, when I was beginning my first steps in semialgebraic geometry and I had received from professor Jacek Bochnak a beautiful handwritten pioneer survey: "Sur la geometrie algebrique reelle" (which later became a joint paper with G. Efroymson[2]), the name of Dubois was, again, among the references.

In those times it was certainly not an everyday event in the Institute of Mathematics to receive visits of foreign professors. Nobody working in real geometry had ever stepped in there (and nobody could have guessed that I was interested in the subject). Our isolation was quite successful. Thus you can imagine my shock when I heard, that morning of 1979, a voice coming from the neighbour office door saying to a colleague of mine: "I am professor Dubois from the University of New Mexico and I would like to consult the library". I jumped from my chair with a sudden inspiration, gained the door and said, with an unreasonable self-security: "You must be Don Dubois". My surprise was only comparable to his surprise when he realized that someone knew him in a country where he had not been aware of anybody studying Real Geometry. In fact, Don was just spending a sabbatical leave, doing some research, alone, at a little village on the Spanish Mediterranean coast; and he was coming to Madrid merely to complete some bibliographical references. The surprise continued when he saw our meager library: of course, it was as impossible for me to have any of his works as it was for him to fulfill the aim of his trip to the capital of Spain.

But who cared about the library: I was anxious to discuss with somebody my doubts and problems in the field and, thus, we spent some hours talking about real geometry together with the two young men (C. Andradas and V. Espino) whom I was trying to get interested in the subject. Don Dubois was then extremely generous with his mathematical knowledge, sharing with us his current work and giving us corrected and personally annotated versions of his

[1] J.J. Risler:"Un theoreme des zeros en geometrie algebrique et analytique reelles", Lecture Notes Math. 409, Springer, 1974.

[2] J. Bochnak, G. Eforymson:"Real Algebraic Geometry and the 17th Hilbert Problem", Math. Annalen 251, 1980.

own papers. We asked him to come back again that year to give a talk at the Department of Algebra in the Universidad Complutense.

All this started a great math collaboration and a wonderful friendship. During the following years his home (and the University of New Mexico) at Albuquerque was periodically invaded by a "Spanish Armada", as he still likes to call the Spanish group of mathematicians working in Real Algebraic Geometry, that started its first steps at that time and that owes so much to him. In fact, we learned from him not only quite a bit of good classic mathematics, but also the basic "elements" of a research career in mathematics (the existence of colloquium talks and international meetings on specialized topics, the importance of the "invisible college" of colleagues working in the same topic with whom one should mantain a fluid communication, the fact that research results should be exposed into papers and the papers submitted, quickly, somewhere....).

He taught all that to us and we taught it to other students in Spain, where such basic facts were not yet part of the scientific tradition in the decade of the seventies. We were enthusiastic students as it corresponds to a very good teacher. The Spanish Real Geometry group was really lucky in receiving the warm support of many colleagues from different countries: French, German, Italian, Polish.... But Don has the merit of being the first, finding the matter in its more rude form.

Early meetings in Real Geometry

Being close to him we were exceptional witnesses of the way he managed to organize, breaking the resistance of the scientific community to opening room for a new subject, the important meetings on Ordered Fields and Real Geometry, held in San Francisco, 1981 (January 7-11) and Boulder, 1983 (July 4-8). There was also another meeting of capital importance for the subject, celebrated in May of 1981 at Rennes (France), that Dubois attended but, this time, without scientific responsibility.

Don's mathematical background was (as we will detail below) basically algebraic rather than geometrical or analytical (in the sense of analytic functions). Nevertheless he had such an open mind and mathematical insight that his invited talks to this first San Franciso meeting were quite representative of what would turn out to be, in the following years, the main research issues in Real Geometry (many of them quite far from Dubois' personal interests): the real spectrum[3] (Coste-Roy), real places (Schülting) and semialgebraic (strong) topology (Andradas, Dubois, Recio, Schwartz, Delzell), the abstract construction of semialgebraic topology (Delfs-Knebusch), the stability index (Merzel), Nash functions (Efroymson) and coherent algebraic sheaves (Tognoli), generalizations of Hilbert's 17th Problem and Theory of Models (Gondard), higher level orderings (Harman), quadratic forms (Bröcker, Dress, Scharlau, Pfister, Shapiro,

[3]the classification and denomination of topics is ours.

Lam), and general theory of ordered fields (Gilmer, Henriksen, Smith, Engler, Viswanathan, Galstad, Mott); plus a collection of open problems collected by Brumfiel (some of them still unsolved). Then, in the 1983 meeting, we must add to this collection of topics the appearance of analytic and semianalytic sets, the topology on real algebraic varieties, specialized issues on real curves (like moduli spaces or Hilbert 16th problem or complete intersections) and, for the first time, algorithmic issues (Arnon, McCallum, Risler). It is interesting to remark that the European talks outnumbered the American by around two-to-one in the Boulder meeting, while the proportion was almost even in the San Francisco session. This speaks clearly for the impulse towards the internationalization of the real geometry school that Dubois fostered in the early stages of this topic development. We should mention also that, several years later, he helped a distinguished European professor in this area (W. Kucharz) to obtain a position at the University of New Mexico, to fill the vacancy left by G. Efroymson.

We can certainly say that these two meetings organized by Dubois greatly helped the creation of an international working group in Real Geometry, allowing the establishment of personal and scientific bonds between many colleagues. For instance, as is mentioned in the Foreword of the most influential book on Real Algebraic Geometry[4], its authors conceived the project of writing the book during the meeting at Boulder. Don obtained, academically speaking, little or no benefit from the success of the Conferences. In fact, he practically retired after the Boulder meeting; the introduction to its Proceedings was his last publication.

We think that no one in his/her sane mind enjoys organizing a scientific conference, although we all agree on the importance of doing it. Don took this job over, but we also believe that Don was particularly reluctant to appear as the chairman, organizer or editor of anything. He was, in the interpersonal relations, the most generous, charming and educated person in the world, when he had in front of him just a few persons. But he was not a manager, not a group leader. He searched for perfection about scientific matters: we just recall how many times our joint papers should be retyped because of minor misprints; or the detailed explanations we had to give him about our contributions in these papers until he finally understood them and, then, he proceeded to change the way they were expressed. That condition is quite incompatible with the required bureaucratic quality of esprit for writing applications to obtain funding for a conference, or for deciding what to do with poorly typed (after returning it to the author for the n-th time) contributions to some Proceedings. We think that he took over the heavy task of organizing the 1981 and 1983 meetings mainly because he found himself in a good position, being the senior scientist of all of us, to promote a subject that started then to obtain some mathematical recognition and that had been his main research activity during a decade (the seventies) of severe difficulties (just take a look to the dates and journals where he published during

[4]J. Bochnak, M. Coste, M.F. Coste-Roy:" Geometrie algebrique reelle". Ergebnisse der Mathematik. Springer. 1987.

that time). That was a wonderful service to the mathematical community and we all thank him for that.

Mathematical work.

Donald Ward Dubois was born in Oklahoma in 1923. He followed sciences and engineering studies at the University of Oklahoma, obtaining his Ph. D. degree in mathematics under the direction of Casper Goffman and A.A. Grau with a dissertation entitled "On partly ordered fields" [1]. After a brief stay at Ohio State University he moved to the University of New Mexico at Albuquerque, where he remained until his retirement. We do not consider ourselves competent even to give a brief account of circa 30 years of mathematical work that includes papers in topics so different as topological fields, number theory, Abelian group theory, algebraic geometry or mathematic education. However we feel a little more at ease to describe below some of his main contributions in Ordered Fields and Real Algebraic Geometry.

Besides his own research activity, Don Dubois devoted efforts to introducing other people to mathematical research or to attracting them to his particular view of Real Geometry. Among the former we can mention the doctoral dissertations written under his direction by Dalton Tarwater (on Abelian groups, ca. 1964), Gail Carns ("Formally real fields", 1968, University of New Mexico), Charles Walter ("Conic Archimedean primes and Galois Groups; Density and non Archimedean Ordered Fields", 1969, University of New Mexico), Art Bukowski ("Branches and Completions for Real Algebraic Curves", 1972, Univ. of New Mexico) and C. Andradas ("Real Places in Function Fields", 1983, Univ. of New Mexico). On the other hand G. Efroymson collaborated closely with Dubois from 1969 until 1971–and continued by himself in the same field until his death; and also T. Recio was Dubois' collaborator from 1980 until Dubois' retirement in 1985.

Don's mathematical career, as we can see it with some perspective, was mostly around Artin-Schreier theory, either in purely algebraic terms or in connection with geometry, with the exception of four early papers on Abelian groups and a detour into mathematical pedagogy from 1972 to 1978. This strange phenomenum of leaving temporarily a successful line of research, while the topic was in expansion, was due to several personal and scientific circumstances that led him to a loss of self-confidence. In particular, according to his own confession, he became disillusioned by the misunderstanding of some of his pioneer steps in Real Geometry (written at the beginning of the 70's, but published much later) by some distinguished classical geometer in California. It seems, also, that this lack of appreciation of his ideas at the time they were published or produced, by many well established research groups was not restricted to his work on Real Geometry. As we will show below (by merely comparing dates of Dubois's writings with dates of comments and references to his work), also some of his results on Ordered Fields that are highly considered in the current developments did not receive attention until much later. There seems to be a constant gap of about 10 years for the true understanding of the relevance of his contributions. Generally speaking, a very good source for details about his work in the subject of Ordered Fields are the Bibliographical and Historical Notes that appear in T.Y. Lam's "Orderings, valuations and quadratic forms"[5] and "The theory of ordered fields"[6]. On the other hand, concerning Real Algebraic Geometry, the Bibliographical Notes of the book by Bochnak, Coste and Roy[7] contain several references to Dubois' contribution and provide all the background needed to understand our own comments.

We could say that, starting from his early work, which stemmed from functional analysis, he mantained as leit-motiv of his research career the study of ordered algebraic structures by means of representations in some ring of functions. For instance, one interesting result attained by Dubois in the middle 60's is the nowadays called "the representation theorem of Kadison[8]-Dubois" ([8],[9]) that, roughly states[9] that any partially ordered archimedean ring $\mathbf{R}$ can be embedded as a subring of a ring $C(X)$ of continuous real functions on some compact space X. Thus, under very mild conditions, it follows from the Stone-Weierstrass theorem that the image of $\mathbf{R}$ is dense in $C(X)$. This theorem generalizes greatly the well known result that every field with an archimedean order admits a unique order embedding into the reals. What makes the theorem so powerful is that the partial orders do not need to contain the squares, and this has been specially useful in the study of orderings of higher level and sums of $2n$-powers (cf. E. Becker's work[10]). Following in this line we also encounter in [12] a first (1971) study of the now called "Holomorphy ring of a field", i.e. the intersection of all real valuation rings of the field. The importance of the holomorphy ring in real geometry as a tool to study birationally invariant properties can be seen, for example, in Schülting's [11] proof of the birational invariance of the number of connected components of a real algebraic variety. A guide to other applications

[5]T.Y. Lam: "Orderings, valuations and quadratic forms". CBSM Regional Conference Series in Mathematics. No. 52. A.M.S. 1983.

[6]T.Y. Lam: "The theory of ordered fields". Ring Theory and Algebra III. (B.McDonald ed.).Lecture Notes in Pure and Applied Math. pp-1–152. Vol. 55, Dekker, New York, 1980.

[7](loc. cit.)

[8]R.V. Kadison: "A representation theory for commutative topological algebra." Memoirs of Amer. Math. Soc. No.7. 1951.

[9]the following comments are taken from E.Becker's paper: "Valuations and real places in the theory of formally real fields." In: Geometrie Algebrique Reelle et Formes Quadratiques. Lecture Notes in Math. vol. 959. pp. 1–40. Springer. 1982.

[10]E. Becker: "The Real Holomorphy Ring and Sums of 2n-th Powers". In: Geometrie Algebrique Reelle et Formes Quadratiques. Lecture Notes in Math. vol. 959. pp. 139–181. Springer. 1982.

[11]H.W.Schulting: "Real holomorphy rings in real algebraic geometry". In: Geometrie algebraique reelle et formes quadratiques. Lecture Notes in Math. vol. 959. pp. 433–442. Springer. 1982.

are the notes of paragraph 9 in Lam's work [12]. Let us say that Dubois' formulation uses essentially Harrison's language of primes[13], a theory that seems to have greatly influenced him. The paper [20], written much later than the work we are commenting here, returns to the old idea of studying ordered structures by means of some functional representation, this time in the style of ringed spaces.

Also inside the general frame of ordered structures, another favourite subject of Don has been Hilbert's 17th problem. In 1967 he published a counterexample [7] stating that the density condition (on a uniquely ordered field with respect to its real closure) can not be dropped in Artin's solution of Hilbert's problem, as was erroneously stated in Lang's "Algebra" [14]. In fact, it was proved later by McKenna[15] that this density is, exactly, the necessary and sufficient condition for Artin's solution to hold true. Further extensions of Dubois counterexample appear in Gamboa-Recio[16] (providing irreducible polynomials which are positive but not sums of squares) or in Schwartz[17] (for a generalization to positiveness over a real variety).

It was also in "an attempt to find an easier solution to Hilbert's problem"[18] that, in 1969, he obtained one of his most celebrated results: "the real nullstellensatz" [10], also proved independently by Risler[19] some months later, although their definitions of real radical look a little bit different. This presence of Hilbert's 17th problem continued in [17] and [23], where he studied different extensions to real varieties, of Robinson's nonnegativity criterion, i.e. the now called positivstellensätze. In these new formulations appears already another key notion in Dubois' work: that of central point of a variety.

In fact, a characteristic of his research has been the interplay between orderings, valuations and real geometry. This is the topic of a whole series of papers, [14], [15], [16], [17], [19], in which he showed (and applied to different contexts) the algebraic characterization of the points of maximun dimension of a real variety: the so-called central points. They are exactly the points of the variety at which can be centered a real place or an order; i.e. those points where the canonical specialization at the point can be extended to a real place. It is precisely this idea of being the "center of" where the name central originated from, although in

[12]T.Y. Lam: "Orderings, valuations and quadratic forms". (loc. cit.)

[13]D. Harrison: "Finite and infinite primes for rings and fields". Memoirs of Amer. Math. Soc. No.68. 1966.

[14]S. Lang : "Algebra". Addison-Wesley, Reading, Mass., 1965.

[15]K. McKenna: "New facts about Hilbert's 17th problem". Lecture Notes in Math. vol. 498. pp 220–230. Springer. 1975.

[16]J.M. Gamboa, T.Recio: "Ordered fields with the dense orbits property". Journal of Pure and Applied Algebra 30, (1983), pp. 237–246.

[17]N. Schwartz: "The strong topology on real algebraic varieties". in: Ordered fields and real algebraic geometry (San Francisco, Calif., 1981). Amer. Math. Soc., Providence, R.I., Series: Contemp. Math., 8, (1982), pp 297–327.

[18]Quoted from Dubois's "A nullstellensatz for ordered fields", Arkiv for Matematik, Band 8 nr13, 1969. pp 111–114

[19]J.J. Risler: "Une characterisation des ideaux des varietes algebriques reelles." C.T. Acad. Sci. Paris 271, 1171-1173, 1970.

the first paper [14], where he dealt with curves, they were called "inner points" since they correspond to the non-isolated points of the curve. Let us remark that this paper and the next two [15], [16], were essentially already presented in October 1971, as a technical report of the University of New Mexico, entitled "Real Algebraic Curves".

In [15], and [16] he works out a theory of real curves over Cantor fields (which later have been named microbial[20] fields) for which he developed some "real analysis" over them, an idea studied again some years later in the well known book of Brumfiel[21] (1979). They are ordered fields at which the notion of convergence of power series can be defined and, therefore, the Puiseux theorem, local decomposition into branches, etc... can be studied. Microbial fields have been mentioned later by R. Robson[22] [23], or M. Knebusch[24] [25], in their efforts to develop some abstract local analytic geometry in relation with the real spectrum. Coming back to the idea of central points, in [19] the birationality of the set of real places is used to study the behaviour of central points of a variety under rational morphisms. Most of his ideas about centrality have now become classic, maybe nowadays expressed in the language of the real spectrum (see, for instance, the dissertation of Saliba[26] or Chapter 7 of the book Bochnak-Coste-Roy[27]).

For the work [11] (and [13] as an added result to the former), the least we can say is that they were, for several years, the building elements of real algebraic geometry, the common reference place for the folklore of the subject in those early days of the seventies. To recall just a result that is in everybody's mind: the *sign changing criterion*[28] (the ideal generated by an irreducible polynomial in n variables is real if and only if it changes sign in the affine space if and only if it has a real non singular point if and only if its zero set is n-1 dimensional). Published at a difficult place for librarians (because both authors were friends of the mathematician Yu-Why Chen in whose honor the paper was written) it was the impossible dream for me to obtain a copy of it from Madrid until I met Don Dubois, that rainy winter day of 1979.

[20]cf. example 4.4, chapter I, in H. Delfs, M. Knebusch: "Locally Semialgebraic Spaces". Lecture Notes in Mathematics. 1173. Springer-Verlag, 1985.

[21]G. Brumfiel: "Partially ordered fields and semi-algebraic geometry". Cambridge University Press. 1979.

[22]R. Robson: "The Ideal Theory of Real Algebraic Curves and Affine Embeddings of Semialgebraic Spaces and Manifolds." Ph. D. Dissertation. Stanford University. 1981.

[23]R. Robson: "Nash wings and real prime divisors". Math. Ann. 273. 177-190. 1986.

[24]M. Knebusch: "An invitation to real spectra". Canadian Mathematical Society. Conference Proceedings. Volume 4, 1984.

[25]M. Knebusch: "Isoalgebraic Geometry: First Steps". Seminaire de Theorie des Nombres, Delange-Pisot-Poitou. Paris (1980-81). In: Progress in Mathematics, pp. 127-140. Birkhauser, 1982.

[26]C.Saliba "Pour une classification des differentes formes de theoreme des zeros et de theoreme des elementes positifs". These. Universite de Rennes I. 1983.

[27](loc. cit.)

[28]see Kapitel II in M. Knebusch, C. Scheiderer "Einführung in die reelle Algebra". Vieweg Studium 63, Aufbaukurs Mathematik, 1989.

Mathematical work of D.W.Dubois.

[1] On partly ordered fields. Dubois, D. W. Proc. Amer. Math. Soc. (1956), 7, 918–930.

[2] A note on division algorithms in imaginary quadratric number fields. Dubois, D. W. Steger, A. Canad. J. Math. (1958),10, 285–286.

[3] Cohesive groups and p-adic integers. Dubois, D. W. Publ. Math. Debrecen (1965), 12, 51–58.

[4] Applications of analytic number theory to the study of type sets of torsionfree Abelian groups. I. Dubois, D. W. Publ. Math. Debrecen (1965), 12, 59–63.

[5] Applications of analytic number theory to the study oftype sets of torsionfree Abelian groups. II. Dubois, D. W. Publ. Math. Debrecen (1966), 13, 1–8.

[6] Modules of sequences of elements of a ring. Dubois, D. W. J. London Math. Soc. (1966), 41, 177–180.

[7] Note on Artin's solution of Hilbert's 17-th problem. Dubois, D. W. Bull. Amer. Math. Soc. (1967), 73, 540–541.

[8] A note on David Harrison's theory of preprimes. Dubois, D. W. Pacific J. Math. (1967), 21, 15–19.

[9] Second note on David Harrison's theory of preprimes. Dubois, D. W. Pacific J. Math. (1968), 24, 57–68.

[10] A nullstellensatz for ordered fields. Dubois, D. W. Ark. Mat. (1969), 8, 111–114 .

[11] Algebraic theory of real varieties. I. Studies and Essays (Presented to Yuwhy Chen on his 60th Birthday, April 1, 1970) Dubois, D. W. Efroymson, G. (1970), Math. Res. Center, Nat. Taiwan Univ., Taipei; pp. 107–135.

[12] Infinite primes and ordered fields. Dubois, D. W. Dissertationes math., Warszawa 69, 40 p. (1970).

[13] A dimension theorem for real primes. Dubois, D.; Efroymson, G. Canadian J. Math. 26, (1974), 108–114.

[14] Real commutative algebra. I: Places. Dubois, D. W. Rev. Mat. Hisp.-Am., IV. Ser. 39, (1979), 57–65.

[15] Real commutative algebra. II: Plane curves. Dubois, D. W.; Bukowski, A. Rev. Mat. Hisp.-Am., IV. Ser. 39, (1979), 149–161.

[16] Real commutative algebra. III: Dedekind-Weber-Riemann manifolds. Dubois, D. W.; Bukowski, A. Rev. Mat. Hisp.-Am., IV. Ser. 40, (1980), 157–167 .

[17] Second note on Artin's solution of Hilbert's 17th problem. Order spaces. Dubois, D. W. Pac. J. Math. 97, (1981), 357–371.

[18] Ordered fields and real algebraic geometry. Proceedings of the Special Session held during the 87th Annual Meeting of the American Mathematical Society, San Francisco, Calif., Jan. 7–11, 1981. Edited by Donald

Ward Dubois and Tomas Recio. Amer. Math. Soc., Providence, R.I. Series: Contemp. Math., 8. (1982).

[19] Order extensions and real algebraic geometry. Dubois, D. W. Recio, T. In: Ordered fields and real algebraic geometry (San Francisco, Calif., 1981). Amer. Math. Soc., Providence, R.I., Series: Contemp. Math., 8, (1982), 265–288.

[20] Subordinate structure sheaves. Dubois, D. W.; Recio, Tomas. Geometrie algebrique reelle et formes quadratiques, Journ. S.M.F., Univ. Rennes 1981, Lect. Notes Math. 959, (1982), 324–342.

[21] General numeration I. Gauged schemes. Dubois, D. W. Rev. Mat. Hisp.-Am., IV. Ser. 42, (1982), 38–50.

[22] General numeration. II: Division schemes. Dubois, D. W. Rev. Mat. Hisp.-Am., IV. Ser. 42, (1982), 139–148.

[23] A note on Robinson's non-negativity criterion. Dubois, D. W.; Recio, T. Fundam. Math. 122, (1984), 71–76.

[24] Ordered fields and real algebraic geometry. Papers from the conference held at the University of Colorado, Boulder, Colo., July 4–8, 1983. Edited by D. W. Dubois. Rocky Mountain J. Math. 14, (1984), no. 4.

[25] Explanation, appreciation and remembrance. Dubois, D. W. Rocky Mt. J. Math. 14, (1984), 729–732.

Universidad de Cantabria, Santander, 39071, Spain.
E-mail address: recio@ccucvx.unican.es

Universidad Complutense, Madrid, 28040, Spain.
E-mail address: andradas@mat.ucm.es

Contemporary Mathematics
Volume **155**, 1994

On Algebraic Structures of Manifolds

S. AKBULUT

Moduli spaces of algebraic structures on manifolds have long been a source of curiosity. Apart from studying them to classify algebraic structures on manifolds there is always an exciting possibility of using them as functors to understand the topology of manifolds themselves as in [**D**]. Here we review the moduli spaces of nonsingular algebraic structures in an elementary way, and then discuss some related facts along with some problems and speculations. The stated propositions are joint work with with H. King.

For any nonsingular (real or complex) algebraic set V let $A(V)$ be the set of nonsingular algebraic subsets of V, and let $A^d(V)$ to be the degree d elements of $A(V)$. If V is real let $A_o(V)$ denote the set of nonsingular components of algebraic subsets of V, and similarly let $A^d_o(V)$ be the degree d elements of $A_o(V)$. If V is complex define $A_{\mathbb{R}}(V)$ to be the elements of $A(V)$ defined over $\mathbb{R}$ and $A^d_{\mathbb{R}}(V) = A_{\mathbb{R}}(V) \cap A^d(V)$. We will also denote the real part of a complex algebraic set V by $V_{\mathbb{R}}$, and denote the complexification of a real algebraic set W by $W_{\mathbb{C}}$. Up to isomorphisim we can always assume that if W is nonsingular then $W_{\mathbb{C}}$ is a nonsingular projective algebraic set [**AK6**, Lemma 2.2.15].

For a compact smooth manifold M as in [**AK4**] we can define:

$$\mathcal{S}\,(M) = \left\{ (V,f)| \begin{array}{l} V \text{ is a nonsingular algebraic set,} \\ f\colon M \to V \text{ is a diffeomorphisim} \end{array} \right\} / \sim$$

where $\sim$ is the equivalence relation $(V,f) \sim (V',f')$ if there is a birational isomorphisim $\varphi\colon V \to V'$ with $\varphi \circ f = f'$. If this space is too big, the standard procedure is to let the group of diffeomorphisim $\mathrm{Diff}(M)$ act on $\mathcal{S}(M)$ by composition on the left and take the quotient $Alg(M) = \mathcal{S}(M)/\mathrm{Diff}(M)$. To be

1991 *Mathematics Subject Classification.* Primary 57R19, 15P05; Secondary: 14C25, 14P25, 57R20, 57R15, 58D29.

This work supported in part by NSF.

This paper is submitted in final form and no version of it will be submitted for publication elsewhere.

consistent with the usual definition of moduli spaces, alternatively we can first divide $\mathcal{S}(M)$ by the identity component $\text{Diff}_o(M)$ of $\text{Diff}(M)$ (this would be the analogue of the Teichmüller space) and then divide the resulting space by the mapping class group $\text{Diff}(M)/\text{Diff}_o(M)$ to get $Alg(M)$.

To topologize $Alg(M)$ we can require that $M \subset \mathbb{R}^n$ and the elements V of $Alg(M)$ be nonsingular algebraic subsets of $\mathbb{R}^n$. In this case we define

$$Alg(M, \mathbb{R}^n) = \{ \, [V, f] \in Alg(M) \mid V \in A(\mathbb{R}^n) \, \}$$

where the square bracket denotes the equivalence class of (V, f) in $Alg(M)$. Here we take the quotient topology induced by the space of imbeddings $M \hookrightarrow \mathbb{R}^n$. This space has many components corresponding to each isotopy class of imbedding of M into $\mathbb{R}^n$. By fixing the isotopy class of an imbedding $f : M \to V \subset \mathbb{R}^n$ we can study a single component $Alg(M, \mathbb{R}^n; f)$ of this space. Hence, by identifying f by the inclusion $M \subset \mathbb{R}^n$ we naturally arrive at the definitions :

$$\begin{aligned} J(M) &= \{ \, V \in A(\mathbb{R}^n) \mid V \text{ is isotopic to } M \, \}/\sim \\ J_o(M) &= \{ \, V \in A_o(\mathbb{R}^n) \mid V \text{ is isotopic to } M \, \}/\sim \\ J_{\mathbb{R}}(M) &= \{ \, V_{\mathbb{R}} \mid V \in A_{\mathbb{R}}(\mathbb{CP}^n) \textit{ and } V_{\mathbb{R}} \text{ is isotopic to } M \, \}/\sim \end{aligned}$$

where $\sim$ denotes the equivalence class under birational isomorphisim. Here we identify $\mathbb{R}^n \subset \mathbb{RP}^n$. Clearly $J_{\mathbb{R}}(M) \subset J(M) \subset J_o(M)$. As above we define the degree d elements of $J_{\mathbb{R}}(M)$, $J(M)$, $J_o(M)$ by $J^d_{\mathbb{R}}(M)$, $J^d(M)$, $J^d_o(M)$ respectively. For a fixed d there are strong restrictions on topological types of manifolds M satisfying $J^d(M) \neq \phi$, e.g. [**V1**], [**V2**]. On the other hand [**AK2**] implies that $J_o(M)$ is nonempty; and $J(M)$ is nonempty if the immersed cobordisim class of M contains an algebraic representative, [**AK3**]. Also in some special cases there are elementary methods available to show that $J_{\mathbb{R}}(M)$ is nonempty, for example:

PROPOSITION 1. *If M is a topological complete intersection, that is if it is an intersection $\cap L_i$ of smooth codimension one submanifolds in general position in $\mathbb{RP}^n$, then $J_{\mathbb{R}}(M) \neq \phi$.*

PROOF. First isotop each L_i to a nonsingular algebraic hypersurface V_i by a small isotopy (this can be done since the group $H_{n-1}(\mathbb{RP}^n; \mathbb{Z}_2)$ is algebraic, e.g. [**AK5**], then change the coefficients of the defining equations of each V_i a little so that the complex solutions become nonsingular and transverse without affecting the isotopy type of $\cap V_i \approx M$. □

For example, this proposition implies that any knot or a link in $\mathbb{RP}^3$ is a real part of a nonsingular complex algebraic curve in $\mathbb{CP}^3$. In general the quotient topology makes these moduli spaces hard to understand. Existence of nonalgebraic homology classes (see below) imply that $J(M)$ is nontrivial in general, [**AK4**]. We now know by [**BK**], and [**B**] that when $M^m \subset \mathbb{R}^{2m+1}$ the set $J(M)$ is uncountable; in fact it contains an imbedded arc.

Even though $A^d(\mathbb{CP}^n)$ is connected $J^d(M)$ is not necessarily connected [**R**]. In the hypersurface case this can be visualized by taking the Verenose imbedding $\lambda : \mathbb{RP}^n \hookrightarrow \mathbb{RP}^N$ which sends a point $[x_0, .., x_n]$ to a point whose coordinates consist of all possible monomials of degree d in $(x_0, .., x_n)$. Then every element $V_d \in A^d(\mathbb{RP}^n)$ corresponds to $\lambda(\mathbb{RP}^n) \cap H_d$ where H_d is a linear subspace of $\mathbb{RP}^N$. Then for example the homology of $\mathbb{RP}^n$ prevents us connecting two submanifolds $\lambda(\mathbb{RP}^n) \cap H_d^i$ diffeomorphic to M, $i = 0, 1$ by a path of submanifolds $\lambda(\mathbb{RP}^n) \cap H_d^t$, $t \in [0, 1]$, each of which is diffeomorphic to M. However any two elements of $J^d(M)$ can be connected inside $J^D(M)$, for some large D [**Na**].

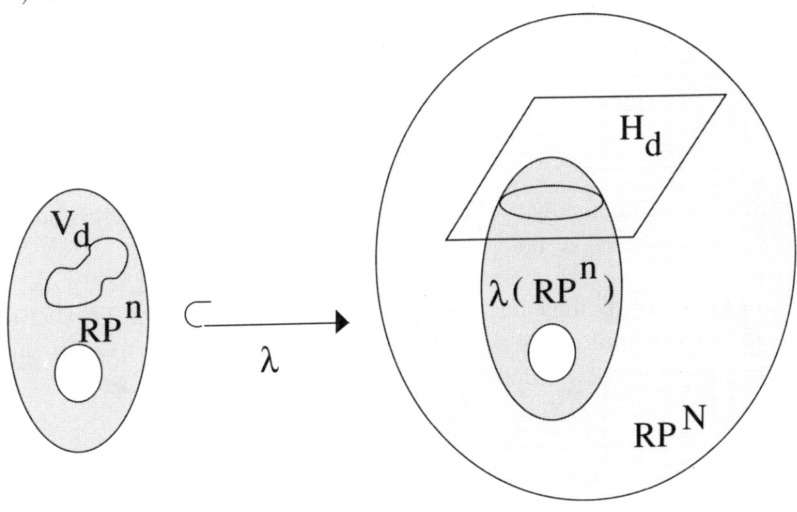

Figure 1.

Now let us review some of properties of nonsingular real algebraic sets. First of all the Grassmanian manifold has a natural real algebraic structure:

Grassmanian:

Recall that the Grassmann manifold $G(n, k)$ of k-planes in $\mathbb{R}^n$ can be identified by the nonsingular algebraic variety (cf. [**AK6**])

$$G(n, k) = \{L \in \mathcal{M}(n) \mid L^2 = L,\ L = L^t,\ \text{trace}(L) = k\}$$

where $\mathcal{M}(n)$ denote the set of $(n \times n)$-matrices with real coefficients, and the identification is given by k-plane $\to$ the matrix of orthogonal projection onto that plane. One of the most important properties of compact real algebraic sets is that their Gauss maps are naturally algebraic. That is if $V \subset \mathbb{R}^n$ is a compact nonsingular real algebraic set then the Gauss map $V \to G(n, k)$, where $k = dim(V)$, which assigns every point of V to the tangent plane at that point, is an entire rational map. In fact more generally if W is a nonsingular real algebraic set with $V \subset W \subset \mathbb{R}^n$ then the map $\alpha : V \to G(n, k)$ given by $\alpha(x) =$ the tangent plane to W which is normal to V at x is an entire rational function (cf. [**AK6**]), where $k = dim(W) - dim(V)$.

Let us recall the proof for $W = \mathbb{R}^n$: For any $y_s \in V$ we pick a set of generators of the ideal of polynomials vanishing at V, $f_1, \dots, f_k \in \mathcal{I}(V)$ so that the gradients ∇f_i are linearly independent at a Zariski open set U_s containing y_s. Let $A(x)$ be the $n \times k$ matrix whose i-th column is $\nabla f_i(x)$ and let $A^t(x)$ be its transpose. Then for $x \in U_s$ we have:

$$\alpha(x) = A(x)\left(A^t(x)A(x)\right)^{-1} A^t(x)\,.$$

By Cramer's rule it is easily seen that $\alpha(x) = P_s(x)/q_s(x)$, where $P_s(x)$ is an $n\times n$ matrix with polynomial entries in x and $q_s(x)$ is a polynomial not vanishing at U_s. We can than extend α to all of M by covering M by open sets $U_1, \ldots U_m$ and defining

$$\alpha = \sum_{s=1}^{m} P_s q_s / q_s^2\,.$$

Notice that α is independent from choices of generators of $\mathcal{I}(V)$, and if $V \in A_d(\mathbb{R}^n)$ then degree$(\alpha) = 4k(d-1)$, where $k = codim(V)$.

PROBLEM 1. By pulling back the universal connection from $G(n,k)$ by the algebraic Gauss map find the relations between the topology, curvature, and the degree of the defining equations of a nonsingular real algebraic set.

Any diffeomorphisim $f : M \to V$ to a nonsingular algebraic set V induces a map between the set of entire rational maps from V to $G(n,k)$ and the set of all maps from M to $G(n,k)$:

$$\theta_f : \mathrm{Rat}(V, G(n,k)) \to \mathrm{Map}(M, G(n,k))\,.$$

Also any two representatives (V_0, f_0) and (V_1, f_1) of $[V,f] \in \mathcal{S}(M)$ induces an isomorphisim $\varphi : \mathrm{Rat}(V_0, G(n,k)) \to \mathrm{Rat}(V_1, G(n,k))$ such that $\theta_{f_1} \circ \varphi = \theta_{f_0}$. This defines a map:

$$\theta : \mathcal{S}(M) \to \text{Subsets of } \mathrm{Map}(M, G(n,k))$$

given by $[M,f] \mapsto \mathrm{image}(\theta_f)$. We can define this map on the level of $Alg(M)$ if we are willing the divide the target by the action of $\mathrm{Diff}(M)$. The image of all rational maps may be too large or even dense; it is more natural to look at the images of degree d entire rational maps, which is finite dimensional:

$$Z_d(V) = \theta_f(\mathrm{Rat}^d(V, G(n,k))\,.$$

PROBLEM 2. Study the topology of $Z_d(V)$ in $\mathrm{Map}(M, G(n,k))$, e.g. study the singularity and non-compactness of this object. Is there a way of associating $Z_d(V)$ a homology cycle in $\mathrm{Map}(M, G(n,k))$? If there is how does this homology class depend on the algebraic structure V?

Having natural homology cycles in $\mathrm{Map}(M, G(n,k))$ is a useful tool in obtaining invariants. For example by imitating [**D**] we can construct a bundle $\xi \longrightarrow M \times \mathrm{Map}(M, G(n,k))$ obtained by pulling back the universal bundle by the evaluation map $ev: M \times \mathrm{Map}(M, G(n,k)) \to G(n,k)$, $ev(x,f) = f(x)$. Then any characteristic class $c \in H^n(M \times \mathrm{Map}(M, G(n,k)))$ of this bundle determines a map:

$$\mu : H_k(M) \to H^{n-k}(\mathrm{Map}(M, G(n,k)))$$

given by the slant product operation $\mu(x) = c/x$. So, if we have a natural homology cycle in $H_{n-k}(\mathrm{Map}(M, G(n,k)))$ depending only on the topology of M, or less ideally on the algebraic structure V of M, we can evaluate $\mu(x)$ on this cycle and get a topological invariant of M, or an invariant of V respectively.

This is the idea of the Donaldson invariants of four dimensional manifolds. In Donaldson's case the homology cycles are given by the subset of anti-self dual connections in the gauge equivalence classes of the space all connections on a fixed bundle over a 4-manifold M, which in turn homotopy equivalent to a component of $\mathrm{Map}(M, G(n,k))$. In this context it is a curious question whether $Z_d(V)$ is related to the set of real stable bundles over V, [**W**]. It likely that in order to get any reasonably nontrivial result from this approach we have to involve the complexification $V_{\mathbb{C}}$. This is one of the lessons we learned from the Nash problem, [**AK2**].

Another natural property of real algebraic sets is that they enable us to define algebraic homology cycles:

Algebraic homology:

Recall [**AK6**], if V is a Zariski open real (or complex) algebraic set and $R = \mathbb{Z}_2$ (or $R = \mathbb{Z}$), then we can define algebraic homology groups $H_*^A(V;R)$ to be the subgroup of $H_*(V;R)$ generated by the compact real (or complex) algebraic subsets of V. It is known that these groups in general do not coincide with the usual homology groups.

When V is real, the resolution theorem implies that $H_*^A(V;\mathbb{Z}_2)$ can equivalently be defined to be the subgroup generated by the classes $g_*([S])$ where $g\colon S \to V$ is an entire rational function, S is a compact nonsingular real algebraic set and $[S]$ is the fundamental class of S. Hence even when V is real, we can define $H_i^A(V;\mathbb{Z})$ to be the subgroup generated by $g_*([S])$ where $g: S \to V$ is an entire rational function from an oriented compact nonsingular real algebraic set and $[S]$ is the fundamental class of S. We define $H_A^*(V;R)$ to be the Poincaré duals of the groups $H_*^A(V;R)$ when defined.

Finally, recall that [**BBK**], for a compact nonsingular real algebraic set V, $H^*_{\mathbb{C}-alg}(V;\mathbb{Z})$ is the subgroup of $H^*(V;\mathbb{Z})$ generated by the restriction of the classes of $H_A^*(V_{\mathbb{C}};\mathbb{Z})$ by the complexification map $i : V \hookrightarrow V_{\mathbb{C}}$ (this is well defined). For convenience we define $H^*_{\mathbb{C}-alg}(V;\mathbb{Z}_2)$ to be the mod 2 reduction of $H^*_{\mathbb{C}-alg}(V;\mathbb{Z})$.

One of the nice applications of the algebraic Gauss map is that all Steifel-Whitney classes of any compact nonsingular real algebraic set V are represented by algebraic subsets. This is because the tangent bundle map $\alpha : V \to G(n,m)$ is entire rational and the Steifel-Whitney classes of $G(n,m)$ are represented by algebraic subsets of $G(n,m)$ (namely the Shubert subvarieties, cf. [**AK1**]). It is well known that the Chern classes of a complex algebraic set are algebraic (e.g. [**F**]); and since $p_k(V) = (-1)^k i^* c_{2k}(V_{\mathbb{C}})$ then Pontryagin classes are in $H^*_{\mathbb{C}-alg}(V;\mathbb{Z})$. The following implies that the Ponryagin classes $p_k(V)$ are also represented by real algebraic subsets.

PROPOSITION 2. *If V is a compact nonsingular real algebraic set then*

$$H^*_{\mathbb{C}-alg}(V;\mathbb{Z}_2) \subset H^*_A(V;\mathbb{Z}_2)\,.$$

PROOF. Let $V \subset V_{\mathbb{C}}$ be the nonsingular projective complexification. Let $\alpha_1 \in H^{2k}_{\mathbb{C}-alg}(V;\mathbb{Z}_2)$ be represented by the restriction of $\alpha_2 \in H^{2k}_A(V_{\mathbb{C}};\mathbb{Z}_2)$. Let $\beta_2 \in H^A_{2m}(V_{\mathbb{C}};\mathbb{Z}_2)$ be the Poincaré dual of α_2 in $V_{\mathbb{C}}$, where $2m = 2n-2k$ and n is the complex dimension of $V_{\mathbb{C}}$. So we can represent β_2 by a complex algebraic subset of $V_{\mathbb{C}}$. In particular if W denotes the underlying real algebraic structure of $V_{\mathbb{C}}$ then $\beta_2 \in H^A_{2m}(W;\mathbb{Z}_2)$. Hence β_2 is represented by $g_*([S])$, where $g : S \to V_{\mathbb{C}}$ is an entire rational function from a compact nonsingular real algebraic set and $[S]$ is the fundamental class.

We can isotop g to a smooth function $g_0 : S \to V_{\mathbb{C}}$ such that it is transverse to V in $V_{\mathbb{C}}$. Then by [**AK6**, Proposition 2.8.8], we can find a nonsingular algebraic set S' and a rational diffeomorphisim $\pi : S' \to S$ and a rational map $g_1 : S' \to V_{\mathbb{C}}$ such that $g_o \circ \pi$ is ϵ-close to g_1, hence g_1 is transverse to V. Clearly $h_*([T])$, where $T = g_1^{-1}(V)$ and $h : T \to V$ is the restriction of g_1, represent the Poincaré dual of α_1 . Hence $\alpha_1 \in H^{2k}_A(V;\mathbb{Z}_2)$. $\square$

Remark. One can show that $H^*_{\mathbb{C}-alg}(V;\mathbb{Z}) \subset H^*_A(V;\mathbb{Z})$, if V is orientable. Also the subgroup of $H^*_{\mathbb{C}-alg}(V;\mathbb{Z}_2)$, consisting of restrictions of the complex algebraic cycles coming from $A_{\mathbb{R}}(V_{\mathbb{C}})$, is the the subgroup of $H^*_A(V;\mathbb{Z}_2)$ generated by cup product squares of the cycles coming from $A(V)$ (this was observed jointly with G. Mikhalkin).

Consider the map:

$$\psi : \mathcal{S}(M) \to \text{ Subgroups of } H^*(M;\mathbb{Z}_2)$$

defined by $(V,f) \longmapsto f^* H^*_A(V;\mathbb{Z}_2)$. By [**BD**] this map is not onto in general, and it depends on the smooth structure of M because by [**AK7**] we can always find a homeomorphisim $f : M \to V$ to a possibly singular algebraic set V with $H^*(M,\mathbb{Z}_2) = f^* H^*_A(V;\mathbb{Z}_2)$. As in the case of θ the map ψ descends to $Alg(M)$ if we divide the image by the action of $Aut(H^*(M;\mathbb{Z}_2))$.

PROBLEM 3. Obtain topological invariants of M from the map ψ.

NOTE ADDED IN PROOF. In [**AK8**] we have shown in general that

$$\bar{H}^{2k}_{\mathbb{C}-alg}(V;\mathbb{Z}_2) = \{\alpha^2 \mid \alpha \in H^k_A(V;\mathbb{Z}_2)\}$$

where $\bar{H}^*_{\mathbb{C}-alg}(V;Z_2)$ is the subgroup of $H^*(V;Z_2)$ generated by the restrictions of cohomology classes of $V_{\mathbb{C}}$, which are the duals of complex algebraic subsets defined over $\mathbb{R}$. One of the amusing corollaries of this result is that there exists closed smooth submanifolds $M \subset \mathbb{R}^n \subset \mathbb{RP}^n$ which can not be isotoped to the real parts of nonsingular complex algebraic subsets of $\mathbb{CP}^n$. □

REFERENCES

[AK1] S. Akbulut and H. King, *The topology of real algebraic sets with isolated singularities*, Annals of Math. **113** (1981), 425–446.

[AK2] S. Akbulut and H. King, *On approximating submanifolds by algebraic sets and a solution to the Nash conjecture*, Invent. Math. **107** (1992), 87–98.

[AK3] S. Akbulut and H. King, *Algebraicity of immersions* (to appear in Topology).

[AK4] S. Akbulut and H. King, *The topology of real algebraic sets*, L'Enseignement Math. **29** (1983), 221–261.

[AK5] S. Akbulut and H. King, *Submanifolds and homology of nonsingular algebraic varieties*, American Journal of Math. (1985), 45–83.

[AK6] S. Akbulut and H. King, *The topology of real algebraic sets*, M.S.R.I. book series (to appear).

[AK7] S. Akbulut and H. King, *All compact manifolds are homeomorphic to totally algebraic real algebraic sets*, Comment. Math. Helv. **66** (1991), 139–149.

[AK8] S. Akbulut and H. King, *Transcendental submanifolds of* $\mathbb{R}^n$ (preprint).

[B] E. Ballico, *An addendum on algebraic models of smooth manifolds*, Geometriae Dedicata **38** (1991), 343–346.

[BD] R. Benedetti and M. Dedo, *Counterexamples to representing homology classes by real algebraic subvarieties up to homeomorphism.*

[BK] J. Bochnak and W. Kucharz, *Nonisomorphic algebraic models of a smooth manifold*, Math. Ann. **290** (1991), 1–2.

[BBK] J. Bochnak, M. Buchner, and W. Kucharz, *Vector bundles over real algebraic varieties*, K-Theory **3** (1990), 271–298.

[D] S. Donaldson, *Connections, cohomology and the intersection forms of 4-manifolds*, Jour. of Diff. Geometry **24** (1986), 275–341.

[F] W. Fulton, *Intersection Theory*, Springer-Verlag, 1984.

[Na] A. Nabutovsky, *Isotopies and non-recursive functions in real algebraic geometry*, Lecture Notes in Math., vol. 1420, Springer-Verlag, Berlin, pp. 194–205.

[R] V.A. Rohlin, *Complex topological characteristics of real algebraic curves*, Uspekhi Math. Nauk **33**(5) (1978), 72–89.

[V1] O. Ya. Viro, *Advances in the topology of real algebraic manifolds during the last six years*, Uspekhi Math. Nauk **41**(3) (1986), 45–67.

[V2] O. Ya. Viro, *Real algebraic plane curves: Constructions with controlled topology*, Leningrad Math. J. **1**(5) (1990), 1059–1134.

[W] S. Wang, *Moduli spaces over manifolds with involutions*, Ph.D. thesis, Oxford University.

MICHIGAN STATE UNIVERSITY, EAST LANSING, MI 48824–0001
E-mail address: akbulut@math.msu.edu

Contemporary Mathematics
Volume **155**, 1994

On Local Uniformization of Orderings

CARLOS ANDRADAS AND JESÚS M. RUIZ

1. Introduction

Let A be an excellent domain with quotient field K. The initial motivation of this paper was the problem of existence of *real* valuation rings V of K having prescribed centers and invariants over A. Here, by *real* we mean that the residue field of V is formally real. However, in studing this one is immediately faced with a problem of uniformization or solving singularities in one way or another. This is of course a fundamental question, so that in the end the paper is devoted mainly to it, relegating the existence of valuations to the last section as an application.

Although real valuations have been studied and used by quite a few number of authors ([**Kn1**], [**Kn2**], [**Lg**], [**Lm**], etc.) the problem of their existence as it is formulated above has not been object of study until the last years. As far as we know, the literature about this question reduces to [**An1**], [**An2**], [**An3**], [**Jd**], [**Kl-Pr**], [**Rb**], [**Be1**], [**Br-Sch**], [**Rz2**]. Except the latter, all these papers deal with function fields of real algebraic varietes, and the techniques that they use are quite distinct: triangulations and real spectra in [**Jd**], model theory in [**Kl-Pr**], Nash wings in [**Rb**], the trace formula in [**Be1**], Hironaka's desingularization in [**Br-Sch**] and commutative algebra methods in [**An1**], [**An2**], [**An3**], more in the spirit of Zariski's local uniformization. The paper [**Rz2**] contains a study of existence of valuation rings in excellent domains by similar methods to those of [**An1**], [**An2**], [**An3**]. This confrontation, Hironaka's global desingularization versus Zariski's local uniformization, is at the basis of this paper. Indeed, since we deal with excellent rings which contain $\mathbf{Q}$, we could have used global desingularization, but we have chosen to develop a more direct approach showing uniformization results by means of quadratic transforms, instead of arbitrary blowing-ups.

1991 *Mathematics Subject Classification.* 14P99 13A18.

Carlos Andradas and Jesús M. Ruiz were partially supported by D.G.I.C.Y.T. PB 89/0379–C02–02.

This paper is submitted in final form and no version of it will be submitted for publication elsewhere.

Since every real valuation ring V of a field K can be obtained from the convex hull of $\mathbf{Q}$ with respect to a suitable ordering of K, and since in some literature the term real valuation is reserved for valuations having as value group a subgroup of $\mathbf{R}$, we rather prefer to formulate the problem in terms of orderings of K, or more precisely of points of the real spectrum $Spec_r(A)$ of A. For the background concerning real spectra we refer to [**B-C-R**], [**Be2**] or the forthcoming [**A-B-R**]. We only remember here that a *prime cone* $\alpha \in Spec_r(A)$ can be seen as a homomorphism $\alpha : A \to \kappa(\alpha)$, where $\kappa(\alpha)$ is a real closed field; the kernel of this homomorphism is called the *support* of α and denoted by $\operatorname{supp}(\alpha)$. We let $\kappa(\operatorname{supp}(\alpha))$ be $A/\operatorname{supp}(\alpha))$. Thus $\kappa(\operatorname{supp}(\alpha)) \subset \kappa(\alpha)$ and we can always take $\kappa(\alpha)$ algebraic over the residue field $\kappa(\operatorname{supp}(\alpha))$. It is also customary to see the elements $f \in A$ as functions on $Spec_r(A)$ by setting $f(\alpha) = \alpha(f)$. This way the sign condition $f > 0$ has a precise meaning, and in fact the sets $\{f > 0\}$ form a subbasis for a topology in $Spec_r(A)$.

Much of the paper is devoted to show uniformization of a given α by means of quadratic transforms. To be more precise, let A be a local excellent domain of dimension d, with maximal ideal $\mathfrak{m}$, residue field k and quotient field K. Then $Spec_r(A)$ contains $Spec_r(k)$ and $Spec_r(K)$. Namely, points of $Spec_r(k)$ can be seen as the points of $Spec_r(A)$ whose support is the maximal ideal of A, and points of $Spec_r(K)$ can be identified with the points of $Spec_r(A)$ whose support is the zero ideal. Therefore we can discuss generations in $Spec_r(K)$ of points in $Spec_r(k)$. Given an ordering α of k, we denote by $\mathcal{G}_\alpha$ the set of all generations of α in $Spec_r(A)$ whose support is (0). Hence $\mathcal{G}_\alpha \subset Spec_r(K)$ and inherits the same Harrison topology as a subset of $Spec_r(A)$ and as a subset of $Spec_r(K)$.

Thus we are lead to the general setting all through this paper: *we have a local excellent domain A and an ordering α in the residue field k of A, and consider the space $\mathcal{G}_\alpha$ of all generations of α in the quotient field K of A.*

Let $\beta \in \mathcal{G}_\alpha$. The notion of the *quadratic transforms of A along $\beta \to \alpha$* is discussed in §2, as well as those of *residue dimension, rank, rational rank of $\beta \to \alpha$*. Then:

DEFINITION 1.1. We say that β *uniformizes* α if there is a quadratic transform $A^{(i)}$ of A along $\beta \to \alpha$ which is regular of dimension d. If we need to stress the upper index i of the quadratic transform, we will say that β uniformizes α in the i-th step. We denote by $\mathcal{U}_\alpha$ the subset of all $\beta \in \mathcal{G}_\alpha$ which uniformize α.

DEFINITION 1.2. We say that β *blows-up* α if there is a quadratic transform $A^{(i)}$ of A along $\beta \to \alpha$ which is regular of dimension d and β specializes in it in the form $\beta = \alpha_d \to \alpha_{d-1} \to \cdots \to \alpha_0$, with $\alpha_l \cap A = \alpha$ for $l = 0, \dots, d-1$. We denote by $\mathcal{B}_\alpha$ the set of all $\beta \in \mathcal{G}_\alpha$ which blow-up α.

Notice that by the very definition, if β uniformizes α then $\dim(A^{(i)}) = \dim(A)$. By the dimension formula [**Mt**, 14.C, theorem 23], this equality of dimensions holds whenever we deal with specializations $\beta \to \alpha$ with residue dimension 0. Also, if β blows-up α, then $\beta \to \alpha$ has rank d and from [**Ab1**] we get that

$\beta \to \alpha$ has residue dimension 0. Therefore, specializations $\beta \to \alpha$ with residue dimension 0 will play a dramatic role in the paper. However not every $\beta \in \mathcal{G}_\alpha$ such that $\beta \to \alpha$ has residue dimension 0 uniformizes α as the following example shows.

EXAMPLE 1.3. Let A be the local ring of Whitney's umbrella at the origin, that is, $A = \mathbf{R}[x,y,z]_{(x,y,z)} = \mathbf{R}[X,Y,Z]_{(X,Y,Z)}/(X^2 - ZY^2)$. Let α be the unique point of $Spec_r(A)$ with $\operatorname{supp}(\alpha) = (x,y,z)$ and let $\beta \in \mathcal{G}_\alpha$ be such that x, y, z are positive and Y is infinitesimal with respect to Z. Then

$$A^{(1)} = A[x', y'] = \mathbf{R}[X', Y', Z]_{(X',Y',Z)}/(X'^2 - ZY'^2)$$

where $x' = x/z$ and $y' = y/z$. Therefore we have $A^{(i)} \simeq A$ for all i.

In view of this example one can ask in a quite natural way, how large or small the sets $\mathcal{U}_\alpha$ and $\mathcal{B}_\alpha$ are. The main results that we will show are the following:

THEOREM 1.4. *The interior of the set* $\mathcal{U}_\alpha$ *is dense in* $\mathcal{G}_\alpha$.

THEOREM 1.5. *The set* $\mathcal{B}_\alpha$ *is dense in* $\mathcal{G}_\alpha$.

The paper is organized as follows. In §§2 and 3 we introduce the notions and algebraic preliminaries needed for the sequel; §4 is devoted to show the existence of enough $\beta \in \mathcal{G}_\alpha$ such that $\beta \to \alpha$ has rank 1 and residue dimension 0. The reason for residue dimension 0 has already been explained. The one for the rank 1 is that we will pass to the completion $\hat{A}$ of A and rank 1 specializations are well behaved with respect to completion as discussed in §3. Sections 5 and 6 are devoted respectively to the proof of Theorems 1.4 and 1.5. Finally, in §7 we apply the previous results to show the existence of different kinds of orderings $\beta \in \mathcal{G}_\alpha$, with preasigned data as residue dimension, rank and rational rank of $\beta \to \alpha$, as well as the existence of specialization chains $\alpha_{j_1} \to \alpha_{j_2} \cdots \to \alpha_{j_r} = \alpha$ with preasigned centers and ranks. As a particular case we single out the following result, where $\mathcal{P}_\alpha$ represents the set of all $\beta \in \mathcal{G}_\alpha$ which *define prime divisors*:

PROPOSITION 1.6. *The set* $\mathcal{P}_\alpha$ *is dense in* $\mathcal{G}_\alpha$.

Actually we will even show that the set $\mathcal{P}_\alpha \cap \mathcal{B}_\alpha$ is dense in $\mathcal{G}_\alpha$. Hereby, $\beta \in \mathcal{G}_\alpha$ *defines a prime divisor* if the unique rank 1 valuation ring containing $V_{\beta\alpha}$ is a prime divisor, that is, its residue dimension over A is $d-1$. The problem of showing the existence of real prime divisors for algebraic varieties was raised by Brumfiel [**Bf**] and solved by different methods in [**An1**], [**An2**], [**Be1**], [**Rb**] and [**Rz3**].

2. Preliminaries

We will use the following standard notations from valuation theory: given a valuation ring V of a field K, we will denote by v the valuation associated to V, by Γ_v its value group, by $\mathfrak{m}_v$ its maximal ideal and by k_v its residue field. If V contains a subring A of K we say that V (or v) *is finite over* A; then we say that V (or v) *is centered at* $\mathfrak{p} = \mathfrak{m}_v \cap A$, which is a prime ideal of A called the *center of* V (or v). Also, we have the usual notions of *rank, rational rank* and *residue dimension over* A (see [**Ab2**], [**Z-S**, Appendix 2]).

Let A be a ring and $\beta \to \alpha$ a specialization in $Spec_r(A)$. We define the *dimension of* $\beta \to \alpha$ as the height of the ideal $\mathfrak{p} = \operatorname{supp}(\alpha)/\operatorname{supp}(\beta)$ in the ring $A/\operatorname{supp}(\beta)$. We attach to $\beta \to \alpha$ a valuation ring $V_{\beta\alpha}$ of the quotient field, $\kappa(\operatorname{supp}(\beta))$, of $A/\operatorname{supp}(\beta)$ in the following way: $V_{\beta\alpha}$ is the convex hull in $\kappa(\operatorname{supp}(\beta))$ of the local ring $(A/\operatorname{supp}(\beta))_{\mathfrak{p}}$ with respect to β, now seen as an ordering of the field $\kappa(\operatorname{supp}(\beta))$. Thus $V_{\beta\alpha}$ is a real valuation ring whose maximal ideal, $\mathfrak{m}_{\beta\alpha}$, is convex with respect to β, and we have the following compatibility condition: if $f, g \in K$ are such that $0 < \beta(f) \leq \beta(g)$, then $v_{\beta\alpha}(f) \geq v_{\beta\alpha}(g)$. Let $\alpha^* \in Spec_r(V_{\beta\alpha})$ be the point defined by the ordering induced by β in the residue field $k_{\beta\alpha}$ of $V_{\beta\alpha}$. To keep the notation consistent, we will denote by β^* the point of $Spec_r(V_{\beta\alpha})$ defined by β. Then $\alpha^* \cap A = \alpha$ and $(k_{\beta\alpha}, \alpha^*)$ is an archimedean extension of (k, α). Moreover, this property of the residue field determines completely $V_{\beta\alpha}$ among the the set of valuation rings of K which are compatible with β. We denote by $v_{\beta\alpha}$ and $\Gamma_{\beta\alpha}$ the valuation and value group associated to the valuation ring $V_{\beta\alpha}$.

We define the *rank of* $\beta \to \alpha$ to be the rank of the valuation ring $V_{\beta\alpha}$. In a similar way we define the *rational rank, residue dimension over A, value group, etc., of* $\beta \to \alpha$ to be the corresponding ones of the valuation ring $V_{\beta\alpha}$.

Assume now that A is local with maximal ideal $\mathfrak{m}$ and residue field k. We will use the standard notation $(A, \mathfrak{m}, k)$ to summarize this situation. Suppose that $\operatorname{supp}(\alpha) = \mathfrak{m}$ and $\operatorname{supp}(\beta) = (0)$, so that in particular A is a domain and β can be seen as an ordering in its quotient field K. We define the *first quadratic transform* $(A^{(1)}, \mathfrak{m}^{(1)}, k^{(1)})$ *of* $(A, \mathfrak{m}, k)$ *along* $\beta \to \alpha$ in the following way: choose any generators $x_1, \dots, x_n$ of $\mathfrak{m}$ with $\beta(x_i) > 0$ for all i; then $A^{(1)} = B^{(1)}_{\mathfrak{p}^{(1)}}$ where

$$B^{(1)} = A[x_2/x_1, \dots, x_n/x_1], \qquad \mathfrak{p}^{(1)} = \mathfrak{m}_{\beta\alpha} \cap B^{(1)}$$

and x_1 is determined by the condition $\beta(x_1) > \beta(x_i)$ for all $i = 2, \dots, n$.

In particular, by the compatibility of β and $v_{\beta\alpha}$ we have $v_{\beta\alpha}(x_1) \leq v_{\beta\alpha}(x_i)$ for all i and $A^{(1)}$ coincides with the first quadratic transform of A along the valuation ring $V_{\beta\alpha}$ in the classical sense of [**Ab2**], [**Sh**]. This shows, in particular, that $A^{(1)}$ does not depend on the choice of the generators $x_1, \dots, x_n$ of $\mathfrak{m}$.

By recurrence we define the *i-th quadratic transform* $(A^{(i)}, \mathfrak{m}^{(i)}, k^{(i)})$ *of* A *along* $\beta \to \alpha$ as the first quadratic transform of $(A^{(i-1)}, \mathfrak{m}^{(i-1)}, k^{(i-1)})$. Obviously $A^{(i)}$ is an essentially finitely generated A algebra with the same quotient field. Therefore, by the dimension formula, [**Mt**, 14.C, Theorem 23], we have $\dim(A) = \dim(A^{(i)}) - \operatorname{tr.deg.}(k^{(i)} : k)$ and $k^{(i)}$ is finitely generated over k. In particular if $\beta \to \alpha$ has residue dimension 0 over A then $k^{(i)}$ is a finite algebraic extension of k and $\dim(A) = \dim(A^{(i)})$.

We will denote by $\beta^{(i)}$ the point in $Spec_r(A^{(i)})$ defined as $\beta^{(i)} = \beta^* \cap A^{(i)}$. Similarly, we will set $\alpha^{(i)} = \alpha^* \cap A^{(i)}$. Thus we have $\beta^{(i)} \to \alpha^{(i)}$, $\beta^{(i)} \cap A = \beta$

and $\alpha^{(i)} \cap A = \alpha$. We will refer to $\alpha^{(i)}$ and $\beta^{(i)}$ as the *i-th quadratic transforms of* α *and* β*, respectively, along* $\beta \to \alpha$.

REMARK 2.2. *a*) Sometimes we will need to deal with a ring A which is not a domain. Then A will be reduced and $\text{supp}(\beta)$ a minimal prime of A. To adapt the notations, we will denote by K the total ring of fractions of A and by $\mathcal{G}_\alpha \subset Spec_r(K)$ the set of all generations of α whose support is a minimal prime of A.

Then, notice that the above definition of first quadratic transform makes also sense whenever x_1 is a regular element of A, that is, it is not a zero divisor, although we need some obvious modifications. In fact, in this case, $B^{(1)}$ is a subring of the total ring of fractions of A, and if we denote by $\pi_{\beta\alpha} : A \to V_{\beta\alpha}$ the canonical homomorphism, then $\pi_{\beta\alpha}$ has a unique extension to a homomorphism

$$\pi_{\beta\alpha}^{(1)} : B^{(1)} \to V_{\beta\alpha}.$$

Then, we define $A^{(1)}$ as $A^{(1)} = {B^{(1)}}_{\mathfrak{p}^{(1)}}$, where $\mathfrak{p}^{(1)} = \left(\pi_{\beta\alpha}^{(1)}\right)^{-1}(\mathfrak{m}_{\beta\alpha})$.

Since for any $f \in B^{(1)}$ we have $x_1^p f \in A$ for a suitable p, it follows that $B^{(1)}$ is reduced. Moreover, it follows that $A^{(1)}/\ker(\pi_{\beta\alpha})A^{(1)}$ is the first quadratic transform of the domain $A/\text{supp}(\beta)$ along $\beta \to \alpha$.

b) We have already mentioned that quadratic transforms do not depend on the choice of the generators. However, different orderings can define the same quadratic transform. For instance, consider $A = \mathbf{R}[t]_{(t)}$. There is a unique choice for α and two for $\beta \to \alpha$, namely $\beta(t)$ positive and infinitesimal with respect to $\mathbf{R}$ or negative and infinitesimal with respect to $\mathbf{R}$. However we obtain always the same valuation ring $V_{\beta\alpha} = A$, and the sequence of quadratic transforms is trivial.

There are examples less trivial than the last one. Actually we will use them later, to change a given ordering in $\mathcal{G}_\alpha$ by another one with better properties, without modifying a finite sequence of quadratic transforms. We give here a precise statement for future reference.

LEMMA 2.3.

a) *Let* $(A, \mathfrak{m}, k)$ *be a local ring,* $\alpha \in Spec_r(k)$ *and* $\beta \in \mathcal{G}_\alpha$. *Suppose that* $\beta \to \alpha$ *has rank* 1 *and residue dimension* 0. *Then for every* $i \geq 0$ *there is an open neighborhood* $U^{(i)}$ *of* β *in* $\mathcal{G}_\alpha$ *such that for all* $\gamma \in U^{(i)}$ *the sequence of quadratic transforms of* A *along* $\gamma \to \alpha$ *is the same until the i-th step.*
b) *Assume that* $\beta \to \alpha$ *has residue dimension* 0. *Let* $A^{(i)}$ *be the i-th quadratic transform of* A *along* $\beta \to \alpha$. *Let* $\gamma \in Spec_r(K)$ *be such that* $\gamma \to \alpha^{(i)}$. *Then* $\gamma \in \mathcal{G}_\alpha$ *and the sequences of quadratic transforms of* A *along* $\gamma \to \alpha$ *and along* $\beta \to \alpha$ *coincide until the i-th step.*

PROOF. *a*) Let $\mathfrak{m} = (x_1, \ldots, x_n)$ and $A^{(1)} = A[x_2/x_1, \ldots, x_n/x_1]_{\mathfrak{p}^{(1)}}$. That means that $\beta(x_i) > 0$ for all i and $\beta(x_1) > \beta(x_i)$ for $i \geq 2$. Moreover, let

$y_1, \dots, y_s$ be a system of generators of $\mathfrak{p}^{(1)}$, positive in β. Since $v_{\beta\alpha}$ has rank one, the value group $\Gamma_{\beta\alpha}$ is archimedean and therefore for every j there is $m_j \geq 1$ such that $v_{\beta\alpha}(x_1) < m_j v_{\beta\alpha}(y_j)$. This implies that $\beta(x_1 - y_j^{m_j}) > 0$. Thus

$$\{x_1 > x_2 > 0, \dots, x_1 > x_n > 0, y_1 > 0, \dots, y_s > 0, x_1 > y_1^{m_1}, \dots, x_1 > y_s^{m_s}\} \cap \mathcal{G}_\alpha$$

is an open neighborhood $U^{(1)}$ of β in $\mathcal{G}_\alpha$, and we claim that for any $\gamma \in U^{(1)}$ the first quadratic transform of A along $\gamma \to \alpha$ is $A^{(1)}$.

Indeed, by the very definition $V_{\gamma\alpha}$ is the convex hull of A with respect to γ and therefore it dominates A. In particular $v_{\gamma\alpha}(x) > 0$ for all $x \in \mathfrak{m}$. Moreover, since $\gamma \in U^{(1)}$, it follows that $v_{\gamma\alpha}(x_1) \leq v_{\gamma\alpha}(x_i)$ and $v_{\gamma\alpha}(x_1) \leq m_j v_{\gamma\alpha}(y_j)$. The first inequality implies that $A[x_2/x_1, \dots, x_n/x_1] \subset V_{\gamma\alpha}$ and the second that $\mathfrak{p}^{(1)} \subset \mathfrak{m}_{\gamma\alpha}$. Now since $v_{\beta\alpha}$ is zero dimensional, $\mathfrak{p}^{(1)}$ is a maximal ideal of $A[x_2/x_1, \dots, x_n/x_1]$, and therefore we get the equality $\mathfrak{p}^{(1)} = m_{\gamma\alpha} \cap A[x_2/x_1, \dots, x_n/x_1]$, showing the claim.

Repeating the argument i times we end up with an open neighbourhood $U^{(i)}$ of β in $\mathcal{G}_\alpha$ that verifies the condition of the statement.

b) By recurrence it is enough to show the case $i = 1$. The fact that $\gamma \in \mathcal{G}_\alpha$ is obvious since $\alpha^{(1)} \cap A = \alpha$. Since $\gamma \to \alpha^{(i)}$ then by the very definition $V_{\gamma\alpha^{(1)}}$ dominates $A^{(1)}$ and therefore A. Now, since $\beta \to \alpha$ has residue dimension 0, $k^{(1)}$ is algebraic over k. On the other hand, $k_{\gamma\alpha^{(1)}}$ is an archimedean extension of $(k^{(1)}, \alpha^{(1)})$, and it follows that $k_{\gamma\alpha^{(1)}}$ is an archimedean extension of (k, α). Altogether we get $V_{\gamma\alpha^{(1)}} = V_{\gamma\alpha}$. In particular we have that $V_{\gamma\alpha}$ dominates $A^{(1)}$ and *b*) follows at once. □

The following general algebraic properties will be essential in the sequel:

LEMMA 2.4. *Let $(A, \mathfrak{m}, k)$ and $(B, \mathfrak{n}, k')$ be local rings, and $A \hookrightarrow B$ a local monomorphism such that $\mathfrak{m}B = \mathfrak{n}$. Let $\mathfrak{m} = (x_1, \dots, x_n)A$, and assume that x_1 is a regular element of A and that B is flat over A. Set*

$$A' = A\big[x_2/x_1, \dots, x_n/x_1\big], \quad \textit{and} \quad B' = B\big[x_2/x_1, \dots, x_n/x_1\big]$$

and let $A' \hookrightarrow B'$ be natural monomorphism. Then B' is flat over A'.

PROOF. Since flatness is preserved by base change, it is enough to show that

$$B\big[x_2/x_1, \dots, x_n/x_1\big] = B \otimes_A A\big[x_2/x_1, \dots, x_n/x_1\big].$$

Let $t_i = x_i/x_1$ for $i = 2, \dots, n$. The inclusion $A[t_2, \dots, t_n] \hookrightarrow A_{x_1}$ gives a monomorphism $B \otimes_A A[t_2, \dots, t_n] \hookrightarrow B \otimes_A A_{x_1} = B_{x_1}$, and it is immediate that the image coincides with $B[x_2/x_1, \dots, x_n/x_1]$. □

LEMMA 2.5. *Let $(A, \mathfrak{m}, k)$ and $(B, \mathfrak{n}, k')$ be local, reduced rings such that*

i) *B dominates A,*

ii) *$\mathfrak{m}B = \mathfrak{n}$,*

iii) *$k' \subset B$ and k' is an algebraic extension of k of characteristic zero, and*

iv) *$\mathfrak{m} = (x_1, \dots, x_n)A$ where x_1 is regular.*

Set

$$A' = (A[x_2/x_1, \dots, x_n/x_1])_{\mathfrak{p}} \quad \textit{and} \quad B' = (B[x_2/x_1, \dots, x_n/x_1])_{\mathfrak{q}},$$

where $\mathfrak{p} \cap A = \mathfrak{m}, \mathfrak{q} \cap B = \mathfrak{n}$ *and* B' *dominates* A'. *Assume that* $B'/\mathfrak{q}B' = k'$. *Then* $\mathfrak{p}B' = \mathfrak{q}B'$, *and this ideal is generated by* n *elements.*

PROOF. Since $B'/\mathfrak{q}B' = k'$ there are $c_2, \dots, c_n \in k'$ such that $x_1, c_2 - x_2/x_1, \dots, c_n - x_n/x_1 \in \mathfrak{q}$. Now note that $B[x_2/x_1, \dots, x_n/x_1]/(x_1, c_2 - x_2/x_1, \dots, c_n - x_n/x_1) = k'$. Hence the elements $x_1, c_2 - x_2/x_1, \dots, c_n - x_n/x_1$ generate a maximal ideal and we conclude $(x_1, c_2 - x_2/x_1, \dots, c_n - x_n/x_1) = \mathfrak{q}$. Let $k_0 \subset A$ be a field such that k is an algebraic extension of k_0, which exists by an easy application of Zorn's lemma. In particular each c_i is algebraic over k_0. Let $P_i(t) \in k_0[t]$ be its minimal polynomial, and assume that it factorizes in $k'[t]$ as $P_i(t) = (t - c_i)Q_i(t)$, with $Q_i(c_i) \neq 0$. Then, in B', we have $P_i(x_i/x_1) = (x_i/x_1 - c_i)Q_i(x_i/x_1)$, and the condition $Q_i(c_i) \neq 0$ implies $Q_i(x_i/x_1) \notin \mathfrak{q}$. It follows that $Q_i(x_i/x_1)$ is a unit in B' and therefore $\mathfrak{q} = (x_1, P_2(x_2/x_1), \dots, P_n(x_n/x_1))$. Since $P_i(x_i/x_1) \in A'$, the proof is complete. □

We finish this section by recalling an algebraic invariant which will be of utmost importance for us in the sequel. Let $(A, \mathfrak{m}, k)$ be a local ring: the *embedding dimension* $\operatorname{emb}(A)$ *of* A is the minimal number of generators of the maximal ideal $\mathfrak{m}$. As is well known, it follows from Nakayama's lemma that

$$\operatorname{emb}(A) = \dim_k (\mathfrak{m}/\mathfrak{m}^2).$$

and that any system of generators of $\mathfrak{m}$ contains another with exactly $\operatorname{emb}(A)$ elements. Also, by dimension theory $\dim(A) \leq \operatorname{emb}(A)$ and the equality holds if and only if the ring A is regular. In this sense, the lower the embedding dimension is, the tamer the singularity of A is. Finally we see from Lemma 2.5 *iv*), that if B is a quadratic transform of A, then

$$\operatorname{emb}(A) \geq \operatorname{emb}(B).$$

This remark contains in fact the whole philosophy of the proof of the uniformization theorem: we will perform quadratic transforms in order to lower the embedding dimension, therefore taming the singularity until it disappears.

3. Quadratic transforms and real strict completions

Our proof of Theorems 1.4 and 1.5 is based on the old idea of passing to the complete case, since then by Cohen's structure theorem we can represent the elements of A as formal power series and understand the effect of quadratic transforms in terms of the descent of multiplicities. However, the property of being complete is lost after quadratic transforms, so that we need to iterate the process, that is, take the completion again, perform one more quadratic transform, etc. On the other hand, by technical reasons it is convenient that the residue field of A is real closed. Thus, we pass first to the real strict henselization and then to the completion, as explained below.

Let us start by recalling some basic facts about completions. Let $(A, \mathfrak{m}, k)$ be an excellent, reduced, equidimensional local ring of dimension d. Then, its $\mathfrak{m}$-adic completion $\hat{A}$ is also excellent, reduced, equidimensional of dimension d and the canonical inclusion $A \hookrightarrow \hat{A}$ is a regular morphism, [**Mt**, 34.C, Theorem 79]. Let $\mathfrak{p}_1, \dots, \mathfrak{p}_r$ be the minimal primes of $\hat{A}$. Every domain $\hat{A}/\mathfrak{p}_i$ is called a *formal branch of A*.

From now on we assume that our local ring $(A, \mathfrak{m}, k)$ is a domain of dimension d. Fix $\alpha \in Spec_r(k)$, which can be seen as a point $\alpha \in Spec_r(A)$ with $\operatorname{supp}(\alpha) = \mathfrak{m}$. Let A_α be the real strict henselization of A at α, cf. [**Al-Ry**], [**A-B-R**]. As is well known, A_α is local, excellent, reduced and equidimensional of dimension d with real closed residue field, namely $\kappa(\alpha)$. Moreover, the canonical monomorphism $A \hookrightarrow A_\alpha$ is regular, and $\mathfrak{m}_\alpha = \mathfrak{m}A_\alpha$, where $\mathfrak{m}_\alpha$ denotes the maximal ideal of A_α. Now consider the $\mathfrak{m}_\alpha$-adic completion $\widehat{A_\alpha}$ of A_α. Then $\widehat{A_\alpha}$ has all properties listed above for A_α and the canonical homomorphism $A_\alpha \hookrightarrow \widehat{A_\alpha}$ is also regular. In particular we get a faithfully flat local homomorphism $A \hookrightarrow \widehat{A_\alpha}$ and $\mathfrak{m}$ generates the maximal ideal $\widehat{\mathfrak{m}}_\alpha$ of $\widehat{A_\alpha}$. We set $C = \widehat{A_\alpha}$, $\mathfrak{n} = \widehat{\mathfrak{m}}_\alpha$ and denote by L the total ring of fractions of C. In this way, we have obtained a couple (A, C) in the conditions of Lemmas 2.4 and 2.5, where C is complete and has real closed residue field, namely $\kappa(\alpha)$. We denote by $\hat{\alpha}$ the unique point of $Spec_r(\kappa(\alpha))$ seen as a point of $Spec_r(C)$.

Let $\beta \in \mathcal{G}_\alpha$. It follows from the going-down theorem for real spectra [**Rz5**] that there is some $\hat{\beta} \in Spec_r(L)$ lying over β, specializing to $\hat{\alpha} = \alpha$, and such that $\operatorname{ht}(\hat{\beta}) = 0$. We point out that this $\hat{\beta}$ is not unique, so that we choose one and fix it for the rest of the section. Since A is a domain, any generators $x_1, \dots, x_n$ of $\mathfrak{m}$ are regular elements of A, and by faithful flatness they are also regular elements of C generating $\mathfrak{n}$. In particular, this shows that we are in the conditions of Remark 2.2 *a*), and it makes sense to consider the first quadratic transform $A^{(1)}$ of A along $\beta \to \alpha$, as well as the first quadratic transform $C^{(1)}$ of C along $\hat{\beta} \to \hat{\alpha}$. We want to study the relationship between them.

We do that under a further assumption: we assume that $\hat{\beta} \to \hat{\alpha}$ has rank one and residue dimension 0. In particular, the latter implies that the residue field $k_{\hat{\beta}\hat{\alpha}}$ of the valuation ring $V_{\hat{\beta}\hat{\alpha}}$ of L is algebraic over $\kappa(\alpha)$ and therefore coincides with it. Moreover, we claim that $V_{\hat{\beta}\hat{\alpha}} \cap K = V_{\beta\alpha}$. Indeed, since $V_{\hat{\beta}\hat{\alpha}}$ is compatible with β and dominates A, we have $V_{\beta\alpha} \subset V_{\hat{\beta}\hat{\alpha}} \cap K$. Moreover, the residue field $k_{\hat{\beta}\hat{\alpha}} = \kappa(\alpha)$ is algebraic over k and in particular archimedean over (k, α). Therefore, we get the equality $V_{\beta\alpha} = V_{\hat{\beta}\hat{\alpha}} \cap K$. Thus, it follows at once that $\beta \to \alpha$ has also rank one and residue dimension 0, and that $C^{(1)}$ dominates $A^{(1)}$. Now, by Lemma 2.4, this extension is flat, and from Lemma 2.5, we get $\mathfrak{m}^{(1)}C^{(1)} = \mathfrak{n}^{(1)}$. Finally, let us denote by C_1 the $\mathfrak{n}^{(1)}$-adic completion $\widehat{C^{(1)}}$ of $C^{(1)}$. Notice that the residue field of C_1 is $\kappa(\alpha)$ which is real closed and therefore C_1 coincides with the completion of the real strict henselization of $C^{(1)}$. We give a special name to this ring:

PROPOSITION AND DEFINITION 3.1. *The ring* $C_1 = \widehat{C^{(1)}}$ *is called the first complete real strict quadratic transform of* C *along* $\hat\beta \to \hat\alpha$. *It holds:*

i) C_1 *is a reduced local ring with residue field* $\kappa(\alpha)$ *which dominates* $A^{(1)}$ *and is flat over it;*

ii) $\mathfrak{n}_1 = \mathfrak{m}^{(1)}C_1$ *is the maximal ideal of* C_1;

iii) $\mathfrak{m}^{(1)}$ *is generated by regular elements;*

iv) $\operatorname{emb}(A^{(1)}) = \operatorname{emb}(C_1)$;

v) $A^{(1)}$ *is regular if and only if* C_1 *is regular.*

PROOF. We have already pointed out, right before the definition, that $A^{(1)} \hookrightarrow C^{(1)}$ is flat and that $\mathfrak{m}^{(1)}C^{(1)}$ is the maximal ideal of $C^{(1)}$. Since C_1 is the completion of $C^{(1)}$, *i)* and *ii)* follow at once. Now *iii)* is an inmediate consequence of *ii)*, and *iv)* follows from *ii)* and the formula $\operatorname{emb}(C_1) = \dim(\mathfrak{n}_1/(\mathfrak{n}_1)^2)$. Finally, we get *v)* directly from *iv)*. □

Moreover, since $\hat\beta \to \hat\alpha$ has rank one, the $\mathfrak{m}_{\hat\beta\hat\alpha}$ topology of $V_{\hat\beta\hat\alpha}$ is finer than the topology defined by the valuation $v_{\hat\beta\hat\alpha}$. Thus, the local homomorphism $\pi^{(1)}_{\hat\beta\hat\alpha}: C^{(1)} \to V_{\hat\beta\hat\alpha}$, is continuous when we equip the first ring with the adic topology and the second with the valuation toplogy. Therefore it extends uniquely to a local homomorphism $\pi_1 : C_1 \to V^*$, where V^* stands for the completion of the valuation $V_{\hat\beta\hat\alpha}$ with its valuation topology. As V^* is an immediate extension of $V_{\hat\beta\hat\alpha}$, [**Ri**], by the Baer-Krull theorem, [**B-C-R**, Proposition 10.1.8, p. 217], there is a specialization $\beta^* \to \alpha^*$ in $Spec_r(V^*)$ lying over $\hat\beta \to \hat\alpha$, with $\operatorname{supp}(\alpha^*) = \mathfrak{m}^*$ and $\operatorname{supp}(\beta^*) = (0)$. Let $\alpha_1 = (\pi_1)^{-1}(\alpha^*)$ and $\beta_1 = (\pi_1)^{-1}(\beta^*)$. By construction we have $A \hookrightarrow A^{(1)} \hookrightarrow C_1/\operatorname{supp}(\beta_1)$, and the chain $\beta_1 \to \alpha_1$ of $Spec_r(C_1)$ lies over $\beta \to \alpha$. In general the dimension of $C_1/\operatorname{supp}(\beta_1)$ may be less that d. However, it holds

LEMMA 3.2. *If* $\hat\beta \to \hat\alpha$ *has rational rank* d, *then* $\dim(C_1/\operatorname{supp}(\beta_1)) = \dim(C_1) = d$, *that is,* $\operatorname{supp}(\beta_1)$ *defines a formal branch of* C_1.

PROOF. We have $\kappa(\operatorname{supp}(\hat\beta)) \subset \kappa(\operatorname{supp}(\beta_1)) \subset L^*$ where L^* is the quotient field of V^*. Now V^* is an immediate extension of $V_{\hat\beta\hat\alpha}$, and therefore so is its restriction W to the field $\kappa(\operatorname{supp}(\beta_1))$. Consequently, W is a valuation ring of rank 1 and rational rank d dominating $C_1/\operatorname{supp}(\beta_1)$. It follows from [**Ab2**, Proposition 2], that $\dim(C_1/\operatorname{supp}(\beta_1)) \geq d$. Next we have $\dim(A) = \dim(A_\alpha) = \dim(C) = \dim(C/\operatorname{supp}(\hat\beta)) = d$. On the other hand, by flatness it follows that $\dim(C_1) \leq \dim(A^{(1)}) \leq \dim(A) = d$. In fact, by the assumption that $\beta \to \alpha$ has residue dimension 0 this last inequality is an equality, but we do not need it here. The lemma is now obvious. □

Next, since by Proposition 3.1 *iii)*, $\mathfrak{m}_1$ is generated by regular elements, and the couple $(A^{(1)}, C_1)$ verifies the conditions of the hypotheses of Lemmas 2.4 and 2.5. Hence we can perform again the quadratic transforms of $A^{(1)}$ and C_1

and compare them to get a similar couple $A^{(2)}$ and C_2. Proceeding this way, we produce two sequences of local rings, $(A^{(i)}, \mathfrak{m}^{(i)}, k^{(i)})$ and $(C_i, \mathfrak{n}_i, \kappa(\alpha))$, and a commutative diagram:

$$\begin{array}{ccccccc} C = C_0 & \longrightarrow & C_1 & \longrightarrow \cdots \longrightarrow & C_i & \longrightarrow \cdots \\ \uparrow & & \uparrow & & \uparrow & \\ A = A_0 & \longrightarrow & A_1 & \longrightarrow \cdots \longrightarrow & A_i & \longrightarrow \cdots \end{array}$$

PROPOSITION AND DEFINITION 3.3. *$(C_i, \mathfrak{n}_i, \kappa(\alpha))$ is called the i-th complete real strict quadratic transform of C along $\hat{\beta} \to \hat{\alpha}$. It holds:*

i) The extension $A^{(i)} \hookrightarrow C_i$ is faithfully flat;
ii) $\mathfrak{n}_i = \mathfrak{m}^{(i)} C_i$, where $\mathfrak{n}_i$ denotes the maximal ideal of C_i;
iii) $\mathfrak{n}_i$ is generated by regular elements;
iv) $\mathrm{emb}(A^{(i)}) = \mathrm{emb}(C_i)$
v) $A^{(i)}$ is regular if and only if C_i is regular.

Therefore, we reduce our initial problem to check the regularity of the ring C_i for some i. To do it we will need some computations, and to that goal we introduce indeterminates. Namely, for each i we may see $Spec_r(C_i)$ as a subspace of $Spec_r(k[[X_1, \dots, X_{e_i}]])$ for some formal power series ring, where $e_i = \mathrm{emb}(C_i)$. Then, roughly speaking, we will realize its quadratic transform as the strict transform inside some quadratic transform of the ring $k[[X_1, \dots, X_{e_i}]]$. Thus, as announced above, we will be able to decide the regularity of the ring $C_i^{(1)}$, (or more precisely of C_{i+1} and therefore also of $A^{(i+1)}$), in terms of the multiplicities of certain elements.

Let us make this precise. For the sake of simplicity we omit the subscript i, of C_i, $\mathfrak{n}_i$, etc., keeping in mind that the process can be started in any C_i. Set $\mathfrak{n} = (x_1, \dots, x_e)$, with $e = \mathrm{emb}(C)$ and all x_j are regular elements. Then, by Cohen's structure theorem, we have $C = R/I$, where $R = \kappa(\alpha)[[X_1, \dots, X_e]]$, $x_i = X_i + I$ and I is a radical ideal of R. We can see α and β as points in $Spec_r(R)$ with $\mathrm{supp}(\alpha) = (X_1, \dots, X_e)$ and $\mathrm{supp}(\beta)$ is an associated prime of I. Since $R_{\mathrm{supp}(\beta)}$ is a regular ring, there exists a valuation ring W of the quotient field L of R which dominates $R_{\mathrm{supp}(\beta)}$ and has residue field $\kappa(\mathrm{supp}(\beta))$. Again by Baer-Krull, there is a generation $\gamma \in Spec_r(R)$ of β with $\mathrm{supp}(\gamma) = 0$, compatible with W, so that $\gamma \to \beta \to \alpha$. In particular we have $V_{\gamma\beta} = W$.

We want to compare the quadratic transforms of R along $\gamma \to \alpha$ and those of C along $\beta \to \alpha$. First of all notice that we have the following commutative diagram, where the vertical arrows are the epimorphisms defined by the places associated to the valuation rings:

$$\begin{array}{lll} \lambda_{\gamma\beta}^{-1}(V_{\beta\alpha}) \subset & V_{\gamma\beta} & \subset \kappa(\mathrm{supp}(\gamma)) = L \\ \downarrow \lambda_{\gamma\beta} & \downarrow \lambda_{\gamma\beta} & \\ V_{\beta\alpha} & \subset \kappa(\mathrm{supp}\,(\beta)) & \\ \downarrow \lambda_{\beta\alpha} & & \\ k_{\beta\alpha} = \kappa(\alpha) & & \end{array}$$

By construction, $\lambda_{\gamma\beta}^{-1}(V_{\beta\alpha})$ is a valuation ring dominating R and compatible with γ. Therefore we have an inclusion $V_{\gamma\alpha} \subset \lambda_{\gamma\beta}^{-1}(V_{\beta\alpha})$. Now, since both have residue field $\kappa(\alpha)$, we get the equality.

Next, assume that $\beta(x_1) > \beta(x_i)$ for $2 \le i \le e$. Then, since $\gamma \to \beta \to \alpha$, it is $\gamma(X_1) > \gamma(X_i)$, and therefore we have a commutative diagram

$$\begin{array}{ccccc} R & \longrightarrow & R[X_2/X_1, \dots, X_e/X_1] & \xrightarrow{\pi_{\gamma\alpha}} & V_{\gamma\alpha} \\ \downarrow & & \downarrow & & \downarrow \lambda_{\gamma\beta} \\ C & \longrightarrow & C[x_2/x_1, \dots, x_e/x_1] & \xrightarrow{\pi_{\beta\alpha}} & V_{\beta\alpha} \end{array}$$

where the vertical arrows are the natural epimorphisms. Thus we get a commutative square

$$\begin{array}{ccc} R & \longrightarrow & R^{(1)} \\ \downarrow & & \downarrow \\ C & \longrightarrow & C^{(1)} \end{array}$$

and taking completions we obtain another

$$\begin{array}{ccccc} R & \longrightarrow & R^{(1)} & \longrightarrow & \widehat{R^{(1)}} = R_1 \\ \downarrow & & \downarrow & & \downarrow \\ C & \longrightarrow & C^{(1)} & \longrightarrow & \widehat{C^{(1)}} = C_1 \end{array}$$

where $R_1 = k[[Y_1, \dots, Y_e]]$, $Y_1 = X_1$, $Y_i = X_i/X_1$ and C_1 is the first complete quadratic transform of C along $\beta \to \alpha$. Remember that $\mathrm{emb}(C_1) \le e$. If we have the equality, we are in the same starting situation and we can repeat the construction. This procedure gives what we call an *embedded sequence of complete quadratic transforms of C along $\gamma \to \beta \to \alpha$ for the embedding dimension e*:

$$\begin{array}{ccccccccc} R = R_0 & \longrightarrow & R_1 & \longrightarrow & \cdots & \longrightarrow & R_i & \longrightarrow & \cdots \\ \downarrow \phi_0 & & \downarrow \phi_1 & & \cdots & & \downarrow \phi_i & & \\ C = C_0 & \longrightarrow & C_1 & \longrightarrow & \cdots & \longrightarrow & C_i & \longrightarrow & \cdots \end{array}$$

Note that all vertical arrows in the diagram are surjective; we set $I_i = \ker(\phi_i)$, which is a reduced ideal.

This sequence is finite if we arrive to some C_s whose embedding dimension is strictly smaller than e, at which moment we can see C_s as a quotient of a formal power series ring in less than e variables, and start the corresponding sequence for the new embedding dimension. We will show in Proposition 5.1, that a suitable choice of γ and β guarantees the finiteness of the above sequence.

We finish this section by pointing out an important fact which is not shown in the previous diagrams. We started our construction by completing the valuation $v_{\beta\alpha}$. Then, as explained before, in every step we have a canonical local homomorphism $C^{(i)} \to V^*$ that extends to $C_i \to V^*$. This way we have mappings $R_i \to C_i \to V^* \to \Gamma_{\hat\beta\hat\alpha} \cup \infty$ for all of which we will use the same notation, namely $v_{\beta\alpha}$ (the construction is canonical and there is little risk of confusion). If we denote by $\mathfrak{p}_i$ (resp. $\mathfrak{q}_i$) the kernel of the homomorphism $C_i \to V^*$ (resp. $R_i \to V^*$), then $\mathfrak{p}_i$ is a minimal prime of C_i and $\mathfrak{q}_i$ the corresponding associated prime of I_i, cf. Lemma 3.2. Thus $R_i/\mathfrak{q}_i = C_i/\mathfrak{p}_i$, and $v_{\beta\alpha}$ is a valuation of the quotient field of that domain. So roughly speaking in producing the sequence of complete quadratic transforms of C and R along $\gamma \to \beta \to \alpha$, we produce also, a sequence of formal branches, which geometrically represent the strict transforms of $Spec(C)$ seen as a subvariety of $Spec(R)$, and that are the geometric objects that will become non-singular at the end of the process.

4. Existence of rank 1 valuations

Rank 1 valuations of residue dimension 0 and rational rank d play, as we have seen in the previous section an important role in the construction of the sequence of quadratic transforms we are going to deal with. In this section we show the existence of enough valuations of this type that verify the extra condition of being discrete until a certain value, a concept that is introduced below and that will be essential in the sequel.

Let k be an ordered field and $\Gamma \subset \mathbf{R}$ a finitely generated additive subgroup of $\mathbf{R}$ of rational rank r. Then $\Gamma = \sum_{i=1}^{r} \xi_i \mathbf{Z}$, where $\xi_1, \dots, \xi_r \in \mathbf{R}$ are rationally independent elements. Moreover, we may assume that $\xi_1 = 1$, so that $\mathbf{Q} \cap \Gamma = \mathbf{Z}$. As is well known [**Fs**] the generalized power series ring $k[[t^\Gamma]]$ is a valuation ring with value group Γ and residue field k. Furthermore, its quotient field $k((t^\Gamma))$ has a unique ordering specializing to the given one in k and such that $t^{\xi_i} > 0$ for all $i = 1, \dots, r$. We fix this ordering in $k((t^\Gamma))$ from now on.

LEMMA 4.1. *Let $A = k[[X_1, \dots, X_d]]$ be the formal power series ring in d variables and K its quotient field. Fix $\alpha \in Spec_r(k)$, and let $f_1, \dots, f_s \in A$ be such that $\emptyset \neq \mathcal{G}_\alpha \cap \{f_1 > 0, \dots, f_s > 0\}$. Let Γ be as above with $2 \le r \le d$. Then there is a local monomorphism $\sigma : A \to \kappa(\alpha)[[t^\Gamma]]$ such that $\sigma(f_i) > 0$ for all i, and the valuation w induced in K by σ verifies the following conditions*:

i) The residue field of w is a finite extension of k.
ii) The value group of w is Γ.
iii) For all $i = 1, \dots, s$ it holds $w(f_i) \in \mathbf{Z}$.

PROOF. By the formal curve selection lemma ([**Ls**], [**Rz4**], [**A-B-R**]) there is a local homomorphism $\tau : A \to \kappa(\alpha)[[t]]$ with $\tau(f_i) > 0$, $i = 1, \dots, s$. In particular, $A/\ker(\tau) \subset \kappa(\alpha)[[t]]$ and therefore, $(A/\ker(\tau))^\nu \subset \kappa(\alpha)[[t]]$, where $(A/\ker(\tau))^\nu$ is the integral closure of $A/\ker(\tau)$. Since $(A/\ker(\tau))^\nu$ has dimension 1, it is a discrete valuation ring, and up to a substitution of the type $t^m u(t) \mapsto t$ with $u(0) \neq 0$, we may assume that its uniformization parameter is t. In particular this means that there are $g_1, g_2 \in A$ such that $\tau(g_1) \neq 0, \tau(g_2) \neq 0$ and $\omega(\tau(g_1)) - \omega(\tau(g_2)) = 1 \in \mathbf{Z}$, where $\omega(x(t))$ represents the order of $x(t)$ as a series.

Since $\kappa(\alpha)[[t]] \subset \kappa(\alpha)[[t^\Gamma]]$, we may consider τ as a map into $\kappa(\alpha)[[t^\Gamma]]$, and, as it happens in $\kappa(\alpha)[[t]]$, the sign of an element of $\kappa(\alpha)[[t^\Gamma]]$ is completely determined by its initial form. This implies, in particular, that for any homomorphism $\tau' : A \to \kappa(\alpha)[[t^\Gamma]]$ such that the initial forms of $\tau(f_i)$ and $\tau'(f_i)$ coincide we also have $\tau'(f_i) > 0$. Pick $\nu_0 \ge 0$ such that if $x_1(t), \dots, x_d(t) \in \kappa(\alpha)[t]$ are truncations of $\tau(X_1),\dots,\tau(X_d)$ of degree $\geq \nu_0$, then the initial forms of $\tau(f_i)$ and $f_i(x_1(t),\dots,x_d(t))$ coincide for every i and the initial forms of $\tau(g_j)$ and $g_j(x_1(t),\dots,x_d(t))$ coincide for $j=1,2$. Let $k' \subset \kappa(\alpha)$ be a finite extension of k containing all the coefficients of $x_1(t),\dots,x_d(t)$.

We define $\sigma : A \to k'[[t^\Gamma]]$ by

$$\sigma(X_1) = x_1(t) + t^{q+1}t^{\xi_1},$$
$$\sigma(X_i) = x_i(t) + t^{s+1}t^{\xi_i} \quad \text{for} \quad 2 \leq i \leq r,$$
$$\sigma(X_j) = x_j(t) + t^{s+1}h_j(t^{\xi_r}) \quad \text{for} \quad r < j \leq r-d,$$

where $q, s > \nu_0$ are integers that will be determined below, and $h_{r+1}(t),\dots,h_d(t)$ are series of $k'[[t]]$ algebraically independent over $k'(t)$; to simplify some computations later, we choose all the $h_j(t)$ of order 1. Since $\xi_1,\dots,\xi_r$ are rationally independent it follows that $t^{\xi_1}, t^{\xi_2},\dots,t^{\xi_r}, h_{r+1}(t^{\xi_r}),\dots,h_d(t^{\xi_r})$ are algebraically independent over k. Notice that for every $f \in A$, $f \neq 0$, we have:

$$\sigma(f) = f(x(t)) + \frac{\partial f}{\partial X_1}(x(t))t^{q+1}t^{\xi_1} + \sum_{i=2}^{r} \frac{\partial f}{\partial X_i}(x(t))t^{s+1}t^{\xi_i} + \\ + \sum_{j=r+1}^{d} \frac{\partial f}{\partial X_j}(x(t))t^{s+1}h_j(t^{\xi_r}) + \dots \tag{4.1.1}$$

where $x(t) = (x_1(t),\dots,x_d(t))$. Therefore, taking q and s large enough the initial forms of $\sigma(f_i)$ and $f_i(x(t))$ coincide, and consequently coincide also with the initial form of $\tau(f_i)$. Thus $\sigma(f_i) > 0, i = 1,\dots,s$. Similarly we can suppose that the initial forms of $\sigma(g_j)$ and $\tau(g_j)$ also coincide.

We claim that σ is a monomorphism. Indeed, let $f \in A$ be such that

$$f(x_1(t) + t^{q+1}t^{\xi_1},\dots,x_q(t) + t^{s+1}h_d(t^{\xi_r})) = 0.$$

The following argument is inspired in **[Mr]** and **[Rz1]**. Assume that $f(X_1,\dots,X_d) \neq 0$. Then also

$$0 \neq f(x_1(t) + t^{q+1}t^{\xi_1}, X_2,\dots,X_d),$$

and replacing X_i by $U_i + x_i(t)$, $i = 2,\dots,d$, we get

$$0 \neq g(t,U) = f(x_1(t) + t^{q+1}t^{\xi_1}, x_2(t) + U_2,\dots,x_d(t) + U_d) \in k'[[t, U_2,\dots,U_d]]$$

(remember that $\xi_1 = 1$). Thus we can write $g = G_m(t,U) + G_{m+1}(t,U) + \dots$ where G_i is a homogeneous form of degree i and $G_m \neq 0$. Then

$$\begin{aligned} 0 &= f(x_1(t) + t^{q+1}t^{\xi_1},\dots,x_q(t) + t^{s+1}h_d(t^{\xi_r})) \\ &= g(t, t^{s+1}t^{\xi_2},\dots,t^{s+1}t^{\xi_{r-1}}, t^{s+1}h_{r+1}(t^{\xi_r}),\dots,t^{s+1}h_d(t^{\xi_r})) \\ &= t^m G_m(t, t^s t^{\xi_2},\dots,t^s t^{\xi_{r-1}}, t^s h_{r+1}(t^{\xi_r}),\dots,t^s h_d(t^{\xi_r})) + \\ &+ t^{m+1}G_{m+1}(t, t^s t^{\xi_2},\dots,t^s t^{\xi_{r-1}}, t^s h_{r+1}(t^{\xi_r}),\dots,t^s h_d(t^{\xi_r})) + \dots \end{aligned}$$

Since the elements $t = t^{\xi_1}, t^{\xi_2}, \ldots, t^{\xi_{r-1}}, h_1(t^{\xi_r}), \ldots, h_d(t^{\xi_r})$ are algebraically independent, it follows easily that $t^{\xi_1}, t^s t^{\xi_2}, \ldots, t^s t^{\xi_{r-1}}, t^s h_1(t^{\xi_r}), \ldots, t^s h_d(t^{\xi_r})$ are algebraically independent too, and we get $G_m = 0$, a contradiction. Hence $f(X_1, \ldots, X_d) = 0$ as a series, as claimed.

Thus we have an inclusion $A \hookrightarrow k'[[t^\Gamma]]$ which gives another of the quotient fields $K \subset k'((t^\Gamma))$. Then $W = K \cap k'[[t^\Gamma]]$ is a valuation ring of K, whose associated valuation will be denoted by w. Clearly W dominates A, its residue field is a subfield of k' (and therefore finite over k), and its value group is a subgroup Γ' of Γ. Let us see that as a matter of fact we have $\Gamma' = \Gamma$.

For every $f \in A$ we have $w(f) = \omega(\sigma(f))$. In particular $v(g_1) - v(g_2) = 1 \in \mathbf{Z}$ and therefore $\mathbf{Z} \subset \Gamma'$. Next, since $x_i(t) \in k'[t]$ for each $i = 1, \ldots, r$, there is a polynomial of minimal degree $\ell_i(X_1, X_i) \in k'[X_1, X_i]$ such that $\ell_i(x_1(t), x_i(t)) = 0$. Let us denote by l_i the degree of ℓ_i. From the equation (3.1.1) we get

$$\sigma(\ell_i) = \frac{\partial \ell_i}{\partial X_1}(x(t))t^{q+1}t^{\xi_1} + \frac{\partial \ell_i}{\partial X_i}(x(t))t^{s+1}t^{\xi_i} + \\ + \sum_{2 \le c+p \le l_i} \frac{1}{(c+p)!} \frac{\partial^{c+p}\ell_i}{\partial X_1^c \partial X_i^p}(x(t))t^{c(q+1)}t^{p(s+1)}t^{p\xi_i}.$$

Moreover, since ℓ_i is of minimal degree we have $\partial\ell_i/\partial X_1(x(t)) \neq 0$ and $\partial\ell_i/\partial X_i(x(t)) \neq 0$. Then if q, s are large enough and $q \ge 2p$ the order of the series $\frac{\partial \ell_i}{\partial X_i}(x(t))t^{s+1}t^{\xi_i}$ is strictly smaller than the others in the above expression, and we conclude

$$w(\ell_i) = \omega(\sigma(\ell_i)) = (s+1) + \xi_i + \omega\left(\frac{\partial \ell_i}{\partial X_i}(x(t))\right).$$

Then $w(\ell_i) \in \xi_i + \mathbf{Z}$, and since $\mathbf{Z} \subset \Gamma'$ we get $\xi_i \in \Gamma'$. Therefore $\Gamma \subset \Gamma'$.

Finally, since the initial forms of $\sigma(f_i)$ and $\tau(f_i)$ coincide, it follows that $w(f_i) \in \mathbf{Z}$, which completes the proof. □

REMARK 4.2. As it is clear from the proof of the lemma, the couple of integers (q, s) can be replaced by any other (q', s') with $q' \ge q$, $s' \ge s$ and $q' \ge 2s'$ so that the statement of the lemma remains true and the values $w(f_i)$, $i = 1, \ldots, s$, are unchanged. This freedom will be of great advantage for the proof of our next result, since it will allow us to keep under control the values of certain number of functions.

Now we introduce the notion of a valuation discrete until a value N:

DEFINITION 4.3. Let A be a domain and let v be a rank 1 valuation of its quotient field K. Assume that $\mathbf{Z} \subset \Gamma_v$, and let $N \in \mathbf{N}$. We say that v is *discrete over A until the value N* if

i) v is finite over A,
ii) Γ_v is finitely generated and $\mathbf{Q} \cap \Gamma_v = \mathbf{Z}$, and
iii) for all $f \in A$ with $v(f) \notin \mathbf{Z}$ it is $v(f) > N$.

The following result shows that these valuations exist.

PROPOSITION 4.4. *Let $(A, \mathfrak{m}, k)$ be a local complete domain of dimension d with quotient field K. Assume that k is real closed and K is formally real. Let $f_1, \dots, f_s \in A$, and consider an integer r with $2 \leq r \leq d$. Then there is a rank one real valuation v of K finite over A and centered at $\mathfrak{m}$, with residue field k, rational rank r and discrete until a value $N \geq v(f_i)$ for all i. In particular, $v(f_i) \in \mathbf{Z}$ for all i.*

PROOF. We have $A = k[[Y_1, \dots, Y_n]]/\mathfrak{p}$ where $\mathfrak{p}$ is a prime ideal of $k[[Y_1, \dots, Y_n]]$. Let $X_1, \dots, X_d$ be analytically independent elements, $X_j = \sum_{i=1}^{n} \lambda_{ij} Y_i$, with $\lambda_{ij} \in k$, so that A is a finite extension of $B = k[[X_1, \dots, X_d]]$. Let $\theta \in A$ be a primitive element of K over the quotient field L of B. Let $P(T) \in B[T]$ be the irreducible polynomial of θ and let β be any total ordering of K. Let $\beta' = \beta \cap L$. By Sturm's theorem the property of $P(T)$ having a root (namely θ) in the real closure of (L, β'), is characterized in terms of a family of elements, $c_1, \dots, c_p \in B$, being positive in β'.

On the other hand, each f_i satisfies a minimal equation over L:

$$f_i^{m_i} + a_{1i} f_i^{m_i - 1} + \dots + a_{m_i i} = 0, \quad a_{ji} \in L. \tag{4.4.1}$$

Since f_i is integral over B and B is integrally closed, a standard argument shows that the a_{ji} belong to B, so that we actually have an equation of integral dependence over B with degree $m_i \leq m = [K : L]$.

Now let $\Gamma = \sum_{i=1}^{r} \mathbf{Z}\xi_i \subset \mathbf{R}$ where $\xi_1 = 1$ and $\xi_1, \xi_2, \dots, \xi_r \in \mathbf{R}$ are rationally independent elements. By Lemma 4.1, there is a local monomorphism $\sigma : B \to k[[t^\Gamma]]$ with $\sigma(\xi_l) > 0$ and $w(a_{ji}) \in \mathbf{Z}$, where $w : L \to \Gamma$ is the valuation induced by σ. Let $M \in \mathbf{N}$ be such that $w(a_{m_i i}) < M$, and assume that the integers q, s used in the proof of the lemma to define σ verify that $\min\{(q+1) + \xi_1, (s+1) + \xi_2, \dots, (s+1) + \xi_r\} > m!M$, which is possible as explained in Remark 4.2.

Sturm's theorem implies that the polynomial $P^\sigma(T) \in k[[t^\Gamma]][T]$ obtained by applying σ to the coefficients of $P(T)$, has some root in the real closure E of $k((t^\Gamma))$. Therefore σ lifts to a monomorphism $\tau : K \to E$. Let V' be the convex hull of $k[[t^\Gamma]]$ in E with respect to the ordering of E. Then V' is a real valuation ring of E, and we set $V = \tau^{-1}(V')$. Thus, V is a real valuation ring of K whose canonical valuation v extends w. Moreover, since A is finite over B we get that v is finite over A and $\mathfrak{m}_v \cap A = \mathfrak{m}$. Also, since $[K : L] = m$, the rank and rational rank of v are equal to those of w, and we have $m!\Gamma_v \subset \Gamma$ [**Z-S**, Corollary to Lemma 3, p. 52]. In particular $\mathbf{Q} \cap \Gamma_v \subset \frac{1}{m!}\mathbf{Z}$ and so $\mathbf{Q} \cap \Gamma_v = \frac{\mu}{m!}\mathbf{Z}$ for some integer μ. Finally the residue field k_v is a finite ordered extension of k, which by assumption is real closed, whence $k_v = k$.

Now we check the assertion about the values of the different elements in the statement. From (4.4.1), we get

$$v(a_{j_1 i} f_i^{m_i - j_1}) = v(a_{j_2 i} f_i^{m_i - j_2})$$

for certain $j_1 > j_2$, and this value is the minimum among the monomials

$v(a_{ji}f_i^{m_i-j})$. Thus

$$v(f_i) = \frac{v(a_{j_1i}) - v(a_{j_2i})}{j_1 - j_2} \in \mathbf{Q} \cap \Gamma_v = \frac{\mu}{m!}\mathbf{Z}.$$

Also, from the same equation we deduce

$$v(f_i) \leq v(a_{m_ii}) \leq M.$$

Next, assume that $v(h) \notin \frac{\mu}{m!}\mathbf{Z}$ for some $h \in A$ and let us see that $v(h) > M$. As explained before, h verifies an equation of the type

$$h^p + b_1h^{p-1} + \ldots + b_p = 0,$$

where $b_p \in B$ and $p \leq m$. We set $\eta = \min\{v(b_ih^{p-i})| \ i = 0, \ldots, p\}$ (with $b_0 = 1$). If we had $v(b_i) \in \mathbf{Z}$ for any i with $v(b_ih^{p-i}) = \eta$, arguing as above, $v(h) \in \frac{\mu}{m!}\mathbf{Z}$, against our assumption. Therefore there is i_0 with $v(b_{i_0}) \notin \mathbf{Z}$ and $\eta = v(b_{i_0}h^{p-i_0})$. Now

$$pv(h) = v(h^p) \geq v(b_{i_0}h^{p-i_0}) = v(b_{i_0}) + (p - i_0)v(h)$$

and $v(h) \geq (1/i_0)v(b_{i_0})$. Thus, since $i_0 \leq p \leq m$, to check that $v(h) > M$ it is enough to check that $v(b_{i_0}) > m!M$. We have $v(b_{i_0}) = \omega(\sigma(b_{i_0}))$, that is, the order of $\sigma(b_{i_0})$ as a series in $k[[t^\Gamma]]$. Thus, using the notations of Lemma 4.1, cf. 4.1.1, we have

$$\sigma(b_{i_0}) = b_{i_0}(x(t)) + \frac{\partial b_{i_0}}{\partial X_1}(x(t))t^{q+1}t^{\xi_1} + \sum_{i=2}^{r} \frac{\partial b_{i_0}}{\partial X_i}(x(t))t^{s+1}t^{\xi_i} + \sum_{j=r+1}^{d} \frac{\partial b_{i_0}}{\partial X_j}(x(t))t^{s+1}h_j(t^{\xi_r}) + \cdots$$

Since $v(b_{i_0}) \notin \mathbf{Z}$, while $\omega(b_{i_0}(x(t))) \in \mathbf{Z}$, the order of $\sigma(b_{i_0})$ cannot be attained in the summand $b_{i_0}(x(t))$, and we deduce

$$v(b_{i_0}) \geq \min\{(q+1) + \xi_1, (s+1) + \xi_2, \ldots, (s+1) + \xi_r\} > m!M$$

(here we use that the series $h_j(t)$ has order 1, so that $h_j(t^{\xi_r})$ has order ξ_r).

To complete the proof we substitute the valuation v by $\frac{m!}{\mu}v$. This equivalent valuation verifies all requirements of the statement, N being the integral part of $\frac{m!}{\mu}M$. □

COROLLARY 4.5. *Let $(A, \mathfrak{m}, k)$ be as in the preceding proposition and denote by α the unique ordering of k. Let $f_1, \ldots, f_r \in A$ and consider an integer r with $2 \leq r \leq d$. Then the set of orderings $\beta \in \mathcal{G}_\alpha \subset Spec_r(K)$ such that $v_{\beta\alpha}$ has rank 1, rational rank r, and is discrete until a value $N \geq v_{\beta\alpha}(f_i)$ for all i, is dense in $\mathcal{G}_\alpha$.*

PROOF. Let $\{g_1 > 0, \ldots, g_s > 0\}$ be any non-empty basic open subset of $\mathcal{G}_a$. We consider $B = A[\sqrt{g_1}, \ldots, \sqrt{g_s}]$ and $L = K(\sqrt{g_1}, \ldots, \sqrt{g_s})$. Then L is formally real and B is a complete local domain with residue field k. Now

Proposition 3.4 applied to B, provides a valuation v of L with the desired rank, rational rank and values for the f_i's, and the residue field of v is k. Let $\beta' \in Spec_r(L)$ be any ordering compatible with v. We claim that $V_{\beta'\alpha}$ is the valuation ring of v. Indeed, $V_{\beta'\alpha}$ is contained in the valuation ring of v and both have residue field k. Now let $\beta = \beta' \cap A$. Then $\beta \in \{g_1 > 0, \dots, g_s > 0\}$ and $\mathbf{Q} \cap \Gamma_{\beta\alpha} = \frac{1}{\mu}\mathbf{Z}$. Replacing $v_{\beta\alpha}$ by $\mu v_{\beta\alpha}$ one easily checks that β satisfies the conditions of the statement. □

5. The uniformization theorem

In this section we prove Theorem 1.4 stated in the introduction. The main step towards that goal is the following announced:

PROPOSITION 5.1. *Let $(A, \mathfrak{m}, k)$ be an excellent local domain of dimension d and quotient field K. Assume that* $\operatorname{emb}(A) > d$. *Let $\alpha \in Spec_r(k)$ be such that $\emptyset \neq \mathcal{G}_\alpha \subset Spec_r(K)$. Let U be a nonempty open set of $\mathcal{G}_\alpha$. Then there exists $\beta \in U$ such that $\beta \to \alpha$ has rank* 1 *and residue dimension* 0, *and* $\operatorname{emb}(A^{(i)}) < \operatorname{emb}(A)$ *for some i.*

PROOF. Let C be the completion of the real strict henselization of A, cf. §3. By Proposition 3.3, $\operatorname{emb}(A) = \operatorname{emb}(C)$. Let this be e. By the real going-down, [**Rz5**], U defines a nonempty open set $\hat{U}$ of $\mathcal{G}_{\hat{\alpha}} \subset Spec_r(C)$. Also note that since C is complete with residue field $\kappa(\alpha)$ we have $\mathcal{G}_\alpha = Spec_r(L)$ where L stands for the total ring of fractions of C.

Let $x_1, \dots, x_e$ be a minimal system of generators of $\mathfrak{n}$. Then $C = R/I$ where $R = k[[X_1, \dots, X_e]]$. Let $\mathfrak{p}$ be a minimal prime of C such that $\hat{U} \cap Spec_r(k(\mathfrak{p})) \neq \emptyset$, and let $\mathfrak{q} \subset R$ be the associated prime ideal of I such that $\mathfrak{p} = \mathfrak{q}/I$. Then by [**Tg**, Proposition 3.6], there is $h \in I$ such that $\partial h/\partial X_j \notin \mathfrak{q}$ for some $j = 1 \dots, e$. Now, by Proposition 4.4, there exists $\hat{\beta} \in \hat{U}$ such that $\operatorname{supp}(\hat{\beta}) = \mathfrak{p}$ and the valuation $v_{\hat{\beta}\hat{\alpha}}$ verifies:

i) $v_{\hat{\beta}\hat{\alpha}}$ has rank 1, rational rank d, residue dimension 0,

ii) $v_{\hat{\beta}\hat{\alpha}}$ is discrete until a value $N \geq \nu_0 2^\eta$, where

$$\nu_0 = \min_j\{v_{\hat{\beta}\hat{\alpha}}(X_j)\} \qquad \eta = \max_j \left\{ v_{\hat{\beta}\hat{\alpha}}\left(\frac{\partial h}{\partial X_j}\right) | \frac{\partial h}{\partial X_j} \notin \mathfrak{q} \right\}$$

iii) $v_{\hat{\beta}\hat{\alpha}}(X_j) \in \mathbf{Z}$ for all j and $v_{\hat{\beta}\hat{\alpha}}(\partial h/\partial X_j) \in \mathbf{Z}$ for all j such that $\partial h/\partial X_j \notin \mathfrak{q}$.

(Here we view $v_{\hat{\beta}\hat{\alpha}}$ as a map $R \to C \to V_{\beta\alpha} \to \Gamma_{\beta\alpha} \cup \infty$.)

We set $\beta = \hat{\beta} \cap A$. By §3 we know that $\beta \to \alpha$ has rank 1 and residue dimension 0. We will show that this β verifies the conditions of the Proposition. Consider the sequence

$$\begin{array}{ccccccccc} C = C_0 & \longrightarrow & C_1 & \longrightarrow & \cdots & \longrightarrow & C_i & \longrightarrow & \cdots \\ \uparrow & & \uparrow & & & & \uparrow & & \\ A = A_0 & \longrightarrow & A_1 & \longrightarrow & \cdots & \longrightarrow & A_i & \longrightarrow & \cdots \end{array} \tag{5.1.1}$$

of complete strict real quadratic transforms of C along $\hat{\beta} \to \hat{\alpha}$. It follows from Proposition 3.3 that $\text{emb}(A^{(i)}) = \text{emb}(C_i)$ for all i. Therefore it is enough to show that for some i, $\text{emb}(C_i) < e$.

To show that, we consider an embedded sequence of complete quadratic transforms of C for the embedding dimension e:

$$\begin{array}{ccccccccc} R = R_0 & \longrightarrow & R_1 & \longrightarrow \cdots \longrightarrow & R_i & \longrightarrow \cdots \\ \downarrow & & \downarrow & & \downarrow & \\ C = C_0 & \longrightarrow & C_1 & \longrightarrow \cdots \longrightarrow & C_i & \longrightarrow \cdots \end{array} \tag{5.1.2}$$

Thus for all i in this diagram except at most, if the sequence is finite, for the very last one we have $\text{emb}(C_i) = e$. We will see that indeed this sequence is finite.

Also, as it was explained at the end of §3, the valuation $v_{\hat{\beta}\hat{\alpha}}$ extends to a map, $R_i \to C_i \to \Gamma_{\hat{\beta}\hat{\alpha}} \cup \infty$ which we denote again by $v_{\hat{\beta}\hat{\alpha}}$. Let $\mathfrak{M}_i$ denote the maximal ideal of R_i and let I_i stand for the kernel of the epimorphism $R_i \to C_i$. Moreover, assume that for each i we have $\mathfrak{M}_i = (X_1^{(i)}, \dots, X_e^{(i)})$. Then we define for all i:

$$\nu_i = \inf\{v_{\hat{\beta}\hat{\alpha}}(f) | f \in \mathfrak{M}_i\} = \min_j\{v_{\hat{\beta}\hat{\alpha}}(X_j^{(i)})\},$$
$$\theta_i = \min\{v_{\hat{\beta}\hat{\alpha}}(\partial f/\partial X_j) \mid f \in I_i, j = 1, \dots, e\}.$$

We will show in several steps that the sequence 5.1.2 is finite, i.e., that $\text{emb}(C_s) < e$ for some s. Indeed, suppose that $\text{emb}(C_i) = e$ (and then also $\text{emb}(C_j) = e$ for $0 \le j \le i$).

FIRST STEP: *It holds*:

iv) $\nu_i \in \mathbf{Z}$.

v) $0 < \nu_i \le \nu_{i-1}$.

vi) *For all* $f \in k[X_1^{(i)}, \dots, X_e^{(i)}]$ *with degree* $\le \rho$, *then either* $v_{\hat{\beta}\hat{\alpha}}(f) \in \mathbf{Z}$ *or* $v_{\hat{\beta}\hat{\alpha}}(f)) \ge \nu_0(2^\eta - \rho(2^i - 1))$.

We work by induction on i. The three assertions are obvious for $i = 0$. For simplicity we set $Y_j = X_j^{(i-1)}$ and $Z_j = X_j^{(i)}$. Thus, we have $R_{i-1} = k[[Y_1, \dots, Y_e]]$, and $R_i = k[[Z_1, \dots, Z_e]]$, and up to renaming the variables we may assume that $Z_1 = Y_1$, $Z_i = (Y_i - c_i Y_1)/Y_1$, where

$$v_{\hat{\beta}\hat{\alpha}}(Y_1) = \min\{v_{\hat{\beta}\hat{\alpha}}(Y_j) | j = 1, \dots, e\} = \nu_{i-1}$$

and $c_i \in k$ is the only element such that

$$v_{\hat{\beta}\hat{\alpha}}(Y_i - c_i Y_1) > v_{\hat{\beta}\hat{\alpha}}(Y_1).$$

Now the second assertion of v) follows at once, since

$$\nu_i \le v_{\hat{\beta}\hat{\alpha}}(Z_1) = v_{\hat{\beta}\hat{\alpha}}(Y_1) = \nu_{i-1}.$$

Next, let $f \in k[Z_1, \dots, Z_e]$ have degree $\leq \rho$. Then $f = \frac{1}{Y_1^\rho} f'(Y_1, \dots, Y_e)$, where $f'(Y_1, \dots, Y_e) \in k[Y_1, \dots, Y_e]$ has degree $\leq 2\rho$. Since

$$v_{\hat{\beta}\hat{\alpha}}(Z_1) = v_{\hat{\beta}\hat{\alpha}}(Y_1) = \nu_{i-1} \in \mathbf{Z},$$

by induction we get that $v_{\hat{\beta}\hat{\alpha}}(f) \notin \mathbf{Z}$ if and only if $v_{\hat{\beta}\hat{\alpha}}(f') \notin \mathbf{Z}$. Again the induction hypothesis implies in this case that

$$v_{\hat{\beta}\hat{\alpha}}(f') \geq \nu_0(2^\eta - 2\rho(2^{i-1} - 1)),$$

and taking into account that $v_{\hat{\beta}\hat{\alpha}}(Y_1) = \nu_{i-1} \leq \nu_0$, we get

$$v_{\hat{\beta}\hat{\alpha}}(f) = v_{\hat{\beta}\hat{\alpha}}(f') - \rho v_{\hat{\beta}\hat{\alpha}}(Y_1) \geq \nu_0(2^\eta - 2\rho(2^{i-1} - 1)) - \rho\nu_0 = \nu_0(2^\eta - (2^i - 1)\rho),$$

showing *vi*). Finally, applying this to Z_j we get that if $v_{\hat{\beta}\hat{\alpha}}(Z_j) \notin \mathbf{Z}$, then

$$v_{\hat{\beta}\hat{\alpha}}(Z_j) \geq \nu_0(2^\eta - 2^i + 1) \geq \nu_0 \geq \nu_{i-1} = v_{\hat{\beta}\hat{\alpha}}(Z_1) \in \mathbf{Z},$$

and $\nu_i = \min\{v_{\hat{\beta}\hat{\alpha}}(Z_j) | j = 1, \dots, e\} \in \mathbf{Z}$, which shows *iv*) and the assertion $\nu_i > 0$ of *v*).

SECOND STEP: It holds:

vii) $0 \leq \theta_0 \leq \eta$.
viii) $\theta_i \geq 0$.
ix) $\nu_i \leq \theta_i$.
x) $\theta_i \leq \theta_{i-1} - \nu_{i-1}$.

The assertions *vii*) and *viii*) are clear from the definitions. To show *i*) suppose that for some $f \in I_i$ we have $\partial f / \partial X_j^{(i)} \notin \mathfrak{M}_i$, that is, $\partial f / \partial X_j^{(i)}(0) \neq 0$. Then, by the implicit function theorem, we get, up to isomorphism, $X_j \in I_i$, and consequently $C_i = k[[X_1^{(i)}, \dots, X_{j-1}^{(i)}, X_{j+1}^{(i)}, \dots, X_e^{(i)}]]/J$ so that $\operatorname{emb}(C_i) \leq e-1$ against our assumption.

Now we will show *x*). Assume that

$$v_{\hat{\beta}\hat{\alpha}}(X_1^{(i-1)}) = \min\{v_{\hat{\beta}\hat{\alpha}}(X_j^{(i-1)}) | j = 1, \dots, e\} = \nu_{i-1}.$$

As above we set $Y_j = X_j^{(i-1)}$ and $Z_j = X_j^{(i)}$, so that $R_{i-1} = k[[Y_1, \dots, Y_e]]$, and $R_i = k[[Z_1, \dots, Z_e]]$, with $Z_1 = Y_1$, $Z_i = (Y_i - c_i Y_1)/Y_1$.

Let $f(Y_1, \dots, Y_e) \in I_{i-1}$, and set

$$f = f_m(Y_1, \dots, Y_e) + f_{m+1}(Y_1, \dots, Y_e) + \dots$$

where f_r is a homogeneous form of degree r. Since $\theta_{i-1} \geq \nu_{i-1} > 0$, we get $\partial f / \partial Y_j \in \mathfrak{M}_{i-1}$, and in particular $m > 1$.

Now set $\tilde{f} = f(Z_1, Z_1(Z_2 + c_2), \dots, Z_1(Z_e + c_e))$. We have $\tilde{f} \in I_i$, and $\tilde{f} = Z_1^m f^{(1)}$, where

$$f^{(1)} = f_m(1, Z_2 + c_2, \dots, Z_e + c_e) + Z_1 f_{m+1}(1, Z_2 + c_2, \dots, Z_e + c_e) + \dots .$$

Since $Z_1 = Y_1$ is not a zero divisor in A_i, we get $f^{(1)} \in I_i$. Let us compute now the values $v_{\hat{\beta}\hat{\alpha}}(\partial f^{(1)}/\partial Z_j)$ for $j = 1, \dots, e$. For $j \geq 2$ we have

$$\frac{\partial \tilde{f}}{\partial Z_j} = \frac{\partial f}{\partial Y_j} Z_1 = Z_1^m \frac{\partial f^{(1)}}{\partial Z_j},$$
$$\frac{\partial f}{\partial Y_j} = Z_1^{m-1} \frac{\partial f^{(1)}}{\partial Z_j},$$

and therefore

$$v_{\hat{\beta}\hat{\alpha}}(\frac{\partial f}{\partial Y_j}) = (m-1) v_{\hat{\beta}\hat{\alpha}}(Z_1) + v_{\hat{\beta}\hat{\alpha}}(\frac{\partial f^{(1)}}{\partial Z_j}).$$

Now, since $v_{\hat{\beta}\hat{\alpha}}(Z_1) = \nu \in \mathbf{Z}$ and $m > 1$, we get

$$v_{\hat{\beta}\hat{\alpha}}(\frac{\partial f^{(1)}}{\partial Z_j}) \leq v_{\hat{\beta}\hat{\alpha}}(\frac{\partial f}{\partial Y_j}) - \nu_{i-1}.$$

Finally,

$$\frac{\partial \tilde{f}}{\partial Z_1} = \frac{\partial f}{\partial Y_1} + \sum_{j \geq 2} \frac{\partial f}{\partial Y_j}(Z_j + c_j) = m Z_1^{m-1} f^{(1)} + Z_1^m \frac{\partial f^{(1)}}{\partial Z_1},$$

and plugging in the value of $\partial f / \partial Y_j$ we get

$$\frac{\partial f}{\partial Y_1} = m Z_1^{m-1} f^{(1)} + Z_1^{m-1} \left(Z_1 \frac{\partial f^{(1)}}{\partial Z_1} - \sum_{j \geq 2} (Z_j + c_j) \frac{\partial f^{(1)}}{\partial Z_j} \right).$$

Now, taking into account that $f^{(1)} \in I_i$, we have:

$$v_{\hat{\beta}\hat{\alpha}}(\frac{\partial f}{\partial Y_1}) \geq (m-1) v_{\hat{\beta}\hat{\alpha}}(Z_1) + \min_j \left\{ v_{\hat{\beta}\hat{\alpha}}(Z_j + c_j) + v_{\hat{\beta}\hat{\alpha}}(\frac{\partial f^{(1)}}{\partial Z_j}) \right\},$$

where we have set $c_1 = 0$, and in the end,

$$\begin{aligned} v_{\hat{\beta}\hat{\alpha}}(\frac{\partial f}{\partial Y_1}) &\geq (m-1) v_{\hat{\beta}\hat{\alpha}}(Z_1) + \min_j \left\{ v_{\hat{\beta}\hat{\alpha}}(\frac{\partial f^{(1)}}{\partial Z_j}) \right\} \geq \\ &\geq \nu_{i-1} + \min_j \left\{ v_{\hat{\beta}\hat{\alpha}}(\frac{\partial f^{(1)}}{\partial Z_j}) \right\}. \end{aligned}$$

In conclusion we have $\theta_i \leq \theta_{i-1} - \nu_{i-1}$, which ends the proof of x).

THIRD STEP: CONCLUSION. Once all these inequalities are proved the conclusion is easy. In fact, we have

$$0 \overset{viii)}{\leq} \theta_i \overset{x)}{\leq} \theta_{i-1} - \nu_{i-1} \overset{v)}{\leq} \theta_{i-1} - 1 \leq \theta_{i-2} - 2 \leq \cdots \leq \theta_0 - i \overset{vii)}{\leq} \eta - i$$

Therefore, $i \leq \eta$, or, in other words, the sequence 5.1.2 has at most a number of terms $\leq \eta$, and in particular it is finite, as we wanted to show. □

COROLLARY 5.2. *Let $(A, \mathfrak{m}, k)$ be an excellent local domain of dimension d and quotient field K. Let $\alpha \in Spec_r(k)$ be such that $\emptyset \neq \mathcal{G}_\alpha \subset Spec_r(K)$. Let U be a nonempty open set of $\mathcal{G}_\alpha$. Then there exists $\beta \in U$ which uniformizes α and such that $\beta \to \alpha$ has rank 1 and residue dimension 0.*

PROOF. We work by induction on $e = \text{emb}(A)$. If $e = d$, then A is regular and the statement asserts only the existence of $\beta \in U$ such that $\beta \to \alpha$ has rank 1 and residue dimension 0. Let C be the completion of the real strict henselization A_α of A, and let $\hat{U} \subset \mathcal{G}_{\hat{\alpha}}$ be the neighbourhood defined by the same equations that define U. By Proposition 4.4, there is $\hat{\beta} \in \hat{U}$ such that $\hat{\beta} \to \hat{\alpha}$ has rank 1 and residue dimension 0. Set $\beta = \hat{\beta} \cap A$. Then β satisfies all the requirements of the statement.

Next suppose $e > d$ and assume the result for rings of dimension d and embedding dimension smaller that e. By Proposition 5.1, there is $\beta \in U$ in the conditions of the statement such that $\text{emb}(A^{(r)}) < e$. Moreover, since $\beta \to \alpha$ has residue dimension 0, we have $\dim(A^{(r)}) = \dim(A) = d$. Also, by Lemma 2.3 *a*) there is a neighbourhood $U^{(r)} \subset U$ of β such that for all $\gamma \in U^{(r)}$ the r-th quadratic transform of A along $\gamma \to \alpha$ coincides with $A^{(s)}$. Now, by the induction hypothesis. there is $\gamma \in U^{(r)}$ which uniformizes $A^{(r)}$, say in the s-th step, and such that $\gamma \to \alpha$ has rank 1 and residue dimension 0. Obviously γ uniformizes A in the $(r + s)$-th step, and the proof is complete. □

Finally we prove Theorem 1.4, which is now an easy consequence of Corollary 5.2.

PROOF OF THEOREM 1.4. Let $(A, \mathfrak{m}, k)$ be a local excellent domain and $\alpha \in Spec_r(k)$. By Corollary 5.2 the set of all $\beta \in \mathcal{U}_\alpha$ such that $\beta \to \alpha$ has rank 1 and residue dimension 0 is dense. Then pick such a β and suppose for instance that the i-th quadratic transform of A along $\beta \to \alpha$ is regular. Then all the orderings γ in the open neighborhood $U^{(i)}$ provided by Lemma 2.3 uniformize α, and so β is an interior point of $\mathcal{U}_\alpha$. The conclusion is obvious. □

REMARK 5.3. We have actually proved a statement stronger than Theorem 1.4. Namely, we have constructed a distinguished dense subset $\mathcal{U}_\alpha^*$ of $\mathcal{G}_\alpha$ and for every point $\beta \in \mathcal{U}_\alpha^*$ and open neighborhood U and an integer $i \geq 0$ such that for all $\gamma \in U$ the sequence of quadratic transforms of A along $\gamma \to \alpha$ is the same until the i-th transform $A^{(i)}$, and this $A^{(i)}$ is a regular ring of dimension d.

6. Proof of Theorem 1.5

In this section we prove the density Theorem 1.5. Before we will show the following result which can be seen as a rectiliniarization property.

PROPOSITION 6.1. *Let $(A, \mathfrak{m}, k)$ be a local excellent domain of dimension d. Let K denote its quotient field and fix $\alpha \in Spec_r(k)$. Let $U \subset \mathcal{G}_\alpha$ be a non-empty open subset of G_α and let $f \in A$. Then there is $\gamma \in U$ such that $\gamma \to \alpha$ has residue dimension 0, some quadratic transform $A^{(r)}$ of A along $\gamma \to \alpha$ is regular* *and* $f = ux^q$ *in* $A^{(r)}$, *where* u *is a unit and* x *belongs to a regular system of parameters of* $A^{(r)}$.

PROOF. By Theorem 1.4 there is $\beta \in U$ such that $\beta \to \alpha$ has rank 1, residue dimension 0 and uniformizes α, say $A^{(s)}$ is regular. Moreover, by Lemma 2.3 a) there is a neighbourhood $U^{(s)} \subset U$ of β such that for any $\gamma \in U^{(s)}$ the sequences of quadratic transforms of A along $\gamma \to \alpha$ and along $\beta \to \alpha$ coincide until the s-th step. Therefore, replacing A, α, and U by $A^{(s)}$, $\alpha^{(s)}$ and $U^{(s)}$, we may assume that A is regular.

Now, suppose $U = \{\beta \in \mathcal{G}_\alpha | h_1(\beta) > 0, \dots, h_s(\beta) > 0\}$, $h_i \in A$. By the real dimension theorem, [**Rz4**], [**A-B-R**], there are $\alpha_d, \alpha_1 \in Spec_r(A)$, such that

- $\alpha_d \to \alpha_1 \to \alpha$,
- $\mathrm{supp}(\alpha_d) = 0$ and $\mathrm{ht}(\mathrm{supp}(\alpha_1)) = d-1$,
- $h_i(\alpha_1) > 0$ for all $i = 1, \dots, s$ and $f(\alpha_1) \neq 0$.

We set $\mathfrak{p} = \mathrm{supp}(\alpha_1)$. Then $A/\mathfrak{p}$ is a local excellent domain of dimension 1, and we consider its quotient field $\kappa(\mathfrak{p})$. Let $\bar{\mathcal{G}}_\alpha \subset Spec_r(k(\mathfrak{p}))$ be the set of generations of α, and correspondingly, let $\bar{U} = \{\gamma \in \bar{\mathcal{G}}_\alpha | \bar{h}_1(\gamma) > 0, \dots, \bar{h}_s(\gamma) > 0\}$ where $\bar{h}$ denotes the class of h (mod $\mathfrak{p}$). Again by Theorem 1.4 there is $\xi_1 \in \bar{U}$ such that the p-th quadratic transform $(A/\mathfrak{p})^{(p)}$ of $A/\mathfrak{p}$ along $\xi_1 \to \alpha$ is regular. Since the ring $A_\mathfrak{p}$ is regular, ξ_1 has a generation ξ_d with $\mathrm{supp}(\xi_d) = 0$, which belongs to U since ξ_1 does.

Obviously $\xi_1 \to \alpha$ has residue dimension 0. Then, arguing as in §3 when we discussed the embedded sequence of quadratic transforms, we see that $V_{\xi_d\alpha}$ is the composite of $V_{\xi_d\xi_1}$ and $V_{\xi_1\alpha}$. In particular $\xi_d \to \alpha$ has residue dimension 0 and we have a commutative diagram

$$\begin{array}{ccccccc}
A = A^{(0)} & \longrightarrow & A^{(1)} & \longrightarrow \cdots \longrightarrow & A^{(p)} \\
\downarrow{\scriptstyle \pi_0} & & \downarrow{\scriptstyle \pi_1} & & \downarrow{\scriptstyle \pi_p} \\
A/\mathfrak{p} = (A/\mathfrak{p})^{(0)} & \longrightarrow & (A/\mathfrak{p})^{(1)} & \longrightarrow \cdots \longrightarrow & (A/\mathfrak{p})^{(p)}
\end{array}$$

where the top line is the sequence of quadratic transforms of A along $\xi_d \to \alpha$ and the bottom one is the sequence of quadratic transforms of $A/\mathfrak{p}$ along $\xi_1 \to \alpha$. Note that the vertical arrows, which are the natural epimorphisms, have as kernel $\mathfrak{p}^{(i)}$ the center of $V_{\xi_d\xi_1}$ in $A^{(i)}$.

Let $\{x_1, x_2, \dots, x_d\}$ be a regular system of parameters of $A^{(p)}$, positive in ξ_d and such that $\mathfrak{p}^{(p)} = (x_2, \dots, x_d)A^{(p)}$. We can write

$$f = c_1 x_1^q + c_2 x_2 + \dots + c_d x_d$$

where $c_1, c_2, \dots, c_d \in A^{(p)}$, and c_1 is a unit in $A^{(p)}$, which is possible because $(A/\mathfrak{p})^{(p)}$ is a discrete valuation ring with uniformization parameter x_1. Next, since $V_{\xi_d\alpha} \subset V_{\xi_d\xi_1}$ the ideal $\mathfrak{p}^{(q)}$ is convex with respect to ξ_d. This implies that $\xi_d(x_i) < \xi_d(x_1^m)$, or in other words that $u_i/u_1^m \in \mathfrak{m}_{\xi_d\alpha}$, for each $i = 2, \dots, d$, and all $m \in \mathbf{N}$. It follows inmediately from this that $\mathfrak{m}^{(p+q)} =$

$(x_1, x_2/x_1^m, \dots, x_d/x_1^m)A^{(p+q)}$. In particular, in $A^{(p+q)}$ we have

$$f = x_1^q\left(c_1 + c_2\frac{x_2}{x_1^q} + \dots + c_d\frac{x_d}{x_1^q}\right).$$

Set $u = c_1 + c_2(x_2/x_1^q) + \dots + c_d(x_d/x_1^q) \in A^{(p+q)}$. Then $c_1 - u \in \mathfrak{m}_{\xi_d\alpha}$ and therefore u is a unit in $A^{(p+q)}$, and we have $f = ux_1^q$. □

We are now ready for the proof of Theorem 1.5.

PROOF OF THEOREM 1.5. Let $U = \{h_1 > 0, \dots, h_s > 0\}$, $h_i \in A$, be an open subset of $\mathcal{G}_\alpha$. We will show that there is $\beta \in U$ which blows-up α. Set $h = \prod_i h_i$. By Proposition 6.1, there is $\gamma \in U$, such that $\gamma \to \alpha$ has residue dimension 0, some quadratic transform $A^{(r)}$ of A along $\gamma \to \alpha$ is regular and we have $h = ux_1^q$, where x_1 belongs to a regular system of parameters $x_1, \dots, x_d$ of $A^{(r)}$ and u is a unit. It follows that for each $i = 1, \dots, s$ we have $h_i = u_i x_1^{q_i}$ where u_i is a unit. We assume that x_i is positive in γ for all $1 \le i \le d$. In particular the classes $\bar{u}_1, \dots .\bar{u}_s$ of $u_1, \dots, u_s$ (mod $\mathfrak{m}^{(r)}$) are positive in the residue field $k^{(r)}$ of $A^{(r)}$.

Consider the domain $D = A^{(r)}[x_2/x_1, \dots, x_d/x_1]$. By [**Ab2**, Lemma 15] it holds:

i) $\mathfrak{n} := \mathfrak{m}^{(r)}D = x_1D$;

ii) $D_\mathfrak{n}$ is a rank one discrete valuation ring;

iii) $D/\mathfrak{n} = k^{(r)}[z_2, \dots, z_d]$, where z_i is the residue class of x_i/x_1 (mod $\mathfrak{n}$);

iv) $\{z_2, \dots, z_d\}$ are algebraically independent over $k^{(r)}$.

We consider the ordering defined by $\alpha^{(r)}$ in the residue field $k^{(r)}$. Since by property *iv)* above $k^{(r)}[z_2, \dots, z_d]$ is a polynomial ring in d variables, we can easily construct a sequence of maximum length $\beta_{d-1} \to \beta_{g-2} \to \dots \to \beta_0 = \alpha^{(r)}$ in $Spec_r(D/\mathfrak{n})$ with $\mathrm{supp}(\beta_{d-1}) = (0)$ and $\beta_i \in \{\bar{u}_1 > 0, \dots, \bar{u}_s > 0\}$ for all i. We identify this chain with the corresponding one in $Spec_r(D)$. Finally, let β_d be the unique generation of β_{d-1} in $Spec_r(D_\mathfrak{n})$ with $\beta_d(x_1) > 0$ and $\mathrm{supp}(\beta_d) = (0)$. Then $\beta_d \to \beta_{d-1} \dots \to \beta_0 = \alpha^{(r)}$ and $\beta_d \in U$. Since $\beta_d \to \alpha^{(r)}$ and $\beta \to \alpha$ was of residue dimension 0, it follows from Lemma 2.3 *b)*, that the sequences of quadratic transforms of A along $\gamma \to \alpha$ and along $\beta_d \to \alpha$ coincide until the r-th step. Moreover, it follows from the construction that the first quadratic transform of $A^{(r)}$ along $\beta_d \to \alpha$ is the ring $A^{(r+1)} = D_{(x_1, x_2/x_1, \dots, x_d/x_1)}$, and in it we have the complete chain $\beta_d \to \beta_{d-1} \to \dots \to \beta_0 = \alpha^{(r)}$. Moreover, since $\mathrm{supp}(\beta_{d-1}) \cap A^{(r)} = \mathfrak{n} \cap A^{(r)} = \mathfrak{m}^{(r)}$ it follows that $\beta_i \cap A = \alpha$ for $i < d$. Altogether this shows that the chain $\beta_d \to \alpha$ blows-up A (namely in the $(r+1)$-th step) and the proof is complete. □

7. Existence of orderings and valuations with preassigned data

In this section we apply the previous results to show different statements on the existence of valuations and orderings $\beta \to \alpha$ with preassigned rank, rational rank and residue dimension. Now A denotes an excellent domain, not necessarily local, K its quotient field and α a point in $Spec_r(A)$. We say that a valuation ring V of K is centered at α in A if V is finite over A, that is, $A \subset V$, $\mathfrak{m}_v \cap A = \mathrm{supp}(\alpha)$

so that $\kappa(\operatorname{supp}(\alpha)) \subset k_v$, and there is a point $\alpha^* \in Spec_r(k_v)$ such that (k_v, α^*) is an ordered extension of $(\kappa(\operatorname{supp}(\alpha)), \alpha)$. Here, as above, we view a point $\alpha \in Spec_r(A)$ as a point in $Spec_r(\kappa(\operatorname{supp}(\alpha)))$. In particular, if $\beta \in Spec_r(K)$ is any ordering compatible with V and specializing to α^* in the residue field k_v, then $\beta \in \mathcal{G}_\alpha$.

We start by showing the existence and density of real prime divisors, as stated in the introduction. We recall that if v is valuation (resp. V is a valuation ring) of K which is finite over A and $\mathfrak{p}$ denotes its center in A, then v (resp. V) is called a *prime divisor over* A if its residue dimension is maximal, that is, $\operatorname{tr.deg.}(k_V : \kappa(\mathfrak{p})) = \dim(A_\mathfrak{p}) - 1$. This in particular implies that v is a rank 1 discrete valuation, cf [**Ab1**], [**Z-S**, vol. II, Appendix 2]. Given a chain $\beta \to \alpha$ in $Spec_r(A)$ with $\operatorname{supp}(\beta) = (0)$ and a prime divisor V centered at α in A and compatible with β, then V is the unique rank 1 valuation ring of K containing $V_{\beta\alpha}$. In this sense V is determined uniquely by $\beta \to \alpha$, and we will say that $\beta \in \mathcal{G}_\alpha$ *defines a prime divisor.* Clearly not every $\beta \in \mathcal{G}_\alpha$ defines a prime divisor and we denote by $\mathcal{P}_\alpha$ the subset of $\mathcal{G}_\alpha$ consisting of those β which do define one. We have the following result from which Proposition 1.6 of the introduction is a direct consequence:

PROPOSITION 7.1. (*Existence and density of real prime divisors*) *Let $U \subset \mathcal{G}_\alpha$ be a non-empty open set. Then there is $\beta \in \mathcal{P}_\alpha \cap U$.*

PROOF. Set $B = A_{\operatorname{supp}(\alpha)}$ and $d = \dim(B)$. By Theorem 1.5 there is $\beta \in U$ which blows-up α. In particular there is a quadratic transform, say $B^{(r)}$ of B such that in $Spec_r(B^{(r)})$ we have a chain $\beta = \beta^{(r)} = \gamma_d \to \gamma_{d-1} \to \dots \to \gamma_0 = \alpha^{(r)}$ with $\operatorname{ht}(\operatorname{supp}(\gamma_i)) = d - i$ and $\gamma_i \cap B = \alpha$ for all $i < d$. It follows immediately that $\beta \to \gamma_{d-1}$ defines a prime divisor, which is obviously the one defined by β. □

As another instance of application of the previous results we have the following about the existence of real valuations of rank one:

PROPOSITION 7.2. *Let $\alpha \in Spec_r(A)$ with $\operatorname{ht}(\operatorname{supp}(\alpha)) = s$, and let $U \subset \mathcal{G}_\alpha$ be a non-empty open set. Let $\Gamma \subset \mathbf{R}$ be a finitely generated ordered subgroup with rational rank $r \geq 2$ and let $d \in \mathbf{N}$ with $d + r \leq s$. Then there exists a valuation v of K centered at α in A, with value group Γ, whose residue field is a finitely generated extension of $\kappa(\operatorname{supp}(\alpha))$ of transcendence degree d and which is compatible with some $\beta \in U$.*

PROOF. Let $B = A_{\operatorname{supp}(\alpha)}$. Then $\dim(B) = s$. By Theorem 1.5, there is $\beta \in U$ which blows-up α. Therefore there is a quadratic transform, say $B^{(j)}$ of B along $\beta \to \alpha$, such that in $Spec_r(B^{(j)})$ we have a chain $\beta = \beta^{(j)} = \gamma_s \to \gamma_{s-1} \to \dots \to \gamma_0 = \alpha^{(j)}$ with $\operatorname{ht}(\operatorname{supp}(\gamma_i)) = s - i$ and $\gamma_i \cap B = \alpha$ for all $i < s$. We set $C = (B^{(j)})_{\operatorname{supp}(\gamma_d)}$. Then C is a regular local ring of dimension $s - d \geq r$ and residue field $k_d = \kappa(\operatorname{supp}(\gamma_d))$ with $\operatorname{tr.deg.}(k_d : k) = d$. Let $\hat{C}$ be the completion of C, $\hat{K}$ its quotient field and $\hat{U} \subset Spec_r(\hat{K})$ the open set defined by U. By the real going-down, [**Rz5**], $\hat{U} \neq \emptyset$. Also since C is regular, $\hat{C}$ is a formal power series ring with residue field k_d, in which we have the ordering γ_d. Then using Lemma 4.1, we find a valuation w of $\hat{K}$ dominating $\hat{C}$, such that w has value group Γ, residue dimension 0, its residue field is finite over k_d and an ordering $\xi \in \hat{U}$ compatible with w. Now, it follows from [**Ab1**, Lemma 12] that the restriction v of w to K, verifies the statement. □

COROLLARY 7.3. *Let $\alpha \in Spec_r(A)$ and $U \subset \mathcal{G}_\alpha$ a non-empty open set. Then there is an open neighbourhood $\bar{U}$ of α in $Spec_r(\kappa(\text{supp}(\alpha)))$ such that for any $\gamma \in \bar{U}$ we have $G_\gamma \cap U \neq \emptyset$ (this again in $Spec_r(A)$).*

PROOF. By considering $B = A[\sqrt{f_1}, \dots, \sqrt{f_r}]$ where $\emptyset \neq \{f_1 > 0, \dots, f_r > 0\} \subset U$ we can forget the condition about U. Now, by Proposition 7.2, there is a valuation v of K centered at α in A and whose residue field k_v is a finitely generated extension of $\kappa(\text{supp}(\alpha))$. Then the canonical mapping $\varepsilon : Spec_r(k_v) \to Spec_r(\kappa(\text{supp}(\alpha)))$ is open, cf. [**E-L-W**]. Since, by the Baer-Krull theorem, any ordering in the residue field k_v can be lifted to an ordering of K, it is enough to take $\bar{U} = \varepsilon(Spec_r(k_v))$ (note that $\alpha \in \varepsilon(Spec_r(k_v))$ because v is centered at α). □

Next we want to give some result on composites of real valuations centered at a given chain of points of $Spec_r(A)$. For that we introduce some extra notation.

Let $\alpha_{m-1} \to \dots \to \alpha_0$ be a chain in $Spec_r(A)$, not necessarily strict. Assume that $V_0 \subset \dots \subset V_{m-1}$ is a chain of valuation rings of K such that for each i, $\text{rank}(V_i) = m - i$ and V_i is centered at α_i in A. Let Γ_i be the value group of V_i, and let r_i and s_i denote respectively the rational rank and and the residue dimension of V_i. Then, cf. [**Z-S**], [**Ab1**], [**Ab2**]:

(*i*) There is a chain of convex subgroups

$$(0) = G_0 \subset G_1 \subset \dots \subset G_{m-1} \subset \Gamma.$$

such that for all $i = 0, \dots, m-1$, $\Gamma_i = \Gamma/G_i$ and $G_i/G_{i-1} \subset \mathbf{R}$ as additive subgroup.

(*ii*) $0 < r_{m-1} < \dots < r_0$ and $s_i \geq 0$, $i = 0, \dots, m-1$.

(*iii*) $s_i + r_i \leq \text{ht}(\text{supp}(\alpha_i))$ and $\text{ht}(\text{supp}(\alpha_i)) + r_i - s_i \leq \text{ht}(\text{supp}(\alpha_{i-1})) + r_{i-1} - s_{i-1}$.

(*iv*) There exists $\beta \in Spec_r(A)$ with $\text{supp}(\beta) = 0$ such that $\beta \to \alpha_{m-1}$.

Our next result shows that under certain circumstances the conditions above are also sufficient.

THEOREM 7.4. *Consider*

a) a chain $\alpha_{m-1} \to \dots \to \alpha_0$ in $Spec_r(A)$,

b) integers $s_0, \dots, s_{m-1}$, and

c) finitely generated ordered groups $\Gamma_0, \dots, \Gamma_{m-1} \subset \mathbf{R}$ with respective rational ranks $r_0, \dots, r_{m-1}$.

Assume that these data verify the conditions (i)–(iv) above and furthermore that $r_i \leq 2 + r_{i-1}$, $i = 1, \dots, m$, *where we have set* $r_m = 0$.

Then there exist a chain $V_0 \subset \dots \subset V_{m-1}$ *of valuation rings of* K, *such that* V_i *is centered at* α_i *in* A, *has value group* Γ_i *and its residue field* k_i *is a finitely generated extension of* $\kappa(\operatorname{supp}(\alpha))$ *of transcendence degree* s_i.

PROOF. We work by induction on m. The case $m = 1$ is Proposition 7.2. Now we apply the induction hypothesis to the chain $\alpha_{m-1} \to \dots \to \alpha_1$, the integers $s_1, \dots, s_{m-1}$ and the groups $\Gamma_1, \dots, \Gamma_{m-1}$. Then there are valuation rings $V_1 \subset \dots \subset V_{m-1}$ of K verifying the conditions of the statement for these data. In particular k_1 is a finitely generated ordered extension of $(\kappa(\operatorname{supp}(\alpha_1)), \alpha_1)$. By **[E-L-W]**, the canonical mapping $\varepsilon : Spec_r(k_1) \to Spec_r(\kappa(\operatorname{supp}(\alpha_1)))$ is open. Let $A_1 = A/\operatorname{supp}(\alpha_1)$. Since $\alpha_1 \to \alpha_0$, we can see them as points $\gamma_1 \to \gamma_0$ in $Spec_r(A_1)$, with supports (0) and $\operatorname{supp}(\alpha_0)/\operatorname{supp}(\alpha_1)$ respectively. Moreover, since V_1 is centered at α_1 we have $\gamma_1 \in \operatorname{Im}(\varepsilon)$. To find the remaining V_0 we distinguish two cases:

FIRST CASE: α_0 *and* α_1 *coincide*. Then $\gamma_0 = \gamma_1$ and $\kappa(\operatorname{supp}(\alpha_0)) = \kappa(\operatorname{supp}(\alpha_1))$. We denote by F this common residue field. We know that k_1 is a finitely generated ordered extension of (F, γ_1) of transcendence degree s_1, say $k_1 = F(z_1, \dots, z_j)$, and the condition *iii*) above implies that $s_0 + (r_1 - r_0) \leq s_1$. Set $B = F[z_1, \dots, z_j]$. By the Artin-Lang theorem, **[B-C-R]**, there is an F-homomorphism $\delta : B \to R$, where R denotes the real closure of (F, γ_1). Thus we get a point of $Spec_r(B)$, also denoted by δ, such that $\dim(\delta) = 0$, $\delta \cap F = \gamma_1$ and $\kappa(\operatorname{supp}(\delta))$ is a finite extension of F. By Proposition 7.2 again, there is a valuation ring W_0 of k_1 centered at δ in B, with value group G_1 (which has rational rank $r_1 - r_0 \geq 2$) and whose residue field k_0 is a finitely generated extension of $\kappa(\operatorname{supp}(\delta))$ of transcendence degree s_0. In particular k_0 is finitely generated over F. Finally, let V_0 be the composite of V_1 and W_0. It is immediate to check that V_0 verifies the conditions of the statement.

SECOND CASE: α_0 *and* α_1 *are different*. Then $\gamma_1 \neq \gamma_0$. Let γ_1^* be an ordering of k_1 extending γ_1 and let H be the convex hull of $(A_1)_{\operatorname{supp}(\alpha_0)/\operatorname{supp}(\alpha_1)}$ in k_1 with respect to γ_1^*. Then H is a valuation ring of k_1 centered at γ_0 in A_1. Now let $z_1, \dots, z_j \in H$ be such that k_1 is the quotient field of $B = A_1[z_1, \dots, z_j] \subset H$, and let γ_0^* be the center of H in B. Then we have a homomorphism

$$A_1/\operatorname{supp}(\gamma_0) \to B/\operatorname{supp}(\gamma_0^*) = (A_1/\operatorname{supp}(\gamma_0))[\bar{z}_1, \dots, \bar{z}_j],$$

where $\bar{z}_i$ denotes the class of z_i mod $(\operatorname{supp}(\gamma_0))$. Furthermore $\kappa(\operatorname{supp}(\gamma_0^*))$ is a finitely generated extension of k_0 of transcendence degree $\leq s_1$. We will modify B so that $\kappa(\operatorname{supp}(\gamma_0^*))$ coincides with k_0. In fact, let $f \in \operatorname{supp}(\gamma_0^*)$ and set $B_1 = A_1[z_1 f, \dots, z_j f, f] \subset H$.

As above, k_1 is the quotient field of B_1, and now we have $\overline{z_i f} = 0$. Therefore if we denote again by γ_0^* the center of H in B_1, we have $B_1/\operatorname{supp}(\gamma_0^*) = A_1/\operatorname{supp}(\gamma_0)$ and consequently $\kappa(\operatorname{supp}(\gamma_0^*)) = k_0$. In particular, $\operatorname{ht}(\operatorname{supp}(\gamma_0^*)) =$

$\mathrm{ht}(\mathrm{supp}(\gamma_0)) + s_1 \geq \mathrm{ht}(\mathrm{supp}(\gamma_0)) + s_1 - s_0$ and we get from *iii*) that $\mathrm{ht}(\mathrm{supp}(\gamma_0^*)) \geq (r_1 - r_0) + s_0$. Now let W_0 be a valuation ring of k_1 centered at γ_0^* in B_1, with value group G_1 and residue field a finitely generated extension of $\kappa(\mathrm{supp}(\alpha_0))$ of transcendence degree s_0. Then the composite of V_1 and W_0 is the valuation ring V_0 we sought. □

It is now immediate from the theorem to produce real valuations with arbitrary ranks. For instance, to get a valuation of rank m centered at $\alpha_0 \in Spec_r(A)$ it is enough to apply Theorem 7.4 to the chain $\alpha_{m-1} \to \ldots \to \alpha_0$ where $\alpha_i = \alpha_0$ for all i. Another direct application of this theorem produces the following result whose proof is omitted:

COROLLARY 7.5. *Let $\alpha_m \to \alpha_{m-1} \to \ldots \to \alpha_0$ be a chain in $Spec_r(A)$ with $\mathrm{supp}(\alpha_m) = 0$ and $f_1, \ldots, f_s \in A$ with $f_1(\alpha_m) > 0, \ldots, f_s(\alpha_m) > 0$. Then for each $r \in \mathbf{N}$ such that $m \leq r \leq \mathrm{ht}(\mathrm{supp}(\alpha_0))$, there is $\beta \in Spec_r(A)$ with $f_1(\beta) > 0, \ldots, f_s(\beta) > 0$, $\mathrm{supp}(\beta) = 0$ and such that the chain $\beta \to \alpha_0$ has rank r and factorizes in exactly m steps $\beta = \gamma_m \to \gamma_{m-1} \to \ldots \to \alpha_0 = \gamma_0$ with $\mathrm{supp}(\gamma_i) = \mathrm{supp}(\alpha_i)$ for all i.*

REFERENCES

[Ab1] S. Abhyankar, *Local uniformization of algebraic surfaces over ground fields of characteristic $p \neq 0$*, Ann. of Math. **63** (1956), 491–521.

[Ab2] ———, *On the valuations centered in a local domain*, Amer. J. Math. **78** (1956), 321–348.

[Al-Ry] M. Alonso and M.-F. Roy, *Real strict localizations*, Math. Z. **194** (1987), 429–441.

[An1] C. Andradas, *Valoraciones reales en cuerpos reales de funciones*, Tesis, Univ. Complutense de Madrid, 1982.

[An2] ———, *Real places in function fields*, Comm. Algebra **15** (1985), 1151–1169.

[An3] ———, *Specialization chains of real valuation rings*, J. of Algebra **124** (1989), 437–446.

[A-B-R] C. Andradas, L. Bröcker, and J. Ruiz, *Real Algebra and Analytic Geometry* (to appear).

[Be1] E. Becker, *The real holomorphy ring and sums of 2n-th powers*, Géométrie algébrique réelle et formes quadratiques, Lecture Notes in Math., vol. 959, Springer-Verlag, Berlin-Heidelberg-New York, 1982, pp. 139–181.

[Be2] ———, *On the real spectrum of a ring and its applications to semialgebraic geometry*, Bull. A.M.S. **15** (1986), 19–60.

[B-C-R] J. Bochnak, M. Coste, and M. F. Roy, *Géometrie algébrique réelle*, Ergeb. Math. Grenzgeb., vol. (3)12, Springer-Verlag, Berlin-Heidelberg-New York, 1987.

[Br-Sch] L. Bröcker and H. W. Schülting, *Valuations of function fields from the geometrical point of view*, J. Reine Angew. Math. **365** (1986), 12–32.

[Bf] G. W. Brumfiel, *Partially ordered rings and semialgebraic geometry*, Lecture Notes in Math., Cambridge Univ. Press, Cambridge, 1979.

[E-L-W] R. Elman, T. Y. Lam, and A. Wadsworth, *Orderings under field extensions*, J. Reine Angew. Math. **306** (1979), 7–27.

[Fs] L. Fuchs, *Partially ordered algebraic systems*, Pergamon Press, 1963.

[Hk] H. Hironaka, *Resolution of singularities of an algebraic variety over a field of characteristic zero*, I, II, Ann. of Math. **79** (1964), 109–326, 205–326.

[Jd] Z. Jadda, *Constructions de places réelles et géométrie semi-algébrique*, Thèse, Univ. de Rennes I, 1986.

[Kl-Pr] F. Kuhlmann and A. Prestel, *On places of algebraic function fields*, J. Reine Angew. Math. **355** (1984), 181–195.

[Kn1] M. Knebusch, *On the uniqueness of real closures and the existence of Real Places*, Comm. Math. Helvet. **47** (1972), 260–269.

[Kn2] ———, *On the extension of Real Places*, Comm. Math. Helvet. **48** (1972), 350–369.

[Lm] T. Y. Lam, *An introduction to real algebra*, Rocky Mountain J. Math. **14** (1984), 767–814.

[Lg] S. Lang, *The theory of real places*, Ann. of Math. **57** (1953), 378–391.

[Ls] G. Lasalle, *Sur le théorème des zeros differentiables*, Singularités d'applications différentiables, Lecture Notes in Math., vol. 535, Springer-Verlag, Berlin-Heidelberg-New York, 1975, pp. 70–97.

[Mt] H. Matsumura, *Commutative algebra*, 2nd edition, Math. Lecture Note Series, vol. 56, Benjamin, London-Amsterdam-Tokyo, 1980.

[Mr] J. Merrien, *Un théorème des zéros pour les idéaux des séries formelles á coefficients réels*, I, C. R. Acad. Sc., Paris **276** (1973), 1055–1058.

[Pr] A. Prestel, *Lectures on formally real fields*, Lecture Notes in Math., vol. 22, Springer-Verlag, Berlin-Heidelberg-New York, 1975.

[Ri] P. Ribenboim, *Théorie des valuations*, 2e édition, les Presses de la Université de Montreal, 1968.

[Rb] R. Robson, *Nash wings and real prime divisors*, Math. Ann. **273** (1986), 177–190.

[Rz1] J. M. Ruiz, *Central orderings in fields or real meromorphic germs*, Manuscripta Math. **45** (1984), 193–214.

[Rz2] ———, *On Hilbert 17 problem and real nullstellensatz for global analytic functions*, Math. Z. **190** (1985), 447–459.

[Rz3] ———, *Cønes locaux et complétions*, I, C. R. Acad. Sc., Paris **302** (1986), 67–69.

[Rz4] ———, *A dimension theorem for real spectra*, J. of Algebra **124** (1989), 271–277.

[Rz5] ———, *A going-down theorem for real spectra*, J. of Algebra **124** (1989), 278–283.

[Sh] D. Shannon, *Monoidal transforms of regular local rings*, Amer. J. Math. **95** (1973), 294–320.

[Tg] J. C. Tougeron, *Idéaux de fonctions différentiables,*, Ergeb. Math. Grenzgeb., vol. (2)71, Springer-Verlag, Berlin-Heidelberg-New York, 1972.

[Z1] O. Zariski, *The reduction of the singularities of an algebraic surface*, Ann. of Math. **40** (1939), 639–689.

[Z2] ———, *Local uniformization on algebraic varieties*, Ann. of Math. **41** (1940), 852–896.

[Z-S] O. Zariski and P. Samuel, *Commutative algebra*, Vol. II, Graduate Text in Math., vol. 29, Springer-Verlag, Berlin-Heidelberg-New York, 1979.

DEPARTAMENTO DE ALGEBRA, UNIVERSIDAD COMPLUTENSE DE MADRID

DEPART. DE GEOMETRÍA Y TOPOLOGÍA, UNIVERSIDAD COMPLUTENSE DE MADRID

Contemporary Mathematics
Volume **155**, 1994

Real Algebraic Geometry Over p-Real Closed Fields

RALPH BERR

Introduction

It is the goal of the present paper to develop the foundations of a real algebraic geometry over a certain class of generalized real closed fields. Before giving a brief outline of the intention of this paper, a few words about the reasons motivating such a development are in order.

To this end we have to start with some remarks about the theory of orderings of higher level invented by E. Becker ([**1**],[**2**],[**4**]). There are two important aspects of this theory. On the one hand, E. Becker's work obviously can be regarded as a generalization of the Artin-Schreier theory of ordered fields and it led to a good structure theory of a large class of formally real fields, namely the generalized real closed fields, i.e. those fields which are real closed with respect to an ordering of higher level. On the other hand, orderings of higher level equally render a refined description of the structure of an ordered field possible. Namely, let K be a field and let $P \subset K$ be an ordering of odd level. Then P is contained in exactly one total order Q of K. From the set of all orderings being contained in Q we gain additional information about the structure of (K, Q). As a matter of fact, this information cannot be recovered from the real closure R of (K, Q), but from the generalized real closed subfields of R. This second aspect finds its formal expression in the notion of a chain signature introduced by N. Schwartz [**14**].

Now, the developments in real algebraic geometry have shown that even the study of varieties over the real numbers is intimately related to the "abstract" theory of ordered fields. At this point it seems natural to ask whether the study of varieties over generalized real closed fields may be useful for real algebraic

1991 *Mathematics Subject Classification.* Primary 14P10; Secondary 54B35, 12J15.
This paper is submitted in final form and no version of it will be submitted for publication elsewhere.

geometry in a similar manner as orderings of higher level for the theory of ordered fields. Evidently, the treatment of this problem requires the development of the foundations of an algebraic geometry over generalized real closed fields. In this paper we will do this for p-real closed fields, i.e. those fields which are real closed with respect to an ordering of level p, where p denotes a prime number.

This work is very much inspired by the developments in real algebraic geometry in the seventies, in particular by the introduction of the real spectrum and by the wellknown applications of model theory to geometrical problems. Consequently we will proceed along the same lines as they are known from real algebraic geometry. So let R be a p-real closed field and V an affine variety over R. In the first section we associate with the ring of regular functions $R[V]$ on V a spectral space $p-Sper_R\, R[V]$, called the admissible p-real spectrum of $R[V]$, and investigate the basic properties of this space. In the second section we point out that $p-Sper_R\, R[V]$ reflects the topological and geometrical structure of $V(R)$ in a suitable manner. The main results of this section correspond to the ultrafilter and finiteness theorem of real algebraic geometry. The second section is devoted to the investigation of the subsets $S \subset V(R)$ being the image of a regular mapping $\varphi : W(R) \longrightarrow V(R)$ and in the last section we will compare different notions of the dimension of a variety over a p-real closed field.

1. The admissible p-real spectrum

Throughout this paper we adopt the following notations:

N	the set of natural numbers,
N_0	$N \cup \{0\}$,
p	a prime number,
R	a fixed p-real closed field,
Q	a fixed total order of R,
w	the natural valuation of (R, Q),
$\alpha(R)$	the reduced order chain $(Q, Q^p, Q^{p^2}, \dots)$,
ω	a fixed element of $Q \setminus Q^p$.

First we will briefly recall some basic facts and definitions concerning p-real closed fields. A field K is p-real closed if and only if K admits a henselian valuation v satisfying the following conditions:

(1.1)
(a) The residue field K_v is real closed.
(b) The value group Γ_v is q-divisible for all primes $q \neq p$.
(c) $\Gamma_v / p\Gamma_v$ is cyclic of order p.

Note that the natural valuation w of (R, Q) satisfies (1.1). Let K be p-real closed. A sequence $\alpha = (\alpha_n)_{n \in N_0}$ of preorderings $\alpha_n \subset K$ is called a reduced order chain of K, if there exists a total order T of K such that $\alpha_n = T^{p^n}$ for all $n \in N_0$. Let $\mathcal{L}_p$ be the extension of the language of fields by a binary relation

$<_n$ for each $n \in N_0$ and let $\alpha = (\alpha_n)_{n \in N_0}$ be a reduced order chain of K. For $a, b \in K$ and $n \in N_0$ set

$$a <_n b :\Leftrightarrow b - a \in \alpha_n \setminus \{0\}.$$

Thus (K, α) is in a natural way a $\mathcal{L}_p$-structure.

Remark (1.2). If p is odd then a p-real closed field K admits an unique reduced order chain, whereas in the case $p = 2$ there are two such chains according to the two different total orders of K (see [**7**, Propositions 1.14 and 1.15]). Thus for $p = 2$ the order chain $\alpha(R)$ of our fixed p-real closed field R depends on the choice of the total order $Q \subset R$.

Next we provide the n-dimensional affine space $A_n(R)$ with the valuation topology induced by the valuation w of R. Note that this topology coincides with the order topology induced by Q as w is compatible with Q. As in the case of real closed fields we call this topology the strong topology. Given an affine variety V over R, we will always assume that the set $V(R)$ of R-rational points is equipped with the restriction of this topology. It is the goal of this section to associate with an affine variety V over R a spectral space reflecting the topological and geometrical structure of $V(R)$ in a suitable manner. This requires some preparation.

Let A be a commutative ring with 1. A p-real point of A is a homomorphism $\varphi : A \longrightarrow (K, \alpha)$, where K is a p-real closed field and α a reduced order chain of K. For $n \in N_0$ set

$$\alpha(\varphi)_n := \varphi^{-1}(\alpha_n)$$

and let $\alpha(\varphi) = (\alpha(\varphi_n))_{n \in N_0}$. A sequence $\alpha = (\alpha_n)_{n \in N_0}$ of preorderings $\alpha_n \subset A$ is called an order chain of A if there exists a p-real point $\varphi : A \longrightarrow (K, \bar{\alpha})$ such that $\alpha = \alpha(\varphi)$. Let $p-Sper\, A$ be the set of order chains of A. A subbasis for a topology on $p-Sper\, A$ is given by the sets

$$D_n(a) = \{\alpha \mid a \in \alpha_n \setminus (\alpha_n \cap -\alpha_n)\}, a \in A, n \in N_0.$$

The topological space $p-Sper\, A$ is called the p-real spectrum of A. With regard to geometrical applications it is important that $p-Sper\, A$ is a spectral space. For this and further results concerning the p-real spectrum see [**7**].

Let V be an affine variety over R and let $R[V]$ be the ring of regular functions on V. For $x \in V(R)$ let $\varphi_x : R[V] \longrightarrow (R, \alpha(R))$ be the evaluation map and let $\alpha(x)_n := \varphi^{-1}(\alpha(R)_n)$. Then $\alpha(x) = (\alpha(x)_n)_{n \in N_0}$ is an order chain of $R[V]$. Thus we have a canonical set-theoretical imbedding $\phi : V(R) \hookrightarrow p-Sper\, R[V]$. Throughout this paper we will identify $V(R)$ with its image in $p-Sper\, R[V]$.

Remark (1.3). Note that in the case $p = 2$ the imbedding $V(R) \hookrightarrow p-Sper\, R[V]$ depends on the choice of the order chain $\alpha(R)$ of R. We will show later on that this causes no problems.

PROPOSITION (1.4). *Let V be an affine variety over R. Then $p-Sper\, R[V]$ induces the strong topology on $V(R)$.*

PROOF. The sets $D_0(f) \cap V(R) = \{x \in V(R) | f(x) >_0 0\}, f \in R[V]$, form a subbasis for the strong topology on $V(R)$. Thus it is sufficient to show that for $n \in N$ the sets

$$D_n(f) \cap V(R) = \{x \in V(R) | f(x) >_n 0\}$$

are open. Recall that $f(x) >_n 0$ means $f(x) \neq 0$ and $f(x) \in Q^{p^n}$. Let $x \in V(R) \cap D_n(f)$. Then there exists an open neighbourhood $U \subset V(R)$ of x such that for all $y \in U$:

$$w((f(y)) = w(f(x)) \in p^n \Gamma_w \text{ and } f(y) \in Q.$$

Since w is a henselian valuation with real closed residue field we get $f(y) \in R^{p^n}$ for all $y \in U$. Now $f(y) \in Q$ implies $f(y) \in Q^{p^n}$. Thus $f(y) >_n 0$ for all $y \in U$. □

But the p-real spectrum of $R[V]$ is not yet the space we are looking for. The reason for this is that the p-real spectrum can be viewed as a space reflecting certain properties of the theories of real closed and p-real closed fields, whereas the geometrical structure of $V(R)$ is only related to the theory of p-real closed fields. So we have to look for a suitable subspace of $p-Sper\, R[V]$. This requires some further notions.

Let A be a commutative R-algebra and let $\alpha \in p-Sper\, A$. Then $\wp_\alpha := \alpha_0 \cap -\alpha_0$ is called the support of α and α induces an order chain α' of the residue field $k(\wp_\alpha)$. We denote by $(k(\alpha), \bar{\alpha})$ the real closure of $(k(\wp_\alpha), \alpha')$ which is unique up to isomorphism (for details see [**7**]). Thus α determines a canonical p-real point

$$\pi_\alpha : A \longrightarrow A/\wp_\alpha \longrightarrow (k(\alpha), \bar{\alpha}).$$

Obviously α is the order chain induced by π_α. Furthermore, the field $k(\alpha)$ is either p-real closed or real closed (see [**7**, sec. 1]). To be precise, $k(\alpha)$ is real closed if and only if $\alpha_0 = \alpha_n$ for all $n \in N$. In this case the induced order chain $\bar{\alpha}$ of $k(\alpha)$ is just the trivial chain $(k(\alpha)^2, k(\alpha)^2, k(\alpha)^2, ...)$. Thus in any case, $(k(\alpha), \bar{\alpha})$ can be considered as a $\mathcal{L}_p$-structure. Finally, we write $R \prec k(\alpha)$ if $(R, \alpha(R))$ is an elementary $\mathcal{L}_p$-substructure of $(k(\alpha), \bar{\alpha})$. Now set

$$p-Sper_R\, A = \{\alpha \in p-Sper\, A | R \prec k(\alpha)\}.$$

We give $p-Sper_R\, A \subset p-Sper\, A$ the subspace topology.

DEFINITION (1.5). Let A be a R-algebra. Then $p-Sper_R\, A$ is called the admissible p-real spectrum of A.

It will turn out that given an affine variety V over R, the space p–$Sper_R\, R[V]$ reflects the topological and geometrical structure of $V(R)$ in a suitable manner. But before carrying out this relationship in detail, we have to state some basic properties of the admissible p-real spectrum.

For $\alpha \in p-Sper_R\, A$ let w_α be the finest henselian valuation of $k(\alpha)$ and let Δ_α be the maximal divisible convex subgroup of $w_\alpha(k(\alpha)^*)$. We denote by $J(\alpha) \subset k(\alpha)$ the valuation ring corresponding to Δ_α. Completely analogous we define the valuation ring $J(R) \subset R$. Note that the definition of the rings $J(R), J(\alpha)$ implies $J(\alpha) \cap R \subseteq J(R)$, as w_α extends w. Applying [**7**, Lemma 2.1 and Theorem 2.3] we see that $R \prec k(\alpha)$ if and only if

$$(1.6) \qquad \begin{array}{ll} (1) & \bar{\alpha}_n \cap R = \alpha(R)_n \text{ for all } n \in N_0, \\ (2) & J(\alpha) \cap R = J(R). \end{array}$$

Note that (1) is equivalent to $\omega \in \alpha_0 \setminus \alpha_1$, where ω denotes the fixed element of $Q \setminus Q^p \subset R$. In the sequel we will need several times the following

LEMMA (1.7). *Let $\alpha \in p-Sper\, A$ such that $\omega \in \alpha_0 \setminus \alpha_1$. Given $a \in \alpha_0 \setminus \wp_\alpha$ and $n \in N_0$ there exists a unique $k \in \{0, \ldots, p^n - 1\}$ such that $\alpha \in D_n(\omega^k a)$.*

PROOF. From $\omega \in \alpha_0 \setminus \alpha_1$ we see that $\bar{\alpha}$ is not the trivial order chain. Thus $k(\alpha)$ is p-real closed and $\bar{\alpha}_0/\bar{\alpha}_n$ is cyclic of order p^n. Now $\omega \in \alpha_0 \setminus \alpha_1$ shows that $\omega\bar{\alpha}_n$ generates $\bar{\alpha}_0/\bar{\alpha}_n$, which implies the claim. □

We still need some further notations. For $a \in A$ and $\alpha \in p-Sper\, A$ let $a(\alpha) := \pi_\alpha(a) \in k(\alpha)$. Thus every $a \in A$ can be viewed as a function

$$a : p-Sper\, A \longrightarrow \coprod_{\alpha \in p\text{-}Sper\, A} k(\alpha)$$

Finally, for $a_1, \ldots, a_k \in A$ let $D_n(a_1, \ldots, a_k) := D_n(a_1) \cap \cdots \cap D_n(a_k)$.

THEOREM (1.8). *$p-Sper_R$ is a contravariant functor from the category of commutative R-algebras into the category of spectral spaces.*

PROOF. Let A be a R-algebra. First we will show that p–$Sper_R\, A$ is a spectral space. Since p–$Sper\, A$ is a spectral space [**7**, Theorem 3.5], it is sufficient to show that $p-Sper_R\, A$ is closed in $p-Sper\, A$. Let $\alpha \in p-Sper\, A \setminus p-Sper_R\, A$. First assume that $\bar{\alpha} \cap R \neq \alpha(R)$. Then $\omega \notin \alpha_0$ or $\omega \in \alpha_1$ as mentioned above. Hence $\alpha \in D_0(-\omega) \cup D_1(\omega) \subset p-Sper\, A \setminus p-Sper_R\, A$. So assume that $\omega \in \alpha_0 \setminus \alpha_1$. But then we have $J(\alpha) \cap R \subset J(R)$, by (1.6). By [**7**, Lemma 2.1], there exist $a, b \in \alpha_0, 1 <_0 t \in J(R)^*$ and $n \in N$ such that

$$1 <_0 \frac{a(\alpha)}{b(\alpha)} <_0 t \text{ and } \neg \left(0 <_n \frac{a(\alpha)}{b(\alpha)}\right).$$

Applying (1.7), we find $k \in \{1, \ldots, p^n - 1\}$ such that $\omega^k \frac{a(\alpha)}{b(\alpha)} \in \bar{\alpha}_n$. Hence we have

$$\alpha \in D_0(ab - b^2, tb^2 - ab) \cap D_n(\omega^k ab^{p^n - 1}) =: U.$$

Since $U \cap p-Sper_R\, A = \emptyset$, we see that $p-Sper_R\, A$ is closed in $p-Sper\, A$.

Now let $\varphi : A \longrightarrow B$ be a R-algebra homomorphism. Since $p-Sper$ is a contravariant functor from the category of commutative rings into the category of spectral spaces [7, Proposition 3.6] it remains to show that the corresponding morphism

$$\varphi^* : p\text{--}Sper\, B \longrightarrow p\text{--}Sper\, A$$

maps $p\text{--}Sper_R\, B$ to $p\text{--}Sper_R\, A$. Let $\alpha \in p\text{--}Sper_R\, B$. Then $\varphi^*(\alpha)$ corresponds to the p-real point

$$\pi_\alpha \circ \varphi : A \longrightarrow B \longrightarrow (k(\alpha), \bar{\alpha})$$

and $k(\varphi^*(\alpha))$ is isomorphic to the relative algebraic closure of $quot(\pi_\alpha(\varphi(A))$ in $k(\alpha)$ (see [7, 4, Proposition 3.6]). Now $R \prec k(\alpha)$ implies $R \prec k(\varphi^*(\alpha))$. Hence $\varphi^*(\alpha) \in p\text{--}Sper_R\, A$. $\square$

Remark (1.9). It follows from (1.2) that in the case $p = 2$ the definition of $p\text{--}Sper_R\, A$ depends on the choice of the order chain $\alpha(R)$. So let $\alpha(R)'$ be the second order chain of R and let

$$X = \{\alpha \in 2\text{--}Sper\;\; A | (R, \alpha(R)') \prec (k(\alpha), \bar{\alpha})\}.$$

We will show that $Y := 2\text{--}Sper_R\, A$ is homeomorphic to X. Let $\alpha \in Y$. Then $k(\alpha)$ admits besides $\bar{\alpha}$ exactly one further reduced order chain $\bar{\alpha}'$. We denote by α' the order chain corresponding to the induced map $\pi'_\alpha : A \longrightarrow (k(\alpha), \bar{\alpha}')$. From (1.6) it follows that $(R, \alpha(R)') \prec (k(\alpha), \bar{\alpha}')$ if and only if $(R, \alpha(R)) \prec (k(\alpha), \bar{\alpha})$. Thus we get a bijective map $\phi : Y \longrightarrow X : \alpha \mapsto \alpha'$. Let $a \in A$. Since $\alpha_n = \alpha'_n$ for all $n \in N$ and $\alpha \in Y$, we have $\phi(Y \cap D_n(a)) = X \cap D_n(a)$ and $\phi^{-1}(X \cap D_n(a)) = Y \cap D_n(a)$. Furthermore, the equalities

$$D_0(a) \cap Y = (D_1(a) \cup D_1(\omega a)) \cap Y \text{ and } D_0(a) \cap X = (D_1(a) \cup D_1(-\omega a)) \cap X$$

imply

$$\phi(D_0(a) \cap Y) = (D_1(a) \cup D_1(\omega a)) \cap X$$

and

$$\phi^{-1}(D_0(a) \cap X) = (D_1(a) \cup D_1(-\omega a)) \cap Y.$$

Thus ϕ is a homeomorphism.

The next result gives a first justification for the notion of the admissible p-real spectrum.

PROPOSITION (1.10). *Let V be an affine variety over R. Then $p\text{--}Sper_R\, R[V]$ is the closure of $V(R) \subset p\text{--}Sper\, R[V]$ with respect to the constructible topology.*

PROOF. Let X be the closure of $V(R)$ with respect to the constructible topology. For every $z \in V(R)$ we have $k(\alpha(z)) = R$, hence $V(R) \subset p\text{--}Sper_R\, R[V]$. In the proof of (1.8) we have seen that $p\text{--}Sper_R\, R[V]$ is closed in $p\text{--}Sper\, R[V]$. Thus it is sufficient to prove $p - Sper_R\, R[V] \subset X$. Let $\alpha \in p - Sper_R\, R[V]$ and let $C \in p - Sper_R\, R[V]$ be constructible with $\alpha \in C$. Furthermore, let $R[V] = R[X_1, \ldots, X_n]/(f_1, \ldots, f_k)$ and for $\beta \in p\text{--}Sper\, R[V]$ we set $x(\beta) =$

$(x_1(\beta), \ldots, x_n(\beta))$, where x_i denotes the image of X_i in $R[V]$. Then we find a quantifier free $\mathcal{L}_p$-formula ϕ with parameters from $R[V]$ such that

$$C = \left\{ \beta \in p\!-\!Sper\, R[V] | k(\beta) \models \bigwedge_{i=1}^{k} f_i(x(\beta)) = 0 \wedge \phi(x(\beta)) \right\}.$$

Now $\alpha \in C$ implies $k(\alpha) \models \exists y \left(\bigwedge_{i=1}^{k} f_i(y) = 0 \wedge \phi(y) \right)$. From $R \prec k(\alpha)$ it follows that there exists $z \in V(R)$ with $R \models \phi(z)$. Hence $\alpha(z) \in C$ and this shows $\alpha \in X$. □

Before we continue the investigation of the connection of an algebraic set $V(R) \subset A_n(R)$ with $p\!-\!Sper_R\, R[V]$ we will state some further properties of the admissible p-real spectrum.

Let A be a R-algebra. For $\alpha \in p\!-\!Sper_R\, A$ set $supp(\alpha) := \wp_\alpha$. Thus we have a map $supp : p\!-\!Sper_R\, A \longrightarrow Spec\ A$. Now [**7**, Proposition 3.7] implies

PROPOSITION (1.11). *The map* $supp : p\!-\!Sper_R\, A \longrightarrow Spec\ A$ *is a morphism of spectral spaces.*

Furthermore, we set $ord(\alpha) := \alpha_0$ for $\alpha = (\alpha_0, \alpha_1, \alpha_2, \ldots) \in p\!-\!Sper_R\, A$. Then $ord(\alpha) \in Sper\ A$ and [**7**, Proposition 3.12] gives

PROPOSITION (1.12). *The map* $ord : p\!-\!Sper_R\, A \longrightarrow Sper\ A$ *is a morphism of spectral spaces.*

Next let $\alpha, \beta \in p\!-\!Sper_R\, A$. Recall that β is called a specialization of α if β lies in the closure of α. We denote this by $\alpha \to \beta$. In the next step we will show that the specializations of a point $\alpha \in p\!-\!Sper_R\, A$ form a chain, i.e. if β, γ are specializations of α, then $\beta \to \gamma$ or $\gamma \to \beta$.

LEMMA (1.13). *Given* $\alpha, \beta \in p\!-\!Sper_R\, A$, *the following statements are equivalent:*

(1) β *is a specialization of* α.
(2) $\alpha_n \subset \beta_n$ *for all* $n \in N_0$.
(3) $\wp_\alpha \subset \wp_\beta$ *and* $\alpha_n \setminus \wp_\beta = \beta_n \setminus \wp_\beta$ *for all* $n \in N_0$.

PROOF. (1) $\Rightarrow$ (3) : By [**7**, Lemma 4.1] we have $\wp_\alpha \subset \wp_\beta$ and $\beta_n \setminus \wp_\beta \subset \alpha_n \setminus \wp_\alpha$, hence $\beta_n \setminus \wp_\beta \subset \alpha_n \setminus \wp_\beta$. Let $a \in \alpha_n \setminus \wp_\beta$. Then $\alpha \in D_n(a) \subset D_0(a)$ and therefore $\beta \in D_0(a)$ as $\alpha \to \beta$. By (1.7) there exists a unique $k \in \{0, \ldots, p^n - 1\}$ such that $\beta \in D_n(\omega^k a)$. Since β is a specialization of α, we get $\alpha \in D_n(a) \cap D_n(\omega^k a)$ which implies $k = 0$, as k is unique. Thus $\beta \in D_n(a)$ which means $a \in \beta_n \setminus \wp_\beta$.
(3) $\Rightarrow$ (2) : Obvious.
(2) $\Rightarrow$ (1) : Let $\beta \in D_n(a) \subset D_0(a)$. Then $\alpha_0 \subset \beta_0$ implies $a \in \alpha_0 \setminus \wp_\beta$. By (1.7) there exists a unique $k \in \{0, \ldots, p^n - 1\}$ such that $\alpha \in D_n(\omega^k a)$. Now $\alpha_n \subset \beta_n$ and $\omega^k a \notin \wp_\beta$ imply $\beta \in D_n(a) \cap D_n(\omega^k a)$. Hence $k = 0$ and $\alpha \in D_n(a)$. □

PROPOSITION (1.14). *The specializations of a point* $\alpha \in p{-}Sper\,A$ *form a chain.*

PROOF. Let β, γ be specializations of α. Since $ord : p{-}Sper_R\,A \longrightarrow Sper\,A$ is continuous, $ord(\beta), ord(\gamma)$ are specializations of $ord(\alpha)$. But in $Sper\,A$ the specializations of a point form a chain. So we may assume w.l.o.g. that $ord(\alpha) \rightarrow ord(\beta) \rightarrow ord(\gamma)$. In particular $\wp_\alpha \subset \wp_\beta \subset \wp_\gamma$. Now $\alpha_n \setminus \wp_\beta = \beta_n \setminus \wp_\beta$ and $\alpha_n \setminus \wp_\gamma = \gamma_n \setminus \wp_\gamma$ imply $\beta_n = \alpha_n \cup \wp_\beta \subset \alpha_n \cup \wp_\gamma = \gamma_n$. Thus $\beta \rightarrow \gamma$, by (1.13). □

This result has some remarkable consequences. Let $p - Max(A)$ be the subspace of closed points of $p{-}Sper_R\,A$. A specialization β of $\alpha \in p{-}Sper_R\,A$ is called maximal if $\beta \in p{-}Max(A)$. Since $p{-}Sper_R\,A$ is a quasicompact T_0-space, the closure of α contains a closed point. Hence to every α there exists a maximal specialization. Now (1.14) implies

COROLLARY (1.15). *Any point* $\alpha \in p{-}Sper_R\,A$ *has a unique maximal specialization* $\alpha_{max} \in p - Max(A)$.

Thus we have a well-defined map

$$spez : p{-}Sper_R\,A \longrightarrow p - Max(A) : \alpha \mapsto \alpha_{max}$$

and from [**10**, Propositions 2 and 3] we get

COROLLARY (1.16).
(1) *The map* $spez : p{-}Sper_R\,A \longrightarrow p - Max(A)$ *is continuous and closed.*
(2) *The space* $p - Max(A)$ *is compact Hausdorff.*

Next let $\varphi : A \longrightarrow B$ be a homomorphism of R-algebras and let $\alpha \in p{-}Sper_R\,A$. We conclude this section by a characterization of the structure of the fibre $\varphi^{*-1}(\alpha)$. We will see that these fibres behave as it is known from the real spectrum.

THEOREM (1.17). *Let* $\varphi : A \longrightarrow B$ *be a homomorphism of R-algebras and let* $\alpha \in p{-}Sper_R\,A$. *Then* $\varphi^{*-1}(\alpha)$ *is homeomorphic to* $p{-}Sper_R\,B \otimes_A k(\alpha)$.

PROOF. First, let $\psi : C \longrightarrow D$ be a homomorphism of R-algebras. Then we write $\psi^* : p{-}Sper_R\,D \longrightarrow p{-}Sper_R\,C$ and $\psi_0^* : p{-}Sper\,D \longrightarrow p{-}Sper\,C$ for the induced morphisms. Now let $\phi : B \longrightarrow B \otimes_A k(\alpha) : b \mapsto b \otimes 1$ and let $Y = \{\gamma \in p{-}Sper\,B \otimes_A k(\alpha) \mid \omega \in \gamma_0 \setminus \gamma_1\}$. By [**7**, Corollary 4.15] the map

$$\phi_{0|Y}^* : Y \longrightarrow \varphi_0^{*-1}(\alpha)$$

is a homeomorphism. Moreover, in the proof of [**7**, Theorem 4.14] it is shown that $k(\phi_0^*(\gamma)) = k(\gamma)$ for all $\gamma \in p{-}Sper\,B \otimes_A k(\alpha)$. Since $p{-}Sper_R\,B \otimes_A k(\alpha) \subset Y$, ϕ_0^* maps $p{-}Sper_R\,B \otimes_A k(\alpha)$ onto $\varphi_0^{*-1}(\alpha) \cap p{-}Sper_R\,B = \varphi^{*-1}(\alpha)$, giving the desired homeomorphism. □

2. Semi-algebraic sets of higher level

Let V be an affine variety over R. It is the goal of this section to investigate the basic properties of the geometrical and topological structure of $V(R)$ and to establish the relationship between V(R) and $p-Sper_R\, R[V]$ mentioned above. The main results of this section correspond to the ultrafilter and finiteness theorem in real algebraic geometry.

For $n \in N_0$ let $\mathcal{S}_n(V)$ be the Boolean lattice generated by the sets

$$\{f >_n 0\} := \{x \in V(R) | f(x) >_n 0\}$$

with $f \in R[V]$. The elements of $\mathcal{S}_n(V)$ are called semi-algebraic sets of level n.

LEMMA (2.1). *A subset $S \in V(R)$ is semi-algebraic of level n if and only if S is the union of finitely many sets of the following kind*

$$\{g = 0, f_1 >_n 0, \ldots, f_k >_n 0\}$$

with $g, f_1, \ldots, f_k \in R[V]$.

PROOF. Let $g \in R[V]$. Since ωQ^{p^n} generates Q/Q^{p^n} we have

$$\{g = 0\} = V(R) \setminus \left\{ \bigcup_{i=0}^{p^n-1} \{\omega^i g >_n 0\} \cup \bigcup_{i=0}^{p^n-1} \{-\omega_i g >_n 0\} \right\}$$

$$V(R) \setminus \{g >_n 0\} = \{g = 0\} \cup \left\{ \bigcup_{i=1}^{p^n-1} \omega^i g >_n 0 \cup \bigcup_{i=0}^{p^n-1} \{-\omega^i g >_n 0\} \right\}.$$

The first equality implies $\{g = 0, f_1 >_n 0, \ldots, f_k >_n 0\} \in \mathcal{S}_n(V)$ and the second implies that every $S \in \mathcal{S}_n(V)$ is of the claimed shape. □

LEMMA (2.2). *For $n < m \in N_0$ we have $\mathcal{S}_n(V) \subset \mathcal{S}_m(V)$.*

PROOF. It is sufficient to show $\{f >_n 0\} \in \mathcal{S}_m(V)$ for $f \in R[V]$. We claim

$$\{f >_n 0\} = \bigcup_{k=1}^{p^{m-n}} \left\{\omega^{kp^n} f >_m 0\right\} =: U.$$

The inclusion $U \subset \{f >_n 0\}$ is obvious. Next let $x \in \{f >_n 0\}$. Then there exists a unique $k \in \{0, \ldots, p^m - 1\}$ such that $\omega^k f(x) >_m 0$. But $f(x) >_n 0$ and $\omega^k f(x) >_n 0$ imply that p^n divides k. □

Thus we have a canonical embedding $i_{nm} : \mathcal{S}_n(V) \hookrightarrow \mathcal{S}_m(V)$ and the lattices $\mathcal{S}_n(V)$ may be regarded as refinements of the usual lattice $\mathcal{S}_0(V)$ of semi-algebraic subsets of V. Let $\mathcal{S}(V)$ be the limit of the direct system $(\mathcal{S}_n(V), i_{nm})$. The elements of $\mathcal{S}(V)$ are called semi-algebraic sets (of higher level).

We are now going to expose the relationship between V and the admissible p-real spectrum of $R[V]$. Let $\mathcal{C}(V)$ be the Boolean lattice of constructible subsets of $p-Sper_R\, R[V]$ and for $n \in N_0$ let $\mathcal{C}_n(V) \subset \mathcal{C}(V)$ be the sublattice generated by the sets $D_n(f)$ with $f \in R[V]$. The elements of $\mathcal{C}_n(V)$ are called constructible of level n. In view of (1.7), the same arguments as in the proof of (2.2) show that $\mathcal{C}_n(V) \subset \mathcal{C}_m(V)$ for $n < m \in N_0$. Furthermore, $\mathcal{C}(V)$ is just the direct limit of the lattices $\mathcal{C}_n(V)$. Since $D_n(f) \cap V = \{f >_n 0\}$, we get a map

$$\pi : \mathcal{C}(V) \longrightarrow \mathcal{S}(V) : C \mapsto C \cap V(R).$$

Let π_n be the restriction of π to $\mathcal{C}_n(V)$. Note that $\pi_n(\mathcal{C}_n(V)) \subset \mathcal{S}_n(V)$.

PROPOSITION (2.3). *The map* $\pi : \mathcal{C}(V) \longrightarrow \mathcal{S}(V)$ *is a lattice isomorphism. In particular, the maps* $\pi_n : \mathcal{C}_n(V) \longrightarrow \mathcal{S}_n(V)$ *are isomorphisms.*

PROOF. Obviously the map π is a lattice homomorphism. The lattice $\mathcal{S}(V)$ is generated by the sets $\{f >_n 0\} = D_n(f) \cap V(R)$. Hence π is surjective. The injectivity of π follows from (1.10). □

As usual we set $\tilde{S} := \pi^{-1}(S)$ for a semi-algebraic set $S \subset V(R)$. So far we have obtained a set-theoretical correspondence between the semi-algebraic subsets of $V(R)$ and the constructible subsets of $p{-}Sper_R\, R[V]$. It remains to show that this correspondence preserves topological properties. The next steps are devoted to this problem.

For the remaining part of this paper we fix a representation

$$R[V] = R[X_1, \ldots, X_m]/(g_1, \ldots, g_k)$$

and set $x = (x_1, \ldots, x_m)$ where x_i denotes the image of the indeterminate X_i in $R[V]$. Let $\mathcal{L}_p(R)$ be the extension of the language $\mathcal{L}_p$ by a constant c_r for each $r \in R$. If $\Phi(t_1, \ldots, t_m)$ is a q.f. $\mathcal{L}_p(R)$-formula with free variables $t_1, \ldots, t_m$, then

$$S(\Phi) := \{z \in V(R) \mid R \models \Phi(z)\}$$

is a semi-algebraic subset of V(R) and

$$X(\Phi) := \{\alpha \in p{-}Sper_R\, R[V] \mid k(\alpha) \models \Phi(x(\alpha))\}$$

is constructible. Moreover, we have $X(\Phi) = \widetilde{S(\Phi)}$. Conversely, if $S \in \mathcal{S}(V)$, then there exists a q.f. $\mathcal{L}_p(R)$ -formula Φ such that $S = S(\Phi)$ and $\tilde{S} = X(\Phi)$. We call such a formula Φ a defining formula for S. Let $(R, \alpha(R)) \prec (L, \alpha)$ be an elementary extension, $S \in \mathcal{S}(V)$ and Φ a defining formula for S. Now set

$$S_L := \left\{ z \in A_m(L) \mid L \models \bigwedge_{i=1}^{k} g_i(z) = 0 \wedge \Phi(z) \right\}.$$

Note that S_L does not depend on the choice of the defining formula Φ as L is an elementary extension of R. For $\alpha \in p{-}Sper_R\, R[V]$ we will write S_α rather than $S_{k(\alpha)}$. As an easy consequence of these definitions we get

LEMMA (2.4). *Let* $S \subset V(R)$ *be semialgebraic. Then* $\tilde{S} = \{\alpha | x(\alpha) \in S_\alpha\}$.

We now come back to the topological problems mentioned above. A semi-algebraic subset $S \subset V$ is called open, if S is open with respect to the strong topology. It follows immediately from the definition of the homomorphism $\pi : \mathcal{C}(V) \longrightarrow \mathcal{S}(V)$ that π maps open-constructible sets onto open semi-algebraic sets. In order to prove the converse direction we need a slightly refined description of the topological structure of the admissible spectrum.

Let $\mathcal{T}_n$ be the topology of $p\!-\!Sper_R\, R[V]$ generated by the sets $D_n(f)$ with $f \in R[V]$. Note that $\mathcal{T}_n \subset \mathcal{T}_m$ holds for $n < m$ and that the topology of $p\!-\!Sper_R\, R[V]$ is the direct limit of the topologies $\mathcal{T}_n$. We call a subset $X \subset p\!-\!Sper_R\, R[V]$ n-closed (resp. n-open) if X is closed (resp. open) with respect to $\mathcal{T}_n$. Let $\mathcal{C}_n(V)^o$ resp. $\mathcal{C}_n(V)_a$ be the set of n-open resp. n-closed constructible sets of level n. Analogously, we denote by $\mathcal{C}(V)^o$ resp. $\mathcal{C}(V)_a$ the open resp. closed constructible subsets of $p\!-\!Sper_R\, R[V]$.

LEMMA (2.5). *Let $C \subset p\!-\!Sper_R\, R[V]$ be a constructible set of level n. Then $C \in \mathcal{C}_n(V)_a$ if and only if C is the union of finitely many sets of the following kind:*

$$\{\alpha | f_1(\alpha) \geq_n 0, \dots , f_k(\alpha) \geq_n 0\}$$

with $f_1, \dots, f_k \in R[V]$.

PROOF. The claim follows from the equality

$$p\!-\!Sper_R\, R[V] \setminus D_n(f) = \bigcup_{i=1}^{p^n-1} \{\alpha \mid \omega^i f(\alpha) \geq_n 0\} \cup \bigcup_{i=0}^{p^n-1} \{\alpha \mid -\omega^i f(\alpha) \geq_n 0\}$$

and the compactness of $p\!-\!Sper_R\, R[V] \setminus C$. $\square$

For $\alpha \in p\!-\!Sper_R\, R[V]$ we denote by $\bar{\{\alpha\}}^n$ the closure of $\{\alpha\}$ with respect to the topology $\mathcal{T}_n$. Now the same arguments as in the proof of (1.13) show

LEMMA (2.6). *For $\alpha, \beta \in p\!-\!Sper_R\, R[V]$ the following statements are equivalent:*

(1) $\beta \in \bar{\{\alpha\}}^n$.
(2) $\alpha_m \subset \beta_m$ *for all* $m \in \{0, \dots, n\}$.
(3) $\wp_\alpha \subset \wp_\beta$ *and* $\alpha_m \setminus \wp_\beta = \beta_m \setminus \wp_\beta$ *for all* $m \in \{0, \dots, n\}$.

LEMMA (2.7). *Let $X \subset Y \subset p\!-\!Sper_R\, R[V]$ be constructible. Then X is n-closed in Y if and only if $\bar{\{\alpha\}}^n \cap Y \subset X$ for all $\alpha \in X$.*

PROOF. It remains to show that the condition is sufficient. Let $\beta \in Y$, $C(\beta) = \{C \in \mathcal{C}_n(V)^o | \beta \in C\}$ and $T = \bigcap_{C \in C\beta)} C$. Assume that β lies in the n-closure of X in Y. Thus $C \cap X \neq \emptyset$ for all $C \in C(\beta)$, which implies that there exists a $\alpha \in T \cap X$. But then $\beta \in \bar{\{\alpha\}}^n$. Hence $\beta \in X$. $\square$

We are now prepared to prove the main result of this section, a finiteness theorem for semi-algebraic sets of higher level. By virtue of the preceeding results we may prove this result in a similar manner as in the case of semi-algebraic sets over real closed fields (see [**11**]).

THEOREM (2.8). *Let $S \subset T \subset V(R)$ be semi-algebraic sets of level n. Then S is closed in T if and only if S is representable as a finite union of sets of the following kind:*

$$T \cap \{f_1 \geq_n 0, \dots, f_k \geq_n 0\}.$$

PROOF. It remains to show that the existence of such a representation is necessary. For this it is sufficient to prove that $\tilde{S}$ is n-closed in $\tilde{T}$ by (2.5). Let $\alpha \in \tilde{S}$ and $\beta \in \overline{\{\alpha\}}^n \cap \tilde{T}$. By (2.7) we have to show that β lies in $\tilde{S}$, i.e. $x(\beta) \in S_\beta$, by (2.4). We consider $R[V]/\wp_\alpha \subset (k(\alpha), \bar{\alpha_0})$ and $R[V]/\wp_\beta \subset (k(\beta), \bar{\beta_0})$ as ordered rings. It follows from (2.6) that there exists an order preserving homomorphism $\varphi : R[V]/\wp_\alpha \longrightarrow R[V]/\wp_\beta$ such that the following diagram commutes:

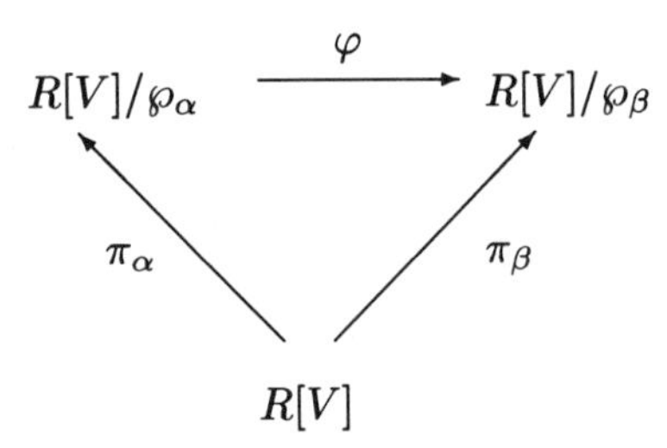

By the real place extension theorem there exists a surjective place $\lambda : k(\alpha) \longrightarrow F \cup \infty$ extending φ, where $k(\wp_\beta) \subset F$. Let B_λ be the valuation ring corresponding to λ and let $\lambda(\alpha) := (\lambda(\bar{\alpha_0} \cap B_\lambda), \lambda(\bar{\alpha_1} \cap B_\lambda), \dots)$. It follows from [**7**, Corollary 1.21] that $(F, \lambda(\alpha))$ is p-real closed and isomorphic to a subfield $(L, L \cap \bar{\alpha})$ of $(k(\alpha), \bar{\alpha})$. So we may assume w.l.o.g. that F is a subfield of $k(\alpha)$. But then F has to be relatively algebraically closed in $k(\alpha)$. Now $R \prec k(\alpha)$ implies $(R, \alpha(R)) \prec (F, \lambda(\alpha)) \prec (k(\alpha), \bar{\alpha})$. Thus the semi-algebraic sets S_F, T_F are well-defined. Next let $y_i := \lambda(x_i(\alpha)) \in F$ and $y := (y_1, \dots, y_m)$. Let $f \in R[V]$. The diagram above and (2.6) imply

$$\lambda(f(\alpha)) = 0 \Leftrightarrow f(\beta) = 0 \text{ and } \lambda(f(\alpha)) >_n 0 \Leftrightarrow f(\beta) >_n 0.$$

Since S, T are semi-algebraic sets of level n, this implies

$$x(\beta) \in S_\beta \Leftrightarrow \lambda(x(\alpha)) = y \in S_F \text{ and } x(\beta) \in T_\beta \Leftrightarrow \lambda(x(\alpha)) = y \in T_F.$$

Thus it remains to prove $y \in S_F$. From $\beta \in \overline{\{\alpha\}}^n \cap T$ it follows that $y \in T$. That S is closed in T is expressible by a $\mathcal{L}_p$-formula Φ. Hence S_F is closed in T_F, as $R \prec F$. For $0 <_0 \varepsilon \in F$ set

$$B_\varepsilon(y) = \{z \in A_m(F) | \sum_{i=1}^m (z_i - y_i)^2 <_0 \varepsilon\},$$

$$B_\varepsilon^\alpha(y) = \{z \in A_m(k(\alpha)) | \sum_{i=1}^m (z_i - y_i)^2 <_0 \varepsilon\}.$$

We have to show that $S_F \cap B_\varepsilon(y)$ is not empty for all $0 <_0 \varepsilon \in F$. From $\alpha \in \tilde{S}$ we get $x(\alpha) \in S_\alpha$. Since $x_i(\alpha) - y_i$ lies in the maximal ideal of B_λ it follows that $x(\alpha) \in B_\varepsilon^\alpha(y)$, hence $B_\varepsilon^\alpha(y) \cap S_\alpha \neq \emptyset$ for all ε. Since the last statement is expressible by a $\mathcal{L}_p(R)$-formula and since $(F, \lambda(\alpha))$ is an elementary substructure of $(k(\alpha), \bar{\alpha})$, we conclude $B_\varepsilon(y) \cap S_F \neq \emptyset$. Thus $y \in S_F$, as S_F is closed in T_F. □

The last result has a couple of consequences. Together with (2.5) it implies

COROLLARY (2.9). *Let $C \subset C_1 \subset p\text{–}Sper_R\, R[V]$ be constructible sets of level n. Then C is closed (resp. open) in C_1 if and only if C is n-closed (resp. n-open) in C_1.*

Furthermore we obtain the following improvement of (2.3):

COROLLARY (2.10). *Let $S \subset T \subset V(R)$ be semi-algebraic of level n. Then S is closed (resp. open) in T if and only if $\tilde{S}$ is closed (resp. open) in $\tilde{T}$.*

In the last step we will prove a description of $p\text{–}Sper_R\, R[V]$ via ultrafilters of semi-algebraic subsets of V. Let $\widehat{\mathcal{S}(V)}$ be the set of ultrafilters of semi-algebraic subsets of V. We provide $\widehat{\mathcal{S}(V)}$ with the topology generated by the sets $\hat{S} = \{F \in \widehat{\mathcal{S}(V)} | S \in F\}$, with $S \subset V(R)$ open semi-algebraic.

THEOREM (2.11). *The admissible spectrum $p-Sper_R\, R[V]$ is canonically homeomorphic to $\widehat{\mathcal{S}(V)}$.*

PROOF. Let $\widehat{\mathcal{C}(V)}$ be the set of ultrafilters of constructible subsets of $p-Sper_R\, R[V]$. We provide $\widehat{\mathcal{C}(V)}$ with the topology generated by the sets $\hat{C} = \{F \in \widehat{\mathcal{C}(V)} | C \in F\}$ with C open-constructible. The map $\pi : \mathcal{C}(V) \longrightarrow \mathcal{S}(V)$ of (2.3) extends uniquely to a map $\hat{\pi} : \widehat{\mathcal{C}(V)} \longrightarrow \widehat{\mathcal{S}(V)}$. From (2.3) and (2.10) it follows that $\hat{\pi}$ is a homeomorphism. For $\alpha \in p\text{–}Sper_R\, R[V]$ let $F(\alpha) \in \widehat{\mathcal{C}(V)}$ be the set of constructible sets containing α. Since $p\text{–}Sper_R\, R[V]$ is a spectral space, the map

$$F : p\text{–}Sper_R\, R[V] \longrightarrow \widehat{\mathcal{C}(V)} : \alpha \mapsto F(\alpha)$$

is a homeomorphism (see [**10**]). Thus $\hat{\pi} \circ F$ is the desired homeomorphism. □

Finally we consider the case that V is irreducible. Let $R(V)$ be the function field of V. Let $\widehat{\mathcal{S}(V)}_d \subset \widehat{\mathcal{S}(V)}$ be the subspace of ultrafilters which contain only Zariski-dense semi-algebraic sets. Furthermore we denote by $\widehat{\mathcal{C}(V)}_d$ the preimage of $\widehat{\mathcal{S}(V)}_d$ with respect to the homeomorphism $\hat{\pi} : \widehat{\mathcal{C}(V)} \longrightarrow \widehat{\mathcal{S}(V)}$.

PROPOSITION (2.12). *Let V be irreducible. Then $p-Sper_R\, R(V)$ is canonically homeomorphic to $\widehat{\mathcal{S}(V)}_d$.*

PROOF. We can identify $p\!-\!Sper_R\, R(V)$ with $\{\alpha \in p\!-\!Sper_R\, R[V] \| \wp_\alpha = (0)\}$. So it is sufficient to show that the homeomorphism

$$F : p\!-\!Sper_R\, R[V] \longrightarrow \widehat{\mathcal{C}(V)} : \alpha \mapsto F(\alpha)$$

maps $p\!-\!Sper_R\, R(V)$ onto $\widehat{\mathcal{C}(V)}_d \cong \widehat{\mathcal{S}(V)}_d$. Let $\alpha \in p\!-\!Sper_R\, R(V)$ and $C \in F(\alpha)$. Since $C \cap p\!-\!Sper_R\, R(V) \neq \emptyset$ it follows from (2.1) and (2.3) that C contains an open-constructible subset C_0 with $\alpha \in C_0$. By [**8**, Proposition 2.1], the open set $C_0 \cap V(R)$ contains a regular point and is therefore Zariski-dense in V [**8**, Corollary 2.3]. Hence $F(\alpha) \in \widehat{\mathcal{C}(V)}_d$. Conversely, let $\alpha \in p\!-\!Sper_R\, R[V]$ such that $F(\alpha) \in \widehat{\mathcal{C}(V)}_d$. We have to show that $\wp_\alpha = (0)$. Let $f \in \wp_\alpha$. Then $Z(f) := \{\beta \in p\!-\!Sper_R\, R[V] \| f(\beta) = 0\}$ lies in $F(\alpha)$. Thus $Z(f) \cap V(R) = \{x \in V(R) | f(x) = 0\}$ is Zariski-dense in V, which implies $f = 0$. □

3. Regular mappings and p-semi-algebraic sets

Let V, W be affine varieties over R. In this section we will investigate the images of regular mappings $\varphi : V(R) \longrightarrow W(R)$. In contrast to semi-algebraic geometry over real closed fields the image of φ need not to be semi-algebraic. We will give a complete description of the images of regular mappings and we will investigate the connection of these sets with certain subsets of the admissible p-real spectrum.

The intention of this section leads to the following definition: a subset $S \subset V(R)$ is called p-semi-algebraic if there exists an affine variety W over R and a regular mapping $\varphi : W(R) \longrightarrow V(R)$ such that $S = \varphi(W(R))$. First note that we have

COROLLARY (3.1). *Any semi-algebraic set $S \subset V(R)$ is p-semi-algebraic.*

PROOF. By (2.1) there exists a representation

$$S = \bigcup_{i=1}^{k} \{g_i = 0, f_{i1} >_n 0, \ldots, f_{ik_i} >_n 0\}.$$

If $p \neq 2$, then set

$$W = \left\{ (x,y) \mid x \in V \wedge \prod_{i=1}^{k} \left(g_i(x)^2 + \sum_{j=1}^{k_i} \left(1 + y_{ij}^{2p^n} f_{ij}(x)\right)^2 \right) = 0 \right\}.$$

If $p = 2$, then one has to replace the terms $1 + y_{ij}^{2p^n} f_{ij}$ by $(1 + y_{ij}^2 f_{ij})(1 + \omega y_{ij}^2 f_{ij})$ resp. $1 + y_{ij} 2^n f_{ij}$ according to $n = 0$ resp. $n \neq 0$. Now let $\pi : W(R) \longrightarrow V(R)$ be the canonical projection. Then $S = \pi(W(R))$, by the choice of W. □

As an immediate consequence we obtain

COROLLARY (3.2). *Let W be an affine R-variety, $T \subset W(R)$ a semi-algebraic subset and $\varphi : W(R) \longrightarrow V(R)$ a regular mapping. Then $\varphi(T)$ is p-semi-algebraic.*

Moreover, this result and general properties of regular mappings imply the following characterization.

COROLLARY (3.3). *A subset $S \subset V(R)$ is p-semi-algebraic if and only if there exists $d \in N$ and a semi-algebraic set $T \subset V(R) \times A_d(R)$ such that $S = \pi(T)$, where $\pi : V(R) \times A_d(R) \longrightarrow V(R)$ denotes the canonical projection.*

We denote by $\mathcal{S}_p(V)$ the set of p-semi-algebraic subsets of $V(R)$. By (3.1) we have $\mathcal{S}(V) \subset \mathcal{S}_p(V)$. But in contrast to the semi-algebraic subsets of $V(R)$ the p-semi-algebraic subsets do not form a Boolean lattice. However, we have

PROPOSITION (3.4). *$\mathcal{S}_p(V)$ is closed under finite unions and intersections.*

PROOF. Let S_1, S_1 be p-semi-algebraic. By (3.3) there exist semi-algebraic sets $T_i \subset A_{d_i}(R) \times V(R)$, $i = 1, 2$, such that $\pi(T_i) = S_i$. Now set

$$M_1 := \{(x, y, z) \in A_{d_1}(R) \times A_{d_2}(R) \times V(R) | (x, z) \in T_1 \vee (y, z) \in T_2\}.$$

$$M_2 := \{(x, y, z) \in A_{d_1}(R) \times A_{d_2}(R) \times V(R) | (x, z) \in T_1 \wedge (y, z) \in T_2\}.$$

Obviously the sets M_i are semi-algebraic and we have $\pi(M_1) = S_1 \cup S_2$ and $\pi(M_2) = S_1 \cap S_2$. □

We are now going to give a complete description of the p-semi-algebraic subsets of $V(R)$. For $f, g \in R[V]$ let

$$\{f/g >_\infty 0\} := \{x \in V(R) | g(x) \neq 0 \wedge 0 <_0 \frac{f(x)}{g(x)} \notin J(R)^*\}.$$

Here $J(R)^*$ denotes the group of units of the valuation ring $J(R) \subset R$. Note that the valuation ring $J(R) \subset R$ corresponds to the maximal divisible convex subgroup $\Delta \subset w(R^*)$, where w denotes the finest henselian valuation of R. Let $1 <_0 z \in R$. Then we have $z \notin J(R)^*$ if and only if there exists $1 <_0 a \in R$ such that $1 <_0 a <_0 z$ and $a \notin R^p$ (see [**7**, Lemma 2.1]). Another equivalent characterization for $z \notin J(R)^*$ is given by $1 <_0 t <_0 z$ for all $1 <_0 t \in J(R)^*$. Given $f, g \in R[V]$, this implies

$$\{f/g >_\infty 0\} = \bigcup_{a \in Q \setminus Q^p} \{\{g^2 <_0 ag^2 <_0 fg\} \cup \{0 <_0 fg <_0 ag^2 <_0 g^2\}\}.$$

(3.5)

$$\{f/g >_\infty 0\} = \bigcap_{0 <_0 t \in J(R)^*} \{\{g^2 <_0 tg^2 <_0 fg\} \cup \{0 <_0 fg <_0 tg^2 <_0 g^2\}\}.$$

Thus we have proven

COROLLARY (3.6). *Let* $f_1, g_1, \dots, f_k, g_k \in R[V]$. *Then the set*

$$\{f_1/g_1 >_\infty 0, \dots, f_k/g_k >_\infty 0\}$$

can be written both as an infinite union and as an infinite intersection of open semi-algebraic sets.

These easy results already allow us to give a complete description of the p-semi-algebraic subsets of $V(R)$.

THEOREM (3.7). *A subset* $S \subset V(R)$ *is p-semi-algebraic if and only if* S *is the finite union of sets of the following shape*

$$T \cap \{f_1/g_1 >_\infty 0, \dots, f_k/g_k >_\infty 0\},$$

where $T \subset V(R)$ *is semi-algebraic and* $f_i, g_i \in R[V]$.

PROOF. First let $S \in \mathcal{S}_p(V)$. By (3.3) there exists a semi-algebraic set $T \subset V(R) \times A_d(R)$ such that $\pi(S) = S$. Let $\Phi(x, y)$ be a defining $\mathcal{L}_p$-formula for T. Then

$$S = \pi(T) = \{x \in V(R) | R \models (\exists y)\Phi(x, y)\}.$$

Now the claim follows from [**7**, Theorem 2.5]. For the converse direction it is sufficient to show that $\{f/g >_\infty 0\}$ is p-semi-algebraic, as $\mathcal{S}_p(V)$ is closed under finite unions and intersections. Let

$$\begin{aligned} T := \{(x_1, x_2, y) \in A_2(R) \times V(R) \quad | \quad & [g^2(y) <_0 x_1 g^2(y) <_0 f(y)g(y) \vee \\ & f(y)g(y) <_0 x_1 g^2(y) <_0 g^2(y)] \wedge \\ & [\bigvee_{i=1}^{p-1} \omega^i x_1 - x_2^p = 0]\}. \end{aligned}$$

Then $\pi(T) = \{f/g >_\infty 0\}$, by (3.5). □

Furthermore, the last two results imply

COROLLARY (3.8). *Let* $S \subset V(R)$ *be p-semi-algebraic. Given* $x \in S$ *and* $y \in V(R) \setminus S$ *there exist semi-algebraic sets* $T_x \subset S, T_y \subset V(R) \setminus S$ *such that* $x \in T_x$ *and* $y \in T_y$.

It remains to describe the subsets of $p\text{--}Sper_R\, R[V]$ corresponding to the p-semi-algebraic sets. To this end we introduce the following notion. A subset $X \subset p\text{–}Sper_R\, R[V]$ is called p-constructible if there exists a polynomial ring $R[V][T] = R[V][T_1, \dots, T_d]$ and a constructible set $C \subset p\text{–}Sper_R\, R[V][T]$ such that $f^*(C) = X$, where $f : R[V] \longrightarrow R[V][T]$ denotes the canonical imbedding. Thus the definition of p-constructible sets is similar to that of p-semi-algebraic sets and we will show that there is a bijective correspondence between both concepts. Before doing so, we have to state two simple properties of p-constructible sets.

Let $\mathcal{C}_p(V)$ be the set of p-constructible subsets. By carrying over the arguments of the proof of (3.4) into the context of the admissible p-real spectrum we obtain

PROPOSITION (3.9). *$\mathcal{C}_p(V)$ is closed under finite unions and intersections.*

Next let W be an affine variety over R and let $\varphi : W(R) \longrightarrow V(R)$ be a regular mapping. Then φ extends to a spectral morphism

$$\Phi : p\text{--}Sper_R\, R[W] \longrightarrow p\text{--}Sper_R\, R[V]$$

via the induced homomorphism $\varphi^* : R[V] \longrightarrow R[W]$ of R-algebras. Thus the following diagram commutes:

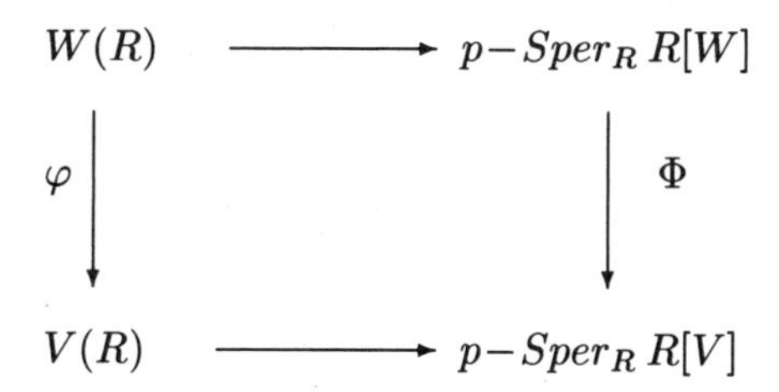

The next result generalizes (3.3).

PROPOSITION (3.10). *Let $\varphi : W(R) \longrightarrow V(R)$ be a regular mapping of affine R-varieties, $\Phi : p\text{–}Sper_R\, R[W] \longrightarrow p\text{–}Sper_R\, R[V]$ the corresonding spectral morphism and $C \subset p\text{–}Sper_R\, R[W]$ a constructible set. Then $\Phi(C)$ is p-constructible.*

PROOF. Choose an embedding $e : W(R) \longrightarrow A_d(R)$ and let $\psi := (e, \varphi) : W(R) \longrightarrow A_d(R) \times V(R)$. Finally let $\pi : A_d(R) \times V(R) \longrightarrow V(R)$ denote the canonical projection. Then the following diagram commutes.

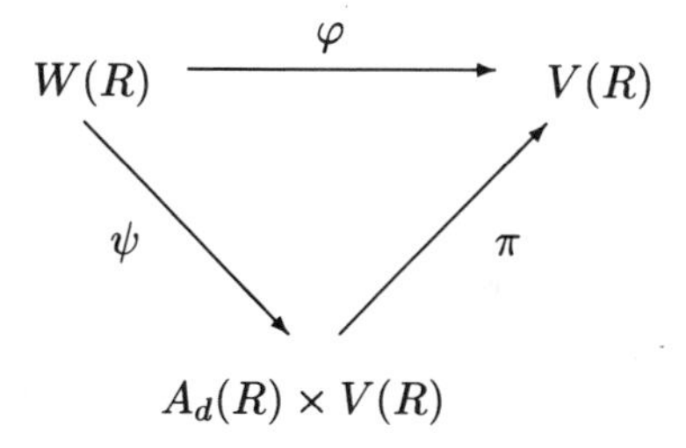

These regular mappings induce morphisms of spectral spaces such that the following diagram commutes.

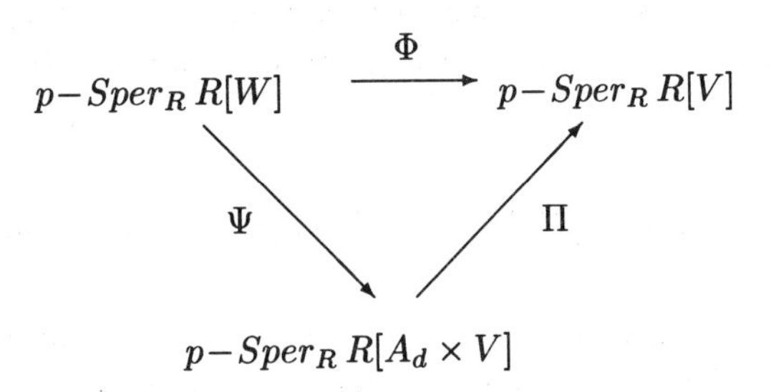

Now let $C \subset p\!-\!Sper_R\, R[W]$ be constructible. One readily checks that $\Psi(C)$ is constructible. Hence $\Phi(C) = (\Pi \circ \Psi)(C)$ is p-constructible. □

Let $S \in V(R)$ be semi-algebraic. Then (1.10) implies that $\tilde{S} \subset p\!-\!Sper_R\, R[V]$ is the closure of S with respect to the constructible topology. This characterization motivates the following assignment: for a p-semi-algebraic set $S \subset V(R)$ let $\tilde{S} \subset p\!-\!Sper_R\, R[V]$ be the closure of S with respect to the constructible topology. The next result already shows that this assignment leads to p-constructible sets.

LEMMA (3.11). *Let W be an affine R-variety, $\varphi : W(R) \longrightarrow V(R)$ a regular mapping and $\Phi : p\!-\!Sper_R\, R[W] \longrightarrow p\!-\!Sper_R\, R[V]$ the corresponding morphism of spectral spaces. Given a semi-algebraic set $S \subset W(R)$ we have*
(a) $\widetilde{\varphi(S)} = \Phi(\tilde{S})$.
(b) $\widetilde{\varphi(S)} \cap V(R) = \varphi(S)$.

PROOF. (a) $\Phi(\tilde{S})$ is proconstructible as Φ is continuous with respect to the constructible topology. Thus it remains to show that $\varphi(S)$ is a dense subset of $\Phi(\tilde{S})$. Let $C \subset p\!-\!Sper_R\, R[V]$ be constructible such that $C \cap \Phi(\tilde{S}) \neq \emptyset$. Then $\Phi^{-1}(C \cap \Phi(\tilde{S})) \supset \Phi^{-1}(C) \cap \tilde{S} \neq \emptyset$. Since $\Phi^{-1}(C) \cap \tilde{S}$ is constructible it contains a rational point $x \in \tilde{S}$. Thus $\Phi(x) = \varphi(x) \in C \cap \varphi(S)$. This proves $\widetilde{\varphi(S)} = \Phi(\tilde{S})$.
(b) Obviously we have $\varphi(S) \subset \widetilde{\varphi(S)} \cap V(R)$. Now the claim follows from (3.8). □

From the above result we see that $\tilde{T}$ is p-constructible for a p-semi-algebraic set $T \subset V(R)$. Thus we have a map $\pi_p : \mathcal{S}_p(V) \longrightarrow \mathcal{C}_p(V) : T \longrightarrow \tilde{T}$. We are now ready to prove the correspondence between p-semi-algebraic and p-constructible sets mentioned above.

PROPOSITION (3.12). *The map $\pi_p : \mathcal{S}_p(V) \longrightarrow \mathcal{C}_p(V)$ is a lattice isomorphism.*

PROOF. From (3.11) (b) it follows that π_p is injective. Next let $X \subset p\!-\!Sper_R\, R[V]$ be p-constructible. By definition, there exists a polynomial ring $R[V][T] = R[V][T_1, \ldots, T_d]$ and $C \subset p\!-\!Sper_R\, R[V][T]$ constructible such that $f^*(C) = X$, where $f : R[V] \longrightarrow R[V][T]$ denotes the canonical imbedding. Now let $S := C \cap V(R) \times A_d(R)$ and let $\pi : V(R) \times A_d(R)$ be the canonical projection. Applying (3.11) we see that $T := X \cap V(R) = \pi(S)$ is p-semi-algebraic with

$\widetilde{\pi(S)} = X$. Hence π_p is bijective. But then it follows immediately from the definition that π_p respects the lattice structure. □

We conclude this section by giving another characterization of p-constructible sets which is sometimes useful.

Let $\mathcal{L}_p^*$ be the extension of the language $\mathcal{L}_p$ by a binary relation $U(,)$. Let (K, α) be a p-real closed field. We denote by $J(K)$ the largest valuation ring of K such that the corresponding valuation satisfies (1.1). For $a, b \in K$ set

$$U(a,b) \Leftrightarrow b \neq 0 \wedge 0 <_0 ab^{-1} \in J(K)^*.$$

In this way (K, α) becomes a $\mathcal{L}_p^*$-structure.

Let $f, g \in R[V]$. By definition we have

$$\{f/g >_\infty 0\} = \{x \in V(R) | R \models \neg U(f(x), g(x))\}.$$

Now let $\mathcal{F}_p(V)$ be the set of q.f. $\mathcal{L}_p^*(R)$-formulas Φ with free variables $x = (x_1, \dots, x_m)$ such that the predicate $U(,)$ occurs only in negated form, i.e. Φ is of the form

$$\bigvee_i \bigwedge_j (\Phi_{ij}(x) \wedge \neg U(f_{ij}, g_{ij}))$$

with $f_{ij}, g_{ij} \in R[V]$ and where the Φ_{ij} are q.f. $\mathcal{L}_p$-formulas with parameters from R.

It follows from (3.7) that a subset $S \subset V(R)$ is p-semi-algebraic if and only if there exists some $\Phi \in \mathcal{F}_p(V)$ such that

$$S = \{x \in V(R) | R \models \Phi(x)\}.$$

In this situation we call Φ a defining formula for S. But there is an essential difference to defining formulas of semi-algebraic sets. Namely, let

$$X(\Phi) := \{\alpha \in p\text{--}Sper_R\, R[V] | k(\alpha) \models \Phi(x(\alpha))\}.$$

If ϕ is a $\mathcal{L}_p$-formula, i.e. S is semi-algebraic, then $\tilde{S} = X(\Phi)$. But if S is p-semi-algebraic this need not to be true, one just gets $X(\Phi) \subset \tilde{S}$. In the sequel we will show how to remove this difficulty.

As before we denote by Δ the largest divisible convex subgroup of $w(R^*)$. For $\alpha \in p\text{--}Sper_R\, R[V]$ let w_α be the unique extension of w to $k(\alpha)$ compatible with $\bar{\alpha}_0$ and let $\Delta(\alpha)$ be the convex hull of Δ in $w_\alpha(k(\alpha)^*)$. Finally let $k(\alpha)(X)$ be a simple transcendental extension. By [7, Lemma 2.2], there exists a p-real closure $(\rho(\alpha), \tilde{\alpha})$ of $k(\alpha)(X)$ such that

(3.13) (1) $\tilde{\alpha} \cap k(\alpha) = \bar{\alpha}$.
(2) The valuation ring $J(\rho(\alpha)) \cap k(\alpha)$ corresponds to $\Delta(\alpha)$.

Note that (2) implies $J(\rho(\alpha)) \cap R = J(R)$. Thus $(R, \alpha(R))$ is an elementary $\mathcal{L}_p$-substructure of $(\rho(\alpha), \tilde{\alpha})$, by (1.6). Finally, given $\Phi \in \mathcal{F}_p(V)$ we set

$$X_p(\Phi) := \{\alpha \in p\text{--}Sper_R\, R[V] | (\rho, \tilde{\alpha}) \models \Phi(x(\alpha))\}.$$

PROPOSITION (3.14). *Let $S \subset V(R)$ be p-semi-algebraic and $\Phi \in \mathcal{F}_p(V)$ a defining formula for S. Then $\tilde{S} = X_p(\Phi)$.*

PROOF. First we will show that $X_p(\Phi)$ is proconstructible. To this end choose $\mathcal{L}_p$-formulas Φ_{ij} and $f_{ij}, g_{ij} \in R[V]$ such that

$$\Phi = \bigvee_i \bigwedge_j (\Phi_{ij} \wedge \neg U(f_{ij}, g_{i,j})).$$

Now let $\alpha \in p{-}Sper_R\, R[V] \setminus X_p(\Phi)$, i.e. $(\rho(\alpha), \tilde{\alpha}) \models \neg\Phi$. Hence

$$(\rho(\alpha), \tilde{\alpha}) \models \bigvee_j \bigwedge_i (\neg\Phi_{ij} \vee U(f_{ij}, g_{ij})).$$

We may assume w.l.o.g. that $(\rho(\alpha), \tilde{\alpha}) \models \bigwedge_i (\neg\Phi_{i1} \vee U(f_{i1}, g_{i1}))$. By (3.13) we have $(\rho(\alpha), \tilde{\alpha}) \models U(f_{i1}, g_{i1})$ if and only if $w_\alpha(f_{i1}/g_{i1}) \in \Delta(\alpha)$. Thus there exist $t_i \in J(R)^*$ such that

$$(k(\alpha), \alpha) \models \bigwedge_i \neg\Phi_{i1} \vee (t_i g_{i1}^2(\alpha) <_0 f_{i1}^2(\alpha) <_0$$

$$g_{i1}^2(\alpha) \vee g_{i1}^2(\alpha) <_0 f_{i1}^2(\alpha) <_0 t_i g_{i1}^2(\alpha)) =: \Psi.$$

Let $C := \{\beta \in p{-}Sper_R\, R[V] | k(\beta) \models \Psi\}$. Then C is constructible with $\alpha \in C$ and $C \cap X_p(\Phi) = \emptyset$. Hence $X_p(\Phi)$ is proconstructible. Since $(R, \alpha(R))$ is an elementary substructure of $(\rho(\alpha), \tilde{\alpha})$, the same arguments as in the proof of (1.10) show that S is dense in $X_p(\Phi)$ with respect to the constructible topology. Hence $\tilde{S} = X_p(\Phi)$. □

As an evident consequence we get

COROLLARY (3.15). *A subset $Y \subset p{-}Sper_R\, R[V]$ is p-constructible if and only if there exists a formula $\Phi \in \mathcal{F}_p(V)$ such that $Y = X_p(\Phi)$.*

4. Central points and the p-real dimension

Let V be an affine variety over R. We denote by $dim_p\, V$ the Krull dimension of $V(R)$ with respect to the Zariski topology and by $dim_p\, R[V]$ the Krull dimension of $p{-}Sper_R\, R[V]$. We call $dim_p\, V$ resp. $dim_p\, R[V]$ the p-real dimension of V resp. $R[V]$. In the first part of this section we will compare this p-real dimensions with the usual dimensions $dim\, V$ and $dim\, R[V]$.

Let $\wp \in Spec\, R[V]$. We denote by $V_\wp \subset V$ the subvariety defined by $\wp$ and by $I(V_\wp(R)) \subset R[V]$ the ideal of regular functions vanishing on $V_\wp(R)$. We call $\wp$ p-real if there exists $\alpha \in p{-}Sper_R\, R[V]$ such that $\wp_\alpha = \wp$, i.e. $\wp$ lies in the image of the morphism $supp : p{-}Sper_R\, R[V] \longrightarrow Spec\, R[V]$. From [**8**, (2.1) and (2.3)] we get the following result which we will use several times.

PROPOSITION (4.1). *Given $\wp \in Spec\, R[V]$, the following statements are equivalent:*

(1) $\wp$ *is p-real.*
(2) $V_\wp(R)$ *contains a regular point.*
(3) $V_\wp(R)$ *is Zariski-dense in* $V_\wp$.
(4) $I(V_\wp(R)) = \wp$.
(5) $p-Sper_R\, R(V_\wp) \neq \emptyset$.

Hence $dim_p\, V$ is equal to the supremum of all integers d such that there exists a chain $\wp_0 \subset \wp_1 \subset \ldots \subset \wp_d$ of distinct p-real prime ideals. Hence $dim_p\, V \leq dim\, V$. Since $p-Sper_R\, R[V]$ is a spectral space, a subset $X \subset p-Sper_R\, R[V]$ is irreducible and closed if and only if X is the set of specializations of a point. But the specializations of a point $\alpha \in p-Sper_R\, R[V]$ form a chain, by (1.14). Thus $dim_p\, R[V]$ is the supremum of all integers d such that there exists in $p-Sper_R\, R[V]$ a chain

$$\alpha^0 \to \alpha^1 \to \cdots \to \alpha^d.$$

By (1.13) we have $supp(\alpha^i) \subset supp(\alpha^j)$ for $i < j$. Since $supp(\alpha^i)$ is p-real, it follows that $dim_p\, R[V] \leq dim_p\, V$. We will show later on that even equality holds. To this end we need some further notations.

Let $\alpha \in p\text{-}Sper_R\, R[V]$ and $B \subset k(\alpha)$ a real valuation ring with $\pi_\alpha(R[V]) \subset B$. We denote by $k(\alpha(B))$ the residue field of B. Let $\lambda : k(\alpha) \longrightarrow k(\alpha(B)) \cup \infty$ be the canonical place associated with B and let

$$\alpha(B)' := (\lambda(\bar{\alpha}_0 \cap B), \lambda(\bar{\alpha}_1 \cap B), \ldots).$$

Since $R \subset B^*$ it follows from [**7**, Corollary 1.12] that $(k(\alpha(B)), \alpha(B)')$ is p-real closed. We denote by $\alpha(B)$ the order chain corresponding to the p-real point $\lambda \circ \pi_\alpha : R[V] \longrightarrow (k(\alpha(B)), \alpha(B)')$.

Proposition (4.2). *Let* $\alpha \in p-Sper_R\, R[V]$ *and let* $B \subset k(\alpha)$ *be a real valuation ring with* $\pi_\alpha(R[V]) \subset B$. *Then* $\alpha(B) \in p\text{-}Sper_R\, R[V]$ *and* $\alpha \to \alpha(B)$.

Proof. By [**7**, Corollary 1.21], there exists a subfield L of R such that

$$\lambda_{|L} : (L, L \cap \bar{\alpha}) \longrightarrow (k(\alpha(B)), \alpha(B)')$$

is an isomorphism. This implies $\alpha(R) = \alpha(B)' \cap R$ and $J(R) = J(k(\alpha)) \cap R \subset J(L) \cap R = J(k(\alpha(B)) \cap R \subset J(R)$. Hence $\alpha(B) \in p-Sper_R\, R[V]$, by (1.6). That $\alpha(B)$ is a specialization of α has been proved in [**7**, Corollary 4.9]. □

Furthermore we will need the notion of a chain signature introduced by N. Schwartz [**14**]. Let $\hat{Z}_p$ be the additive group of p-adic integers and let $Z^* = \{1, -1\}$. Let K be a field. A homomorphism $f : K^* \longrightarrow Z^* \times \hat{Z}_p$ is called a chain signature if there exists a total order $P \subset K^*$, a valuation v compatible with P and a homomorphism $f_1 : v(K^*) \longrightarrow \hat{Z}_p$ such that $f = sign_p \times (f_1 \circ v)$. Given a chain signature f of K and $n \in N_0$, we set

$$\alpha_n(f) := f^{-1}(1 \times p^n \hat{Z}_p) \cup \{0\}.$$

Then $\alpha(f) := (\alpha(f)_n)_{n\in N_0}$ is an order chain of K and conversely, for every $\alpha \in p-Sper\,K$ there exists a chain signature f of K such that $\alpha = \alpha(f)$ (see [**14**] or [**7**]). In this situation we say that f induces the order chain α. For the remaining part of this paper we fix a homomorphism

$$\bar{g}_R : w(R^*) \longrightarrow \hat{Z}_p$$

such that the chain signature $g_R := sign_Q \times (\bar{g}_R \circ w)$ induces our fixed order chain $\alpha(R)$ of R.

The next two results serve as preliminaries for the proof of the equality $dim_p\,V = dim_p\,R[V]$.

LEMMA (4.3). *Let $R[X] = R[X_1, \ldots, X_n]$ be a polynomial ring over R. Given $z \in A_n(R)$, there exists in $p-Sper_R\,R[X]$ a chain*

$$\alpha^0 \to \alpha^1 \to \cdots \to \alpha^n = z$$

of specializations.

PROOF. Let $P \subset R(X)$ be the total order being determined by

$$R \ll (X_n - z_n)^{-1} \ll (X_{n-1} - z_{n-1})^{-1} \ll \cdots \ll (X_1 - z_1)^{-1}$$

and $Q \subset P$. Let v be the unique extension of the valuation w to $R(X)$ compatible with P. Then the value group $v(R(X)^*)$ is isomorphic to the lexicographically ordered group

$$w(R^*) \oplus Zv(X_n - z_n) \oplus Zv(X_{n-1} - z_{n-1}) \oplus \cdots \oplus Zv(X_1 - z_1),$$

where $v(X_i - z_i) \ll v(X_j - z_j)$ for $i > j$. For $i \in \{1, \ldots, n\}$ let $f_i : Zv(X_i - z_i) \longrightarrow \hat{Z}_p$ be the canonical imbedding. Then

$$f := sign_P \times ((\bar{g}_R \oplus (\oplus f_i)) \circ v)$$

is a chain signature of $R(X)$ extending g_R. Let (R_0, α^0) be a p-real closure of $(R(X), f)$, i.e. $(R_0, \bar{\alpha}^0)$ is p-real closed with $\bar{\alpha}^0 \cap R(X) = \alpha(f)$. Let α^0 be the order chain of $R[X]$ corresponding to the canonical imbedding

$$\pi : R[X] \longrightarrow R(X) \longrightarrow (R_0, \bar{\alpha}^0).$$

Then we have $\alpha^0 \cap R = \alpha(f) \cap R = \alpha(R)$ and $(R, \alpha(R)) \prec (R_0, \bar{\alpha}^0)$ as R is archimedean closed in $(R_0, \bar{\alpha}^0_0)$. Thus $\alpha^0 \in p-Sper_R\,R[V]$. It follows from the choice of the total order $P \subset R(X)$ that for all $i \in \{1, \ldots, n\}$ the homomorphism

$$\varphi_i : R[X_1, \ldots, X_n] \longrightarrow R[X_{i+1}, \ldots, X_n] : X_k \longmapsto z_k, k \leq i$$

extends to a place $\lambda_i : R_0 \longrightarrow R_i \cup \infty$ compatible with the total order $\bar{\alpha}^0_0$. Let $B_i \subset R_0$ be the valuation ring of λ_i and let $\alpha^i := \alpha(B_i)$. By (4.1) we have $\alpha^0 \to \alpha^i$ for all. Since the specializations of α^0 form a chain we get

$$\alpha^0 \to \alpha^1 \to \cdots \to \alpha^n$$

and it follows from the definition of the homomorphism $\varphi_n : R[X] \longrightarrow R$ that $\alpha_n = z$. $\square$

LEMMA (4.4). *Let $S \subset V(R)$ be p-semi-algebraic and suppose that there exists $\alpha \in \tilde{S}$ with $\wp_\alpha = (0)$. Then there exists an open-constructible set $U \subset S$.*

PROOF. By (2.1) and (3.7) it is sufficient to prove the claim for p-semi-algebraic sets of the form $S = S_1 \cap S_2$ with

$$S_1 = \{g = 0, h_1 >_n 0, \dots, h_k >_n 0\} \text{ and } S_2 = \{f_1/g_1 >_\infty 0, \dots, f_s, g_s >_\infty 0\}.$$

Since there exists $\alpha \in \tilde{S} \subset \tilde{S}_1$ with $\wp_\alpha = (0)$, we have $S_1 = \{h_1 >_n 0, \dots, h_k >_n 0\}$. Thus $\tilde{S}_1$ is open-constructible. By (3.6) there exists an open semi-algebraic set $T \subset S_2$. Now set $U := \tilde{S}_1 \cap \tilde{T}$. Then U is open-constructible with $U \subset \tilde{S}$. $\square$

Let us call an affine variety V over R p-real if $V(R)$ is Zariski-dense in V. Thus V is p-real if and only if every minimal prime ideal $\wp \subset R[V]$ is p-real, by (4.1). Moreover, this is equivalent to the property that every irreducible component W of V contains a R-rational regular point. We are now ready to state one of the main results of this section.

THEOREM (4.5). *Let V be an irreducible p-real variety over R. Then dim_p $R[V] = dim\, R[V]$ and $dim_p V = dim\, V$.*

PROOF. As $dim_p R[V] \leq dim_p V \leq dim\, V = dim\, R[V]$, it is sufficient to prove $dim_p R[V] = dim\, R[V]$. Let $d = dim\, R[V]$ and let $Y_1, \dots, Y_d \in R[V]$ be algebraically independent over R such that $R[V]$ is integral over $A = R[Y_1, \dots, Y_d]$. Now let

$$\pi^* : p\!-\!Sper_R\, R[V] \longrightarrow p\!-\!Sper_R\, A$$

be the morphism induced by the canonical imbedding $\pi : A \hookrightarrow R[V]$. Then $Im\, \pi^*$ is p-constructible, by (3.10). Hence there exists a p-semi-algebraic set $S \subset A_d(R)$ with $\tilde{S} = Im\, \pi^*$. Since V is p-real, there exists $\beta \in p\!-\!Sper_R\, R[V]$ with $\wp_\beta = (0)$, by (4.1). Let $\alpha = \pi^*(\beta) \in \tilde{S}$. Then $\wp_\alpha = (0)$. Thus (4.4) implies that there exists an open-constructible set $U \subset \tilde{S}$. Hence we find $z \in U \cap A_d(R)$. By (4.3) there exists in $p\!-\!Sper_R\, A$ a chain

$$\alpha^0 \to \alpha^1 \to \cdots \to \alpha^d = z$$

of specializations. By [**7**, Theorem 4.8], there exists for each $i \in \{1, \dots, d\}$ a valuation ring $B_i \subset k(\alpha^0)$ with $\alpha_i = \alpha(B_i)$. Furthermore we have $\alpha^0 \in U \subset \tilde{S}$, as U is open. Choose $\beta^0 \in p\!-\!Sper_R\, R[V]$ with $\pi^*(\beta^0) = \alpha^0$. Since $R[V]$ is integral over A, we have $k(\beta^0) = k(\alpha^0)$ and $\pi_{\beta^0}(R[V]) \subset B_i$ for all i. Applying (1.14) and (4.2) we see that we get the chain

$$\beta^0 \to \beta^0(B_1) \to \cdots \to \beta^0(B_d)$$

of distinct specializations in $p\!-\!Sper_R\, R[V]$. $\square$

COROLLARY (4.6). *For any affine variety V over R we have $dim_p\, R[V] = dim_p\, V$.*

PROOF. Note that we have

$$dim_p\, V = Max\{dim_p\, V_\wp | \wp \in Spec\, R[V] p\text{-real}\}$$

$$dim_p\, R[V] = Max\{dim_p\, R[V_\wp] | \wp \in Spec\, R[V] p\text{-real}\}.$$

Thus the claim follows from (4.5). □

Next we assume that V is a p-real variety over R. Then every minimal prime ideal of $R[V]$ is p-real. Hence (4.5) implies

COROLLARY (4.7). *For any p-real affine variety V over R we have $dim_p\, V = dim\, V$.*

Given a p-semi-algebraic set $S \subset V(R)$, we denote by $S^Z \subset V$ the Zariski-closure of S in V and we call $dim\, S := dim_p\, S^Z$ the dimension of S. Note that S^Z is a p-real variety. Thus $dim\, S = dim\, S^Z$ as well. In the next step we will show that $dim\, S = dim\, \tilde{S}$, where $dim\, \tilde{S}$ denotes the Krull dimension of $\tilde{S} \subset p\!-\!Sper_R\, R[V]$. To this end we need the following

LEMMA (4.8). *Let V be an irreducible affine variety over R with $d = dim_p\, V$ and let $x \in V(R)$ be regular. Then there exists in $p\!-\!Sper_R\, R[V]$ a chain*

$$\alpha^0 \to \alpha^1 \to \cdots \to \alpha^d = x$$

of distinct specializations.

PROOF. Since $V(R)$ contains a regular point, V is p-real, by (4.1). Thus we have

$$d = dim\, \mathcal{O}_x = dim\, R[V].$$

Let

$$\varphi_x : \mathcal{O}_x \longrightarrow R$$

be the evaluation map corresponding to x. By [**3**, Lemma 1.4], φ_x extends to a place

$$\lambda : R(V) \longrightarrow R \cup \infty$$

such that the valuation ring B of λ is discrete of rank d. Let $B = B_d \subset B_{d-1} \subset \ldots \subset B_1 \subset R(V)$ be the corresponding chain of distinct valuation rings. By [**14**, Lemma 9], there exists a chain signature f of $R(V)$ compatible with λ such that g_R is the pushdown of f with respect to λ. Let $(R_0, \bar{\alpha}^0)$ be a p-real closure of $(R(V), f)$. Note that $(R, \alpha(R)) \prec (R_0, \bar{\alpha}^0)$, by the choice of f. Let α^0 be the order chain given by the p-real point $\pi_0 : R[V] \longrightarrow (R_0, \bar{\alpha}^0)$. For $i \in \{1, \ldots, d\}$ let $C_i \subset R_0$ be the unique extension of B_i to R_0 compatible with $\bar{\alpha}^0_0$ and let $\alpha^i := \alpha(C_i)$. Since g_R is the pushdown of f, we have $\alpha^d = x$. Using (1.14) and (4.2) we see that

$$\alpha^0 \to \alpha^1 \to \cdots \to \alpha^d = x. \quad \square$$

PROPOSITION (4.9). *Suppose $S \subset V(R)$ be p-semi-algebraic. Then $dim\, S = dim\, \tilde{S}$.*

PROOF. Let $d = dim\, S$. Since S^Z is p-real and $\widetilde{S^Z} \cong p\text{–}Sper_R\, R[S^Z]$, we have $dim\, S = dim\, \widetilde{S^Z}$. Hence $dim\, \tilde{S} \leq d$. Let $W \subset S^Z$ be an irreducible component with $dim\, W = d$. We choose a finite representation $S \cap W = \bigcup_{i=1}^k S_i$, where the sets S_i are finite intersections of sets of the following kind:

$$W \cap \{f = 0\}, W \cap \{f >_n 0\}, W \cap \{f/g >_\infty 0\}.$$

Since W is irreducible and $S \cap W$ Zariski-dense in W, there exists $i \in \{1, \ldots, k\}$ such that S_i is Zariski-dense in W. We may assume w.l.o.g. that $i = 1$. Then S_1 contains a regular point x of W and in $p\text{–}Sper_R\, R[W] \subset p\text{–}Sper_R\, R[V]$ there exists a chain

$$\alpha^0 \to \alpha^1 \to \cdots \to \alpha^d = x$$

of specializations, as we have shown above. Since S_1 is Zariski-dense in W, we find a representation

$$S_1 = W \cap \{f_1 >_n 0, \ldots, f_s >_n 0\} \cap \{g_1/h_1 >_\infty 0, \ldots, g_t/h_t >_\infty 0\}.$$

Hence S_1 is an open subset of $W(R)$, by (3.6). In particular, there exists an open semi-algebraic set $U \subset S_1$ with $x \in U$. But then the chain $\alpha^0 \to \alpha^1 \to \cdots \to \alpha^d = x$ lies in $\tilde{U} \subset \tilde{S}$. Thus $dim\, \tilde{S} = dim\, S$. □

Next we will give a further characterization of the dimension of a p-semi-algebraic set. For $\alpha \in p\text{–}Sper_R\, R[V]$ we call $dim\, \alpha := dim\, R[V]/\wp_\alpha$ the dimension of α. Note that we have $dim\, \alpha = dim_p\, V_{\wp_\alpha}$, as $\wp_\alpha$ is p-real.

PROPOSITION (4.10). *Let $S \subset V(R)$ be p-semi-algebraic. Then $dim\, S = Max\{dim\, \alpha | \alpha \in \tilde{S}\}$.*

PROOF. Let $d = dim\, S$ and $d^* = Max\{dim\, \alpha | \alpha \in \tilde{S}\}$. Then (4.9) implies $d \leq d^*$. Next let Φ be a defining formula for S and let $\alpha \in \tilde{S}$. Hence $(\rho(\alpha), \tilde{\alpha}) \models \Phi(x(\alpha))$, by (3.15). Let $I(S) \subset R[V]$ be the ideal of S. For $f \in I(S)$ we have

$$(R, \alpha(R)) \models \forall x(\Phi(x) \Rightarrow f(x) = 0).$$

But then $(\rho(\alpha), \tilde{\alpha}) \models \forall x(\Phi(x) \Rightarrow f(x) = 0)$, as $(R, \alpha(R)) \prec (\rho(\alpha), \tilde{\alpha})$. Hence $\rho(\alpha) \models f(x(\alpha)) = 0$ which implies $f \in \wp_\alpha$. Thus $I(S) \subset \wp_\alpha$. Since S^Z and $V_{\wp_\alpha}$ are p-real varieties we get

$$dim\, \alpha = dim\, V_{\wp_\alpha} \leq dim\, S^Z = dim\, S = d. \qquad \square$$

Now we turn to the investigation of the central points of $V(R)$. Let $Reg_R(V)$ be the set of regular points of $V(R)$. A point $x \in V(R)$ is called central if x lies in the closure of $Reg_R(V)$ with respect to the strong topology. We denote by $Cent_R(V) \subset V(R)$ the set of central points of $V(R)$. The following results are completely analoguous to the well known results about central points of varieties over real closed fields.

PROPOSITION (4.11). *Let V be a irreducible p-real variety of dimension d. Given $z \in V(R)$, then the following statements are equivalent:*

(1) *z is a central point.*

(2) *z is the specialization of some $\alpha \in p{-}Sper_R\, R(V)$.*

(3) *For any open p-semi-algebraic $S \subset V(R)$ with $z \in S$ we have $dim\, S = d$.*

PROOF. (1) $\Rightarrow$ (2): Let $C \subset p{-}Sper_R\, R[V]$ be open-constructible with $z \in C$. Then C contains a regular point x. By (4.8) we find in $p{-}Sper_R\, R[V]$ a chain

$$\alpha^0 \to \alpha^1 \to \cdots \to \alpha^d = x.$$

Since V is p-real we have $dim_p\, R[V] = dim\, R[V]$. Thus $\wp_{\alpha^0} = (0)$, i.e. α^0 lies in $C \cap p{-}Sper_R\, R(V)$. Applying the compactness of $p{-}Sper_R\, R[V]$, we find

$$\alpha \in p{-}Sper_R\, R(V) \cap (\bigcap C),$$

where C ranges over all open-constructible sets containing z. Hence $\alpha \to z$.
(2) $\Rightarrow$ (3): Let $S \subset V(R)$ be an open p-semi-algebraic set containing z. Then there exists an open semi-algebraic neighbourhood $U \subset S$ of z. By assumption there exists $\alpha \in p{-}Sper_R\, R(V) \cap \tilde{U}$. Note that we have $dim\, \alpha = dim\, R[V] = d$. Now (4.10) implies

$$d = dim\, \alpha \leq dim\, S \leq dim\, V = d.$$

(3) $\Rightarrow$ (1) : Let U be an open semi-algebraic neighbourhood of z. Since dim $Sing(V) < d = dim\, U$, we have $U \cap Reg_R(V) \neq \emptyset$. □

COROLLARY (4.12). *$p{-}Sper_R\, R(V)$ is a dense subspace of $\widetilde{Reg_R(V)}$.*

PROOF. Obviously we have $p - Sper_R\, R(V) \subset \widetilde{Reg_R(V)}$. Next let $C \subset \widetilde{Reg_R(V)}$ be open-constructible. Choose any $x \in Reg_R(V) \cap C$. By (4.11), there exists $\alpha \in p{-}Sper_R\, R(V)$ such that $\alpha \to x$. Hence $\alpha \in C$, as C is open. □

We conclude this paper with a few comments. The results we have obtained so far are quite analogous to the basic results in real algebraic geometry. However, there are some essential differences. For example one can show that the space $p - Max(R[V])$ is totally disconnected. Thus a subset $Z \subset p - Sper_R\, R[V]$ is a connected component if and only if Z is the set of generalizations of a closed point. (For the case $p = 2$ see [**6**].) Nevertheless there are applications of the results presented here to real algebraic geometry, as will be shown in a forthcoming paper. For example, the study of varieties over p-real closed fields yields a description (in the language of usual real algebraic geometry) of the sums of $2n$-th powers in function fields over real closed fields.

REFERENCES

1. E. Becker, *Hereditarily pythagorean fields and orderings of higher level*, Lecture Notes, Instituto de Mathematica Pura e Aplicade, Rio de Janeiro, 1978.
2. ———, *Summen n-ter Potenzen in Körpern*, J. reine angew. Math., **307/308** (1979), 8–30.
3. ———, *Valuations and real places in the theory of formally real fields*, Géométrie Algébrique Réelle et Formes Quadratiques, Lecture Notes, Springer **959**, 1–40.
4. ———, *Extended Artin-Schreier theory of formally real fields*, Rocky Mt. J. Math. **14** (1984), 881–897.
5. ———, *On the real spectrum of a ring and its application to semialgebraic gemoetry*, Bull. of the AMS **15** (1986), 1-9-60.
6. R. Berr, *Réelle algebraische Geometrie höherer Stufe*, Technische Berichte der Universität Passau, MIP 1916, 1989.
7. ———, *The p-real spectrum of a commutative ring*, Comm. in Algebra **20** (10) (1982), 3055–3103.
8. ———, *Null- and Positivstellensätze for generalized real closed fields* (submitted).
9. J. Bochnak, M. Coste, M. F. Roy, *Géométrie algébrique réelle*, Ergebnisse der Mathematik und ihrer Grenzgebiete, 3. Folge Band **12**.
10. M. Carral and M. Coste, *Normal spectral spaces and their dimensions*, J. Pure Appl. Alg **30** (1983), 227–235.
11. L.v.d. Dries, *Some applications of a model theoretic fact to (semi-)algebraic geometry*, Indag. Math **44** (1982), 397–401.
12. S. Prieß-Crampe, *Angeordnete Strukturen: Gruppen, Körper, projektive Ebenen*, Ergebnisse der Mathematik und ihrer Grenzgebiete **98**.
13. G. E. Sacks, *Saturated model theory*, W. A. Benjamin, Inc., Reading, MA, 1972.
14. N. Schwartz, *Chain signatures and real closures*, J. reine angew. Math. **347** (1984), 1–20.

FACHBEREICH MATHEMATIK, UNIVERSITÄT DORTMUND, POSTFACH 50 05 00, 4600 DORTMUND, FRG

E-mail address: UMA004@DDOHRZ11.BITNET

Contemporary Mathematics
Volume **155**, 1994

On the Reduction of Semialgebraic Sets by Real Valuations

LUDWIG BRÖCKER

Introduction

Consider the family of ellipses $S_t \subset \mathbb{R}^2$, where $t \in \mathbb{R}^*$ and

$$S_t = \{t^2 X_1^2 + X_2^2 = t^2\} .$$

We make three observations:

1) As t tends to 0, the ellipses S_t do not tend (in any reasonable sense) to $\{X_2^2 = 0\}$, which is obtained by inserting $t = 0$ into the defining equation of S_t.
2) In a reasonable sense, for $t \to 0$, S_t tends to $I = \{X_2 = 0\,,\ X_1^2 \leq 1\}$, which is semialgebraic.
3) For $t \to 0$, S_t tends to I with multiplicity 2.

Of course, which kinds of limits we consider, and also what multiplicity means, have to be defined precisely. In this article we choose a purely algbraic approach. So instead of $\mathbb{R}$, consider a nonarchimedean real-closed field R. Then R admits a canonical real valuation ring V with maximal ideal $\mathfrak{m}$ and real-closed residue field contained in $\mathbb{R}$. For a $\in V$, let $\bar{a}$ be its residue class in $\mathbb{R}$. Now let $t \in \mathfrak{m}$ and consider the ellipse

$$S = \{t^2 X_1^2 + X_2^2 = t^2\} ,$$

which is now defined over R. There is a natural map $S \to \mathbb{R}^2$, which we get by throwing all elements of S componentwise to their residue classes. Let $\bar{S} :=$ image of S. In fact, in our case

$$\bar{S} = I = \{X_2 = 0\,,\ X_2^2 \leq 1\} ,$$

1991 *Mathematics Subject Classification.* Primary:14P10 Secondary:14C17, 13F30, 12D15, 12J10.

This paper is submitted in final form and no version of it will be submitted for publication elsewhere.

which is the limit set we considered before.

The first goal of this paper is the investigation of this kind of reduction map $S \to \bar{S}$ in a more general setting. Then we extend such maps to the real spectrum, which allows us to define multiplicities in a purely algebraic way. In fact, these multiplicities do not live on ordinary points but on orderings.

More important than the reduction of sets is the reduction of semialgebraic chains. This, in turn, is constructed by means of multiplicities. As a result, we get a reduction of homology groups. (For simplicity, we restrict ourselves to coefficients in $\mathbb{Z}/2\mathbb{Z}$.) Finally, we investigate the relation between reduction and intersection, which culminates in a nice formula (Th. 5.3).

Applications to limits in families of semialgebraic sets and a result on non-approximability of properly semialgebraic sets by algebraic sets will be studied in a forthcoming article [**Br**].

Of course, connections between limits and reductions by real valuations have been discovered before: They appear in Robinson's Non-Standard Analysis [**Ro**, 9.4] and also in the works of van den Dries [**V.d.D**] and others.

However, rather than applying more non-standard constructions, here we just open another page of the dictionary between ordinary Real Algebra and Semialgebraic Geometry.

1. The reduction map

From now on we fix the following notation: R is a real-closed field and $V \subset R$ is a non-trivial real valuation ring with residue field $\bar{R}$, value group Γ_v^*, value map $v\colon R \to \Gamma_v = \Gamma_v^* \cup \infty$ and maximal ideal $\mathfrak{m}_v$. Note that V is henselian, Γ_v is divisible and $\bar{R}$ is real-closed. For $a \in V$ we denote by $\bar{a}$ its residue class in $\bar{R}$. More generally, for $x = (x_1, \dots, x_n) \in V^n$ we set $\bar{x} := (\bar{x}_1, \dots, \bar{x}_n)$. The map

$$V^n \to \bar{R}^n ; \quad x \to \bar{x}$$

is called the *reduction map.* Let $S \subset V^n$ be any subset. We write

$$\bar{S} = \left\{ \bar{x} \in \bar{R}^n \middle| \, x \in S \right\}$$

for the image of S under the reduction map.

With respect to the Euclidean topology in R^n and in $\bar{R}^n$, given by the standard scalar product $<,>$, we shall use the following notation: $\mathrm{Int}(A)$ = interior of A, $\mathrm{Cl}(A)$ = closure of A, $\mathrm{Bd}(A)$ = boundary of A. For the Zarisky-topology, we use the subscript Z, so $\mathrm{Cl}_Z(A)$ denotes the Zariski-closure of A in R^n or $\bar{R}^n$, respectively. For any two sets A, B, we write

$$A \triangle B := (A \cup B) \setminus (A \cap B) .$$

For a subset $T \subset R^n$ and $0 < \varepsilon \in R$, we write:

$$U_\varepsilon(T) := \left\{ x \in R^n \mid \exists y \in T : \; < x-y, \, x-y > \, < \varepsilon^2 \right\} ,$$

$$B_\varepsilon(T) := \mathrm{Cl}\big(U_\varepsilon(T)\big) \quad \text{and} \quad S_\varepsilon(T) := \mathrm{Bd}\big(U_\varepsilon(T)\big) .$$

For a semialgebraic set S of dimension p, we denote by S^* the part of pure dimension p in S. So S^* is again semialgebraic and closed in S.

We may wonder when $\bar{S} \subset \bar{R}^n$ is semialgebraic if $S \subset V^n$ is semialgebraic. In fact, this holds, and we are going to consider a more general result. This, in turn, is a consequence of a theorem of Delon [**De**, Cor. 2.23]. For the convenience of the reader we present a direct geometrical proof for our situation. Then we look at properties of S, which are stable under the reduction map. These are not so numerous.

Consider the first-order language of value ordered fields, that is, the language of ordered fields, extended by the atomic formulas

$$v\big(p(X)\big) > v\big(q(X)\big),$$

where $p(X)$ and $q(X)$ are polynomials with constant coefficients in the variables X. The fundamental fact we need is the following:

THEOREM 1.1 (Cherlin, Dickman). *The theory of real-closed fields with real valuation admits elimination of quantifiers in the language of valued ordered fields.*

PROOF. [**C-Di**]. □

For the sequel, we also fix a section

$$s\colon \bar{R} \to R$$

for the reduction map. It is easily seen that such a section always exists. Let

$$R_1 := s(\bar{R}).$$

Correspondingly, for a polynomial $f \in R[X]$, $X = (X_1, \dots, X_n)$, we denote by $s(f) \in R_1[X]$ the polynomial which we get by applying s to the coefficients of f. Thus a quantifier-free formula Φ over $\bar{R}$ naturally yields a formula $s(\Phi)$ over R_1.

Now let $S' \subset \bar{R}^n$ be semialgebraic, say $S' = \big\{x \in \bar{R}^n \,\big|\, \Phi(x)\big\}$. From S' we get $S_1 = \{x \in R_1^n \mid s(\Phi)(x)\}$ and $S = \{x \in R^n \mid s(\Phi)(x)\}$. By model completeness of the theory of real-closed fields, S does not depend on the particular formula Φ which describes S'. With this notation we get

PROPOSITION 1.2. $\overline{S \cap V^n} = \mathrm{Cl}(S')$.

PROOF. It is easily seen that $\overline{S \cap V^n} \subset \mathrm{Cl}(S')$. Conversely, assume $x \in \mathrm{Cl}(S')$. Applying the sections s, we get $x_1 = s(x) \in \mathrm{Cl}(S_1) \subset R_1^n$. This means

$$R_1 \vDash \forall \varepsilon > 0 \; \exists y : \; s(\Phi)(y) \wedge \|x_1 - y\| < \varepsilon .$$

Again, by model completeness of the class of real-closed fields, we get

$$R_1 \vDash \forall \varepsilon > 0 \; \exists y : \; s(\Phi)(y) \wedge \|x_1 - y\| < \varepsilon .$$

In particular, we may choose $\varepsilon \in \mathfrak{m}_v$. Then $y \in S$ and $\bar{y} = \bar{x}_1 = x$. □

NOTATION 1.3.

a) Any two sets $A, B \subset \bar{R}^n$ are called *generically equal*, written $A \underset{g}{=} B$, if $\dim(\mathrm{Cl}_Z(A \triangle B)) < n$.

b) A polynomial $f \in R[X]$, $X = (X_1, \ldots, X_n)$ is called *normalized* if $f \in V[X]$, $f \notin \mathfrak{m}_v[X]$.

c) For $f \in V(X)$ we denote by $\bar{F} \in \bar{R}[X]$ the polynominal which is obtained by reducing the coefficients.

Let $S \subset V^n$ be semialgebraic. We saw already that we do not get $\bar{S}$ just by finding a formula for S in terms of normalized polynomials f_i and using the same formula after passing to $\bar{f}_i$. However, one has

LEMMA 1.4. *Let $S \subset R^n$ be basic, say $S = \{f_1 > 0, \ldots, f_m > 0\}$ where $f_i \in R[X]$ is normalized. Then $\overline{S \cap V^n} \underset{g}{=} \{\bar{f}_1 > 0, \ldots, \bar{f}_m > 0\}$.*

PROOF. Let $x \in V^n$, $i \in \{1, \ldots, m\}$ such that $\bar{f}_i(\bar{x}) \neq 0$. Then $\mathrm{sign}(f_i(x)) = \mathrm{sign}(\bar{f}_i(\bar{x}))$. □

Next, assume that $S \subset R^n$ is definable (in the language of ordered valued fields). By Th. 1.1, apart from order inequalities, we have to consider conditions of the form $v(f) \geq v(g)$ or $v(f) > v(g)$. Again, by scaling we may assume that $f, g \in V[X]$, but at least one of $f, g \notin \mathfrak{m}_v[X]$. In this case we get

LEMMA 1.5. *Let $S = \{x \in V^n, v(f) \ ? \ v(g)\}$, where ? stands for $\geq$ or $>$. Then $\bar{S} = \emptyset$ or $\bar{S} \underset{g}{=} \bar{R}^n$.*

PROOF. This is easily seen by inspection of the 6 cases according to whether f or g or both are not in $\mathfrak{m}_v[X]$ and ? stands for $\geq$ or $>$. □

LEMMA 1.6. *Let $S \subset V^n$ be definable. Then there exists a semialgebraic set $S' \subset \bar{R}^n$ such that $\bar{S} \underset{g}{=} S'$.*

PROOF. In view of Th. 1.4 and the preceding lemmas, it is enough to show the following: If $\bar{S}_1 \underset{g}{=} S'_1$ and $\bar{S}_2 \underset{g}{=} S'_2$, then $\overline{S_1 \cap S_2} \underset{g}{=} S'_1 \cap S'_2$ (the corresponding fact for unions being clear). Since we have obviously the inclusion "$\subset$", it remains to show that for

$$S = \{f_1 > 0, \ldots, f_m > 0, \ v(g_1) \geq v(h_1), \ldots, v(g_r) \geq v(h_r)\},$$

all f_i, g_i, h_i normalized, we get $\overline{S \cap V^n} \underset{g}{=} \{\bar{f}_1 > 0, \ldots, \bar{f}_m > 0\}$ or $\emptyset$, which one sees similarly as in Lemma 1.4 and Lemma 1.5. □

Now we are well prepared to show the main result of this section.

THEOREM 1.7. *Let $S \subset V^n$ be definable (in the language of valued ordered fields). Then $\bar{S} \subset \bar{R}^n$ is semialgebraic.*

PROOF. We proceed by induction on $\dim_Z(\bar{S}) := \dim\ \mathrm{Cl}_Z(\bar{S})$.

1) $\dim_Z(\bar{S}) = 0$. Obvious.

2) Assume that $\dim_Z(\bar{S}) = m > 0$ and that the claim is true for $\dim_Z(\bar{S}) < m$. Let $W' = \mathrm{Cl}_Z(\bar{S}) = \{f = 0\}$, $f \in \bar{R}[X]$ and $W = \{s(f) = 0\} \subset R^n$.

Then, by Prop. 1.2, $\overline{W \cap V^n} = W'$. Now W is defined over R_1. Let $W_1 = W \cap R_1^n$.

We find a finite semialgebraic decomposition: $W_1 = \cup W_{1\alpha}$ and for each $W_{1\alpha}$ a linear subspace $L_{1\alpha}$ such that the orthogonal projection $\pi_\alpha : R_1^n \to L_{1\alpha}$ induces a semialgebraic isomorphism

$$W_{1\alpha} \to \pi_\alpha(W_{1\alpha}) \subset L_{1\alpha}.$$

Moreover,

$$\dim\bigl(\pi_\alpha(W_{1\alpha})\bigr) = \dim\bigl(W_{1\alpha})\bigr) = \dim(L_{1\alpha}).$$

Using the corresponding formulas in $\bar{R}$, we get a decomposition $W' = \cup W'_\alpha$ and linear subspaces $\bar{L}_\alpha \subset \bar{R}^n$ such that the orthogonal projection $\bar{\pi}_\alpha : \bar{R}^n \to \bar{L}_\alpha$ induces a semialgebraic isomorphism

$$W'_\alpha \to \bar{\pi}_\alpha(W'_\alpha) \subset \bar{L}_\alpha.$$

On the other hand, by Tarski's principle, the formulas which define the $W_{1\alpha}$ also define a decomposition: $W = \cup W_\alpha$ such that $W_{1\alpha} = W_\alpha \cap R_1^n$ and we get $\pi_\alpha : W_\alpha \to L_\alpha$, where

$$L_\alpha = L_{1\alpha} \otimes_{R_1} R$$

with the corresponding properties. Now consider

$$\tilde{S} := \{x \in W \mid \exists y \in S : \bar{y} = \bar{x}\}.$$

Then $\tilde{S}$ is a definable subset of W and $\bar{\tilde{S}} = \bar{S}$. So it remains to show that $\overline{\tilde{S} \cap W_\alpha}$ is semialgebraic. For this, consider the commutative diagram

$$\begin{array}{ccc} \tilde{S} \cap W_\alpha & \longrightarrow & \overline{\tilde{S} \cap W_\alpha} \\ \downarrow{\scriptstyle \pi_\alpha} & & \downarrow{\scriptstyle \bar{\pi}_\alpha} \\ L_\alpha & \longrightarrow & \bar{L}_\alpha. \end{array}$$

Now $\pi(\tilde{S} \cap W_\alpha)$ is definable in L_α. Hence, by Lemma 1.6, there is a semialgebraic set $S' \subset \bar{L}_\alpha$ such that

$$\overline{\pi_\alpha(\tilde{S} \cap W_\alpha)} \underset{g}{=} S'.$$

On the other hand, $\overline{\tilde{S} \cap W_\alpha} \subset \bar{W}_\alpha = \mathrm{Cl}(W'_\alpha)$ (see Prop. 1.2). Since

$$\bar{\pi}_\alpha : W'_\alpha \to \bar{\pi}_\alpha(W'_\alpha) \subset \bar{L}_\alpha$$

is a semialgebraic isomorphism, we get

$$\dim\left(\overline{\tilde{S} \cap W_\alpha} \,\triangle\, T'\right) < \dim W'$$

for $T' = \bar{\pi}_\alpha^{-1}(S')$.

We may assume that $T' \subset \overline{\tilde{S} \cap W_\alpha}$. So let

$$\overline{\tilde{S} \cap W_\alpha} \setminus T' \subset Z',$$

where Z' is Zariski closed and $\dim(Z') < \dim W'$. Let $U = \{x \in V^n | \ \bar{x} \in Z'\}$. Then $\tilde{S} \cap W_\alpha \cap U$ is definable. Hence, by induction

$$\overline{\tilde{S} \cap W_\alpha \cap U} := U'$$

is semialgebraic. On the other hand,

$$\overline{\tilde{S} \cap W_\alpha} = U' \cup Z'. \quad \square$$

2. Properties of the reduction map

First of all, the reduction map is continuous. Concerning the functionality one has to be a bit careful: Let $S, T \subset V^n$ be definable and let $f: S \to T$ be a continuously definable map. The latter means that the graph of f is definable in $V^N \times V^n$. Then in general there is no map $\bar{f}$ such that the diagram

$$\begin{array}{ccc} S & \longrightarrow & T \\ \downarrow & & \downarrow \\ \bar{S} & \xrightarrow{\bar{f}} & \bar{T} \end{array}$$

commutes (where the vertical arrows denote the reduction map). However, $\bar{f}$ exists and it is semialgebraic continuous; if f is *moderate*, that means f fulfills the following strong Lipschitz condition: *There exists* $\lambda \in V$, $\lambda > 0$ *such that* $\| f(x) - f(y) \| \leq \lambda \| x - y \|$ *for all* $x, y \in S$.

Obviously, any semialgebraic set $S' \subset \bar{R}^n$ is of the form $S' = \bar{S}$ where $S \subset V^n$ is definable, since the preimage $\{x \in V^n | \bar{x} \in S'\}$ is already definable. On the other hand, we have

PROPOSITION 2.1. *Let* $S \subset R^n$ *be semialgebraic. Then* $\overline{S \cap V^n}$ *is closed.*

PROOF. Let $x' \in \mathrm{Cl}(\overline{S \cap V^n})$ and $x \in V^n$ such that $\bar{x} = x'$. By Tarski's principle there exists $y \in \mathrm{Cl}(S)$ such that $\| x - y \|$ is minimal. Assume $\| x - y \| \notin \mathfrak{m}_v$. Then, for $\rho = \frac{1}{2} \| x - y \|$, $B_\rho(x) \cap S = \emptyset$, where $B_\rho(x)$ denotes the closed ball with midpoint x and radius ρ. It follows that

$$B_{\bar{\rho}}(x') \cap \bar{S} = \emptyset.$$

Contradiction. $\square$

On the other hand, the reduction map shares many properties with ordinary polynomial maps of semialgebraic sets. For instance, if $S \subset V^n$ is semialgebraic, for the Betti numbers b_i one has $b_0(\bar{S}) \leq b_0(S)$, whereas the higher Betti numbers may decrease or increase under the reduction. Here we use standard notations from the algebraic topology of semialgebraic sets over real-closed fields [**D-Kn**], [**Kn**], [**B-C-R**, 11.7]. The example of the flat ellipse $\{t^2 X_1^2 + X_2^2 = t^2\}$, $t \in$

$\mathfrak{m}_v$, shows that in general algebraic sets are no longer algebraic after reduction. However, by Lemma 1.4, basic sets remain at least generically basic. Prop. 1.2 shows that, for any closed semialgebraic set $S' \subset \bar{R}^n$, there is a semialgebraic set $S \subset R^n$ such that $S' = \overline{S \cap V^n}$. If $\dim(S') < n$ one has even

PROPOSITION 2.2. *Let $S' \subset \bar{R}^n$ be semialgebraic and closed such that* $\dim(S') = m < n$. *Then there exists a bounded smooth hypersurface $H \subset R^n$ such that $S' = \overline{H \cap V^n}$.*

We need some preparations. For $\varepsilon \in R$, $\varepsilon > 0$, and $A, B \subset R^n$ we say that A and B are ε-close if $A \subset U_\varepsilon(B)$ and $B \subset U_\varepsilon(A)$. Again, for $\rho \in R$, $\rho > 0$, denote by $B_\rho = B_\rho^n = B_\rho^n(0)$ the ball $\{x \in R^n | < x, x > \leq \rho^2\}$ and correspondingly, $S_\rho = S_\rho^n = \mathrm{Bd}(B_\rho)$.

LEMMA 2.3. *Let $f_0 \in R[X]$ and ε , $\rho \in R$, ε , $\rho > 0$. Then there exists $f \in R[X]$ such that* $\mathrm{grad}(f) \neq 0$ *on $\{f = 0\}$ and $\{f = 0\} \cap B_\rho$ is ε-close to $\{f_0 = 0\} \cap B_\rho$.*

PROOF. We choose disjoint closed subsets $D_1, \ldots, D_m \subset \{f_0 = 0\} \cap B_\rho$ such that each D_i admits an open neighborhood U_i where $U_i \cap \{f_0 = 0\}$ is smooth and $D_1 \cup \cdots \cup D_m$ is $\varepsilon/2$–close to $\{f_0 = 0\} \cap B_\rho$. Then we find semialgebraic tubular maps

$$t_i : D_i \times B^{k(i)} \to B_\rho\,, \quad k(i) = n - \dim(D_i)$$

(see [**B-C-R**, 8.9]). We can also assume that $f_0(t(y,s)) \neq 0$ for $s \neq 0$ and that $\| t_i(y,0) - t_i(y,s) \| < \varepsilon/2$ for $y \in D_i$ and $s \in B_1^{k(i)}$. Now replace f_0 by f_0^2 and let

$$\mu := \min_i \left\{ f_0(t(y,s)) | (y,s) \in D_i \times S_1^{k(i)} \right\}.$$

Then $\mu > 0$. By Sard's theorem [**B-C-R**, 9.5] we find $\delta \in]0, \mu[$ such that $\mathrm{grad}(f) \neq 0$ on $\{f = 0\}$ for $f = f_0 - \delta$. Let $(y,s) \in D_i \times S_1^{k(i)}$. Then $f_0(t(y,0)) = 0$ and $f_0(t(y,s)) > \delta$, so $f(t(y, \lambda s)) = 0$ for $\lambda \in]0,1[$. This shows that $\{f_0 = 0\} \cap B_\rho$ is ε-close to $\{f_0 = 0\} \cap B_\rho$. $\square$

PROOF OF PROP. 2.2. Let $S' = \{x \in \bar{R}^n | \Phi(x)\}$ and $S = \{x \in R^n | s(\Phi)(x)\}$. By the finiteness theorem [**B-C-R**, 2.7.1], Φ can be chosen in such a way that S is a finite union of basic sets, say $B = \{f_0 = 0\ g_1 \geq 0, \ldots, g_s \geq 0\}$. Let $\rho \in R$, $\rho > 0$, $\rho \notin V$. We may assume that B is bounded by ρ, just after adjoining the additional inequality, say $g_s = < x, x > - \rho^2$. Let $\varepsilon \in \mathfrak{m}_v$. According to Lemma 2.3, we make $\{f_0 = 0\}$ smooth such that the new $\{f_0 = 0\}$ is ε-close to the old one inside $B_{2\rho}$. Consider again a semialgebraic tubular map

$$t : (\{f_0 = 0\} \cap B_\rho) \times B_1^1 \to B_{2\rho}$$

such that $\| t(y,0) - t(y,s) \| < \varepsilon$ for $y \in \{f_0 = 0\} \cap B_\rho$ and $s \in B_1^1$. Next, by scaling, we may assume that $f_0^2(x) > g_1(x)$ for all $x \in B_{2\rho}$ such that $x = t(y,s)$ with $y \in \{f_0 = 0\} \cap B_\rho$ and $\frac{1}{2} \leq |s| \leq 1$. Let $f_1 := g_1 - f_0^2$. As in Lemma 2.3, we see that $\{f_1 = 0\} \cap B_\rho$ is 2ε-close to $\{f_0 = 0\ ,\ g_1 \geq 0\} \cap B_\rho$. Now we repeat the same procedure, replacing f_0 by f_1 and g_1 by g_2 until we end up with a function f_s such that $\{f_s = 0\} \cap B_\rho$ is $2s\varepsilon$-close to $B \cap B_\rho$.

Finally, let $S = B_1 \cup \cdots \cup B_t$, B_i be basic closed and let h_i be a polynomial for $i = 1, \ldots, s$ such that B_i , h_i is like B , f_s above. We set $h := h_1 \cdots h_t$. Then we modify it again such that it becomes smooth and the new zero set is ε-close to the old one in $B_{2\rho}$. Now $H := \{h = 0\}$ does the job. □

The preceding proofs also yield the following

PROPOSITION 2.4. *Let $S \subset R^n$ be semialgebraic,* dim $S < n$. *Then there exists $d \in \mathbb{N}$, only depending on S, such that for all $\epsilon > 0$, $\rho > 0$ there exists a polynomial $f \in R(x)$ with* $\deg(f) \le d$ *such that* $\operatorname{grad}(f) \neq 0$ *on* $\{f = 0\}$ *and* $\{f = 0\} \cap B_\rho(0)$ *is ε-close to $S \cap B_\rho(0)$.*

Consider the small circle $S = \{X_1^2 + X_2^2 = t^2\} \subset R^2$, $t \in \mathfrak{m}_v$. Then $\bar{S}$ is just a single point. In particular, $\dim(\bar{S}) < \dim(S)$. Since it might be not completely obvious that the dimension cannot increase, I shall show this below.

PROPOSITION 2.5. *Let $S \subset R^n$ be semialgebraic. Then* $\dim(\overline{S \cap V^n}) \le \dim(s)$.

PROOF. Let $Q = Q^n \subset V^n$ be a cube, defined over R_1, such that $\dim(\bar{Q}) = n$. First, consider the following special case.

CLAIM. If $S \subset Q$ and $\bar{S} \supset \operatorname{Int}(\bar{Q})$, then $\operatorname{Int}(S) \neq \emptyset$. This will be shown by induction on n.

1) $n = 1$: Obvious.
2) $n > 1$: We write $Q = Q^n = Q^{n-1} \times I$.

Consider the segments $s_a = \{(a,t) | t \in I\}$ for $a \in Q^{n-1}$. There exists k such that $b_0(s_a \cap S) < k$ for all $a \in Q^{n-1}$. Now we subdivide I and thus Q into $2k$ equal closed parts Q_i, $i = 1, \ldots, 2k$, such that $Q_i \cap Q_{i+1} \simeq Q^{n-1}$.

By construction, for each $a \in Q^{n-1}$, there exists at least one $i \in \{1, \ldots, 2k\}$ such that

$$s_a \cap S \cap Q_i = s_a \cap Q_i \text{ or } \emptyset .$$

So let

$$P_i = \{a \in Q^{n-1} | s_a \cap S \cap Q_i = s_a \cap Q_i\}$$

and

$$R_i = \{a \in Q^{n-1} | s_a \cap S \cap Q_i = \emptyset\} .$$

By elimination of quantifiers, P_i and R_i are semialgebraic and $Q^{n-1} = P_1 \cup \cdots \cup P_{2k} \cup R_1 \cup \cdots \cup R_{2k}$. So $\bar{Q}^{n-1} = \bar{P}_1 \cup \cdots \cup \bar{P}_{2k} \cup \bar{R}_1 \cup \cdots \cup \bar{R}_{2k}$.

If one of the $\bar{R}_i$ admits an interior point, then by induction there is a cube $Q_0^{n-1} \subset Q^{n-1}$, Q_0^{n-1} defined over R_1, such that $S \cap (Q_0^{n-1} \times I) \cap Q_i = \emptyset$ contradicting the assumption that $\bar{S} \supset \operatorname{Int}(\bar{Q})$. Hence at least one $\bar{P}_i$ admits an interior point and so does P_i by induction. But then $(P_i \times I) \cap Q_i \subset S$, which proves the claim.

Now for the general case assume $\dim(\bar{S}) = 1$. Let L' be a linear subspace of $\bar{R}^n$, $\dim(L') = 1$, and π the orthogonal projection: $\bar{R}^n \to L'$. We can achieve that $\dim\big(\bar{\pi}(\bar{S})\big) = 1$ too. As before, we take the corresponding subspace L in R^n, which yields a commutative diagram:

$$\begin{array}{ccc} S \cap V^n & \xrightarrow{\pi} & L \\ \downarrow & & \downarrow \\ \bar{S} & \xrightarrow{\bar{\pi}} & \bar{L}\,. \end{array}$$

By the claim, $\dim\big(\pi(S \cap V^n)\big) \geq 1$. Hence $\dim(S \cap V^n) \geq 1$ too. $\square$

3. The reduction map on the real spectrum

We consider the same situation as before, but now we are going to extend the reduction map to the real spectrum. First, recall the tilde-correspondence between semialgebraic sets and constructible sets in the real spectrum [**B-C-R**, 7.2], [**Kn-Sch**, III, §5]. So let $\mathrm{Sper}(R[X])$ be the real spectrum of $R[X]$, $X = (X_1, \ldots, X_n)$. Then for a semialgebraic set $S \subset R^n$, defined by a quantifier-free formula Φ, the same formula Φ defines the constructible set $\tilde{S} \subset \mathrm{Sper}(R[X])$. By the model completeness of the theory of real-closed fields, $\tilde{S}$ does not depend on the particular formula Φ.

NOTATION 3.1. For a definable set $S \subset V^n$, we denote by $\tilde{S}$ the set

$$\bigcap_{S \subset T\,,\,T \text{ semialgebraic}} \tilde{T}\,.$$

As for semialgebraic sets, we have a canonical embedding $S \to \tilde{S}$. Now let $\alpha \in \tilde{V}^n$. We define $\bar{\alpha} \subset \mathrm{Sper}(\bar{R}(X))$ as follows: Let φ be the ultrafilter associated to α [**B-C-R**, 7.2], [**Kn-Sch**, III, §5].

PROPOSITION AND DEFINITION 3.2. *The family* $\bar{\varphi}_0 = \{\overline{S \cap V^n} | S \in \varphi\}$ *generates a unique ultrafilter* $\bar{\varphi}$ *of semialgebraic sets in* $\bar{R}^n$. *The corresponding element* $\bar{\alpha} \in \mathrm{Sper}\big(\bar{R}[X]\big)$ *is defined to be the image of* α *under the reduction map.*

PROOF. Obviously $\bar{\varphi}_0$ is a filter basis. Let $\mathrm{Cl}_Z(\overline{S \cap V^n}) = W'$ for all sufficiently small $S \in \varphi$, and assume that $\bar{\varphi}_1 \neq \bar{\varphi}_2 \supset \bar{\varphi}_0$ for ultrafilters $\bar{\varphi}_1$, $\bar{\varphi}_2$. Let $\bar{\alpha}_1$ and $\bar{\alpha}_2$ be the corresponding elements in $\mathrm{Sper}(W')$. We may assume that $\dim(\alpha_1) = \dim(\alpha_2) = \dim(W')$. Therefore neither $\bar{\alpha}_1 \to \bar{\alpha}_2$ nor $\bar{\alpha}_2 \to \bar{\alpha}_1$ where the arrow means specialization. So we find

$$f \subset \bar{R}[X] \text{ with } f\big(\bar{\alpha}_1\big) > 0 \text{ and } f\big(\bar{\alpha}_2\big) < 0\,.$$

But $s(f)(\alpha) \neq 0$. If $s(f)(\alpha) > 0$, then

$$\{f \geq 0\} \in \bar{\varphi}_0 \quad \text{(see Prop. 1.2), thus} \quad \{f \geq 0\} \in \bar{\varphi}_1$$

and $\{f < 0\} \in \bar{\varphi}_1$. Contradiction. Correspondingly, $s(f)(\alpha) < 0$ leads to a contradiction. $\square$

From the construction it follows that $\dim(\bar{\alpha}) \leq \dim(\alpha)$ (see Prop. 2.5). In case that $\dim(\bar{\alpha}) = \dim(\alpha)$, we shall give a more algebraic description of the reduction.

For an arbitrary set $A \subset \tilde{V}^n$, again we set $\bar{A} := \{\bar{\alpha} | \alpha \in A\}$. Then the identity below is immediately clear.

PROPOSITION 3.3. *Let $S \subset V^n$ be definable. Then*

$$\bar{\tilde{S}} = \tilde{\bar{S}}\,.$$

Now let us look at the fibers under the reduction map of the real spectrum. We want to show that these are finite under the extracondition that the dimensions are preserved. For this, let us consider the following construction with fixed notations for the rest of this section.

Let $W \subset R^n$ be algebraic, $\dim(W) = k$, and let $\bar{W}_1, \ldots, \bar{W}_r$ be the irreducible components of $\mathrm{Cl}_Z(\overline{W \cap V^n})$ for which $\dim(\bar{W}_i) = k$. Let L' be a k-dimensional linear subspace of $\bar{R}^n$, say $L' = \{l_1 = 0, \ldots, l_{n-k} = 0\}$ where the l_i are linear forms, and let $L = \{s(l_1) = 0, \ldots, s(l_{n-k}) = 0\} \subset R^n$. We can choose L' in such a way that the orthogonal projections $\pi\colon W \to L$ and $\pi\colon \bar{W}_i \to L' = \overline{L \cap V^n}$ are finite (in fact, we need only generically finite). In the first step we shall define for $i = 1, \ldots, r$ a set of valuations v_{ij} of $R(W)$ such that v_{ij} extends v and $\{v_{ij}\}$ depends only on v, the embedding $W \subset R^N$ and i. For this let

$$V[W] := \{f + I(W) \in R[S] \| f \in V[X]\}$$

and

$$\mathfrak{p}_i := \left\{f + I(W) \in R[W] \| f \in V[X] \text{ and } \bar{f}(x) = 0 \quad \forall x \in \bar{W}_i\right\}.$$

Then $\mathfrak{p}_i$ is a prime ideal in $V[W]$ and $V[W]/\mathfrak{p}_i \cong \bar{R}[\bar{W}_i]$. Set $A_i = V[W]_{\mathfrak{p}_i}$ and let $\tilde{A}_i$ be the integral closure of A_i in $R(W)$. In general, $A_i \neq \tilde{A}_i$ and also $\tilde{A}_i$ is not a valuation ring in $R(W)$. However, one has

PROPOSITION 3.4. *Under the above notations, $\tilde{A}_i$ is semilocal with maximal ideal $\mathfrak{m}_{ij}$, $j = 1, \ldots, s(i)$, where $\mathfrak{m}_{ij} \cap V[W] = \mathfrak{p}_i$. The localization $(\tilde{A}_i)_{\mathfrak{m}_{ij}}$ is the valuation ring of a valuation v_{ij} of $R(W)$ extending v. The residue field of v_{ij} is a finite extension of $\bar{R}(\bar{W}_i)$ and the value group of v_{ij} is divisible for $j = 1, \ldots, s_i$.*

PROOF. We may assume that $l_i = X_{n+1-i}$, that is $R[L] = R[X']$ with $X' = (X_1, \ldots, X_k)$. Let $B = V[X']_{\mathfrak{m}_v[X']}$, that is, the valuation ring of the canonical extension b of v to $R(X') = R(L)$. The residue field of b is $\bar{R}(X')$ and the value group of b is the same for v, so it is divisible. It is well known that the integral closure $\tilde{B}$ in $R(W)$ of B is semilocal and that the localizations $\tilde{B}_{\mathfrak{m}_\lambda}$ at the maximal ideals $\mathfrak{m}_\lambda$ of $\tilde{B}$ are just the valuationrings of the extensions of b to $R(W)$ [**Bour**, Chap. 6, §8.3 Remarque]. Now A_i dominates B, hence $\tilde{B} \subset \tilde{A}_i$. Consider any valuation of $W(X)$ which dominates A_i. Then it dominates B too. So it is one of the finitely many extensions of b to $W(X)$. This yields the assertion (see [**Bour**, Chap. 6, §7.1]). $\square$

We keep the above notations and consider an ordering $\alpha \in \text{Sper}(R(W))$, which is compatible with one of the valuations v_{ij}. Then α induces an ordering α'_{ij} on the residue field $\tilde{A}_i/\mathfrak{m}_{ij}$, which restricts to an ordering α' on $A_i/p_iA_i \cong V[W]/\mathfrak{p}_i \cong \bar{R}[\bar{W}_i]$. With these notations, we get

PROPOSITION 3.5. *The ordering $\alpha \in Sper(R(W))$ is compatible with one of the valuations v_{ij} if and only if $\alpha \in \tilde{V}^n$ and* $\dim(\alpha) = \dim(\bar{\alpha}) = k$. *In that case,* $\bar{\alpha} = \alpha' \in \text{Sper}(\bar{R}(\bar{W}_i))$.

PROOF. Assume that $\alpha \in \text{Sper}(R(W))$ is compatible with one of the valuations v_{ij}. Then in particular $1 + p >_\alpha 0$ for all $p \in \mathfrak{p}_iA_i$. First, we have to show that $\alpha \in \tilde{V}^n$. Let φ be the ultrafilter corresponding to α and let $T \in \varphi$ be closed. If $\alpha \notin \tilde{V}^n$, we may choose T such that $T \cap V^n = \emptyset$. Let $g(X) := < X, X >$ (with respect to the euclidean standard metric). Then g takes its minimum value τ on T where $\tau > 0$, $\tau \notin V$. Choose $r \in R$ with $0 < r < \tau$ and $r \notin V$. Then $p := -r^{-1}g \in \mathfrak{p}_i \subset \mathfrak{p}_iA_i$, but $1 + \mathfrak{p}$ is negative on T. Contradiction. Next assume that $\dim(\bar{\alpha}) < k = \dim(\alpha)$ or $\bar{\alpha} \notin \text{Sper}(\bar{R}[\bar{W}_i])$. Again, we want to lead both cases to a contradiction. For this consider a polynomial $f \in V[X]$, $f \notin \mathfrak{p}_i$, such that $\bar{f}$ vanishes on $\text{supp}(\bar{\alpha})$. This exists by assumption. Consider also a closed-bounded semialgebraic set $T \in \varphi$, where φ is the ultrafilter corresponding to α. Then $f(x) \in \mathfrak{m}_v$ for all $x \in T \cap V^n$. We want that $f(x) < \rho \in \mathfrak{m}_v$ for all $x \in T \cap V^n$ and some bound $\rho > 0$. This is clear, if $T \subset V^n$. Otherwise, $T \subset \lambda V^n$ for some $\lambda \in R$. Then consider the dilatation $\delta: x \mapsto \lambda^{-1}(x)$ and choose $g \in V[X]$ such that $\bar{g}$ vanishes on $\overline{\delta(T)}$, but not on $\overline{\delta(W)}$. Now let $f(x) := g(\lambda^{-1}x)$ for $x \in R^n$. Then $f \in V[X]$ and $f(x) \in \mathfrak{m}_v$ for all $x \in T$. Now $p := -\rho f^{-2} \in \mathfrak{p}_iA_i$, but $1 + p < 0$ on T. Contradiction.

For the converse, consider the commutative diagram

$$\begin{array}{ccc} W & \longrightarrow & \overline{W \cap V^n} \\ \pi \downarrow & & \bar{\pi} \downarrow \\ L \cap V^n & \longrightarrow & \bar{L}, \end{array}$$

where the horizontal arrows are reductions and the vertical arrows are corresponding orthogonal projections. By assumption and construction, α, $\bar{\alpha}$, $\pi(\alpha) =: \beta$ and $\bar{\beta} = \bar{\pi}(\bar{\alpha})$ are all of maximal dimension k. First, we show that β is compatible with the canonical valuation b of $R(L)$. In fact, if $g \in \mathfrak{m}_b$, then $\bar{g}$ is well-defined and vanishing outside a set of dimension $< k$ in $\overline{L \cap V^n}$. So clearly $1 + g >_\beta 0$. Now let C be the convex hull of B in $R(W)$ with respect to α. Thus C defines a valuation c which extends b and which is compatible with α. We have to show that c coincides with one of the v_{ij}. But $\bar{\alpha} \in \text{Sper}(\bar{R}(\bar{W}_i))$ for some $i \in \{1, \dots, r\}$ and for this i turns out that $C \supset A_i$, which proves the assertion. It remains to show that in case $\dim(\alpha) = \dim(\bar{\alpha}) = k$ one has $\bar{\alpha} = \alpha'$. For this, let $\alpha \in \text{Sper}(\bar{R}(\bar{W}_i))$ and $f \in V[X]$, $f \notin \mathfrak{p}_i$. By what we have seen in the second part of the proof, $f >_\alpha 0$ if and only if $\bar{f} >_{\alpha'} 0$. On the other hand, since $\overline{\{f>0\}} \underset{g}{=} \{\bar{f}>0\}$ from $f >_\alpha 0$ we get $\bar{f} >_{\bar{\alpha}} 0$. This settles the proof. □

The preceding results enable us to define multiplicities, as considered in the introduction, in a purely algebraic way.

DEFINITION 3.6. Under the above notations, let $\gamma \in \operatorname{Sper}(\bar{R}(\bar{W}))$, $\dim(\gamma) = k$. Then the number

$$\mu_W(\gamma) = \mu(\gamma) := \#\{\alpha \in \operatorname{Sper}(\bar{R}(\bar{W})) | \bar{\alpha} = \gamma\}$$

is called the multiplicity of γ.

PROPOSITION 3.7. *Under the above notations, for each $i \in \{1, \dots, r\}$ there exists a nondegenerate quadratic form φ_i over $\bar{R}(\bar{W}_i)$ such that*

$$\mu(\gamma) = \operatorname{sign}_\gamma(\varphi_i)$$

for all $\varphi \in \operatorname{Sper}(\bar{R}(\bar{W}_i))$. Here the right-hand side is the signature of φ_i with respect to γ.

PROOF. Since all valuations v_{ij} have divisible groups, the preimages α with $\bar{\alpha} = \beta$ are uniquely defined by the orderings they induce on the residue fields of the v_{ij}. Therefore, φ_i is just given by the trace form of the $\bar{R}(\bar{W}_i)$-algebra $\tilde{A}_i/\mathfrak{p}_i\tilde{A}_i$, which is finite dimensional and separable of dimension, say m_i. □

COROLLARY 3.8. *For any two $\gamma_1, \gamma_2 \in \operatorname{Sper}(\bar{R}(\bar{W}_i))$, one has $\mu(\gamma_1) \equiv \mu(\gamma_2)$* mod 2.

This element in $\mathbb{Z}/2\mathbb{Z}$ will be denoted by $\mu_W(W_i)$.

PROOF. In fact, $\mu(\gamma_1) \equiv \mu(\gamma_2) \equiv m_i \mod 2$. □

REMARKS AND EXAMPLE 3.9.

a) By Prop. 3.7, multiplicities are always finite.
b) Consider the flat ellipse $W = \{t^2X_1^2 + (X_2 - 2t)^2 - t^2\} \subset V^2$ for $t \in \mathfrak{m}_v$. Then $\bar{W} = \{X_1^2 \leq 1\} \subset \bar{R}^1$ and $\bar{W}_1 = \operatorname{Cl}_Z(\bar{W}) = \bar{R}^1$. For $\alpha \in \operatorname{Sper}(\bar{R}^1(X_1))$, we have

$$\mu(\alpha) = 2 \qquad \text{if } \alpha \in \tilde{\bar{W}},$$

$$\mu(\alpha) = 0 \qquad \text{if } \alpha \notin \tilde{\bar{W}}.$$

This coincides with the visual meaning of multiplicities.
c) Note that for points $\beta \in \tilde{\bar{W}}$ with $\dim(\beta) < \dim(W)$, multiplicities are not defined. In general, these have infinite fibers under the reduction map. However, by Prop. 3.7 we see that there is a partition of $\operatorname{Cl}_Z(\bar{W})$ into semialgebraic subsets:

$$\operatorname{Cl}_Z W = S_1 \cup \dots \cup S_m,$$

unique up to sets of dimension less than $\dim(W)$, such that the multiplicity is constant on S_i, that is $\mu(\gamma_1) = \mu(\gamma_2)$ for $\gamma_1, \gamma_2 \in S_i$ with

$$\dim(\gamma_1) = \dim(\gamma_2) = \dim W .$$

d) The latter can be extended to a more general situation. Let $S \subset R$ be semialgebraic, $\dim S = k$. Let $W = \mathrm{Cl}_(S)$. We may assume that W is irreducible. Consider the first case where S is basic, say $S = \{f_1 \geq 0, \dots, f_m \geq 0\}$. Now let $\gamma \in \overline{\widetilde{S \cap V^n}}$, $\dim(\gamma) = k$, say $\gamma \in \tilde{\bar{W}}_i$ (using the former notations). Then

$$\mu_S(\gamma) = \mu(\gamma) := \{\alpha \in (S \cap V^n)^{\sim} \mid \bar{\alpha} = \gamma\}$$

again is given by the sign of a quadratic form, namely

$$2^m \mu(\gamma) = \mathrm{sign}_\gamma \mathrm{Tr}_*(\varphi) ,$$

where the right-hand side means the following: first take the Pfister form $\varphi = < 1, f_1 > \otimes \cdots \otimes < 1, f_m >$ over $R(W)$. For each valuation v_{ij}, let φ_{ij} be the unique residue form of φ. Next, let $\mathrm{Tr}_{ij}(\varphi_{ij})$ be the transfer with respect to the trace of the finite extension $\tilde{A}_i/m_{ij}$ of $A_i/\mathfrak{p}_i$ [**Schar**, Ch. 2, §5], then

$$\mathrm{Tr}_*(\varphi) := \sum_j \mathrm{Tr}_{ij}(\varphi_{ij}) .$$

(For the use of trace forms including same historical background, see also [**Be-W**].) Of course, from this we get a partition as in c) for the multiplicities with respect to reduction maps $S \cap V^n \to \bar{S}$, where $S \subset R^n$ is an arbitrary semialgebraic set.

e) Let $W = L$ be a linear subspace of $\mathbb{A}^n$ and let $\alpha \in \mathrm{Sper}(\bar{R}(\bar{L}))$ with $\dim(\alpha) = \dim \overline{L \cap V^n} = \dim L$. Then $\mu_L(\alpha) = 1$.

PROPOSITION 3.10. *Let $S \subset R^n$ be semialgebraic of dimension k and let $L = L^k$ a k-dimensional linear subspace of R^n such that*

i) $\mathrm{Cl}_Z(S)$ *is finite over L.*
ii) $\mathrm{Cl}_Z(\overline{S \cap V^n})$ *is finite over $\overline{L \cap V^n}$.*
iii) $\dim(\overline{S \cap V^n}) = k$.

Then there are semialgebraic sets $T_1, \dots, T_m \subset \bar{L}$ and a definable set $D \subset (L \cap V^n)$ such that the following holds.

a) $\bar{\pi}(\overline{S \cap V^n}) = T_1 \cup \cdots \cup T_m$.
b) $\dim\, T_i = k$ *for $i = 1, \dots, m$ and* $\dim\, (T_i \cap T_j) < k$ *for $i \neq j$.*
c) *For $\alpha \in \tilde{T}_i$, $\dim\, (\alpha) = k$, the number*

$$t_i := \sum\nolimits_{\beta \in \widetilde{\overline{S \cap V^n}},\, \bar{\pi}(\beta) = \alpha} \mu_S(\beta)$$

depends only on i.

d) $\dim(\overline{D \cap V^n}) < k$.
e) $\#\{\pi^{-1}(x) \cap S \cap V^n\} = t_i$ *for all $x \in (L \cap V^n) \setminus D$ with $\bar{x} \in \mathrm{Int}(T_i^*)$.*

(As before, π and $\bar{\pi}$ denote the orthogonal projections onto L and $\bar{L}$. Recall that T_i^* denotes the p-dimensional part of T_i.)

PROOF. By part d) of the proceeding remark, we find $T_1, \ldots, T_m \subset \bar{R}^n$ for which a), b) and c) hold. Now let D be the set of $x \in L \cap V^n$ for which $\bar{x} \in \mathrm{Int}(T_i)$ for some i, but $\#\left\{\pi^{-1}(x) \cap S \cap V^n\right\} \neq t_i$. We have to show that $\dim(\bar{D}) < k$. Assume that not. Then for some $i \in \{1, \ldots, m\}$ there exists $\alpha' \in \left(\bar{D} \cap T_i\right)^{\sim}$ with $\dim(\alpha') = k$.

Consider the commutative diagram

$$\begin{array}{ccc} S \cap V^n & \longrightarrow & \overline{S \cap V^n} \\ \pi \big\downarrow & & \bar{\pi} \big\downarrow \\ L \cap V^n & \longrightarrow & \overline{L \cap V^n}\,. \end{array}$$

Let $\beta_1, \ldots, \beta_r$ be the elements in $\widetilde{\overline{S \cap V^n}}$ for which $\tilde{\bar{\pi}}(\beta_i) = \alpha'$. By part e) of the preceding remark, there is a unique $\alpha \in (L \cap V^n)^{\sim}$ with $\bar{\alpha} = \alpha'$. Actually, $\alpha \in \tilde{D}$. Now let $\{\gamma_1, \ldots, \gamma_s\}$ be the set of all orderings in $(S \cap V^n)^{\sim}$ for which $\bar{\gamma}_i \in \{\beta_1, \ldots, \beta_r\}$. By construction, $s = t_i$. On the other hand, $\{\gamma_1, \ldots, \gamma_s\}$ is the set of all extensions of α in $(S \cap V^n)^{\sim}$. Hence there are points $x \in D$ with $\#(\pi^{-1}(x)) = t_i$. Contradiction. $\square$

4. Reduction and homology

Let R be a real-closed field (for a moment we do not consider the valuation v of R). We fix also $\mathbb{Z}/2\mathbb{Z}$ as a ring of coefficients. This will be enough for our purposes and it simplifies the notations considerably.

DEFINITION 4.1. A semialgebraic set $S \subset R^n$ is called a p-chain, if S is bounded and closed in R^n and of pure dimension p. For a subset $X \subset R^n$, we denote by $C_r(X)$ the set of all p-chains S with $S \subset X$.

4.2 Homology. $C_p(X)$ forms a $\mathbb{Z}/2\mathbb{Z}$–module

$$S_1 + S_2 := 0 \text{ if } S_1 = S_2$$
$$S_1 + S_2 := \mathrm{Cl}(S_1 \triangle S_2) \text{ otherwise.}$$

We get a boundary operator ∂ as follows: Take any semialgebraic triangulation of the p-cycle S. This defines a singular p-chain σ with $|\sigma| = S$, where $|\sigma|$ denotes the carrier of σ. Now we set $\partial S := |\partial\sigma|$. This is independent of the triangulation. Then clearly $(C(X), \partial)$ is a complex. Now let $A \subset X \subset R^n$ be semialgebraic. As usual we set

$$C_q(X/A) := \{S \in C_q(X) | \partial S \subset A\}$$

and

$$B_q(X/A) := \{S \in C_q(X) | \exists T \in C_{q+1}(X) : \partial T + S \in C_q(A)\}\,.$$

These are $\mathbb{Z}/2\mathbb{Z}$–modules. So we get

$$H_q(X/A) = C_q(X/A)/B_q(X/A).$$

This group coincides (for coefficients in $\mathbb{Z}/2\mathbb{Z}$) with that defined in [**B-C-R**, 11.7] and for $R = \mathbb{R}$ with the usual relative homology group (see also [**D**]). It shares with the latter its essential properties as long as one stays in the semialgebraic category, but serious effort has to be spent to show this [**D**]. However, let us indicate how one gets the functionality directly. So let $X, Y \subset \mathbb{R}^n$ be semialgebraic, $f: X \to Y$ semialgebraic continuous, and let $S \subset C_q(X)$. If $\dim(f(S)) < q$ we set $f_*(S) = 0$. Otherwise there are q-chains $T_1, \ldots, T_k \subset Y$ with the following properties:

a) $f(S)^* = T_1 + \cdots + T_k$ (remember that $f(S)^*$ denotes the p-dimensional part of $f(S)$).
b) For all $\alpha \in \tilde{T}_i$ with $\dim(\alpha) = q$, there are exactly r_i orderings $\beta \in \tilde{S}$ which $\tilde{f}(\beta) = \alpha$.

Then we set

$$f_*(S) := \sum_{i=1}^{k} r_i T_i,$$

for which one gets the fundamental property

$$f_* \circ \partial = \partial \circ f_*.$$

We omit the proof here, since it is simiilar to the one we give for the reduction below (though the present case would be easier).

Now we come back to the situatiion where R admits a real valuation with residue field $\bar{R}$.

4.3 Reduction of chains. Let $S \subset V^n$ be a q-chain. We define a q-chain $r(S) \subset \bar{R}^n$ as follows. If $\dim(\bar{S}) < q$ we set $r(S) = 0$. Otherwise there are q-chains $T_1, \ldots, T_k \subset \bar{R}^n$ with the following properties:

a) $\bar{S}^* = T_1 + \cdots + T_k$
b) For all $\alpha \in \tilde{T}_i$ with $\dim(\alpha) = q$, there are exactly r_i orderings $\beta_i \in \tilde{S}$ with $\bar{\beta}_i = \alpha$.

Then we set

$$r(S) := \sum_{i=1}^{k} r_i T_i.$$

So the definition of $r(S)$ is quite analogous to that of $\tilde{f}(S)$. Clearly $r(S+T) = r(S) + r(T)$ for any two q-chains $S, T \subset V^n$.

THEOREM 4.4. *Let $S \subset V^n$ be a q-chain and let $f: S \to V^n$ be semialgebraic moderate* (see §2). *Then*

a) $\partial \circ r(S) = r \circ \partial(S)$.
b) $\bar{f}_* \circ r(S) = r \circ f_*(S)$.

Consequently, for semialgebraic pairs $A \subset X$ and $B \subset Y$ in V^n and a moderate semialgebraic map $f\colon (X, A) \to (Y, B)$, the diagram

$$\begin{array}{ccc} H_q(X,A) & \xrightarrow{f_*} & H_q(Y,B) \\ {\scriptstyle r}\downarrow & & {\scriptstyle r}\downarrow \\ H_q\left(\bar{X},\bar{A}\right) & \xrightarrow{\bar{f}_*} & H_q\left(\bar{Y},\bar{B}\right) \end{array}$$

commutes.

PROOF. b) Let $\alpha \in \left(\overline{f(S)^*}\right)^\sim$, $\dim(\alpha) = q$, and let $\alpha_1, \dots, \alpha_k$ be the orderings in $\tilde{S}$ for which $\overline{\tilde{f}(\alpha_i)} = \alpha$ or equivalently $\tilde{\bar{f}}(\bar{\alpha}_i) = \alpha$. Then $\alpha \in (r \circ f_*(S))^\sim$ if and only if k is odd if and only if $\alpha \in \bar{f}_* \circ r(S)$.

a) We need the following

CLAIM. There are q-chains $S_1, \dots, S_k \subset V^n$ such that $S = S_1 + \dots + S_k$ and for each S_i either

i) $\dim(\bar{S}_i) < q$, or

ii)

(1) $\dim(\bar{S}_i) = q$

(2) $\bar{S}_i = \bar{S}_i^*$

(3) $\mu_{S_i}(\alpha) = 1$ for all $\alpha \in \tilde{\bar{S}}_i$

(4) $\overline{\partial S_i} = \overline{\partial S_i^*} = \partial \bar{S}_i$

(5) $\mu_{\partial S_i}(\beta) = 1$ for all $\beta \in \left(\partial \bar{S}_i\right)^\sim$.

First we show that b) follows from the claim. So we may assume that S is of type i) or ii). But in the latter case there is nothing to prove. So let $\dim\left(\bar{S}\right) < q$. Again, if $\dim\left(\bar{S}\right) < q - 1$, we are done.

For the case $\dim\left(\bar{S}\right) = q - 1$, assume that there exists $\alpha \in \widetilde{\overline{\partial S}}$ with $\dim(\alpha) = q - 1$ and $\mu_{\partial S}(\alpha) = k \equiv 1 \mod 2$. We find $x_0 \in V^n$, $\rho \in V$, $\bar{\rho} > 0$ and a linear subspace $L \subset R^n$, $\dim L = q - 1$ such that $\mu_{\partial S}(\beta) = k$ for all $\beta \in \overline{U_\rho(x_0)} \cap \overline{\partial S}$ with $\dim(\beta) = q - 1$. Moreover, assume that L is chosen according to Prop. 3.10 and $\bar{\pi}$ is injective on $\overline{U_\rho(x_0)} \cap \overline{\partial S}$. Then this proposition says that there are points $x \in L$ for which $\pi^{-1}(x) \cap S_\rho(x_0) \cap S = \emptyset$ and $\pi^{-1}(x) \cap B_\rho(x_0) \cap \partial S = k$. But this is impossible since the class of $\partial S \cap B_\rho(x_0)$ is zero in $H_{q-1}\left(B_\rho(x_0)/\left(S_\rho(x_0) \setminus S\right), \mathbb{Z}/2\mathbb{Z}\right)$ (compare §5).

For the proof of the claim, we proceed by induction on n, the case $n = 1$ being obvious.

For $n > 1$, consider the projection $\pi\colon R^n \to L$ where $L = L^{n-1}$ is a linear subspace of R^n, say $L = R^{n-1}$, and take a semialgebraic partition $: \pi(S) = D_1 \cup \dots \cup D_m$, which defines a cylindrical decomposition of S (called *saucissonage* in [**B-C-R**, 2.3]). Let $T_1, \dots, T_k$ be the collection of those D_i for which $\dim(T_i) = \dim \pi(S)$. We may assume that $T_i = T_i^*$, ∂S is finite over L and also that S is finite over L if $\dim(S) < n$. Then $\dim(T_i) = \dim(S)$ if $\dim(S) < n$ and $\dim(T_i) = n - 1$ if $\dim(S) = n$. Moreover, $\pi(S) = \pi(S)^* = T_1 + \dots + T_k$. Taking a sufficiently fine partition of $\pi(S)$, we may assume that the induction hypothesis holds for $\pi(S)$ and $T_1, \dots, T_k$, and that for each leaf F_{ij} over T_i (loc. cit.) the projection $\pi: F_{ij} \to T_i$ is a moderate isomorphism. We may also assume that $k = 1$, so $\pi(S) = T = T_1$.

Let $F_1, \dots, F_r$ be the leaves over T. If T is of type i), that is $\dim(\bar{T}) < \dim(T)$, then also $\dim(\bar{S}) < \dim(S)$ and we are done. So let t be type ii). If $\dim(S) < n$, then $S = F_1 + \dots + F_r$, so we may assume $S = F_1 = F$. We have a commutative diagram

$$\begin{array}{ccc} S & \longrightarrow & \bar{S} \\ {\scriptstyle \pi}\downarrow & & {\scriptstyle \bar{\pi}}\downarrow \\ T & \longrightarrow & \bar{T}, \end{array}$$

where π and $\bar{\pi}$ are isomorphisms. Therefore the properties ii) also hold for S.

Finally, let $\dim(S) = n$. Let $\varphi_i: T \to F_i$ be the inverse of $\pi: F_i \to T$. Here we may assume that S is the slice between F_1 and F_2, that is

$$S = \{(x', x_n) | x' \in T, \ \varphi_1(x) \le x_n \le \varphi_2(x)\},$$

where $x' = (x_1, \dots, x_{n-1})$. So $\partial S \subset F_1 \cap F_2 \cup \pi^{-1}(\partial T)$. Let $D := \{x' \in T | \varphi_2(x') - \varphi_1(x') \in \mathfrak{m}_v\}$. Then D is definable. It is easily seen that S is of type ii) if $\dim(\bar{D}) < n-1$. So let $\dim(\bar{D}) = n-1$ and let $\bar{D} := \{y \in \bar{R}^{n-1} | \Phi(y)\}$ where Φ is a quantifier-free formula for $\bar{D}$. We set $D_1 := \{x' \in R^n | s\,\Phi(x')\}$. (Remember from §1 that $s: \bar{R} \to R_1 \subset R$ denotes a fixed section which extends to formulas.) Also by §1 we get $\dim(\overline{D \setminus D_1}) < n-1$. But $\dim(S \cap \pi^{-1}(D_1)) \le n-1$. So by what we have shown we may replace S by $S \cap \pi^{-1}(T')$ for $T' := \mathrm{Cl}(T \setminus D_1)$. Splitting T' again into a sum of chains of type i) or ii), we get into the cases we considered before. □

REMARK 4.5. Let $S = W(R)^* \subset R^n$ be a semialgebraic chain where W is an affine real algebraic variety. Then S is a cycle (see [**B-C-R**, 11.3]). Such a cycle S is called algebraic. Note that $\deg_2(S)$, the degree of $s \mod 2$ is well defined. Now let $S \subset V^n$. Then, by Remark 3.9 c), $r(S)$ is again an algebraic cycle in $\bar{R}^n$, and, using Prop. 3.10, one gets $\deg_2(S) = \deg_2(r(S))$.

Similarly, $f_*(S)$ is algebraic and $\deg_2(f_*(S)) = \deg_2(S)$ for any polynomial map $f: R^n \to R^n$.

5. Reduction and intersection

We keep the notations of §1. Let $S \subset R^n$ be semialgebraic. As we mentioned before, for a point x of $\overline{S \subset V^n}$ in general a multiplicity with respect to S is not defined, since the fiber under the reduction map is infinite. However, one could look at semialgebraic subsets $T \subset R^n$, $\dim(S) + \dim(T) = n$ such that $\dim(S \cap T) = 0$ and $x \in \overline{S \cap V^n} \cap \overline{T \cap V^n}$ and then count the number of points $y \in (S \cap V^n) \cap (T \cap V^n)$ for which $\bar{y} = x$. It turns out that in mod 2 there is a nice formula for this number, which connects the intersection behavior of semialgebraic sets with reduction and multiplicities.

To start with, let us consider a classical situation of the Algebraic Intersection Theory. Let Y^p, Z^q be algebraic, reduced in $I\!P^n$ (defined over a real-closed field R in our case) such that $p+q=n$ and $\dim(Y\cap Z)=0$. For $x\in Y\cap Z$ the local intersection number $I(x,Y,Z)$ is defined by

$$I(x,Y,Z)=\sum_{i=0}^{n}(-1)^i \text{ length } \operatorname{Tor}_i^A(A/I(Y)\,,\ A/I(Z))\,,$$

where $A=\mathcal{O}_{x,I\!P^n}$ and $I(Y)\,,\ I(Z)$ are the defining ideals of Y and Z respectively [**Har**, p. 427]. Then by Bezout's theorem

$$\sum_{x\in Y\cap Z} I(x,Y,Z)=\deg(Y)\deg(Z)\,.$$

Let

$$I_2(x,Y,Z):=I(x,Y,Z) \mod 2 \text{ in } \mathbb{Z}/2\mathbb{Z}\,.$$

The significance of this comes from the fact that

$$\sum_{\substack{x\in Y\cap Z\\ x \text{ real}}} I_2(x,Y,Z)\equiv\deg(Y)\deg(Z) \mod 2\,.$$

We are going to present a topological interpretation of the local mod 2–intersection numbers $I_2(x,Y,Z)$. This is done along the usual lines. Let us recall:

5.1 Topological intersection. For the sequel, let A be a ring of coefficients (commutative with unit). Let η be a singular p-chain and ζ a singular q-chain in $\mathbb{R}^n$, $p+q=n$, such that $|\partial\eta|\cap|\zeta|=\emptyset=|\eta|\cap|\partial\zeta|$ (where $|\ |$ denotes the carrier of the chain). If these conditions hold, we call η and ζ *admissible.* Then an intersection number $\eta\circ\zeta$ is defined with the following properties [**Do**, VII, §4], [**S-T**, §73].

a) If $|\eta|\subset Y\,,\ |\partial\eta|\subset B\,;B\subset Y\subset\mathbb{R}\,,\ |\zeta|\subset Z\,,\ |\partial\zeta|\subset C\,,\ C\subset Z\subset\mathbb{R}$ and $Y\cap C=\emptyset=Z\cap B$, then $\eta\circ\zeta$ depends only on the homology-classes of η in $H_p(Y,B)$ and ζ in $H_q(Z,C)$.

b) $\zeta\circ\eta=\eta\circ\zeta$ and $(\eta+\eta')\circ\zeta=\eta\circ\zeta+\eta'\circ\zeta$ if the corresponding pairs of cycles are admissible.

c) If $|\eta|\cap|\zeta|=\emptyset$, then $\eta\circ\zeta=0$.

d) Let x be an isolated point in $|\zeta|\cap|\eta|$. Write $\eta=\eta_0+\eta_1$, $\zeta=\zeta_0+\zeta_1$ such that $x\notin|\eta_1|\cup|\zeta_1|$, $|\eta_0|\cap|\zeta_0|=x$ and η_0, ζ_0 are admissible. Then $\eta_0\circ\zeta_0$ only depends on x,η,ζ and defines the *local intersection number* $i\ (x,\eta,\zeta)$.

e) If $|\eta|\cap|\zeta|$ is finite, then $\eta\circ\zeta=\sum_{x\in|\eta|\cap|\zeta|} i(x,\eta,\zeta)$.

f) If η, ζ intersect transversally in x, then $i(x,\eta,\zeta)=\pm1$.

g) Let $f\colon\mathbb{R}^n\to\mathbb{R}^n$ be a homeomorphism. Then $\eta\circ\zeta=\pm(f^*\eta)\circ(f^*\zeta)$.

If $\mathbb{R}$ is replaced by an arbitrary real-closed field R, one has a corresponding intersection number, using Algebraic Topology over arbitrary real-closed fields (see §4, **[D-Kn]**, **[Kn]**, **[B-C-R**, 11.7]). Of course, in that case all sets and all maps under consideration have to be semialgebraic. So again we fix a real-closed field R and also $A = \mathbb{Z}/2\mathbb{Z}$ as a ring of coefficients. In order to indicate the latter, we write $\eta \underset{2}{\circ} \zeta$ instead of $\eta \circ \zeta$ for admissibole semialgebraic chains $\eta, \zeta \subset \mathbb{R}^n$ and also $i_2(x, \eta, \zeta)$ for $i(x, \eta, \zeta)$.

As a first application for topological intersection numbers, defined over an arbitrary real-closed field, we consider the situation of the beginning of this section. We assume further that Y and Z are real varieties. Let $\eta := Y(R)^*$ and $\zeta = Z(R)^*$. Then η and ζ are contained in $I\!P^n(R)$, but not in any $R^n \subset I\!P^n(R)$ in general. However, $i_2(x, \eta, \zeta)$ is still defined for $x \in \eta \cap \zeta$. In this situation we have

PROPOSITION 5.2. *Let $x \in \eta \cap \zeta$. Then*

$$I_2(x, Y, Z) = i_2(x, \eta, \zeta).$$

PROOF (sketch). One has

$$\sum_{y \in Y(R) \cap Z(R)} I_2(y, Y, Z) = \deg_2(Y) \deg_2(Z) = \eta \vee \zeta,$$

where $\eta \vee \zeta$ is the cup product of the homology classes of η and ζ in $H_n(\mathbb{P}_n(R), \mathbb{Z}/2\mathbb{Z}) \simeq \mathbb{Z}/2\mathbb{Z}$ after identification with cohomology classes via Poincaré duality. On the other hand

$$\eta \vee \zeta = \sum_{y \in \eta \cap \zeta} i_2(y, \eta, \zeta).$$

So we are done if $\#(\eta \cap \zeta) = 1$. To achieve this, merely consider a projection $\pi : S^n \to \mathbb{P}^n$ such that $\pi(S^n)$ is a small ball around x. We get

$$I(\pi^{-1}(Y), \pi^{-1}(Z)) = 2I(x, Y, Z) = i(\pi^{-1}(\eta), \pi^{-1}(\zeta)) = 2i(x, \eta, zeta) \mod 4,$$

and thus the claim. □

Of course, one should also compare algebraic and topological intersection for more general real algebraic varieties over R, but we do not enter into this. Our topic is reduction, and since our results here are essentially of local nature, the above illustrations may be sufficient.

So we return to the situation that R admits a real valuation v with residue field $\bar{R}$.

THEOREM 5.3 (Intersection-Reduction Formula). *Let S and T be semialgebraic chains in V^n of dimensions p and q respectively, with $p + q = n$. Assume that $\bar{S} \cap \overline{\partial T} = \emptyset$ and $\overline{\partial S} \cap \bar{T} = \emptyset$. Then S, T and $r(S), r(T)$ are admissible and $S \underset{2}{\circ} T = r(S) \underset{2}{\circ} r(T)$.*

First, we show a special case:

LEMMA 5.4. *Let S and T be semialgebraic chains in V^n of dimension p and q respectively, with $p+q=n$. Let $x_0 \in V^n$ such that the following conditions hold:*

i) *$\bar{S}$ and $\bar{T}$ intersect transversally in $\bar{x}_0$.*
ii) *The multiplicity of $\bar{S}$ is* constant $= s$ *and the multiplicity of $\bar{T}$ is* constant $= t$ *in a neighborhood of $\bar{x}_0$.*
iii) *$\bar{x}_0 \notin \overline{\partial S} \cup \overline{\partial T}$.*

Then there exists $\rho \in V$ with $\bar{\rho} > 0$ such that

a) *$S' := \mathrm{Cl}(S \cap U_\rho(x_0))$ and $T' := \mathrm{Cl}(T \cap U_\rho(x_0))$ are admissible and $(\partial S' \cup \partial T') \subset S_\rho(x_0)$.*
b) *$\overline{S' \cap T'} = \{\bar{x}_0\}$.*
c) *$S' \underset{2}{\circ} T' = st \mod 2$.*

PROOF. The assumption implies that $\bar{S}$ and $\bar{T}$ are smooth in a neighborhood of $\bar{x}_0$. We may assume that $x_0 = 0$ and we choose $\rho \in V$ such that $\bar{\rho} > 0$ and

$$B_\rho(x_0) \cap (\partial S \cup \partial T) = \emptyset .$$

For such a ρ we get a) and b). We choose linear subspaces $L = L^p \subset R^n$ for S and $M = M^q \subset R^n$ for T according to Prop. 3.10. We may assume that $\bar{L}$ and $\bar{M}$ are the tangent spaces at $\bar{X}_0$ of $\bar{S}$ and $\bar{T}$ respectively, and also that $L \perp M$.

Choosing ρ sufficiently small (but still $\bar{\rho} > 0$), we find $\varepsilon \in V$, $\bar{\varepsilon} > 0$, such that the following conditions hold.

1) $S' \subset U_\varepsilon(L)$, $T' \subset U_\varepsilon(M)$.
2) $S_\rho(x_0) \cap U_\varepsilon(L) \cap U_\varepsilon(M) = \emptyset$.

Now $S' \underset{2}{\circ} T'$ depends only on the class of T' in

$$G := H_q\left(B_\rho(x_0)/\left(S_\rho(x_0) \setminus U_\varepsilon(L)\right),\quad \mathbb{Z}/2\mathbb{Z}\right) .$$

Using the exact homology sequence for the couple $(B_\rho(x_0),\ (S_\rho(x_0) \setminus U_\varepsilon(L)))$ and [**Gr**, §18], we get $G = \mathbb{Z}/2\mathbb{Z}$ where the class of $M' := M \cap B_\rho(x_0)$ is a generator. Let also $L' := L \cap B_\rho(x_0)$. So $S' \underset{2}{\circ} T' = S' \underset{2}{\circ} \mu M'$ for some $\mu \in \mathbb{Z}/2\mathbb{Z}$.

Correspondingly, we may replace S' by $\lambda L'$ for some $\lambda \in \mathbb{Z}/2\mathbb{Z}$. So $S' \underset{2}{\circ} T' = \lambda L' \underset{2}{\circ} \mu M' = \lambda\mu \left(L' \underset{2}{\circ} M'\right) = \lambda\mu$.

But by Prop. 3.10, $\mu = L' \underset{2}{\circ} \mu M' = L' \underset{2}{\circ} T' \equiv t \mod 2$, and correspondingly $\lambda \equiv s \mod 2$. □

PROOF OF Th. 5.3. Obviously, S and T are admissible and, using Th. 4.4, we see that also $r(S)$ and $r(T)$ are admissible. By assumption, we find $\varepsilon \in V$ with $\bar{\varepsilon} > 0$ such that

$$S \cap U_\varepsilon(\partial T) = 0 , \quad \partial S \cap U_\varepsilon(T) = \emptyset$$
$$r(S) \cap U_{\bar{\varepsilon}}\left(\partial r(T)\right) = \emptyset , \quad \partial r(S) \cap U_\varepsilon\left(r(T)\right) = \emptyset .$$

For $a \in V^n$ with $\overline{< a, a >} < \bar{\varepsilon}^2$, let $T_a := T + a$.

Then $\bar{T}_a = T + \bar{a}$, $r(T_a) = r(T) + \bar{a}$ and correspondingly $\partial r(T_a) = r\partial(T_a) = \partial r(T) + \bar{a}$. Consequently, S, T_a and $r(S)$, $r(T_a)$ are still admissible with $S \underset{2}{\circ} T_a = S \underset{2}{\circ} T$ and $r(S) \underset{2}{\circ} r(T_a) = r(S) \underset{2}{\circ} r(T)$. So we may replace T by T_a. But by a Sard-Bertini argument [**B-C-R**, 9.5], we can achieve that $\bar{S}$ intersects $\bar{T}$ transversally at finitely many points. Now the claim follows from Lemma 5.4. □

REFERENCES

[Be-W] E. Becker and T. Wörmann, *On trace forms and applications to real algebraic geometry* (to appear in this volume).

[B-C-R] I. Bochnak, M. Coste, and M. F. Roy, *Géometrie Algébrique réele*, Springer, Berlin, Heidelberg, and New York, 1987.

[Bour] N. Bourbaki, *Algébre Commutative*, Chap. 6, Hermann, Paris, 1964.

[Br] L. Bröcker, *Families of semialgebraic sets and limits*, Real Algebraic Geometry, M.Coste, L. Mahé, M-F. Roy (Eds), Springer Lecture Notes in Mathematics.

[C-Di] G. Cherlin and M. Dickmann, *Real closed rings II. Model theory*, Annals of Pure and Applied Logic **25** (1983), 213–231.

[D] H. Delfs, *Kohomologie affiner semialgebraischer Räume*, Diss. Univ. Regensburg, 1980.

[D-Kn] H. Delfs and M. Knebusch, *On the homology of algebraic varieties over real closed fields*, J. reine Angew. Math. **335** (1982), 122–163.

[De] F. Delon, *Quelques propriétés des corps valués en théorie del molèles*, Thèse, Paris, 1982.

[Do] A. Dold, *Lectures on algebraic topology*, Springer, Berlin, Heidelberg and New York, 1972.

[Gr] M. Greenberg, *Lectures on algebraic topology*, Benjamin, New York and Amsterdam, 1967.

[Har] R. Hartshorne, *Algebraic geometry*, Springer, New York, Heidelberg and Berlin, 1977.

[Kn] M. Knebusch, *Weakly semialgebraic spaces*, Lecture Notes in Math., vol. 1367, Springer, Berlin and Heidelberg, 1989.

[Kn-Sch] M. Knebusch and C. Scheiderer, *Einführung in die reele Algebra*, Vieweg, Braunschweig and Wiesbaden, 1989.

[Ro] A. Robinson, *Introduction to model theory and the metamathematics of algebra*, North-Holland, Amsterdam, 1965.

[Schar] W. Scharlau, *Quadratic and hermitian forms*, Springer, Berlin, Heidelberg, New York and Tokyo, 1985.

[S-T] H. Seifert and W. Threlfall, *Lehrbruch der Topologie*, Teubner, Leipzig, 1934.

[v.d.D] L. van den Dries, *Tarski's problem and Pfaffian functions*, Logic Colloquium 1984 (J. B. Paris, A. J. Wilkie, and G. M. Wilmers, eds.), North-Holland, Amsterdam, 1986, pp. 59–90.

MATHEMATISCHES INSTITUT, EINSTEINSTR. 62, W–4400 MÜNSTER, GERMANY

Contemporary Mathematics
Volume **155**, 1994

Orderings for Noncommutative Rings

THOMAS C. CRAVEN

Introduction and notation

The goal of this paper is to present the beginnings of a theory of real algebraic geometry for noncommutative rings. For a basic introduction to the commutative theory, see Lam [**L**]. The word *field* will be used in this paper to mean a (generally noncommutative) skewfield; we shall specify a commutative field when we need to. R will denote a noncommutative ring with 1. We shall define a concept of ordering for R which we show behaves properly with respect to orderings of "residue fields" and generalizes the usual concepts of orderings for fields and commutative rings. In the final section, we take a brief look at the real spectrum. The complications of this theory can be avoided for special classes of noncommutative rings such as Ore domains (see, for example, [**P**]).

Let $\mathcal{M}(R)$ denote the set of all square matrices over R. The notation $(a|A)$ will be used to denote an augmented matrix with a as the first column and A as the remainder of the matrix. The basic concepts for noncommutative algebraic geometry which we use are due to P. M. Cohn [**Co1**], [**Co2**], [**Co3**] and we shall generally use his notation. In addition to the usual matrix operations, we will use $A \dotplus B$ to denote the matrix $\begin{bmatrix} A & 0 \\ 0 & B \end{bmatrix}$. If two matrices A and B are identical except for one row (or one column), then $A \nabla B$ denotes the matrix identical to A and B in all entries except that the entries in the special row (column) are added. Note that for commutative rings R, we have $\det(A \nabla B) = \det A + \det B$. For this reason, the operation ∇ is referred to as the *determinantal sum.* Cohn has shown that, in analogy to the residue field of a localization of a commutative ring at a prime ideal, one can study *epic R-fields.* These are fields generated by the homomorphic image of R inside them. They are determined, not by a prime ideal, but by a "prime matrix ideal" $\wp$. We shall denote the corresponding epic

---

1991 *Mathematics Subject Classification.* Primary 14P99; Secondary 16W80.
This paper is submitted in final form and no version of it will be submitted for publication elsewhere.

R-field by $K(\wp)$; the matrix ideal $\wp$ is called the *singular kernel* of the mapping from R to $K(\wp)$.

An $n \times n$ matrix A is called *nonfull* if $A = BC$, where B is $n \times r$, C is $r \times n$ and $r < n$. A *matrix ideal* is a subset $\wp$ of $\mathcal{M}(R)$ which contains all nonfull matrices, is closed under $\dotplus$ and $\triangledown$ when defined, and contains a matrix A whenever it contains $A \dotplus 1$. A matrix ideal $\wp$ is called a *prime matrix ideal* if it is a proper subset of $\mathcal{M}(R)$ and satisfies $A \dotplus B \in \wp \implies A \in \wp$ or $B \in \wp$.

The real theory

We now look at how orderings can be included in the algebraic geometry developed by Cohn. We thank Jim Madden for pointing out that some of these ideas were developed earlier (and independently) by Gábor Révész [**R**]. His approach and goals are somewhat different than ours, but a maximal matrix cone in his paper can be seen to be equivalent to a *matrix ordering* as defined below. Our approach will be far more natural for anyone familiar with the commutative theory. Also, we provide a proof for the fundamental result showing that the matrix orderings of R with center $\wp$ are in one-to-one correspondence with the orderings of $K(\wp)$; there appears to be a large gap in Révész proof [**R**, Prop. 6 and Corollary]. (See Theorem 5 below.)

The fundamental idea is to think of how the Dieudonné determinant would behave if it were defined for rings. We shall see that for fields, the Dieudonné determinant induces a one-to-one correspondence between matrix orderings and ordinary orderings.

We write 1 and -1 for the matrices $[1]$ and $[-1]$.

DEFINITION. A *matrix preordering* of R is a subset $T \subseteq \mathcal{M}(R)$ satisfying

(P1) $A, B \in T$ implies $A \triangledown B \in T$ if it is defined.
(P2) $A, B \in T$ implies $A \dotplus B \in T$.
(P3) $A \dotplus A \in T$ for all $A \in \mathcal{M}(R)$.
(P4) $-1 \notin T$.
(P5) $A \dotplus 1 \in T$ implies $A \in T$.
(P6) T contains all nonfull matrices.

DEFINITION. A *matrix ordering* of R is a subset $P \subseteq \mathcal{M}(R)$ satisfying

(O1) $A, B \in P$ implies $A \triangledown B \in P$ if it is defined.
(O2) $A, B \in P$ implies $A \dotplus B \in P$.
(O3) $A \dotplus A \in P$ for all $A \in \mathcal{M}(R)$.
(O4) $-1 \notin P$.
(O5) $P \cup -P = \mathcal{M}(R)$, where $-P = \{\, B \in \mathcal{M}(R) \mid \exists A \in P,\ A \triangledown B \text{ nonfull} \,\}$.
(O6) $\wp = P \cap -P$ is a prime matrix ideal. It will be called the *center* of P.

It is not hard to show that the last two items in the definition of matrix ordering imply the last two items in the definition of matrix preordering. The set $-P$ can be defined in several different ways. The one above seems natural since the nonfull matrices fulfill a roll much like that of zero. In particular, a square matrix over a field is nonfull iff its Dieudonné determinant is zero. Two equivalent ways of writing $-P$ are included in our list of consequences of these definitions, both of which are much more useful in practice.

Other properties of matrix preorderings and orderings:

(1) For any matrix preordering T, we have $\{ B \in \mathcal{M}(R) \mid \exists A \in T,\ A \triangledown B \text{ nonfull} \} = \{ (-a|A') \mid (a|A') \in T \}$.

PROOF. One containment is clear since $(0|A')$ is nonfull. For the converse, we may assume $A = (a|A') \in T$, $B = (b|A')$ so that $\triangledown$ is defined. As $(a+b|A') = A \triangledown B$ is nonfull, (P6) implies that $A \triangledown B \in T$, and that $(-a-b|A') \in T$. Since $(a|A') \in T$, (P1) implies $(-b|A') = (-a-b|A') \triangledown (a|A') \in T$, whence B lies in the right hand side as desired. □

(2) For any matrix preordering T, we have $\{ -1 \dotplus A \mid A \in T \}$ equal to the two sets in (1). (This follows from the previous result and the fact that $1 \in T$.) This set will be denoted by $-T$.

(3) If A is a nonfull matrix, then $A \in T \cap -T$ (by (P6)).

(4) If $A \in T$, the result of adding any right multiple of one column (or left multiple of a row) to another column (row) again lies in T. (Use (P1), (P6) and follow [**Co1**, p. 397].)

(5) If $A \in T$, then interchanging two columns (rows) and changing the sign of one of them results in a matrix which again lies in T. (Use (P1), (P6) and follow [**Co1**, p. 397].)

(6) For $A, B \in \mathcal{M}(R)$ and C of the appropriate size, $A \dotplus B \in T$ iff $\begin{bmatrix} A & 0 \\ C & B \end{bmatrix} \in T$ (or equivalently, $\begin{bmatrix} A & C \\ 0 & B \end{bmatrix} \in T$). (Use (P2) and follow [**Co1**, pp. 397–398].)

(7) If $A \in T$, the result of any even permutation of rows or columns again lies in T by (5).

(8) If $A, B \in T$ of the same size, then $AB \in T$. More generally, $AB \in T \iff BA \in T$ for any $A, B \in \mathcal{M}(R)$ of the same size. (Same argument as [**Co1**, p. 398].)

(9) If $A \triangledown B \in T \cap -T$ and $A \in T$, then $B \in -T$. (See proof of (1).)

(10) If $A \triangledown B \in T \cap -T$, then $A \dotplus B \in -T$. (This follows from (9).)

(11) If $A \in T, B \in -T$, then $A \dotplus B \in -T$.

One would like to be able to show that a maximal matrix preordering is a matrix ordering. This would allow a generalization of the Artin-Schreier criterion for existence of an ordering. This remains an open problem. One can easily handle matrix preorderings only with the prior assumption that $T \cap -T$ is a prime matrix ideal. This is essentially what Révész has done and what we do with matrix orderings below.

Let D be a field. We shall assume that its characteristic is zero since we are only interested in ordered fields. Of special interest is the Dieudonné determinant (cf. [**A**, IV.1], defined to be zero for singular square matrices, and defined modulo commutators otherwise. (Note that commutators are always squares, hence positive in any ordering!) Two properties of the Dieudonné determinant make it work for matrix orderings. The first is that $\det(A \bigtriangledown B) \subseteq \det A + \det B$ whenever $A \bigtriangledown B$ is defined [**A**, Theorem 4.5]. The second is the following proposition, in which $GL_n\, D$ is the general linear group of nonsingular $n \times n$ matrices and $D^\times$ is the multiplicative group of nonzero elements of D.

LEMMA 1. *Let $n \geq 1$ be an integer. The group homomorphism $GL_n\, D \to D^\times / D^{\times 2}$ induced by the Dieudonné determinant is surjective with kernel equal to the subgroup $S_n\, D$ generated by all squares of $n \times n$ matrices.*

PROOF. We have the special linear group $SL_n\, D$ equal to the commutator subgroup

$$[GL_n\, D, GL_n\, D]$$

by [**D**, p. 138] and, in particular, contained in $S_n\, D$. Modulo $SL_n\, D$, each matrix has the form $\begin{bmatrix} d & & & \\ & 1 & & \\ & & \ddots & \\ & & & 1 \end{bmatrix}$ by [**A**, Theorem 4.1]. Thus the mapping

$$d \mapsto \begin{bmatrix} d & & & \\ & 1 & & \\ & & \ddots & \\ & & & 1 \end{bmatrix}$$

gives an inverse to $GL_n\, D / S_n\, D \to D^\times / D^{\times 2}$. □

From these results, we obtain the characterization of matrix orderings for fields as an immediate corollary.

THEOREM 2. *Let D be a field.*

(a) *Let P_0 be an ordering of D. Then P_0 extends to a matrix ordering P of D given by $A \in P \iff \det A \subset P_0$.*
(b) *Any matrix ordering P of D induces an ordering P_0 defined by $a \in P_0 \iff [a] \in P$.*
(c) *There is a one-to-one correspondence between orderings and matrix orderings of D.*

□

REMARK. In a similar manner, the ordinary determinant for commutative rings gives a one-to-one correspondence between orderings (in the sense of [**L**]) and matrix orderings.

Rings and epic R-fields

Let R be a ring and let $\wp$ be a prime matrix ideal of R. Our goal in this section is to establish the fundamental result that the matrix orderings of R with center $\wp$ are in one-to-one correspondence with the orderings of the residue field $K(\wp)$. The proofs depend heavily on the construction of $K(\wp)$ detailed in [**Co2**, Chapter 4] and [**Co1**, Chapter 7]. Basically, this involves forming the localization $R_\wp$, a local ring with a homomorphism $R \to R_\wp$ with the property that every square matrix over R not in $\wp$ becomes invertible over $R_\wp$. The elements of $R_\wp$ can be characterized by the fact that they are precisely the components of solutions of matrix equations $Au + a = 0$ where A lies in the image of $\mathcal{M}(R) \setminus \wp$, and a is a column vector defined over the image of R [**Co1**, Theorem 7.1.2]; [**Co2**, Theorem 4.2.1]. In fact, this can be done so that the element of $R_\wp$ is the first component of the solution and $a = e_1$, the first column of an identity matrix.

Now let P be a matrix ordering of R with center $\wp$. We define a subset P_D of $D = K(\wp)$ as follows: for $x \in D$, choose $u \in R_\wp$ with image x under the canonical mapping to the residue field; let $A \in \mathcal{M}(R) \setminus \wp$ be such that $A \begin{bmatrix} u \\ u_2 \\ \vdots \\ u_n \end{bmatrix} = r \in R^n$. Set $A_1 = [r \quad a_2 \quad \dots \quad a_n]$, where a_i is the ith column of A. Define $x \in P_D$ if and only if $A \dotplus A_1 \in P$.

THEOREM 3. *P_D is an ordering of $K(\wp)$.*

PROOF. We first show that the definition of P_D is independent of the choice of A and u. Consider two liftings of x, say u, v, with $A \begin{bmatrix} u \\ u' \end{bmatrix} = r$ and $B \begin{bmatrix} v \\ v' \end{bmatrix} = s$, where u', v', r, s are column vectors. Set $M = \left[\begin{array}{c|cc} A & a_1 & 0 \\ \hline 0 & \multicolumn{2}{c}{B} \end{array}\right]$, so that

$$M \begin{bmatrix} u - v \\ u' \\ v \\ v' \end{bmatrix} = \begin{bmatrix} r \\ s \end{bmatrix}.$$

With notation as above, $M_1 = M \begin{bmatrix} u - v & 0 \\ 0 & I \end{bmatrix} \begin{bmatrix} 1 & 0 \\ u' & I \end{bmatrix}$, which becomes singular over D; thus $M_1 \in \wp$. Now $M_1 = N_1 \nabla N_2$, where $N_1 = \left[\begin{array}{c|cc} A_1 & a_1 & 0 \\ \hline 0 & \multicolumn{2}{c}{B} \end{array}\right]$ and $N_2 = \left[\begin{array}{c|c|cc} 0 & a_2 \dots a_n & a_1 & 0 \\ \hline s & 0 & \multicolumn{2}{c}{B} \end{array}\right]$. By (10) of the additional properties, we have $N_1 \dotplus N_2 \in -P$. Switching columns 1 and $n+1$ of N_2 yields

$$\left[\begin{array}{c|c} \begin{array}{c|cc} A_1 & a_1 & 0 \\ \hline 0 & \multicolumn{2}{c}{B} \end{array} & 0 \\ \hline 0 & \begin{array}{cc|c} \multicolumn{2}{c|}{A} & 0 \\ \hline b_1 & 0 & B_1 \end{array} \end{array}\right] \in P.$$

Using (6) we obtain $A_1 \dotplus A \dotplus B_1 \dotplus B \in P$, whence $A \dotplus A_1 \in P \iff B \dotplus B_1 \in P$.

Next we check that P_D is an ordering. The facts that $-1 \notin P_D$ and $P_D \cup -P_D = D$ follow immediately from the corresponding facts for P. If $x \in P_D \cap -P_D$, then $A \dotplus A_1 \in P \cap -P$, which implies that A_1 is in $\wp$ since A cannot be; i.e., A_1 becomes singular over D. Now, taking Dieudonné determinants, the equation

$$A_1 = A \begin{bmatrix} u & 0 \\ u' & I \end{bmatrix} = A \begin{bmatrix} u & 0 \\ 0 & I \end{bmatrix} \begin{bmatrix} 1 & 0 \\ u' & I \end{bmatrix}$$

yields $x = 0$. Finally, we check that P_D is closed under multiplication and addition. Let $x, y \in P$ and lift them to $u, v \in R_\wp$ satisfying $A \begin{bmatrix} u \\ u' \end{bmatrix} = e_1$ and $B \begin{bmatrix} v \\ v' \end{bmatrix} = e_1$, where $A \dotplus A_1$ and $B \dotplus B_1$ lie in P. Then $M \begin{bmatrix} uv \\ u'v \\ v \\ v' \end{bmatrix} = \begin{bmatrix} 0 \\ e_1 \end{bmatrix}$, where $M = \left[\begin{array}{c|cc} A & -e_1 & 0 \\ \hline 0 & \multicolumn{2}{c}{B} \end{array}\right]$. In this case $M_1 = \left[\begin{array}{c|c|cc} 0 & a_2 \dots a_n & -e_1 & 0 \\ \hline e_1 & 0 & \multicolumn{2}{c}{B} \end{array}\right]$. Switching columns 1 and $n+1$ and changing the sign of the resulting first column shows that $M_1 \in P$ iff $\left[\begin{array}{cc|c} \multicolumn{2}{c|}{A_1} & 0 \\ \hline -b_1 & 0 & B_1 \end{array}\right] \in P$; i.e., iff $A_1 \dotplus B_1 \in P$. Also $M \in P \iff A \dotplus B \in P$, so, since $A \dotplus A_1 \dotplus B \dotplus B_1 \in P$, we have $M \dotplus M_1 \in P$ and hence $xy \in P_D$. Changing M to $\left[\begin{array}{c|cc} A & -a_1 & 0 \\ \hline 0 & \multicolumn{2}{c}{B} \end{array}\right]$, we obtain $M \begin{bmatrix} u+v \\ u' \\ v \\ v' \end{bmatrix} = \begin{bmatrix} e_1 \\ e_1 \end{bmatrix}$. In this case, $M_1 = \left[\begin{array}{cc|cc} \multicolumn{2}{c|}{A_1} & -a_1 & 0 \\ \hline e_1 & 0 & \multicolumn{2}{c}{B} \end{array}\right]$, which can be rewritten as $N_1 \triangledown N_2$, where $N_1 = \left[\begin{array}{c|cc} A_1 & -a_1 & 0 \\ \hline 0 & \multicolumn{2}{c}{B} \end{array}\right]$ and $N_2 = \left[\begin{array}{c|c|cc} 0 & a_2 \dots a_n & -a_1 & 0 \\ \hline e_1 & 0 & \multicolumn{2}{c}{B} \end{array}\right]$. Clearly $N_1 \in P \iff A_1 \dotplus B \in P$. Also $N_2 \in P \iff A \dotplus B_1 \in P$, which can be seen by switching columns 1 and $n+1$ and changing a sign. Since $A \dotplus A_1$ and $B \dotplus B_1$ lie in P, we obtain $M \dotplus N_1$ and $M \dotplus N_2$ in P, so that $M \dotplus M_1 = M \dotplus (N_1 \triangledown N_2) = (M \dotplus N_1) \triangledown (M \dotplus N_2) \in P$. □

Lifting orderings from $K(\wp)$ to R is much simpler.

THEOREM 4. *Let $D = K(\wp)$ be an epic R-field with matrix ordering P_D. Let $\phi \colon \mathcal{M}(R) \to \mathcal{M}(D)$ be the homomorphism induced by the canonical mapping from R to the residue field D. Set $P \subset \mathcal{M}(R)$ equal to $\phi^{-1}(P_D)$. Then P is a matrix ordering of R.*

PROOF. Let $A, B \in P$ with $A \triangledown B$ defined. Then $\phi(A) \triangledown \phi(B)$ is defined. Since P_D is a matrix ordering, $\phi(A) \triangledown \phi(B) \in P_D$, hence $A \triangledown B \in P$. Conditions (O2) and (O3) are verified similarly. Condition (O4) is clear since $\phi([-1]) = [-1]$. The final two conditions require the fact that $\phi^{-1}(-P_D) = -P$. To see this, one need only use the definition of minus given in (2) and the fact that ϕ preserves $\dotplus$. Condition (O6) then follows immediately, since $\wp = \{ \phi^{-1}(A) \mid A \text{ is singular } \} =$

$\phi^{-1}(P_D \cap -P_D) = \phi^{-1}(P_D) \cap \phi^{-1}(-P_D) = P \cap -P$. For condition (O5), let $A \in \mathcal{M}(R)$. Then $\phi(A) \in P_D \cup -P_D$, so that $A \in \phi^{-1}(P_D \cup -P_D) = P \cup -P$. $\square$

We now come to our main theorem, which shows that the definitions we have made really behave properly.

THEOREM 5. *The constructions of Theorems* 3 *and* 4 *induce a one-to-one correspondence between matrix orderings of a ring R with center $\wp$ and orderings of the residue field $D = K(\wp)$.*

PROOF. We shall show that the given constructions are inverses of each other. We begin with an ordering P_0 of D; let P_0 also denote its unique extension to a matrix ordering as given by Theorem 2. Let P be the matrix ordering of R given by the construction of Theorem 4 and let P_D be the ordering of D induced by P as in Theorem 3. We claim that $P_0 = P_D$. Choose $0 \neq a \in P_0$. The element a has a defining matrix equation $A \begin{bmatrix} u \\ u' \end{bmatrix} = e_1$, where $\bar{u} = a$. From the equation $A_1 = A \begin{bmatrix} u & 0 \\ 0 & I \end{bmatrix} \begin{bmatrix} 1 & 0 \\ u' & I \end{bmatrix}$, we obtain $a[D, D] = (\det A)^{-1} \det A_1 = \det(A \dotplus A_1)(\det A)^{-2}$, which implies that $\det(A \dotplus A_1) \subset P_0$ and thus that $A \dotplus A_1 \in P$. By definition, $a \in P_D$. Conversely, let $0 \neq a \in P_D$. Then, with the matrix notation above, $A \dotplus A_1 \in P$, hence $(\det A)(\det A_1) \subset P_0$. As above, $a[D, D] = (\det A)^{-1} \det A_1$, which is a subset of P_0, hence $a \in P_0$.

For the other composition of constructions, let P be a matrix ordering of R with $P \cap -P = \wp$. Let P_D be the induced ordering of D given by Theorem 3; we have seen that it extends uniquely to a matrix ordering of D via the Dieudonné determinant. Let P_1 denote the inverse image of this matrix ordering, a matrix ordering of R. We know from the first part of the proof that P_1 also induces the ordering P_D on the residue field. To see that P_1 is actually P, we proceed by induction on the number of rows n of a matrix $A \in P$. We may assume that $A \notin \wp$. If $n = 1$, we can write $A = [a]$, $a \in R$. Using the equation $[a] \cdot [a] = [a^2]$, our defining condition for P_D shows that, writing A also for the image in $\mathcal{M}(D)$, $\det A = \bar{a}$ is positive iff $\begin{bmatrix} a & 0 \\ 0 & a^2 \end{bmatrix} \in P$; as $A = [a]$ and $[a^2]$ both lie in P, we obtain $\det A \in P_D$ and thus $A \in P_1$. Now assume that $n \geq 2$ is minimal such that $A \notin P_1$. Since A becomes invertible over $R_\wp$, the equation $A \begin{bmatrix} u \\ u' \end{bmatrix} = e_1$ can be solved for $u \in R_\wp$ and $u' \in R_\wp^{n-1}$. Now the sign of $\bar{u}$ (with respect to P_D) equals the sign of $A \dotplus A_1$ with respect to both P and P_1, where $A_1 = \begin{bmatrix} 1 & * \\ 0 & A_0 \end{bmatrix}$ has the usual meaning, and A_0 is the $(n-1) \times (n-1)$ matrix formed by deleting the first row and column of A. Since A has opposite signs with respect to P and P_1, this must also be true of A_1. But A_1 and A_0 have the same sign with respect to any matrix ordering by (6), whence A_0 must have opposite signs with respect to the two matrix orderings P and P_1. As A_0 has only $n-1$ rows, this contradicts the induction hypothesis. $\square$

COROLLARY 6. *Let D be an epic R-field with singular kernel $\wp$. If two matrices in $\mathcal{M}(R)$ have images in $\mathcal{M}(D)$ with the same Dieudonné determinant (or even determinants which are congruent modulo sums of products of squares), then the matrices have the same sign with respect to every matrix ordering of R with center $\wp$.*

REMARK. Let T be a matrix preordering such that $\wp = T \cap -T$ is a prime matrix ideal. The construction and proof of Theorem 3, applied to T, show that T induces a preordering on $K(\wp)$. This preordering is contained in an ordering of $K(\wp)$ which, by Theorem 4, pulls back to a matrix ordering of R containing T.

The real spectrum

The set of orderings X_D of a field D has been studied by several authors, even in the noncommutative case (cf. [**Cr**]). As in the commutative case, this can be generalized to rings, namely to the set of all matrix orderings X_R of a ring R. This set, together with the topology defined below, will be called the *real spectrum* of R. We assume that the set X_R is nonempty.

The topology on X_R is the one generated by taking, as a subbasis for the open sets, all sets of the form

$$H(A) = \{ P \in X_R \mid A \notin -P \}, \qquad A \in \mathcal{M}(R).$$

This will be called the *Harrison topology.*

LEMMA 7. *Let T be a matrix preordering. Assume that T satisfies*

$$A \dotplus B \in -T \implies A \in T \text{ or } B \in T.$$

Then T is a matrix ordering.

PROOF. Since $A \dotplus (-1 \dotplus A) \in -T$, we have either $A \in T$ or $A \in -T$; thus $T \cup -T = \mathcal{M}(R)$. We must show that $\wp = T \cap -T$ is a prime matrix ideal. From the definition of matrix preordering, we know that $\wp$ contains the nonfull matrices, is closed under ∇ and satisfies $A \dotplus 1 \in \wp \implies A \in \wp$. Now let $A \in \wp, B \in \mathcal{M}(R)$. We know that $B \in T \cup -T$; assume $B \in T$. Then $A \dotplus B \in T$ and $A \dotplus B \in -T$. Similarly if $B \in -T$. Therefore $\wp$ is a matrix ideal. To see that $\wp$ is prime, assume $A \dotplus B \in \wp$ and $A \notin \wp$. We must show that $B \in \wp$. Without loss of generality, $A \notin T$. Since $A \dotplus B \in -T$ and $A \dotplus (-1 \dotplus B) \in -T$, the hypothesis implies $B \in T$ and $-1 \dotplus B \in T$, whence $B \in \wp$. □

With the lemma above, we can give a proof that the real spectrum is compact (but not Hausdorff) which is very similar to the proof in the commutative case.

THEOREM 8. *The real spectrum X_R is compact.*

PROOF. Let $Y = \{0,1\}^{\mathcal{M}(R)}$, the set of all functions from $\mathcal{M}(R)$ to the discrete two point space $\{0,1\}$. We give Y the product topology, so it is compact by Tychonoff's theorem. For any matrix ordering $P \in X_R$, we can define a function $f_P\colon \mathcal{M}(R) \to \{0,1\}$ as the characteristic function of $P \setminus (-P)$. The mapping $P \mapsto f_P$ is easily seen to give an embedding of X_R into Y. The standard basis for the product topology on Y consists of all sets of the form

$$H_{\epsilon_1,\ldots,\epsilon_n}(A_1,\ldots,A_n) = \{\, f \in Y \mid (\forall i) f(A_i) = \epsilon_i \,\} \qquad (A_i \in \mathcal{M}(R),\ \epsilon_i \in \{0,1\})$$

Note that $H_{1,\ldots,1}(A_1,\ldots,A_n) \cap X_R = \bigcap H(A_i)$ is open in X_R. Thus if we can show that X_R is closed in Y, we will know that it is compact with respect to the finer Tychonoff topology induced from Y and hence is compact with respect to the Harrison topology as well. Let $f \in Y \setminus X_R$. We show that X_R is closed by constructing a basic open set containing f and missing X_R. Since $f \notin X_R$, the set $T = \{\, -1 \dotplus A \mid f(A) = 0 \,\}$ must violate one of the six conditions in the definition of matrix preordering or the condition given in the previous lemma. Each of these leads immediately to the desired basic open set; e.g. for condition (P1), there exist matrices $A, B \in T$ with $A \bigtriangledown B$ defined, but not in T; thus $f \in H_{0,0,1}(-1 \dotplus A, -1 \dotplus B, -1 \dotplus (A \bigtriangledown B))$, an open set which contains no matrix ordering. Similarly, for (P6), there exists a nonfull matrix $A \notin T$, so $f \in H_1(-1 \dotplus A)$; since $-1 \dotplus A$ is again nonfull, $f_P(-1 \dotplus A)$ must be zero for any matrix ordering P. □

As in the commutative case, we have the immediate corollary (cf. [**Cr**]) that for any field D, the space of orderings X_D is a Boolean space (compact, Hausdorff and totally disconnected) since the sets $H(A)$ and $H(-1 \dotplus A)$ are complements, hence both closed and open.

REFERENCES

[A] E. Artin, *Geometric algebra*, Interscience Publishers, New York, 1957.

[Co1] P. M. Cohn, *Free rings and their relations*, 2nd ed., Academic Press, London, 1985.

[Co2] P. M. Cohn, *Skew field constructions*, London Math. Soc. Lecture Notes No. 27, Cambridge Univ. Press, New York, 1977.

[Co3] P. M. Cohn, *Principles of non-commutative algebraic geometry*, Rings and Geometry (R. Kaya, et al., eds.), D. Reidel Publishing Co., Hingham, MA, 1985.

[Cr] T. Craven, *Witt rings and orderings of skew fields*, J. Algebra **77** (1982), 74–96.

[D] P. Draxl, *Skew fields*, London Math. Soc. Lecture Notes No. 81, Cambridge Univ. Press, New York, 1983.

[L] T. Y. Lam, *An introduction to real algebra*, Rocky Mountain J. Math. **14** (1984), 767–814.

[P] V. Powers, *Higher level orders on noncommutative rings*, J. Pure Appl. Algebra **67** (1990), 285–298.

[R] G. Révész, *Ordering epic R-fields*, Manuscripta Math. **44** (1983), 109–130.

DEPARTMENT OF MATHEMATICS, UNIVERSITY OF HAWAII, HONOLULU, HI 96822

E-mail address: tom@kahuna.math.hawaii.edu

Contemporary Mathematics
Volume **155**, 1994

Nonexistence of Analytically Varying Solutions to Hilbert's 17th Problem

CHARLES N. DELZELL

ABSTRACT. Let R be a real closed field, and K a subfield. If $f \in K[X] := K[X_1, \dots, X_n]$ is positive semidefinite (psd) over R, then $f = \sum p_i r_i^2$, with $0 \leq p_i \in K$ and $r_i \in K(X)$ (Artin, Kreisel, Henkin, A. Robinson). In the eighties we showed that (a) the p_i and the coefficients of the r_i could be chosen to be continuous $\mathbf{Z}$-R-piecewise-polynomial functions g of the coefficients of f, depending only on n and $d := \deg f$, but (b) the g cannot all be chosen to be polynomial functions (except when $d \leq 2$). Now we improve (b) by proving (1.3) that for even $d \geq 4$ and $K = R = \mathbf{R}$, the g cannot even be taken to be real analytic functions. For this, we first extend the transversal zeros theorem from the case of real polynomials [**CKLR**] to that of germs of real analytic functions (2.2). Then we use the latter to show that the set $P := \{ (a,b) \in \mathbf{R}^2 \mid X_1^4 + aX_1^2 + b \text{ is psd in } X_1 \}$ $(= \{ (a,b) \in \mathbf{R}^2 \mid 4b \geq a^2 \vee (a \geq 0 \wedge b \geq 0) \})$ is not a *basic* closed semianalytic set (specifically, for every open neighborhood V of $(0,0)$ and for every finite set $\{p_i\}$ of real analytic functions on V, $P \cap V \neq \{ (a,b) \in V \mid \forall i\ p_i(a,b) \geq 0 \}$); it had been known since 1979 that P was not basic semi*algebraic* [**D1**].

1. Introduction

Before explaining Hilbert's 17th problem, we recall a similar result which is easier to state: In 1921 Siegel [**Si**] generalized Lagrange's 4-square theorem by proving that if z is a totally positive element in a number field K (i.e., $z \geq 0$ in every ordering of K; equivalently, $z \in \sum K^2 :=$ {sums of squares in K}), then z is a sum of 4 squares in K. In [**D4**] we showed that if K is embedded in $\mathbf{C}$, then there exist functions g_1, g_2, g_3, g_4 analytic in $L := \mathbf{C} \setminus$ {the non-positive real axis} such that $z = g_1(z)^2 + \cdots + g_4(z)^2$ and $g_i(L \cap \sum K^2) \subseteq K$. Heilbronn [**He**] had

1991 *Mathematics Subject Classification.* 12D15 14P15 32B20.

The author was supported in part by the Humboldt Foundation, the Louisiana Board of Regents (Educational Quality Support Fund), and NSF.

This paper is submitted in final form. An abstract was included in *AMS Abstracts* **12**, Issue 73 (January 1991), #863-14-743, p. 47.

already proved the special case where $K = \mathbf{Q}$, providing an analytically varying version of Lagrange's theorem, and answering a question of Kreisel; our proof was a straightforward extension of Heilbronn's.

The purpose of the present paper is to show (1.3) that there is no such analytically varying version of the sum-of-squares representation in Hilbert's 17th problem [**Hi**], which we now explain. Let R be a real closed field, K be a subfield (with the inherited order), $X := (X_1, \dots, X_n)$ be indeterminates, and $f \in K[X]$.

QUESTION 1.1. *If f is positive semidefinite* ("psd") *over R (i.e., $\forall x \in R^n$, $f(x) \geq 0$), then does $f = \sum p_i r_i^2$, with $0 \leq p_i \in K$ and $r_i \in K(X)$?*

If $R = \mathbf{R}$ and $K = \mathbf{Q}$ or $\mathbf{R}$, then we can absorb the p_i into the r_i; the resulting question is Hilbert's 17th problem, answered affirmatively by Artin in 1926 [**Ar**]. The general question was answered affirmatively in approximately 1955 by Henkin, Kreisel, and A. Robinson, using model theory and proof theory (see [**D1**] for historical references). These logicians proved more, as we now explain using more notation:

Fix $d \in \mathbf{N}$; now let $f \in \mathbf{Z}[C; X]$ be the general polynomial of degree d in X with coefficients $C := (C_1, \dots, C_{\binom{n+d}{n}})$:

$$f = \sum_{|\alpha| \leq d} C_\alpha X^\alpha,$$

where $\alpha = (\alpha_1, \dots, \alpha_n) \in \mathbf{N}^n$, $|\alpha| = \sum \alpha_i$, $X^\alpha = X_1^{\alpha_1} \cdots X_n^{\alpha_n}$, and C_α is a re-indexing of the C's. Write $P_{nd} = \{ c \in R^{\binom{n+d}{n}} \mid f(c; X)$ is psd in X over $R \}$ = a closed, convex cone. They constructed finitely many $p_{ij} \in \mathbf{Z}[C]$ and $r_{ij} \in \mathbf{Q}(C; X)$ such that

$$\forall j \quad f = \sum_i p_{ij} r_{ij}^2 \text{ and} \tag{1.1.1}$$

$$\forall c \in P_{nd}\ \exists j\ \forall i\ p_{ij}(c) \geq 0 \text{ and } r_{ij}(c; X) \text{ is defined.} \tag{1.1.2}$$

Therefore the number and degrees of the r_i in (1.1) may be bounded by a function of n and d which is independent of c, answering a question of Artin; and the p_i and the coefficients of the r_i are "piecewise-polynomial"—in particular, piecewise-analytic—functions g of c.

For the purpose of giving an idea of the proof of the main theorem (1.3) below, we point out another consequence of their representation: Write $q_{ij} \in K[C]$ for the sum of the squares of the X-coefficients of the denominator of r_{ij}, $U_j = \{ c \in R^{\binom{n+d}{n}} \mid \forall i\ q_{ij}(c) > 0 \}$, and $W_j = \{ c \in R^{\binom{n+d}{n}} \mid \forall i\ p_{ij}(c) \geq 0 \}$; then

$$P_{nd} = \bigcup_j (U_j \cap W_j) \tag{1.1.3}$$

($\supseteq$ by (1.1.1) and $\subseteq$ by (1.1.2)).

A set such as U_j (respectively, W_j), which is the intersection of the positivity (resp., nonnegativity) sets of finitely many polynomials in $K[C]$, is called a *basic open* (resp., *basic closed*) *semialgebraic set*; a set such as $U_j \cap W_j$, which is the intersection of a basic open and a basic closed semialgebraic set, is called a *basic semialgebraic set*; finally, a set such as P_{nd}, which is the union of finitely many basic semialgebraic sets, is called a *semialgebraic set*.

For an open subset V of a real analytic manifold M, write $\mathcal{O}_V$ for the ring of real analytic functions on V. We say that $S \subseteq M$ is a *basic open* (respectively, *basic closed*), *semianalytic set* if for each $c \in M$ there are an open neighborhood V of c and $q_1, \ldots, q_{k_c} \in \mathcal{O}_V$ such that $S \cap V = \{ b \in V \mid \forall i \ q_i(b) > 0 \}$ (resp., $\{ b \in V \mid \forall i \ q_i(b) \geq 0 \}$). Similarly, we say that S is *basic semianalytic* if it is, locally, the intersection of a basic open and a basic closed semianalytic set, and that S is *semianalytic* if it is, locally, a finite union of basic semianalytic sets. Thus in case $K = R = \mathbf{R}$, the U_j, W_j, and $U_j \cap W_j$ above are basic open, basic closed, and basic semianalytic subsets of $M := \mathbf{R}^{\binom{n+d}{n}}$, respectively.

(The fact that P_{nd} is closed, mentioned before (1.1.1), is obscured by the representation (1.1.3); the finiteness theorem for closed semialgebraic sets (proved in approximately 1979, independently by Bochnak and Efroymson, Coste and Coste-Roy, Delzell, van den Dries, and Recio—see [**D3**] for references) provides a better representation, by asserting that every closed semialgebraic set is the union of finitely many basic closed semialgebraic sets; and Łojasiewicz [**Ł**, p. 98] had proved the analog for open semianalytic sets, which is equivalent to the statement for closed semianalytic sets (see also [**Mc**, Theorem 2], and [**BM**, (2.9)]).)

Since the piecewise-polynomial functions g in the logicians' solution to the 17th problem (mentioned after (1.1.2)) are usually not continuous, their solution was computationally inadequate when $K = R = \mathbf{R}$, where most elements can be presented only as approximations (by, say, rational numbers). Intuitionistic logic gives small changes in the logical laws which do ensure continuity of functions constructed, and for a wide range of topologies. After the contributions of model- and proof-theory in the 50's to the 17th problem, Kreisel [**K**] asked whether intuitionistic logic could also contribute, by determining whether there exists a continuous solution, i.e., one in which the g are continuous. So far, intuitionistic logic has contributed little, and real algebraic geometry much, as we now review.

QUESTION 1.2. *From what subrings of* $R^{P_{nd}} := \{ g : P_{nd} \to R \}$ *may we choose the* p_i *and the coefficients of the* r_i *in* (1.1)?

(a) On the positive side, we showed in [**D3**] that we may choose the g to be continuous semialgebraic (roughly, real algebraic) functions, and in 1988 we showed [**D5**] that they may even be chosen to be continuous piecewise-polynomial functions (even suprema of infima of finitely many integer polynomials). And for $d \leq 2$, we may even use (not merely piecewise-)polynomial functions, and avoid denominators in X [**D2**]. (b) On the negative side, we had shown in [**D1**] that for even $d \geq 4$, the g cannot all be chosen to be polynomial functions; equivalently, that j in (1.1.1) and (1.1.2) (or (1.1.3)) must in general range over an index set with more than 1 element, or, again equivalently, that P_{nd} is not a basic semialgebraic set. In the present paper we improve (b) by proving our

MAIN THEOREM 1.3. *For even $d \geq 4$ and $K = R = \mathbf{R}$, the p_i and the coefficients of the r_i in* (1.1) *cannot be taken to be real analytic functions defined on an open neighborhood of P_{nd}; i.e., there is no analytically varying solution to Hilbert's 17th problem.*

This follows (by the discussion after (1.1.3)) from the assertion that P_{nd} is not a basic semianalytic set. In fact, setting most $C_\alpha = 0$ or 1, we see that it suffices to show that the 2–dimensional cross-section

$$P := \{ (a,b) \in \mathbf{R}^2 \mid X_1^4 + aX_1^2 + b \text{ is psd} \}$$

is not basic semianalytic. It is an easy exercise using the quadratic formula to show that

$$P = \{ (a,b) \in \mathbf{R}^2 \mid 4b \geq a^2 \vee (a \geq 0 \wedge b \geq 0) \}$$

(see the Figure). Our task is now to show that for every open neighborhood $V \subseteq \mathbf{R}^2$ of $(0,0)$ and for every finite subset $\{p_i'\} \subset \mathcal{O}_V$,

$$P \cap V \neq \{ (a,b) \in V \mid \forall i \; p_i'(a,b) \geq 0 \}. \qquad (1.3.1)$$

We shall prove this in §3, after proving the transversal zeros theorem for germs of real analytic functions in §2. Finally, in §4, we review some of the existing literature about and around these topics.

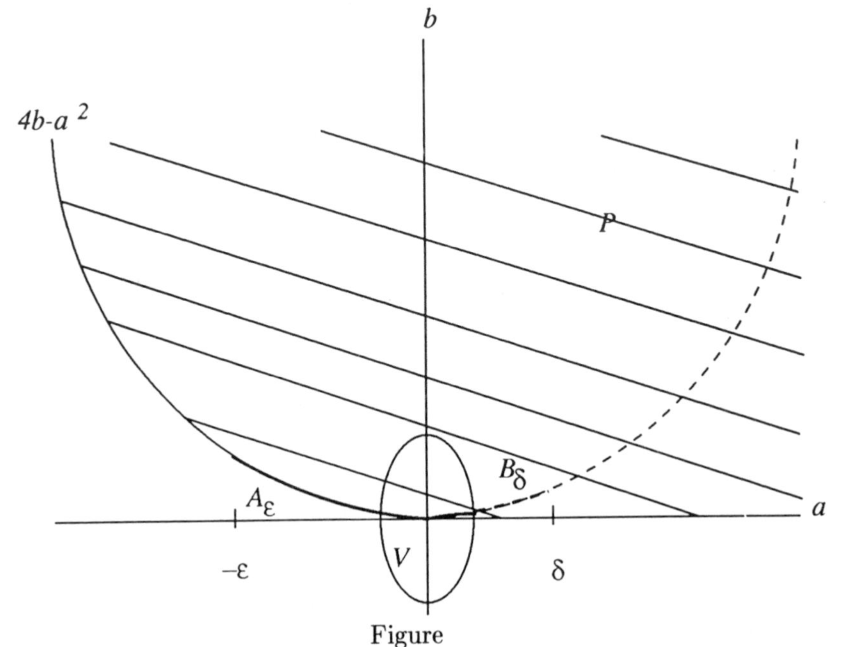

Figure

I am grateful to William Adkins, Carlos Andradas, Manfred Knebusch, James Madden, Jesús Ruiz, and Gilbert Stengle for helpful conversations on this subject. I thank Alexander Prestel for his hospitality in Konstanz, where most of this work was done in 1990. I presented these results in the Special Year in Real Algebraic Geometry and Quadratic Forms at UC Berkeley in 1990, and at the Special Session on Real Algebraic Geometry at the AMS meeting in San Francisco in 1991; I am grateful to Tsit-Yuen Lam and Robert Robson for organizing the former, and to William Jacob for organizing the latter.

2. The transversal zeros theorem for germs of real analytic functions

We begin this section by explaining the notions of—local—dimension and transversal zeros.

Again let V be an open subset of a real analytic manifold M. Let $m = \dim M$. For $q \in \mathcal{O}_V$ write $Z(q) = \{ y \in V \mid q(y) = 0 \}$. For $c \in M$, write $\mathcal{O}_c$ for the local UFD of germs at c of real analytic functions on M. For $\mathbf{q} \in \mathcal{O}_c$ write $\mathbf{Z}(\mathbf{q})$ for the germ of $Z(q)$ at c, where (q, U) is any representative of $\mathbf{q}$.

Let S be a semianalytic set in M, with germ $\mathbf{S}$ at c. There are two basic approaches to defining $\dim S$ and $\dim \mathbf{S}$ ($= \dim_c S$, the local dimension of S at a point $c \in S$). (1) The first approach is algebraic, and begins by defining $\dim \mathbf{S}$ to be the dimension of the smallest real analytic set germ $\mathbf{N}$ in M at c containing $\mathbf{S}$ (e.g., [**Mc**, §3] or [**Ru2**, (2.5)]). The latter is defined as the maximum of the dimensions of the irreducible components $\mathbf{K}$ of $\mathbf{N}$ (e.g., [**N**, Definition III.1.3]). The dimension of such a $\mathbf{K}$, in turn, can be defined in two ways, both algebraic: (a) as the Krull dimension of the prime ideal P of functions in $\mathcal{O}_c$ which vanish on $\mathbf{K}$ (= the Krull dimension of the ring $\mathcal{O}_c/P$) (e.g., [**Mc**, §3]), or (b) the number of distinguished coordinates in any "regular system of coordinates for P" (e.g., [**N**, Definition III.1.3]; [**Ru2**] gives a similar definition in his (1.3)). It is well-known that (a) and (b) are equivalent, say by the going-up theorem combined with the fact ([**T**, Theorem II.1.5]) that the Krull dimension of $\mathcal{O}_c$ is m. ([**Ru2**] proves a related result, (4.4).) Once $\dim \mathbf{S}$ ($= \dim_c S$) has been defined, one could go on to define $\dim S = \max\{ \dim_c S \mid c \in S \}$ (as Narasimhan did for real or complex analytic, but not semianalytic, sets; McEnerney, on the other hand, dealt only with *germs* of sets and functions). (In the semialgebraic situation, where S is a semialgebraic subset of a (real) algebraic variety, [**DK**, §13, Definition 1] and [**CKLR**, §1] went in the opposite order: they defined

$$\dim_c S = \min\{ \dim(S \cap V) \mid V \text{ is an open semialgebraic neighborhood of } c \}, \tag{2.1.1}$$

after first defining $\dim(S \cap V)$ to be the dimension of the Zariski closure of $S \cap V$.)

(2) The second approach, followed by [**Ł**] and [**BM**, 2.12], is "topological" or "geometric." They defined global rather than local dimension, by constructing a "locally finite semianalytic stratification $\bigcup A_k$ of M compatible with S"; they then defined $\dim S = \max\{ \dim A_k \mid A_k \subseteq S \}$, where $\dim A_k$ is the familiar dimension of a real analytic manifold; they showed that this is independent of the choice of stratification. Using this, one could go on to define $\dim_c S = \min\{ \dim(S \cap V) \mid V$ is an open semianalytic neighborhood of $c \}$, by analogy with (2.1.1).

The two definitions, (1) and (2), agree, using (a) Łojasiewicz's theorem [**BM**, Theorem 2.13] that the dimension (according to (2)) of S equals the dimension (also according to (2)) of the smallest real analytic set containing S, and (b) Proposition III.1.6 of [**N**].

All we shall need to know about dimension in this paper is (a) that $\dim S$ is the smallest k such that S contains a homeomorphic copy of a nonempty open subset of $\mathbf{R}^k$ (by definition (2)), (b) the consequence that if $M = \mathbf{R}^m$, V is open semianalytic, and $\pi : V \to \mathbf{R}^{m-1}$ is the projection, then $\dim(S \cap V) \geq \dim \pi(S \cap V)$, (c) the further consequence (using the identity theorem for real analytic functions, whose statement and proof are obvious analogs of the identity theorem for complex analytic functions, (I.A.6) of [**GR**]) that $\mathbf{0} \neq \mathbf{q} \in \mathcal{O}_c \Rightarrow \dim \mathbf{Z}(\mathbf{q}) < m$, and (d) for $\mathbf{p}$ and $\mathbf{q}$ relatively prime in $\mathcal{O}_c$, $\dim[\mathbf{Z}(\mathbf{p}) \cap \mathbf{Z}(\mathbf{q})] < m-1$, where $m = \dim M$ (by Definition 1(a)).

Like [**DK**, §13] and [**CKLR**, (1.2)] in the semialgebraic case, we call S *pure of dimension* p if for all $c \in S$, $\dim_c S = p$ (compare [**GR**, (III.C.10) and (V.B)], who define "pure" in the complex analytic case).

We say that $q \in \mathcal{O}_V$ *changes sign* or *is indefinite* on a subset U of V if $\exists y, z \in U$ such that $q(y)q(z) < 0$. Write $Z_t(q) = \{ y \in V \mid q$ changes sign in every neighborhood of y in $V\}$ ($\subseteq Z(q)$); the points of $Z_t(q)$ are called the *transversal zeros* of q. Since $Z_t(q) = \overline{\{c \in V \mid q(c) > 0\}} \cap \overline{\{c \in V \mid q(c) < 0\}}$, $Z_t(q)$ is semianalytic (using the fact, first proved in [**Ł**, p. 76], that the closure of a semianalytic set is semianalytic).

A set germ (at c) is called *semianalytic* if one of its representatives is, and a semianalytic set germ $\mathbf{S}$ is called *basic open*, *basic closed*, or *basic* if one of its representatives is. For a point y in a representative S of $\mathbf{S}$, the *local dimension* $\dim_y \mathbf{S}$ of $\mathbf{S}$ at y is defined to be $\dim_y S$ (this is independent of the representative S containing y). $\mathbf{S}$ is called *pure of dimension* p if one of its representatives is. For $\mathbf{q} \in \mathcal{O}_c$ write $\mathbf{Z}_t(\mathbf{q})$ (the germ of *transversal zeros* of $\mathbf{q}$), for the germ of $Z_t(q)$, where (q, V) is any representative of $\mathbf{q}$; the choice of (q, V) is irrelevant here. (For example,—the germ (at any point) of—$Y_1^2 + Y_2^2$ has no transversal zeros.) By the previous paragraph, $\mathbf{Z}_t(\mathbf{q})$ is semianalytic.

TRANSVERSAL ZEROS THEOREM FOR $\mathcal{O}_c$ (2.2).

(1) $\mathbf{Z}_t(\mathbf{q})$ *is either empty or pure of dimension* $m-1$,

(2) $\emptyset \neq \mathbf{Z}_t(\mathbf{q}) \subseteq \mathbf{Z}(\mathbf{s}) \Rightarrow \mathbf{q}$ *and* $\mathbf{s}$ *have a common non-unit factor in* $\mathcal{O}_c$, *and*

(3) *if* $\mathbf{q} = \mathbf{t}_1^{e_1} \cdots \mathbf{t}_l^{e_l}$ *is a(n essentially unique) factorization of* $\mathbf{q}$ *into powers of pairwise-nonassociate irreducible elements* $\mathbf{t}_j \in \mathcal{O}_c$ $(e_j \geq 1)$, *then* $\mathbf{Z}_t(\mathbf{q}) = \cup\{ \mathbf{Z}_t(\mathbf{t}_j) \mid e_j$ *is odd*$\} = \bigcup_j \mathbf{Z}_t(\mathbf{t}_j^{e_j})$.

PROOF. Since the theorem is local, it is no trouble to reduce to the case that $M = \mathbf{R}^m$ (this facilitates our use of projections below). Pick a representative (q, V) of $\mathbf{q}$ with V connected.

(1) (Simplified by William Adkins.) Suppose $\mathbf{Z}_t(\mathbf{q})$ is nonempty, and let $y \in Z_t(q)$. Then $q \neq 0$. Since V is connected, the identity theorem implies

$m > \dim_y Z(q) \geq \dim_y Z_t(q)$. It remains to show that $\dim_y Z_t(q) \geq m-1$. Let B be a neighborhood of y in $\mathbf{R}^m$, which we may (without loss of generality) assume is convex; we must show that $\dim(B \cap Z_t(q)) \geq m-1$. Since $y \in Z_t(q)$, there exist $z, w \in B$ such that $q(z)q(w) < 0$. Choose coordinates $Y_1, \ldots, Y_m$ so that the Y_m-axis is parallel to the line through z and w. Let $\pi : B \to \mathbf{R}^{m-1}$ be projection. (At this point **[CKLR]** used a "central" projection (2.3), which, unfortunately, was not well-defined; however, this mistake could be easily corrected.) Therefore there is a neighborhood $V' \subseteq \pi(B)$ of $\pi(z)$ $(= \pi(w))$ such that for all $v' \in V'$, $q(v', Y_m)$, regarded as a function of Y_m, changes sign on $\pi^{-1}(v')$ (using the convexity of B); i.e., such that for all $v' \in V$, $\pi^{-1}(v') \cap Z_t(q) \neq \emptyset$. Thus $\pi(B \cap Z_t(q)) \supseteq V'$. Hence $m-1 = \dim \pi(B \cap Z_t(q)) \leq \dim(B \cap Z_t(q))$.

(2) If $\mathbf{q}$ and $\mathbf{s}$ were relatively prime in $\mathcal{O}_c$, then by Lemma II.E.17 of **[GR]**, we could choose coordinates Y and construct $\mathbf{t}$ and $\mathbf{v} \in \mathcal{O}_c$ such that $\mathbf{g} := \mathbf{tq} + \mathbf{vs}$ is a nonzero element of $\mathcal{O}'_{c'} :=$ the local UFD of germs at $c' := (c_1, \ldots, c_{m-1})$ of real analytic functions in the variables $Y' := (Y_1, \ldots, Y_{m-1})$ (actually, one needs to change $\mathbf{C}$ to $\mathbf{R}$ throughout—the proof of—that lemma; this is no problem). By (1), we can further adjust coordinates so that, in addition, $\pi(Z_t(q))$ contains a nonempty open subset of $\mathbf{R}^{m-1}$ (where $\pi : \mathbf{R}^m \to \mathbf{R}^{m-1}$ is projection). Since $\mathbf{g} \in \mathcal{O}'_{c'}$ and $\mathbf{s} = \mathbf{0}$ on $\mathbf{Z}_t(\mathbf{q})$, $\mathbf{g} = \mathbf{0} \in \mathcal{O}'_{c'}$, contradiction.

(3) To prove the last equation, note that if e_j is even, then $\mathbf{Z}_t(\mathbf{q}_j^{e_j}) = \emptyset$. We prove the earlier equation in 2 steps: ($\subseteq$): This direction would hold even if the $\mathbf{t}_j$ were reducible, for $\mathbf{q}$ cannot change sign unless at least 1 of the $\mathbf{t}_j^{e_j}$'s does, too, and for such a j, e_j must be odd. ($\supseteq$): We may assume that e_1 is odd. It suffices to show $\mathbf{Z}_t(\mathbf{q}) \supseteq \mathbf{Z}_t(\mathbf{t}_1)$. Let V be a neighborhood of c within which some representatives q and t_k of $\mathbf{q}$ and all the $\mathbf{t}_k$ are defined. We must show that $V \cap Z_t(q) \supseteq V \cap Z_t(t_1)$. We claim that the (Zariski open) subset $U := V \cap Z_t(t_1) \setminus [V \cap Z(t_1) \cap Z(t_2^{e_2} \cdots t_l^{e_l})]$ of $V \cap Z_t(t_1)$ is dense. Indeed, assume (as we may) that $V \cap Z_t(t_1) \neq \emptyset$; pick $y \in V \cap Z_t(t_1)$ and pick a neighborhood N of y in M. Since $\dim[V \cap Z(t_1) \cap Z(t_2^{e_2} \cdots t_l^{e_l})] < m-1 = \dim[V \cap Z_t(t_1) \cap N]$ (by (2.2.1)), $V \cap Z_t(t_1) \cap N$ meets U, proving the claim. Since $Z_t(q)$ is closed in V, it therefore suffices to show $U \subseteq Z_t(q)$. So pick $y \in U$. In some neighborhood of y, $t_2^{e_2} \cdots t_l^{e_l}$ is strictly definite (either positive or negative), and t_1 (hence also $t_1^{e_1}$) is indefinite, whence q is indefinite there; i.e., $y \in Z_t(q)$. □

3. Conclusion of the proof of the main theorem

We now return to the proof of (1.3.1). Let $M = \mathbf{R}^2$. For $\varepsilon, \delta > 0$, write $A_\varepsilon = \{ (a, a^2/4) \in \mathbf{R}^2 \mid -\varepsilon < a < 0 \}$ and $B_\delta = \{ (a, a^2/4) \in \mathbf{R}^2 \mid 0 \leq a < \delta \}$ (see the Figure). Write $\mathbf{A}$ and $\mathbf{B}$ for the germs of A_1 and B_1 at $(0,0)$, respectively.

LEMMA 3.1. *For a finite subset* $\{\mathbf{q}_i\} \subset \mathcal{O}_{(0,0)}$, *if* $\mathbf{A} \subseteq \bigcup_i \mathbf{Z}_t(\mathbf{q}_i)$, *then for some* i, $\mathbf{A} \subseteq \mathbf{Z}_t(\mathbf{q}_i)$.

PROOF. Since each A_ε is infinite and $\{\mathbf{q}_i\}$ is finite, there exists an i (which we now fix), a representative (q_i, U) of $\mathbf{q}_i$, and a sequence $z_1, z_2, \ldots$ of negative reals which converges to 0, such that for all l, $(z_l, z_l^2/4) \in Z_t(q_i)$. Since $Z_t(q_i)$ is semianalytic, there exist finitely many g_j, h_{jk} analytic (and not identically 0) on a neighborhood V of $(0,0)$ such that $Z_t(q_i) \cap V = \{ (a,b) \in V \mid \exists j \ [g_j(a,b) = 0$ and $\forall k \ h_{jk}(a,b) > 0] \}$. Since j ranges over a finite set, there exist j and infinitely many l such that $g_j(z_l, z_l^2/4) = 0$ and for all k, $h_{jk}(z_l, z_l^2/4) > 0$. Since a (real) analytic function of 1 variable can vanish infinitely often on a bounded proper subinterval of its domain only if it is identically 0, we conclude that for this j and for some $\varepsilon > 0$, $g_j = 0$ and $\forall k$, $h_{jk} > 0$ throughout A_ε. Therefore $\mathbf{A} \subseteq \mathbf{Z}_t(\mathbf{q}_i)$. $\square$

To prove (1.3.1), note that if it were false, then $\mathbf{A} \subseteq \bigcup_i \mathbf{Z}_t(\mathbf{p}'_i)$. By (3.1), there exists i (which we fix) such that $\mathbf{A} \subseteq \mathbf{Z}_t(\mathbf{p}'_i)$. Let $\mathbf{p}'_i = \mathbf{t}_1^{e_1} \cdots \mathbf{t}_l^{e_l}$ be a factorization of $\mathbf{p}'_i$. By (2.2.3), $\mathbf{A} \subseteq \cup\{ \mathbf{Z}_t(\mathbf{t}_j) \mid e_j \text{ is odd} \}$. By (3.1) again, there exists j (which we fix) such that e_j is odd and $\mathbf{A} \subseteq \mathbf{Z}_t(\mathbf{t}_j) \subseteq \mathbf{Z}(\mathbf{t}_j)$. Thus $t_j(z, z^2/4) = 0$ on some interval $(-\varepsilon, 0)$, hence also on some interval $[0, \delta)$, by the identity theorem. I.e., $\mathbf{Z}(\mathbf{t}_j) \supseteq \mathbf{Z}(\mathbf{a}^2 - 4\mathbf{b}) = \mathbf{Z}_t(\mathbf{a}^2 - 4\mathbf{b})$. By (2.2.2), $\mathbf{t}_j$ is an associate of $\mathbf{a}^2 - 4\mathbf{b}$. Therefore $\mathbf{Z}_t(\mathbf{t}_j) = \mathbf{Z}(\mathbf{t}_j)$, so $\mathbf{B} \subseteq \mathbf{Z}_t(\mathbf{t}_j) \subseteq \mathbf{Z}_t(\mathbf{p}'_i)$, proving (1.3.1), after all. $\square$

Returning to (1.2), we ask, "Can the p_i and the coefficients of the r_i be chosen to be C^∞ functions, again at least for $K = R = \mathbf{R}$?"

4. Notes on the literature

The reduction of (1.3) to (1.3.1) was the same as the 1979 reduction [**D1**] of the algebraic version of (1.3) to the statement that P is not basic semi*algebraic*. The latter fact is best proved by the transversal zeros theorem for polynomials, of which an excellent exposition was given by Choi, Knebusch, T.-Y. Lam, and Reznick [**CKLR**]. In 1979, however, this theorem was still folklore to real algebraic geometers, so the original proof of the fact that P is not basic semialgebraic was based, instead, on the somewhat less informative "sign-changing" theorem for polynomials (2.7) in Dubois and Efroymson [**DE**]: an irreducible polynomial changes sign iff it generates a real ideal.

The present situation is somewhat similar. We thought that these proceedings would be a good occasion to publish an account of (2.2), not just because we needed it for (1.3.1), but because it was folklore. For example, a slightly weaker form of (2.2.2) had been proved by Risler in his Proposition 1 [**Ri1**] (= Proposition 3 [**Ri2**] and Proposition 4.2 [**Ri3**]): for sign-changing $\mathbf{q}$, irreducible in $\mathcal{O}_c$, and for $\mathbf{s} \in \mathcal{O}_c$, if $Z(\mathbf{q}) \subseteq \mathbf{Z}(\mathbf{s})$, then $\mathbf{s} \in (\mathbf{q})$. Adkins [**Ad**] had proved a global theorem (4.6) related to (2.2) and the Dubois-Efroymson sign-changing theorem: for a connected real analytic manifold M with $H^1(M, \mathbf{Z}_2) = 0$, and for an irreducible $q \in \mathcal{O}_M$, if q changes sign, then q generates a real ideal in $\mathcal{O}_M$ (and conversely if $\dim M = 2$ (4.7)). And Ruiz [**Ru2**] gave an exposition of many topics in the general area of (2.2).

C. Andradas and J. Ruiz have pointed out to me another approach to (1.3.1).

Define the subset

$$\sigma_1 = \left\{ f \in \mathcal{O}_V \;\middle|\; \text{either } \begin{array}{l} \text{(a) } f(0,0) > 0, \text{ (b) } f(0,0) = 0 \text{ and } f > 0 \\ \text{on some } B_\delta, \text{ or (c) } f = 0 \text{ on some } B_\delta, \text{ and} \\ f \geq 0 \text{ just above } B_\delta \end{array} \right\},$$

where B_δ is as before (3.1). In a similar way we get subsets $\sigma_2, \sigma_3, \sigma_4$, defined by replacing "$B_\delta$" with "$A_\varepsilon$," and/or "above" with "below." Each σ_i extends to an ordering $\geq_i$ of the field $\mathcal{M}_V$ of meromorphic functions on V, by declaring $f/g \geq_i 0$ iff $fg \in \sigma_i$. The 4 orderings $\geq_1, \geq_2, \geq_3, \geq_4$ constitute a *fan*, i.e., a set of 4 orderings on a field such that the "product" of any 3 of them is the fourth; to see this, one needs (2.2.3), or something like it. Now it is obvious that any (Harrison) "basic" (cl)open (constructible) set of orderings of a field (such as $\mathcal{M}_V$) defined by finitely many inequalities (such as the $p_i' \geq 0$ in (1.3.1)) cannot contain three elements of a fan (such as $\geq_1, \ldots, \geq_4$) without containing the fourth; thus the abstract set $\widetilde{P}$ of orderings which corresponds to the concrete semianalytic set P cannot be basic. Finally, one concludes that P itself is not basic, by the semianalytic version of the Artin-Lang correspondence (cf. [**Ru1**, (4.1) and (4.2)], or [**Ru3**]).

One last remark on the literature. The 1979 result that P is not basic semialgebraic, and the present result (1.3.1), are "qualitative," and rely on ideas going back at least to 1970 (namely, (2.4) of Dubois and Efroymson [**DE**], supplemented, in the analytic case, by the Weierstrass preparation theorem). Around 1983 L. Bröcker initiated the study of *quantitative* aspects of basic semialgebraic sets. The main results, found by Bröcker and Scheiderer in 1988 [**B**], are that basic open and basic closed semialgebraic sets can be described by n strict, and $n(n+1)/2$ nonstrict inequalities, respectively (and these estimates are sharp); as to the number of basic sets required to represent an arbitrary semialgebraic set, the known estimates are less satisfactory. Shortly thereafter, Andradas, Bröcker, and Ruiz showed that the situation is the same in the analytic case. For recent accounts, see [**ABR**], [**B**], [**Ma**], and [**Sch**].

References

[Ad] W. A. Adkins, *A real analytic nullstellensatz for two dimensional manifolds*, Boll. U.M.I. (5) **14-B** (1977), 888–903; MR **58** (1979), 6323; Zbl. **444** (1981), 14019.

[ABR] C. Andradas, L. Bröcker, and J. Ruiz, *Minimal generation of basic open semianalytic sets*, Invent. Math. **92** (1988), 409–430; MR **89f**, 32016; Zbl. **655**, 32011.

[Ar] E. Artin, *Über die Zerlegung definiter Funktionen in Quadrate*, Abh. Math. Sem. Hamburg **5** (1927), 100–115.

[BM] E. Bierstone and P. D. Milman, *Semianalytic and subanalytic sets*, Publ. Math. I.H.E.S. **67** (1988), 5–42; MR **98k**, 32011; Zbl. **674** (1990), 32002.

[B] L. Bröcker, *On basic semi-algebraic sets*, Expos. Math. **9** (1991), 289–334.

[CKLR] M.-D. Choi, M. Knebusch, T.-Y. Lam, and B. Reznick, *Transversal zeros and positive semidefinite forms*, Géométrie Algébrique Réelle et Formes Quadratiques (J.-L. Colliot-Thélène, M. Coste, L. Mahé, and M.-F. Roy, eds.), Proceedings, Rennes 1981, Lect. Notes in Math., vol. 959, Springer, 1982, pp. 273–298; MR **84b**, 10027; Zbl. **506** (1983), 10019.

[DK] H. Delfs and M. Knebusch, *Semialgebraic topology over a real closed field II: Basic theory of semialgebraic spaces*, Math. Z. **178** (1981), 175–213; MR **82m**, 14011; Zbl. **461** (1982), 14005.

[D1] C. N. Delzell, *Case distinctions are necessary for representing polynomials as sums of squares*, Proc. Herbrand Symp., Logic Coll., 1981 (J. Stern, ed.), North Holland, 1982, pp. 87–103; MR **86i**, 11015; Zbl. **502** (1983), 03032.

[D2] ______, *Continuous sums of squares of forms*, L. E. J. Brouwer Cent. Symp. (A. S. Troelstra and D. van Dalen, eds.), North Holland, 1982, pp. 65–75; MR **85g**, 03086; Zbl. **527** (1984), 10017.

[D3] ______, *A continuous, constructive solution to Hilbert's 17th problem*, Invent. Math. **76** (1984), 365–384; MR **86e**, 12003; Zbl. **547** (1985), 12017; also reviewed by I. Stewart, *The power of positive thinking*, Nature **315** (1985), 539; abstract in AMS Abstracts **2**(1) (1981), #783-12-28.

[D4] ______, *Analytic right-inverses for quadratic forms over number fields*, Bull. London Math. Soc. **17** (1985), 449–452; MR **87b**, 11029; Zbl. **595** (1987), 10014; abstract in AMS Abstracts **3**(2) (1982), #792-12-269, under title *Analytic version of Siegel's theorem on sums of squares.*

[D5] ______, *Continuous, piecewise-polynomial functions which solve Hilbert's 17th problem*, J. f. reine u. angew. Math. (1993), abstract in AMS Abstracts **10**(3), Issue 63 (1989), #849-14-160, 208–209, under title *A sup-inf-polynomially varying solution to Hilbert's 17th problem.*

[DE] D. W. Dubois and G. Efroymson, *Algebraic theory of real varieties. I.*, Studies and Essays Presented to Yu-Why Chen on his 60th Birthday, Taiwan Univ., 1970, pp. 107–135; MR **43**(5) (1972), 6203; Zbl. **216**(1) (1971), 54.

[GR] R. C. Gunning and H. Rossi, *Analytic Functions of Several Complex Variables*, Prentice-Hall, Englewood Cliffs, NJ, 1965; MR **31**(5) (1966), 4927; Zbl. **141**(1) (1968), 86–87.

[He] H. Heilbronn, *On the representation of a rational as a sum of four squares by means of regular functions*, Bull. London Math. Soc. **39** (1964), 72–76; MR **28** (1964), 3003; Zbl. **131** (1967), 19–20.

[Hi] D. Hilbert, *Mathematische Probleme*, Göttinger Nachrichten (1900), 253–297; Archiv der Math. u. Physik (3rd ser.) **1** (1901), 44–53, 213–237; trans. by M. W. Newson, Bull. Amer. Math. Soc. **8** (1902), 437–479, reprinted in *Mathematical developments arising from Hilbert problems*, Proc. Symp. in Pure Math., F. Browder (ed.), **28**, AMS (1976), 1–34; Zbl. **326** (1977), 00002; MR **54** (1977), 7158.

[K] G. Kreisel, *Review of Goodstein*, #A1821, MR **24A** (1962), 336–337.

[Ł] S. Łojasiewicz, *Ensembles semi-analytiques*, Lecture Note #A66.765, I.H.E.S., 1965.

[Ma] M. Marshall, *Minimal generation of constructible sets in the real spectrum of a ring* (preprint).

[Mc] J. McEnerney, *Trim stratification of semianalytic sets*, manuscr. math. **25** (1978), 17–46; MR **58** (1979), 11487; Zbl. **402** (1979), 32005.

[N] R. Narasimhan, *Introduction to the Theory of Analytic Spaces*, Lect. Notes Math., vol. 25, Springer, 1966; MR **36**(1) (1968), 428; Zbl. **168**(1) (1969), 60–61.

[Ri1] J.-J. Risler, *Un théorème des zéros en géométrie analytique réelle*, C.R. Acad. Sc. Paris (Series A & B) **274** (1972), 1488–1490; MR **45** (1973), 3754; Zbl. **236** (1972), 14001.

[Ri2] ______, *Un théoreme des zéros en géométries algébrique et analytique réelles*, Lect. Notes Math, vol. 409, Fonctions de Plusieurs Variables Complexes: Seminaire François Norguet, Octobre 1970–Decembre 1973, Springer, 1974, pp. 522–531; MR **51**(1) (1976), 489; Zbl. **296** (1975), 14014.

[Ri3] ______, *Le théorème des zéros en géométries algébrique et analytique réelles*, Bull. Soc. math. France **104** (1976), 113–127; MR **54** (1977), 5226; Zbl. **328** (1977), 14001.

[Ru1] J. M. Ruiz, *Central orderings in fields of real meromorphic function germs*, manuscr. math. **46** (1984), 193–214; MR **85c**, 58011; Zbl. **538** (1985), 14018.

[Ru2] ______, *Basic properties of real analytic and semianalytic germs*, Publ. Inst. Recherche Math. Rennes, Fasc. 4: Algèbre, 1986, pp. 29–51; MR **89h**, 32016; Zbl. **634**

(1988), 32007.

[Ru3] ———, *The Basic Theory of Power Series*, Vieweg, 1993.

[Sch] C. Scheiderer, *Stability index of real varieties*, Invent. Math. **97** (1989), 467–483; MR **90g**, 14011; Zbl. **715** (1991), 14049.

[Si] C. L. Siegel, *Darstellung total positiver Zahlen durch Quadrate*, M. Zeit. **11** (1921), 246–275; Ges. Abh. **1** (1966), 47–76; MR **33** (1967), 5441; Zbl. **143** (1968), 1.

[T] J. C. Tougeron, *Idéaux de fonctions différentiables*, Ergeb. Math. u. I. Grenzgebiete, vol. 71, Springer, 1972; MR **55**(6) (1973), 13472; Zbl. **251** (1973), 58001.

DEPARTMENT OF MATHEMATICS, LOUISIANA STATE UNIVERSITY, BATON ROUGE, LOUISIANA 70803–4918

E-mail address: mmdelz@lsuvax.sncc.lsu.edu

Contemporary Mathematics
Volume **155**, 1994

A Combinatorial Geometric Structure on the Space of Orders of a Field II.

M. A. DICKMANN

1. Introduction. Summary of Results

In the extensive literature on formally real fields the set $\chi(K)$ of all total orders on a formally real (abbreviated, f.r.) field K has hitherto been considered as a topological space, endowed with the so-called *Harrison topology* generated by the family of sets

$$H_K(a_1, \ldots, a_n) = \{P \in \chi(K) \,|\, a_1, \ldots, a_n \in P\}\ ,$$

for all finite sequences $a_1, \ldots, a_n \in K$, as a base of open (in fact, clopen) sets. (N.B.: We identify a total order $\leq$ on a field K with its *positive cone*: $P = \{x \in K \,|\, x \geq 0\}$.)

The purpose of this paper is to introduce another—combinatorial—way of looking at $\chi(K)$ and develop at some length the ensuing theory. This yields new (and, we hope, interesting) results, but a part of our task consists also in recasting in terms of the new concepts a part of the existing theory; the most fruitful aspects of this reinterpretation lie in the link with the combinatorial theory of quadratic forms, developed essentially by Marshall and Bröcker (see [**11**], [**12**], [**13**], [**14**], [**4**], [**5**]).

An informal notion of an "independent" set of orders has been around for some time, especially in some papers by Bröcker, see [**4**, p. 149]. However, this notion never was used as any more than a terminology. It never was the object of systematic investigation using the concepts and tools of the theory where it belongs, namely the theory of matroids. The idea of checking whether this vague notion of independence—or rather the corresponding notion of closure—did satisfy the matroid axioms occurred to me in November 1988. The immediate positive answer provided the impetus to continue. In a few months I obtained most of the results of Part I, to be published elsewhere. The link with the

---

1991 *Mathematics Subject Classification.* 05B35 51D20 12D15 12J15 12J25.

This paper is submitted in final form and no version of it will be submitted for publication elsewhere

work of Marshall which constitutes the bulk of the present Part II came later, in December 1989, after a conversation with E. Becker.

This Introduction is intended to provide a summary of the basic notions introduced and principal results proved in Part I—some of which are used here—and to convey the intuition behind them.

NOTATION. We adhere to standard notation used in both matroid theory—cf. White [**19**], [**20**], Welsh [**18**] or Aigner [**1**]—and the theories of ordering spaces and valuations, as in Lam [**10**]. We only depart from the latter on the following point: consistent use of the field operations of $\mathbf{Z}_2$ and of additive notation for valuations forces a change in the usual, multiplicative notation for the sign function in favor of additive notation. Thus, we will write:

$$\mathrm{sgn}_P(a) = \begin{cases} 0 & \text{if } a \in P \\ 1 & \text{if } a \notin P \end{cases}$$

for $P \in \chi(K)$ and $a \in K$. Consequently, $\mathrm{sgn}_P(ab) = \mathrm{sgn}_P(a) + \mathrm{sgn}_P(b)$. Since quadratic form signatures are not used in this paper, this convention does not involve risks of clash with other notation used in [**10**].

We begin by defining an operator $c\ell_K$ on subsets of the set $\chi(K)$.

DEFINITION 1.1.1. Let K be a f.r. field. For $\emptyset \neq \mathcal{X} \subseteq \chi(K)$ and $P \in \chi(K)$,

$P \in c\ell_K(\mathcal{X})$ iff there is a finite subset $\mathcal{X}' \subseteq \mathcal{X}$ so that $\bigcap \mathcal{X}' \subseteq P$, and

$c\ell_K(\emptyset) = \emptyset$.

Note that $\bigcap \mathcal{X}' \subseteq P$ simply means that any element of K positive under each order in $\mathcal{X}'$ is also positive in P. We write $c\ell$ instead of $c\ell_K$ when no confusion is possible.

THEOREM 1.1.2.

(a) *The operator $c\ell_K$ satisfies the axioms of a closure operator defining a combinatorial geometry on $\chi(K)$* (cf. White [**19**, p. 300]). *Further, $c\ell_K$ verifies:*
(b) $c\ell(\{P, Q\}) = \{P, Q\}$, *i.e., "lines consist of two points."*
(c) *"A point in the closure of a set is in the closure of some finite subset."*
(d) (*Local finiteness*) *"The closure of a finite set is finite."*

The last two finiteness conditions are particularly important when $\chi(K)$ is infinite, a case which by no means we want to exclude. Moreover, it is easily checked that the closure of an n-element set has cardinality $\leq 2^{n-1}$. The possible cardinalities that the closure of a finite set may take on have been determined, as a function of its rank, by Bröcker [**5**].

Next, a well-known theorem of Bröcker (cf. Lam [**9**, Thm. 10.5]) is used to characterize the *free matroids* (i.e. those whose closure operator is trivial) of type $\chi(K)$ in terms of the field K:

THEOREM 1.1.8. *$\chi(K)$ is a free matroid iff the field K has the strong approximation property* (SAP).

With these basic facts established, the first question which comes to mind is: in which of the best known classes of matroids considered by combinatorists do the geometries $\chi(K)$ lie? We give a very satisfactory answer to this question in §1.2:

THEOREM 1.2.2. *For every* (f.r.) *field K, the matroid $\chi(K)$ is binary.*

Here $\chi(K)$ *may be infinite* (in which case the coordinatizing vector space over $\mathbf{Z}_2$ is infinite-dimensional). This result is proved in two steps:
—First, if $M = \chi(K)$ is finite (or, more generally, M is a finite flat of $\chi(K)$), we use the following well-known criterion (cf. White [**20**, Thm. 2.2.1(8)]): a (finite) matroid is binary iff no coline (= flat of corank 2) is contained in four or more hyperplanes (= flats of corank 1). The verification of this criterion is carried out by means of "dual basis" elements.
—Second, in order to pass from the finite to the infinite case, we use an ultraproduct construction to put together the binary coordinatizations of finite flats of $\chi(K)$ given by the previous step.

The existence of an obvious map

$$\begin{aligned} \chi(K) &\longrightarrow \operatorname{Hom}(K^\times/\Sigma K^{\times^2}, \mathbf{Z}_2) \\ P &\longrightarrow \operatorname{sgn}_P \end{aligned}$$

raises the question whether this is a matroid representation of $\chi(K)$ into the dual of the $\mathbf{Z}_2$–vector space $K^\times/\Sigma K^{\times^2}$. This is actually the case, as shown in Theorem 1.2.9.

Another important result is proved by the same type of "dual base" argument as in the previous theorem, namely:

THEOREM 1.2.6. *Every circuit of $\chi(K)$ has even cardinality.*

(Matroids with this property are called *bipartite.* A *circuit* is a minimal dependent set.)

Notice that the results mentioned above show that circuits are finite (cf. 1.1.2(c)), of cardinality ≥ 4 (since any three elements are independent, by (1.1.2(b)). A corollary to Theorem 1.2.6 is that finite spaces of orders of the form $\chi(K)$ have a coordinatization by odd vectors of $\mathbf{Z}_2^n$, that is, vectors with an odd number of non-zero coordinates (Corollary 1.2.8).

Once we know that the matroids $\chi(K)$ are binary, the next basic question suggested by matroid theory is: which of them are unimodular? A point of caution is in order here: what do we mean by a unimodular, possibly infinite, matroid? We shall take it to mean that *every finite flat is unimodular* in the usual sense.

In fact, there is a simple example of a non-unimodular matroid $\chi(K)$, namely for $K = \mathbf{R}(X, Y, Z)$; Example 1.3.2. This is shown as follows: since it is known how all orders of a rational function field $k(X)$ which extend a given order of k are constructed (cf., f.ex., Dickmann [**7**, Ch. I, §5]), by placing the elements X, Y, Z, in suitable positions with respect to one another, we construct eight orders $P_0, \ldots, P_7$ of $\mathbf{R}(X, Y, Z)$ whose dependencies can be explicitly computed. The resulting configuration is a three-dimensional cube (= 3–dimensional affine space) over $\mathbf{Z}_2$:

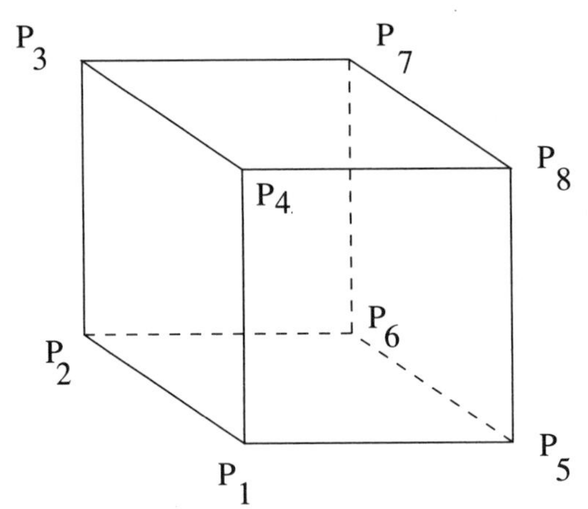

The presence of this configuration as a minor in $\chi(\mathbf{R}(X, Y, Z))$ makes it non-unimodular, by Tutte's famous excluded minor characterization of unimodular (binary) matroids: Fano's plane $\mathbf{F}$ is obtained from the cube by contraction of any (one) vertex, and its dual, $\mathbf{F}^*$, by deletion (cf. White [**20**, Thm. 3.1.1]).

Which is, then, an example (at least a non-trivial one) of a field K so that $\chi(K)$ *is* unimodular? Familiarity with the example above shows that the argument proving non-unimodularity depends essentially on having three variables, since the cube has rank four; it breaks down for $\mathbf{R}(X, Y)$. Another candidate is $\mathbf{Q}(X)$. (We exclude $\mathbf{R}(X)$ since it is unimodular for trivial reasons: it is SAP and hence its closure operator is trivial.)

Before explaining the general solution to this problem, let us take a further look at the cube above. Such a configuration is an example of a notion crucial in quadratic form theory: that of a *fan.* This notion is defined as follows:

DEFINITION. Let K be a f.r. field. A *fan* of K is a preorder T such that any multiplicative subgroup S of $K^\times$ so that $T \subseteq S$, $[K^\times : S] = 2$, and $-1 \notin S$, is an order.

The following characterization of fans bring us to our point:

PROPOSITION. (Bröcker [**4**, p. 149]). *A preorder T of K is a fan iff for every three different orders P_0, P_1, P_2 containing T, the closure $c\ell_K(\{P_0, P_1, P_2\})$ has cardinality four (this is the largest cardinality it can have).*

In this case, $c\ell_K(P_0, P_1, P_2)$ is necessarily a four-element circuit of $\chi(K)$. In our geometric language it is more telling to redefine this notion as follows: a subset

$\mathcal{F} \subseteq \chi(K)$ is a *fan* iff any three elements of $\mathcal{F}$ are contained in a four-point circuit contained in $\mathcal{F}$.

It is an easy matter to show that there is a unique fan of rank n for every integer $n \geq 1$; it has 2^{n-1} elements. In terms of the coordinatization by odd vectors mentioned above, the fan of rank n is exactly the geometry of *all* odd vectors of $\mathbf{Z}_2^n$ (Proposition 1.3.5). Concrete examples: (i) If K is a field with a unique order, then the fan of rank $n \geq 2$ is isomorphic to the matroid $\chi(K((X_1, \ldots, X_{n-1})))$. (ii) If L is a Rolle field with 2^n orders, $\chi(L)$ is the fan of rank $n+1$. Lam [**10**, Ch. 5] contains a comprehensive analysis of fans (in their disguise as preorders) and their role in quadratic form theory.

Returning to the problem of characterizing the fields K such that $\chi(K)$ is unimodular, we obtain a very satisfactory solution in terms of the so-called *stability index* of K:

THEOREM 1.3.8. *Let K be a* f.r. *field. Then $\chi(K)$ is unimodular iff the reduced stability index of K is at most* 2.

One-half of the proof, the implication ($\Longleftarrow$), is based on Bröcker's local-global principle for stability indices [**3**, Satz 3.19], which reduces the question to an analysis of the structure of the order spaces $\chi(\overline{K_v})$ of the residue fields $\overline{K_v}$ of K, for all real valuations v of K. For the other implication, ($\Longrightarrow$), we use a characterization of the (reduced) stability index in terms of fans, cf. [**10**, Thm. 13.7] and [**17**]:

$$\operatorname{st}(K) = \max\{n \in \mathbf{N} \mid \chi(K) \text{ contains a fan of cardinality } 2^n\}$$

($\operatorname{st}(K) = \infty$ if $\chi(K)$ contains fans of arbitrarily large finite cardinality). Thus, if $\operatorname{st}(K) \geq 3$, then $\chi(K)$ contains a fan of at least 8 elements, hence a fan of 8 elements, i.e. the affine cube above. By Tutte's criterion again, $\chi(K)$ is not unimodular.

Thus, the space of orders of the fields $\mathbf{Q}(X)$ and $\mathbf{R}(X, Y)$, which have stability index 2, are unimodular matroids (for an explicit "drawing" of $\chi(\mathbf{Q}(X))$, see below). Since SAP-fields are exactly the fields of stability index ≤ 1 [**3**, Satz 3.201], Theorem 1.3.8 may be considered as an extension of Theorem 1.1.8 to the next stability index level. Theorem 1.3.8 yields:

COROLLARY 1.3.9.

(1) *Let K be a* f.r. *field such that* $\operatorname{st}(K) \leq 2$. *Let F be any field. Then:*
   (a) *Every finite subset of $\chi(K)$ has a coordinatization over F.*
   (b) *If F is finite, $\chi(K)$ has a coordinatization over F.*
   (c) *Every finite subset of $\chi(K)$ has a coordinatization over $\mathbf{Q}$ given by a totally unimodular matrix.*

   In particular, this is the case if $K = \mathbf{R}(X, Y)$.

(2) *If K is a* f.r. *field with* $\operatorname{st}(K) \geq 3$, *then $\chi(K)$ is coordinatizable only over fields of characteristic* 2.

The same technique can be used to improve the preceding result in a way which further underlines the sharp dichotomy between the spaces of orders of 2–stable fields (stability index ≤ 2) and those with a larger stability index. The result is as follows:

THEOREM 1.3.10. *Let K be a* f.r. *field. Each of the following conditions on $\chi(K)$ is equivalent to $st(K) \leq 2$:*

(a) *$\chi(K)$ is a graphical matroid.*
(b) *$\chi(K)$ is a cographical matroid.*
(c) *$\chi(K)$ is a planar graphical matroid.*
(d) *$\chi(K)$ is a series-parallel matroid.*

As in the case of unimodular matroids, the meaning of these notions for infinite $\chi(K)$ is that every finite flat has the stated property. Well-known excluded minor characterizations for each of these classes of matroids (cf. White [**19**, pp. 146–147]) are used in the proof.

The point in the proof of Theorems 1.3.8 and 1.3.10 is that Bröcker's local-global principle, together with Theorem 1.1.8, implies that the space $\chi(K)$ has a very simple structure if the field K is 2–stable. Indeed, each connected component of $\chi(K)$ (in the matroid sense) is either a single point or a (possibly infinite) slab of the form:

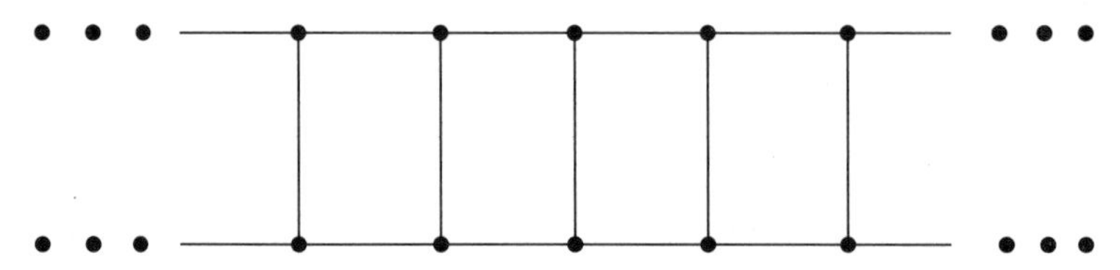

A particularly transparent example is that of $\chi(\mathbf{Q}(X))$, where the components are:

(i) Single points, corresponding to Archimedean orders (equivalently, transcendental cuts on $\mathbf{Q}$), or to orders of the form a^+, a^-, for $a \in \mathbf{Q}$, obtained by placing X infinitesimally near a, to the left ($= a^-$) or to the right ($= a^+$).
(ii) Slabs as above containing $2r$ points (i.e., of rank $r+1$) for $r \geq 2$. For each such integer there are countably many such slabs, classified by the irreducible polynomials $F \in \mathbf{Q}[X]$ with exactly r real roots, say $\alpha_1, \ldots, \alpha_r$. The $2r$ orders are:

$$\alpha_i^+ = \{P/Q \mid P, Q \in \mathbf{Q}[X], Q \neq 0 \text{ and } \exists \varepsilon > 0 (PQ \lceil (\alpha_i, \alpha_i + \varepsilon) > 0)\},$$
$$\alpha_i^- = \{P/Q \mid P, Q \in \mathbf{Q}[X], Q \neq 0 \text{ and } \exists \varepsilon > 0 (PQ \lceil (\alpha_i - \varepsilon, \alpha_i) > 0)\},$$

for $i = 1, \ldots, r$. These are precisely the liftings of the order of $\mathbf{Q}$ along the F-adic valuation of $\mathbf{Q}(X)$.

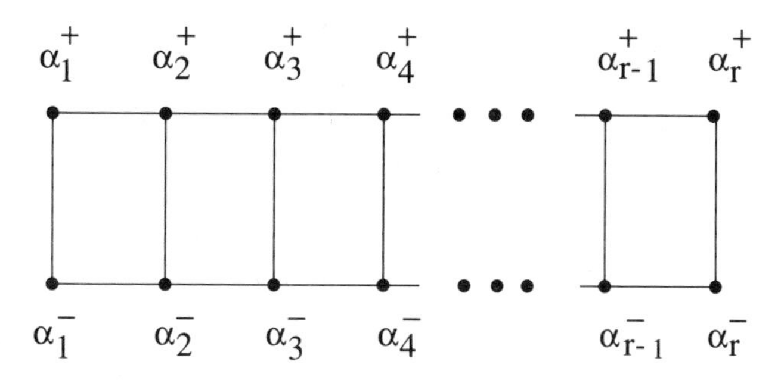

A (non-trivial) component of $\chi(\mathbf{Q}(X))$

Any inclusion of fields: $i : K \hookrightarrow F$ induces a dual map $\rho : \chi(F) \to \chi(K)$ given by the restriction of orders from F to K. From a combinatorial point of view, ρ is a *strong map* of the geometry $\chi(F)$ into $\chi(K)$, i.e. a closure-preserving map ($P \in c\ell_F(\mathcal{X}) \Rightarrow \rho(P) \in c\ell_K(\rho[\mathcal{X}])$; cf. White [**19**, Prop. 8.1.3]. The interpretation of a well-known result from valuation theory (cf. Lam [**10**, Prop. 3.17]) in our geometric language yields:

FACT 1.4.2. *Let $\langle K, v\rangle \subseteq \langle F, w\rangle$ be an immediate extension of valued fields, where w is a real valuation. Let $\chi(K, v)$ denote the set of all orders of K compatible with v; similarly for F, w. Then $\rho\lceil\chi(F, w) : \chi(F, w) \to \chi(K, v)$ is a matroid isomorphism. In particular, if $\widehat{K}^v$ denotes the henselization of $\langle K, v\rangle$, then ρ is an isomorphism of $\chi(\widehat{K}^v)$ onto $\chi(K, v)$.*

As a motivation for our next theme, we recall a result of Baer and Krull (cf. Lam [**10**, Thm. 3.10 and Notes on Ch. 3]). Let v be a real valuation of a f.r. field K. Two objects are associated to each $P \in \chi(K, v)$:

1) The push-down order $\overline{P}$ of P on the residue field $\overline{K_v}$, defined by:

$$\overline{P} = \{x/M_v \mid x \in A_v \cap P\}\,.$$

Incidentally, we remark that $^{-}$: $\chi(K, v) \to \chi(\overline{K_v})$ $[P \mapsto \overline{P}]$, is also a strong map.

2) A character $P^* \in \mathrm{Hom}(\Gamma_v/2\Gamma_v, \mathbf{Z}_2)$, essentially induced by the sign function on $K^\times/\Sigma(K^\times)^2$; Γ_v denotes the value group of v. Actually, P^* is in the dual of the $\mathbf{Z}_2$-vector space $\Gamma_v/2\Gamma_v$, but is not intrinsically defined.

The formal definition of P^* is as follows. Denote by v' the composition of the valuation $v : K^\times \to \Gamma_v$ and the canonical quotient map $\Gamma_v \to \Gamma_v/2\Gamma_v$. Let us choose a system of representatives $\overline{a} = \{a_i \mid i \in I\} \subseteq K^\times$ such that $\{v'(a_i) \mid i \in I\}$ forms a $\mathbf{Z}_2$-base of $\Gamma_v/2\Gamma_v$. We define $P^*_{\overline{a}}$ by specifying its action on the chosen base, as follows:

$$P^*_{\overline{a}}(v'(a_i)) = \mathrm{sgn}_P(a_i) = \begin{cases} 0 & \text{if } a_i \in P \\ 1 & \text{if } a_i \notin P \end{cases}$$

and extending to all of $\Gamma_v/2\Gamma_v$ by linearity.

The result we have in mind is:

THEOREM. (Baer-Krull). *The map*

$$g_{\overline{a}} : \chi(K,v) \longrightarrow \chi(\overline{K_v}) \times \mathrm{Hom}(\Gamma_v/2\Gamma_v, \mathbf{Z}_2)$$
$$P \longrightarrow \langle \overline{P}, P^*_{\overline{a}} \rangle$$

is a bijection. A similar result holds with the subgroup $2\Gamma_v$ $(= v[\Sigma(K^\times)^2])$ *replaced by the subgroup* $v[T]$, *where* T *is an arbitrary preorder of* K.

A refinement of the technique employed in proving surjectivity of this map (lifting a residual order along a character $\chi \in \mathrm{Hom}(\Gamma_v/2\Gamma_v, \mathbf{Z}_2)$) suffices to prove the following result:

THEOREM 1.4.5. (The rank formula.) *Let* v *be a real valuation of a field* K. *Let* $\mathcal{X} \subseteq \chi(K,v)$, $T = \bigcap\{P \mid P \in \mathcal{X}\}$, $\overline{\mathcal{X}} = \{\overline{P} \mid P \in \mathcal{X}\}$. *Then:*

$$r_K(\mathcal{X}) = r_{\overline{K_v}}(\overline{\mathcal{X}}) + \dim_{\mathbf{Z}_2}(\,\mathrm{Hom}(\Gamma_v/v[T], \mathbf{Z}_2)) = r_{\overline{K_v}}(\overline{\mathcal{X}}) + \dim_{\mathbf{Z}_2}(\Gamma_v/v[T])\,.$$

A further question suggested by the foregoing results is whether the set-theoretic bijection established by the Baer-Krull theorem has a geometric interpretation. Since $\chi(\overline{K_v})$ and $\mathrm{Hom}(\Gamma_v/2\Gamma_v, \mathbf{Z}_2)$ each have a natural matroid structure, we may ask, for instance, whether they induce a matroid structure on the cartesian product which makes the Baer-Krull map $g_{\overline{a}}$ into a matroid isomorphism. The answer is positive whenever $\chi(K,v)$ is finite; more generally, we show:

THEOREM 1.4.6. (Geometric form of the Baer-Krull theorem.) *Let* T *be a preorder of* K *and* v *a valuation fully compatible with* T. *Assume* $\chi(K,T)$ *is finite. Let* $d = \dim_{\mathbf{Z}_2}(\mathrm{Hom}(\Gamma_v/v[T], \mathbf{Z}_2))$. *Let* $\overline{a} = \{a_1, \ldots, a_d\} \subseteq K^\times$ *be such that* $\{v'(a_1), \ldots, v'(a_d)\}$ *is a basis of the* $\mathbf{Z}_2$*-vector space* $\Gamma_v/v[T]$. *The set* $\chi(\overline{K_v}, \overline{T}) \times \mathrm{Hom}(\Gamma_v/v[T], \mathbf{Z}_2)$ *can be endowed with a matroid structure (defined in terms of that of the factors) so that the Baer-Krull map* $g_{\overline{a}}(P) = \langle \overline{P}, P^*_{\overline{a}} \rangle$ $(P \in \chi(K,T))$ *is a matroid isomorphism.*

The present Part II is devoted to the decomposition theory of the matroids $\chi(K)$. The main tool used here is the approximation theorem for V-topologies, a very efficient instrument in dealing simultaneously with orders and valuations.

Matroids are naturally split in connected components by the *circuit-connectivity* relation: for $p, q \in M$,

$$p \underset{c}{\sim} q \text{ iff either } p = q \text{ or there is a circuit containing both } p \text{ and } q.$$

The circuit axioms (see White [**19**, p. 301–302]) imply that $\underset{c}{\sim}$ is an equivalence relation; the equivalence classes modulo $\underset{c}{\sim}$ are the *connected components* of M. A corresponding external operation of *direct sum* of a (possibly infinite) family of matroids can be defined in such a way that a matroid is the direct sum of its connected components (considered themselves as matroids under the induced closure operator); see §2.17.

In [**5**] Bröcker introduced a natural way of splitting the spaces $\chi(K)$; for $P, Q \in \chi(K)$,

$$P \underset{v}{\sim} Q \quad \text{iff} \quad \text{either } P = Q \text{ or there is a non-trivial valuation of } K \text{ compatible with both } P \text{ and } Q\,.$$

Since Archimedean orders are characterized by the fact that only the trivial valuation is compatible with them, the class $\mathrm{Arch}(K)$ of such orders, if non-empty, gets split into singletons modulo $\underset{v}{\sim}$. The approximation theorem for V-topologies implies that the classes $\underset{v}{\sim}$ are flats of $\chi(K)$. Furthermore, it implies:

THEOREM. *Let K be a* f.r. *field. Then:*

(a) (Corollary 2.11) *The operator $c\ell_K$ is trivial on* $\mathrm{Arch}(K)$, *i.e.* $c\ell_K(\mathcal{X}) = \mathcal{X}$, *for* $\mathcal{X} \subseteq \mathrm{Arch}(K)$.
(b) (Corollary 2.12) $\mathrm{Arch}(K)$ *is a separator of* $\chi(K)$ (cf. White [**19**, pp. 175–176]). *Hence*

$$\chi(K) \simeq \mathrm{Arch}(K) \oplus (\chi(K) - \mathrm{Arch}(K))\,.$$

(c) (Corollary 2.13) *Every* $P \in \mathrm{Arch}(K)$ *is an isthmus of* $\chi(K)$ (cf. White [**19**, pp. 128–130]).

The connection between the relations $\underset{c}{\sim}$ and $\underset{v}{\sim}$ is as follows:

PROPOSITION 2.14. *Circuit-connectivity implies Bröcker (valuation)-equivalence.*

PROPOSITION 2.18. *Every Bröcker class of $\chi(K)$ is the direct sum of the connected components of $\chi(K)$ contained in it.*

Circuit-connectivity admits a characterization in valuation-theoretic terms; namely:

THEOREM 2.25. *Let $P_1, P_2 \in \chi(K)$. Then:*

$$P_1 \underset{c}{\sim} P_2 \quad \text{iff} \quad P_1 = P_2 \text{ or there is } P_3 \neq P_1, P_2 \text{ and a (necessarily non-trivial) valuation } v \text{ compatible with } P_1, P_2, P_3, \text{ such that } \Gamma_v \text{ is not 2–divisible.}$$

The main technical tool used in proving this result (and also Theorem 2.19 below) is the Baer-Krull theorem mentioned above.

This characterization suggests that circuit-connectivity is a finer relation than Bröcker-equivalence. We verify that this is the case in a non-trivial way by constructing a field K where $\chi(K)$ is Bröcker-connected (= a single $\underset{v}{\sim}$-class), has cardinality 8, but splits into two 4-element connected components (Example 2.28).

One of our main results concerning circuit-connectivity in order spaces is:

THEOREM 2.19. *Let K be a* f.r. *field, and $P, Q \in \chi(K)$. Then:*

$P \mathrel{\underset{c}{\sim}} Q$ iff $P = Q$ or there is a 4–element circuit containing both P and Q.

This turns out to be a crucial property of the geometries $\chi(K)$. In order to understand its significance (and its origin), let us recall that Marshall [**11**] considered, in the context of a generalization of the spaces $\chi(K)$ called by him *abstract order spaces*, a relation equivalent to that on the right-hand side of the statement of Theorem 2.19. We shall denote this relation by $\underset{m}{\sim}$; obviously, it is reflexive and antisymmetric. Showing that it is transitive is a non-trivial matter requiring the use of yet poorly understood notions pertaining to the combinatorics of quadratic forms. Of course, Theorem 2.19 yields at once:

COROLLARY 2.23. (Marshall [**10**, Thm. 3.2].) *The relation $\underset{m}{\sim}$ is an equivalence relation.*

The point of Theorem 2.19 is that it offers an elucidation of Marshall's relation $\underset{m}{\sim}$ and puts the heart of his work in a natural and, hopefully, fruitful perspective. Of course, Marshall's work takes place in the more general context of abstract order spaces, and is of a more difficult technical nature due to the absence of valuation-theoretic tools. However:

1) In spite of their usefulness (for example, in the investigation of the stability index of real varieties and related matters, cf. [**17**]), it is not known whether abstract order spaces yield anything different from the concrete spaces $\chi(K)$, at least as far as isomorphism types are concerned (in the finite case they do not, see [**11**, Thm. 4.10]).
2) Our combinatorial geometric approach works equally well for Marshall's abstract order spaces; in fact, I do not (yet?) know of any property of these abstract spaces which is not shared by the concrete order spaces of fields.
3) Our approach helps to elucidate other, ill understood, points of Marshall's work, as well (for example, the Basic Lemma 3.1 of [**11**]).

The interaction of the combinatorial geometric approach presented here with Marshall's work is the subject of joint work in progress with A. Lira.

I wish to thank E. Becker for calling my attention to Marshall's work; A. Prestel for explanations concerning Bröcker's local-global principle which led to the proof of Theorem 1.3.8; M. Las Vergnas for useful advice on matroid theory; A. Engler for suggesting Example 2.28 and Corollary 2.29; and D. Marker for his interest on the ideas of this paper.

NOTE. Having realized too late that the "Ragsquad Special Year" in which I participated was partially financed by the U.S. National Security Agency, I take this opportunity to express, once more, my total opposition to the involvement of military or "security" agencies in financing scientific activities. I have always—in particular in this case—refused to receive funds from such agencies.

2. Decompositions of $\chi(K)$

Various ways of splitting a space of orders have been considered by Bröcker [**5**] and Marshall [**11**], [**13**]. The aim of this section is to study the relationship between these decompositions and the natural splitting of $\chi(K)$ into connected components in the matroid-theoretic sense. We employ a combination of techniques belonging to valuation theory and to matroid theory.

NOTATION.

(1) v_0 denotes the *trivial valuation* of a field K, i.e. the valuation whose ring A_{v_0} is K; hence $M_{v_0} = \{0\}$, and $\Gamma_{v_0} = \{0\}$.
(2) Valuations of a field K are partially ordered by inclusion of their valuation rings:
$$v \preceq w \qquad \text{iff} \qquad A_v \subseteq A_w\,.$$
(This is the reverse of the usual notation.) v_0 is the coarsest (i.e. the $\preceq$-largest) valuation. The expression "*the smallest valuation*" with a certain property means smallest with this property for the order $\preceq$; this expression is used only when such smallest valuation exists.
(3) We denote by $\mathcal{T}_v$ the topology on K determined by the valuation v; thus, $\mathcal{T}_{v_0}$ is the discrete topology. Similarly, $\mathcal{T}_P$ denotes the topology on K determined by the order P.
(4) $\mathrm{Arch}(K)$ stands for the set of all *archimedean* orders of K.

We list some standard results used later.

FACTS 2.1.

(1) *The following are equivalent for $P \in \chi(K)$:*
  (a) $P \in \mathrm{Arch}(K)$.
  (b) *v_0 is the only valuation compatible with P.*
(2) (a) *Let $P, Q \in \mathrm{Arch}(K)$. Then, $P \neq Q$ iff $\mathcal{T}_P \neq \mathcal{T}_Q$.*
  (b) *Let $P \in \mathrm{Arch}(K)$, $Q \in \chi(K) - \mathrm{Arch}(K)$. Then, $\mathcal{T}_P \neq \mathcal{T}_Q$.*
(3) *Let v be a non-trivial valuation of K, and $P \in \chi(K)$. If v and P are compatible, then $\mathcal{T}_v = \mathcal{T}_P$.*
(4) (a) *Two valuations compatible with a given order P are comparable under the relation $\preceq$.*
  (b) *If $v \preceq w$ and v is compatible with $P \in \chi(K)$, then so is w.*
(5) *The following are equivalent for arbitrary valuations v_1, v_2:*
  (a) *There is a non-trivial valuation $v \succeq v_1, v_2$.*
  (b) $\mathcal{T}_{v_1} = \mathcal{T}_{v_2}$.

PROOF. Exercise. For (2) use the Approximation Theorem below. A hint for the proof of (3) is given in Dickmann [**7**, Ex. 38]. For (5), see Bourbaki [**2**, Ch. 6, §7, Prop. 3, p. 136]. □

DEFINITION 2.2. (V-topology.) A V-topology on a ring R is a ring topology (sum and product continuous) such that for any $A, B \subseteq R$, if $0 \notin \overline{A}$ and $0 \notin \overline{B}$, then $0 \notin \overline{A \cdot B}$. ($\overline{A}$ denotes the closure of A, and $A \cdot B = \{xy \mid x \in A,\ y \in B\}$.)

EXAMPLES. The typical examples of V-topologies on a field K are the topologies defined by an order or by a valuation of K.

The fundamental result on V-topologies is (see [**16**]):

THEOREM. (The approximation theorem; Stone.)
Let $\mathcal{T}_1, \ldots, \mathcal{T}_n$ be pairwise distinct V-topologies on a field K. If $U_1, \ldots, U_n$ are non-empty sets open for $\mathcal{T}_1, \ldots, \mathcal{T}_n$, respectively, then $\bigcap_{i=1}^n U_i \neq \emptyset$.

For the reader's convenience we include here the statement of some results from Part I which we will need later. Recalling the setup for the Baer-Krull theorem laid down in §1, we have

LEMMA 2.3. *Let T be a preorder of K and v a valuation of K fully compatible with T. Let $\overline{a} = \{a_i \mid i \in I\} \subseteq K^\times$ be such that $\{v'(a_i) \mid i \in I\}$ is a $\mathbf{Z}_2$-basis of $\Gamma_v/v[T]$ (where $v'(b) = v(b)/v[T]$). Then every element of $K^\times$ is representable in the form:*

(1) $tu \prod_{i \in I} a_i^{n_i}$,
where $t \in T$, $u \in U_v$, $\{n_i \mid i \in I\} \subseteq \mathbf{Z}_2$ almost all zero.

(2) *This representation is unique modulo T; if*

$$t_1 u_1 \prod_{i \in I} a_i^{n_i} = t_2 u_2 \prod_{i \in I} a_i^{m_i} ,$$

where t_j, u_j, n_i and m_i are as in (1), then
  (a) *$n_i = m_i$ for all $i \in I$;*
  (b) *$u_1 = (t_1^{-1} t_2) u_2$.*

In particular, $u_1 \equiv u_2 \pmod{T}$; hence u_1, u_2, have same sign under every order of $\chi(K, T)$.

REMARK. It follows that there are well-defined maps depending on $\overline{a} = \{a_i \mid i \in I\}$, $f_{\overline{a}} : K^\times/T^\times \to U_v/T \cap U_v$ and $n = n_{\overline{a}} : K^\times/T^\times \to \mathbf{Z}_2^I$, so that for every $x \in K^\times$ (where $\overline{x} = x/T^\times$):

- $\{i \in I \mid n(\overline{x})(i) = 1\}$ is finite, and
- $\overline{x} = f_{\overline{a}}(\overline{x}) \cdot \prod_{i \in I} a_i^{n(\overline{x})(i)}$.

LEMMA 2.4. *Let T, v and $\overline{a}$ be as in Lemma 2.3. Then, for every $P \in \chi(K, T)$ and $x \in K^\times$:*

$$P^*_{\overline{a}}(v'(x)) = \mathrm{sgn}_P(x) + \mathrm{sgn}_{\overline{P}}(f_{\overline{a}}(\overline{x})/M_v) .$$

LEMMA 2.5. *Let T be a preorder of K and v a valuation fully compatible with T. Let $P_1, \ldots, P_n \in \chi(K, T)$ and $Q \in c\ell_{\overline{K_v}}(\overline{P_1}, \ldots, \overline{P_n})$. Then, there is $P \in \chi(K, T)$ so that $\overline{P} = Q$ and $P \in c\ell_K(P_1, \ldots, P_n)$.*

Bröcker [**5**] considered the binary relation $\underset{v}{\sim}$ on $\chi(K)$, defined in §1, which we will call *Bröcker-equivalence* or *valuation-equivalence*. It is an equivalence relation (cf. 2.1(4)) whose equivalence classes will be called the *Bröcker classes* of $\chi(K)$. Clearly, $\underset{v}{\sim}$ splits $\mathrm{Arch}(K)$ into singletons (2.1(2)). We also have:

FACT 2.6. *$P \underset{v}{\sim} Q$ iff $\mathcal{T}_P = \mathcal{T}_Q$. In others words, there is a unique V-topology associated to each Bröcker class of $\chi(K)$, and different Bröcker classes determine different topologies.*

PROOF. Left as an exercise using 2.1 (1)–(5). □

The following proposition will play a crucial role in the sequel.

PROPOSITION 2.7. *Let $\mathcal{B}_1, \ldots, \mathcal{B}_r$ be distinct Bröcker classes of $\chi(K)$, and let $\mathcal{T}_1, \ldots, \mathcal{T}_r$ denote the topologies associated to them. For $j = 1, \ldots, r$ let $\mathcal{X}_j = \{P_1^j, \ldots, P_{k_j}^j\}$ be an arbitrary, non-empty finite subset of $\mathcal{B}_j$. Let $P \in c\ell_K(\bigcup_{j=1}^r \mathcal{X}_j)$. Then,*

(1) *There is a unique $i \in \{1, \ldots, r\}$ such that $\mathcal{T}_P = \mathcal{T}_i$.*
(2) *The following are equivalent for arbitrary $i \in \{1, \ldots, r\}$:*
(a) *$\mathcal{T}_P = \mathcal{T}_i$.* (b) *$P \in \mathcal{B}_i$.* (c) *$P \in c\ell_K(\mathcal{X}_i)$.*
Hence,
(3) *$P \in c\ell_K(\mathcal{X}_i)$ for a unique $i \in \{1, \ldots, r\}$.*

PROOF. (3) is immediate from (1) and (2).

(1) Assume $\mathcal{T}_P \neq \mathcal{T}_i$ for $i = 1, \ldots, r$. Let

$$U_0 = (-\infty, 0)_P, \quad \text{and} \quad U_i = \bigcap_{j=1}^{k_i} (0, +\infty)_{P_j^i} \qquad (i = 1, \ldots, r),$$

where, for $a, b \in K \cup \{\pm\infty\}$, $(a, b)_Q = \{x \in K \mid a <_Q x <_Q b\}$. Since the interval $(0, +\infty)_{P_j^i}$ is open in $\mathcal{T}_i$ for each $j = 1, \ldots, k_i$, then $U_0, U_1, \ldots, U_r$ are non-empty open sets in the topologies $\mathcal{T}_P, \mathcal{T}_1, \ldots, \mathcal{T}_r$, respectively. By the approximation theorem, $\bigcap_{i=0}^r U_i \neq \emptyset$, which contradicts the assumption $P \in c\ell(\bigcup_{j=1}^r \mathcal{X}_j)$. Uniqueness follows at once from Fact 2.6.

(2) (b) $\Longrightarrow$ (a) is obvious by 2.6.

(c) $\Longrightarrow$ (b). If $\mathcal{X}_i$ contains an archimedean order, P_i, then $\mathcal{X}_i = \{P_i\}$ by 2.6, and (c) entails $P = P_i$ by Theorem 1.1.2(a). Otherwise, let v be a non-trivial valuation compatible with each order in $\mathcal{X}_i$ (exists because $\mathcal{X}_i \subseteq \mathcal{B}_i$ is finite). By (c), $1 + M_v \subseteq \bigcap_{j=1}^{k_i} P_j^i \subseteq P$, i.e., v is compatible with P. Hence $P \underset{v}{\sim} P_j^i$ for $j = 1, \ldots, k_i$, and $P \in \mathcal{B}_i$.

(a) $\Longrightarrow$ (c). By induction on $r \geq 1$ we shall prove:

(†) Fix $i \in \{1, \ldots, r\}$. Given arbitrary, non-empty finite families $\mathcal{X}_j = \{P_1^j, \ldots, P_{k_j}^j\} \subseteq \chi(K)$, $j = 1, \ldots, r$, such that
(α) Every order in $\mathcal{X}_j$ defines the same topology $\mathcal{T}_j$ on K, and
(β) The topologies $\mathcal{T}_j$ are distinct for different indices j, then,

$P \in c\ell(\bigcup_{j=1}^r \mathcal{X}_j)$ and $\mathcal{T}_P = \mathcal{T}_i$ imply $P \in c\ell(\mathcal{X}_i)$.

Clearly, this establishes the required implication. □

PROOF OF (†). If $r = 1$ there is nothing to be proved. Assume that $r > 1$ and that (†) holds for all possible situations of the same kind involving $r-1$ classes, i.e.,

$$(*)\ Q \in c\ell\Big(\bigcup_{j=1}^{r-1} \mathcal{Y}_j\Big) \text{ and } \mathcal{T}_Q = \mathcal{T}_k \text{ imply } Q \in c\ell(\mathcal{Y}_k),$$

for every order Q, $k \in \{1, \ldots, r-1\}$ and $r-1$ non-empty finite families $\mathcal{Y}_1, \ldots, \mathcal{Y}_{r-1}$ of arbitrary cardinality satisfying the requirements (α) and (β) of (†).

Replacing $\bigcup_{j=1}^{r} \mathcal{X}_j$ by any of its bases we may assume, without loss of generality, that this set is independent. Since $r \geq 1$ we may choose $\ell \in \{1, \ldots, r\}$, $\ell \neq i$. Now we proceed by a subsidiary induction on k_ℓ, the cardinality of $\mathcal{X}_\ell$. Let $\mathcal{X} = \bigcup_{\substack{j=1 \\ j\neq \ell}}^{r} \mathcal{X}_j$.

$k_\ell = 1$) Assume $P \notin c\ell(\mathcal{X})$; then $P \in c\ell(\mathcal{X} \cup \{P_1^\ell\}) - c\ell(\mathcal{X})$, and the exchange property implies $P_1^\ell \in c\ell(\mathcal{X} \cup \{P\})$. Since $\mathcal{T}_P = \mathcal{T}_i$, the $r-1$ classes $\mathcal{X}_i \cup \{P\}$ and $\mathcal{X}_j$, $j \in \{1, \ldots, r\} - \{i, \ell\}$, satisfy the requirements (α), (β) of (†). By (1) we get $\mathcal{T}_{P_1^\ell}$ $(= \mathcal{T}_\ell) = \mathcal{T}_k$ for some $k \in \{1, \ldots, r\}, k \neq \ell$, contradicting assumption (β). It follows that $P \in c\ell(\mathcal{X}) = c\ell\Big(\bigcup_{\substack{j=1 \\ j\neq \ell}}^{r} \mathcal{X}_j\Big)$, and the induction hypothesis (*) shows that $P \in c\ell(\mathcal{X}_i)$.

$k_\ell > 1$) Assume that, whenever $\mathcal{Y}$ is a subset of $\mathcal{X}_\ell$ of cardinality $k_\ell - 1$ the statement (*) holds with $\bigcup_{j=1}^{r-1} \mathcal{Y}_j$ replaced by $\Big(\bigcup_{j=1}^{r-1} \mathcal{Y}_j\Big) \cup \mathcal{Y}$, where $\mathcal{Y}_1, \ldots, \mathcal{Y}_{r-1}, \mathcal{Y}$ satisfy the requirements (α), (β) of (†). Suppose that

$$P \in c\ell(\mathcal{X} \cup \{P_1^\ell, \ldots, P_{k_\ell}^\ell\}) - c\ell(\mathcal{X} \cup \{P_1^\ell, \ldots, P_{k_\ell - 1}^\ell\}).$$

By the exchange property, $P_{k_\ell}^\ell \in c\ell(\mathcal{X} \cup \{P\} \cup \{P_1^\ell, \ldots, P_{k_\ell-1}^\ell\})$. Since the classes $\mathcal{X}_i \cup \{P\}$, $\mathcal{X}_j$, $j \in \{1, \ldots, r\} - \{i, \ell\}$, and $\{P_1^\ell, \ldots, P_{k_\ell-1}^\ell\}$ verify the conditions (α), (β), the induction hypothesis can be applied (with $\mathcal{Y} = \{P_1^\ell, \ldots, P_{k_\ell-1}^\ell\}$, and $\mathcal{Y}_1, \ldots, \mathcal{Y}_{r-1}$ the other $r-1$ classes); since $\mathcal{T}_{P_{k_\ell}^\ell} = \mathcal{T}_\ell$, we conclude $P_{k_\ell}^\ell \in c\ell(\{P_1^\ell, \ldots, P_{k_\ell-1}^\ell\})$ which contradicts our assumption that $\mathcal{X}_\ell$ is an independent set. This shows that $P \in c\ell(\mathcal{X} \cup \{P_1^\ell, \ldots, P_{k_\ell-1}^\ell\})$, and the induction hypothesis (on k_ℓ) proves, then, that $P \in c\ell(\mathcal{X}_i)$. □

We collect now several consequences of Proposition 2.7.

PROPOSITION 2.8.

(a) *Let $\mathcal{X} \subseteq \chi(K)$ and $P \in c\ell_K(\mathcal{X})$. Let $\{\mathcal{B}_i \,|\, i \in I\}$ be an enumeration (without repetitions) of all Bröcker classes of $\chi(K)$. Then, there is a* *unique* $i_0 \in I$ *such that* $P \in c\ell_K(\mathcal{X} \cap \mathcal{B}_{i_0})$.

(b) *An arbitrary union of Bröcker classes is a flat. In particular,*

(c) *Arch*(K) *is a flat of* $\chi(K)$.

PROOF.

(a) By the finiteness property 1.1.2(c) (see §1), there are $i_1, \ldots, i_r \in I$ and finite sets $\mathcal{X}_j \subseteq \mathcal{B}_{i_j} \cap \mathcal{X}$ $(j = 1, \ldots, r)$ so that $P \in c\ell_K(\mathcal{X}_1 \cup \cdots \cup \mathcal{X}_r)$. From Proposition 2.7(3) we get $P \in c\ell(\mathcal{X}_{j_0})$, for a unique $j_0 \in \{1, \ldots, r\}$.

(b) Use (a) with $\mathcal{X} = \bigcup_{s \in S} \mathcal{B}_s$, where $\{\mathcal{B}_s \mid s \in S\}$ is a given collection of Bröcker classes. □

COROLLARY 2.9. *Let* $\{\mathcal{B}_s \mid s \in S\}$ *be a collection of Bröcker classes, and* $P \in c\ell_K(\bigcup_{s \in S} \mathcal{B}_s)$. *If* $P \in \mathrm{Arch}(K)$, *then* $\mathcal{B}_{s_0} = \{P\}$ *for some* $s_0 \in S$.

PROOF. By 1.1.2(c) and Proposition 2.7(1), $\mathcal{T}_P = \mathcal{T}_{s_0}$ for some $s_0 \in S$. Then 2.1(2) yields $\mathcal{B}_{s_0} = \{P\}$. □

PROPOSITION 2.10. *Let* $\mathcal{X} \subseteq \chi(K)$. *Then*

$$\mathrm{Arch}(K) \cap \mathcal{X} = \mathrm{Arch}(K) \cap c\ell(\mathcal{X}) = c\ell(\mathrm{Arch}(K) \cap \mathcal{X}) .$$

PROOF. Call $\mathcal{A}$, $\mathcal{B}$, $\mathcal{C}$ the three terms above, in order from left to right. Clearly, we have: $\mathcal{A} \subseteq \mathcal{B}$, $\mathcal{A} \subseteq \mathcal{C}$, $\mathcal{C} \subseteq c\ell(\mathcal{X})$. The inclusion $\mathcal{C} \subseteq \mathrm{Arch}(K)$ holds because $\mathrm{Arch}(K)$ is a flat (2.8(c)). Hence we have $\mathcal{A} \subseteq \mathcal{C} \subseteq \mathcal{B}$. The inclusion $\mathcal{B} \subseteq \mathcal{A}$ follows from Corollary 2.9 by splitting $\mathcal{X}$ into Bröcker classes. □

COROLLARY 2.11. *The closure operator of* $\chi(K)$ *is trivial on* $\mathrm{Arch}(K)$: $\mathcal{X} \subseteq \mathrm{Arch}(K)$ *implies* $c\ell_K(\mathcal{X}) = \mathcal{X}$.

PROOF. Immediate from the equality of the first and third terms in 2.10. □

COROLLARY 2.12. $\mathrm{Arch}(K)$ *is a separator of* $\chi(K)$, *i.e.*

$$\chi(K) = \mathrm{Arch}(K) \oplus \big(\chi(K) - \mathit{Arch}(K)\big) .$$

PROOF. White [**19**, Prop. 7.6.4] shows that the second equality of 2.10 is a convenient cryptomorphism for the notion of a separator in terms of closure. □

COROLLARY 2.13. *Any* $P \in \mathrm{Arch}(K)$ *is an isthmus of* $\chi(K)$.

PROOF. White [**19**, Prop. 7.2.2] shows that the implication

$$P \in c\ell_K(\mathcal{X}) \Longrightarrow P \in \mathcal{X} ,$$

which follows from the first equality of 2.10 is a cryptomorphism for the notion of an isthmus. □

A further consequence of Proposition 2.8 shows that the circuit-connectivity relation, $\underset{c}{\sim}$, is finer than that of Bröcker-equivalence, $\underset{v}{\sim}$.

PROPOSITION 2.14. *Let* $P, Q \in \chi(K)$. *Then,*

$$P \underset{c}{\sim} Q \quad \text{implies} \quad P \underset{v}{\sim} Q\,.$$

PROOF. Assume $P \underset{c}{\sim} Q$ but not $P \underset{v}{\sim} Q$. Let $\mathcal{C} = \{P_1, \ldots, P_n\}$ be a circuit of $\chi(K)$ containing P and Q, say $P = P_1$, $Q = P_n$; in particular, $Q \in c\ell(P_1, \ldots, P_{n-1})$. Assume $\mathcal{X} = \{P_1, \ldots, P_{n-1}\}$ is split into Bröcker classes $\mathcal{X}_1, \ldots, \mathcal{X}_r$, with $P \in \mathcal{X}_1$, say. By assumption, $r \geq 2$ and $Q \notin \mathcal{X}_1$. Proposition 2.8(a) gives $Q \in c\ell(\mathcal{X}_{j_0})$ for a unique $j_0 \in \{2, \ldots, r\}$. Since $\overline{\overline{\mathcal{X}_1}} \geq 1$, we have $\overline{\overline{\mathcal{X}_{j_0}}} < n-1$, and $\mathcal{X}_{j_0} \cup \{Q\}$ is a dependent set of cardinality $\leq n-1$ contained in $\mathcal{C}$, absurd. $\square$

COROLLARY 2.15. *Every connected component of* $\chi(K)$ *is contained in a single Bröcker class. Hence, every Bröcker class is a disjoint union of connected components.*

Corollary 2.15 shows that every finite connected set $\{P_1, \ldots, P_n\}$, $n \geq 2$, — e.g., a circuit—is also contained in a single (non-trivial) Bröcker component, and hence there is a non-trivial valuation of K compatible with the orders $P_1, \ldots, P_n$. We note:

FACT 2.16. *Let* $\{P_1, \ldots, P_n\}$ *be a (circuit-)connected subset of* $\chi(K)$ *of cardinality* ≥ 2. *Let* v *be the smallest valuation compatible with* $P_1, \ldots, P_n$. *Then* Γ_v *is not* 2*-divisible.*

PROOF. If $\Gamma_v = 2\Gamma_v$, the quotient map $\bar{\ } : \chi(K, v) \to \chi(\overline{K_v})$ is a matroid isomorphism (check!); hence the set $\{\overline{P_1}, \ldots, \overline{P_n}\}$ is also connected. By Corollary 2.15 there is a proper valuation ring $\overline{B}$ of $\overline{K_v}$ compatible with $\overline{P_1}, \ldots, \overline{P_n}$. Lifting $\overline{B}$ to K we get a valuation ring $B \subset A_v$, contradicting the choice of v ($B = \pi^{-1}[\overline{B}]$, where $\pi{:}A_v \to \overline{K_v}$ is the canonical map). $\square$

Before proceeding further, however, we want to improve on the foregoing results by showing that the partition of Corollary 2.15 is actually a *direct sum decomposition* in the matroid-theoretic sense. Since direct sums of infinitely many factors are seldom considered in the literature on matroid theory, a comment is in order at this point; the basic facts, however, are essentially the same as for finite direct sums.

2.17. Digression; infinite direct sums of matroids. The definition of direct sum is the same as for finitely many summands. We denote by $|M|$ the set underlying a matroid M.

DEFINITION. Let $\{M_i \mid i \in I\}$ be an arbitrary non-empty family of matroids (of any cardinality) such that $|M_i| \cap |M_j| = \emptyset$, for $i \neq j$. An operator on subsets of the set $|M| = \bigcup_{i \in I} |M_i|$ is defined as follows:

$$p \in c\ell_M(X) \quad \text{iff} \quad \text{there is } i \in I \text{ (necessarily unique) so that } p \in c\ell_{M_i}(X \cap |M_i|)\,.$$

Straightforward checking shows that $c\ell_M$ satisfies the closure axioms. The matroid M thus defined is denoted by $\bigoplus_{i\in I} M_i$. Note that the M_i's are flats of M.

The following cryptomorphic descriptions of direct sum are routine checking:

FACT A. *Let* $M = \bigoplus_{i\in I} M_i$. *Then:*

(a) $I \subseteq |M|$ *is independent iff for all* $i \in I$, $I \cap |M_i|$ *is independent in* M_i.
(b) $C \subseteq |M|$ *is a circuit in* M *iff there is* $i \in I$ *so that* $C \subseteq |M_i|$ *and* C *is a circuit in* M_i.

The next fact is used below:

FACT B.

(a) *A matroid* M *is the direct sum of its connected components (endowed with the matroid structure induced by* M*).*
Further,
(b) *Let* $F \subseteq M$ *be a flat, and* $\{M_i \mid i \in I\}$ *be the connected components of* M. *Then* $F = \bigoplus_{i\in I}(F \cap M_i)$, *where both sides are endowed with the matroid structure induced by* M.

From 2.15, 2.8(b) and 2.17(B) we obtain at once:

PROPOSITION 2.18.

(a) $\chi(K)$ *is the direct sum of its connected components.*
(b) *Every Bröcker class of* $\chi(K)$ *is the direct sum of the connected components of* $\chi(K)$ *contained in it.*
Hence:
(c) $\chi(K)$ *is the direct sum of its Bröcker classes.*

Now we turn to the fundamental characterization of the circuit-connectivity relation for the geometries $\chi(K)$.

THEOREM 2.19. *Let* $P_1, P_2 \in \chi(K)$. *Then:*

$$P_1 \underset{c}{\sim} P_2 \quad \text{iff} \quad P_1 = P_2 \text{ or there is a four-element circuit containing } P_1 \text{ and } P_2.$$

As mentioned in §1, the binary relation on $\chi(K)$ defined by the right-hand side of this equivalence was first considered by Marshall [**11**], [**13**], in the context of *abstract spaces of orderings*, a generalization of the spaces $\chi(K)$. Theorem 2.19 will follow from Propositions 2.20 and 2.21.

NOTATION.

(a) The setting of the Baer-Krull theorem will be used throughout the proof of Propositions 2.20–2.22. In each of these proofs a certain preorder T_0 of

K will be considered. Given $P \in \chi(K,T_0)$ we shall write P^* for the $\mathbf{Z}_2$-character $P^*_{\overline{a}}$ of ${}^{\Gamma_v}/_{v[T_0]}$, with respect to a fixed set $\overline{a}$ of representatives, as defined in §1.

(b) Given a valuation v of K and preorders T of K and S of $\overline{K_v}$ such that $S \supseteq \overline{T} = \{{}^{x}/M_v \,|\, x \in T\}$, the *wedge product* $T \wedge S$ is the set $T \cdot \pi^{-1}[S]$, where $\pi : A_v \to \overline{K_v}$ is the quotient map. $T \wedge S$ is a preorder of K fully compatible with v, and $\overline{T \wedge S} = S$; cf. Lam [**10**, 3.2].

PROPOSITION 2.20. *Let $P_1, P_2, P_3 \in \chi(K)$ be three distinct orders; let v be a non-trivial valuation compatible with them. If $\overline{P_1} = \overline{P_2} = \overline{P_3}$ $(= Q)$ in $\overline{K_v}$, then there is $P_4 \in \chi(K,v)$ so that $\{P_1,\dots,P_4\}$ is a circuit of $\chi(K)$ and $\overline{P_4} = Q$.*

PROOF. Let $T = P_1 \cap P_2 \cap P_3$ and $T_0 = T \wedge Q$. Using the Baer-Krull theorem get $P_4 \in \chi(K,T_0)$ so that $\overline{P_4} = Q$ and $P^*_4 = P^*_1 + P^*_2 + P^*_3$; this is possible since $\overline{P_i} = Q = \overline{T_0}$ for $i = 1,2,3$. Since $P_4 \supseteq T_0 \supseteq T = P_1 \cap P_2 \cap P_3$, we need only prove that $P_4 \notin \{P_1,P_2,P_3\}$. If, for example, $P_4 = P_1$, we get $P^*_2 + P^*_3 = \mathbf{O}$, i.e. $P^*_2 = P^*_3$. By showing that this implies $P_2 = P_3$ we get a contradiction.

By Lemma 2.4,

$$P^*_i(v'(x)) = \mathrm{sgn}_{P_i}(x) + \mathrm{sgn}_Q({}^{f_{\overline{a}}(\overline{x})}/M_v)\ ,$$

for $x \in K^\times$, $i = 2,3$. Let $x \in P_2$; then:

$$\mathbf{O} = (P^*_2 + P^*_3)(v'(x)) = \mathrm{sgn}_{P_3}(x) + 2\,\mathrm{sgn}({}^{f_{\overline{a}}(\overline{x})}/M_v) = \mathrm{sgn}_{P_3}(x)\ ,$$

i.e., $x \in P_3$. Hence $P_2 \subseteq P_3$, and $P_2 = P_3$.

Similarly, $P_4 = P_2$ implies $P_1 = P_3$, and $P_4 = P_3$ implies $P_1 = P_2$. □

PROPOSITION 2.21. *Let $P_1, P_2, P_3 \in \chi(K)$ be three different orders, and let v be a non-trivial valuation compatible with them. If $\{\overline{P_1}, \overline{P_2}, \overline{P_3}\}$ has cardinality two $(= \{Q,R\}$, say), then there is $P_4 \in \chi(K,v)$ so that $\{P_1,\dots,P_4\}$ is a circuit of $\chi(K)$ and $\overline{P_4} \in \{Q,R\}$. Further, each of Q,R is the push-down of exactly two orders among $P_1,\dots,P_4$.*

PROOF. Assume, for instance, that $\overline{P_1} = \overline{P_2} = Q$ and $\overline{P_3} = R$. Let $S = Q \cap R$, $T = P_1 \cap P_2 \cap P_3$, and $T_0 = T \wedge S$. Clearly, $S \supseteq \overline{T}$. Using the Baer-Krull theorem we get $P_4 \in \chi(K,T_0)$ so that $\overline{P_4} = R$ and $P^*_4 = P^*_1 + P^*_2 + P^*_3$; in particular, $P_4 \supseteq T = P_1 \cap P_2 \cap P_3$. Only $P_4 \notin \{P_1,P_2,P_3\}$ remains to be shown. Since $\overline{P_4} = R$, $\overline{P_1} = \overline{P_2} = Q$, we have $P_4 \neq P_1, P_2$. If $P_4 = P_3$, the argument used in the proof of Proposition 2.20 yields $P_1 = P_2$, a contradiction. □

Next we deal with the case of orders which push down to different residual orders.

PROPOSITION 2.22. *Let $P_1, P_2, P_3 \in \chi(K)$ be three different orders, and let v be a non-trivial valuation compatible with them. Assume P_1, P_2, P_3 push down to three different orders of $\chi(\overline{K_v})$, say $\overline{P_i} = Q_i$, $i = 1,2,3$. Then the following are equivalent:*

(i) *There is $P_4 \in \chi(K)$ so that $\{P_1,\dots,P_4\}$ is a circuit of $\chi(K)$.*

(ii) *There is* $Q_4 \in \chi(\overline{K_v})$ *so that* $\{Q_1, \ldots, Q_4\}$ *is a circuit of* $\chi(\overline{K_v})$.

PROOF. (i) $\Rightarrow$ (ii). It is easily checked that if $\{P_1, \ldots, P_4\}$ is a dependent set, so is $\{\overline{P_1}, \ldots, \overline{P_4}\}$. Since any 3–element set of $\chi(\overline{K_v})$ is independent (Theorem 1.1.2(b)), $\{\overline{P_1}, \ldots, \overline{P_4}\}$ is a 4–element circuit. Set $Q_4 = \overline{P_4}$.

(ii) $\Rightarrow$ (i). Lemma 2.5 applied to $T = P_1 \cap P_2 \cap P_3$. □

Now we are ready for

PROOF OF THEOREM 2.19. Only the implication ($\Rightarrow$) needs proof. Assume $P_1, P_2 \in \chi(K)$, $P_1 \underset{c}{\sim} P_2$, $P_1 \neq P_2$. Let $\mathcal{C} = \{P_1, \ldots, P_n\}$ be a circuit containing P_1, P_2. By 2.16, the smallest valuation v compatible with $P_1, \ldots, P_n$ satisfies $|\Gamma_v / 2\Gamma_v| \geq 2$.

If for some $i_0 \geq 3$ the set $\{\overline{P_1}, \overline{P_2}, \overline{P_{i_0}}\}$ has cardinality 1 or 2, the result follows from Propositions 2.20 or 2.21, respectively, applied to P_1, P_2, P_{i_0}.

Suppose, then, that $\overline{P_i} \neq \overline{P_1}, \overline{P_2}$ for $i \geq 3$, and $\overline{P_1} \neq \overline{P_2}$. The Baer-Krull theorem entails that, for $Q \in \chi(\overline{K_v})$, the set $\{P \in \chi(K, v) \mid \overline{P} = Q\}$ has same cardinality as $\mathrm{Hom}(\Gamma_v/2\Gamma_v, \mathbf{Z}_2)$. Since, by the choice of v, this set has cardinality ≥ 2, there is $R_3 \in \chi(K, v)$ such that $\overline{R_3} = \overline{P_1}$ and $R_3 \neq P_1$. Using Proposition 2.20 on P_1, P_2, R_3 we get $P_4 \in \chi(K, v)$ so that $\{P_1, P_2, R_3, P_4\}$ is a circuit. □

COROLLARY 2.23. (Marshall [**13**, Thm. 2.31].) *Let* $\underset{m}{\sim}$ *denote the binary relation on* $\chi(K)$ *defined by the right-hand side of the statement of Theorem 2.19 (Marshall-equivalence). Then* $\underset{m}{\sim}$ *is an equivalence relation.*

COROLLARY 2.24. *Let* v *be a real valuation of* K, *and* $\mathcal{F} \subseteq \chi(K, v)$ *be a fan. Then* $\overline{\mathcal{F}} = \{\overline{P} \mid P \in \mathcal{F}\}$ *is a fan of* $\chi(\overline{K_v})$.

PROOF. Obvious, if $v = v_0 =$ the trivial valuation. If v is non-trivial use 2.22 (i) $\Rightarrow$ (ii). □

By use of the same kind of argument as in the preceding results we obtain a characterization of circuit connectivity in valuation-theoretic terms.

THEOREM 2.25. *Let* $P_1, P_2 \in \chi(K)$. *Then:*

$P_1 \underset{c}{\sim} P_2$ iff $P_1 = P_2$ or there is $P_3 \neq P_1, P_2$ and a (necessarily non-trivial) valuation v compatible with P_1, P_2, P_3, such that Γ_v is not 2–divisible .

PROOF.

($\Longrightarrow$) This is an immediate consequence of 2.16 and the fact that circuits have cardinality ≥ 4 (see Theorem 1.2.6, ff.).

($\Longleftarrow$) Assume that $P_1 \neq P_2$ and that the right-hand side of the statement holds. If the set $\{\overline{P_1}, \overline{P_2}, \overline{P_3}\}$ has cardinality 1 or 2, Propositions 2.20 and 2.21, respectively, show that $\{P_1, P_2, P_3\}$ is contained in a 4–element circuit of $\chi(K)$; hence $P_1 \underset{c}{\sim} P_2$.

Assume that the residual orders $\overline{P_1}, \overline{P_2}, \overline{P_3}$ are pairwise distinct. Since Γ_v is not 2-divisible, $\mathrm{Hom}(\Gamma_v/2\Gamma_v, \mathbf{Z}_2)$ has cardinality ≥ 2. Using the Baer-Krull theorem get $P_3' \in \chi(K,v)$ so that $P_3' \neq P_1$ and $\overline{P_3'} = \overline{P_1}$. Now apply Proposition 2.21 to get a 4–element circuit of $\chi(K)$ containing P_1, P_2, P_3'; it follows $P_1 \underset{c}{\sim} P_2$. □

COROLLARY 2.26. *Let v be a valuation of K such that Γ_v is not 2–divisible. Then the set $\chi(K,v)$ is (circuit-)connected, unless it has cardinality 2.*

The proofs of Theorems 2.19 and 2.25 also yield:

COROLLARY 2.27. *Let $P_1, P_2 \in \chi(K)$. Then:*

$$P_1 \underset{c}{\sim} P_2 \quad \text{iff} \quad \begin{array}{l} P_1 = P_2 \text{ or there is a four-element circuit } \{P_1, \ldots, P_4\} \\ \text{containing } P_1, P_2, \text{ and a valuation } v \text{ of } K \text{ such that} \\ \Gamma_v \text{ is not 2–divisible and } \overline{P_3} = \overline{P_1}, \overline{P_4} = \overline{P_2}\ . \end{array}$$

Theorem 2.25 makes clear that the circuit-connectivity relation is, in general, finer than Bröcker-equivalence. A trivial example is any chain-closed field K, or $K = \mathbf{R}((X))$: $\chi(K)$ has two elements, is Bröcker-connected (cf. Dickmann [**6**]), but has two connected components. We present now a non-trivial example of the same situation.

EXAMPLE 2.28. (Engler). A field K with Bröcker-connected $\chi(K)$ having several connected components of cardinality ≥ 4.

For definiteness we shall construct K with $\overline{\overline{\chi(K)}} = 8$ and two 4–element connected components. By Theorem 2.25 the following requirements ought to be fulfilled:

(i) K has two $\preceq$-incomparable valuations v_1, v_2, such that Γ_{v_i} is not 2–divisible, and four orders compatible with each v_i $(i = 1,2)$.
(ii) The smallest valuation $v \succeq v_1, v_2$ is non-trivial and Γ_v is 2–divisible.

We start with a field F having two archimedean orders—e.g., $F = \mathbf{Q}(\sqrt{2})$—and built K by a suitable 2–step lifting.

Let $G_1, G_2 \in F[X]$ be two relatively prime polynomials of degree 1, say $G_i = X - a_i$, $a_i \in F$. Let w_i' denote the G_i-adic valuation of $F(X)$; w_1', w_2' are independent valuations. Let

$$\begin{aligned} \langle L_i, w_i'' \rangle &= \text{ Henselization of } \langle F(X), w_i' \rangle, \\ L &= L_1 \cap L_2\ , \\ w_i &= w_i'' \lceil L \qquad\qquad (i = 1,2)\ . \end{aligned}$$

A result of Endler [**8**, Cor. 2, Prop. 3] shows:

(a) $\langle L, w_i \rangle$ is an immediate extension of $\langle F(X), w_i' \rangle$ $(i = 1,2)$.
(b) Every non-trivial valuation $w \neq w_1, w_2$ of L has divisible group Γ_w and algebraically closed residue field $\overline{L_w}$.

Further, for $i = 1, 2$, each order P of F lifts to exactly two orders of $F(X)$ compatible with w_i', namely $P_{a_i^+}$ and $P_{a_i^-}$ (see Dickmann [**7**, Ch. I, §5]). By (a) and Fact 1.4.2, the same holds of the field L:

(c) L has eight orders, four of them compatible with each of the valuations w_1, w_2. Moreover, by (b) none is compatible with any other non-trivial valuation of L.

Finally, we set $K = L((Y^{\mathbf{Q}}))$, the field of formal power series in one variable Y with coefficients in L and exponents in $\mathbf{Q}$ (or in any other 2–divisible group). Let v be the canonical valuation of K: for $s \in K$, $v(s)$ is the least element in the support of s. Lifting the rings A_{w_i} to K we get valuations v_i such that for $i = 1, 2$:

(d) $A_{v_i} \subseteq A_v$ and A_v is the only valuation ring containing A_{v_1}, A_{v_2}.
(e) Γ_{v_i} is not 2–divisible,

since $\Gamma_v = \mathbf{Q}$, $\chi(K)$ is isomorphic to $\chi(L)$.

Starting with a field F having n archimedean orders, and using m pairwise relatively prime polynomials of degree 1, this construction yields a field K with Bröcker-connected space $\chi(K)$ splitting into m connected components, each of cardinality $2n$.

A. Engler has also observed that the construction of the preceding example provides a general method for constructing finite connected components of $\chi(K)$. This is a consequence of the results above:

COROLLARY 2.29. *Let $\mathcal{B}$ be a connected component of $\chi(K)$ of finite cardinality ≥ 2. Then there is a valuation v of K such that Γ_v is not 2–divisible and $\mathcal{B} = \chi(K, v)$.*

PROOF. Since $\mathcal{B}$ is finite of cardinality ≥ 2, Fact 2.16 shows that the group Γ_v of the smallest valuation compatible with every order of $\mathcal{B}$ is not 2–divisible and, of course, $\mathcal{B} \subseteq \chi(K, v)$. By Corollary 2.26, $\chi(K, v)$ is connected; since $\mathcal{B}$ is a connected component—i.e., a maximal connected subset—$\mathcal{B} = \chi(K, v)$. □

Marshall [**13**] has considered yet another decomposition of the spaces $\chi(K)$, the so-called *canonical P-structure* of $\chi(K)$, obtained by considering real places $\alpha{:}K \to \mathbf{R} \cup \{\infty\}$ of K. We shall not deal with this type of decomposition here beyond the remark that it splits $\chi(K)$ in fans, and that every other fan of $\chi(K)$ intersects *at most two* members of the decomposition.

REFERENCES

1. M. Aigner, *Combinatorial theory*, Springer-Verlag, New York, 1979.
2. N. Bourbaki, *Algèbre Commutative.* XXX, Chs. 5–6, Actual. Sci. Ind., Hermann, Paris, 1964.
3. L. Bröcker, *Zur Theorie der quadratischen Formen über formal reellen Körpern*, Math. Ann. **210** (1974), 233–256.
4. ———, *Characterizations of fans and hereditarily pythagorean fields*, Math. Z. **151** (1976), 149–163.

5. ———, *Über die Anzahl der Anordnungen eines kommutativen Körpers*, Arch. Math. **29** (1977), 458–464.
6. M. A. Dickmann, *The model theory of chain-closed fields*, J. Symb. Logic **53** (1988), 921–930.
7. ———, *Model-theoretic methods in real algebraic geometry*, Math. Library, North-Holland Publ. Co. (to appear).
8. O. Endler, *On henselizations of valued fields*, Bol. Soc. Brasil. Mat. **4** (1973), 97–109.
9. T. Y. Lam, *The theory of ordered fields*, Ring Theory and Algebra III (B. McDonald, ed.), Lect. Notes Pure Appl. Math., vol. 55, M. Dekker, New York, 1980, 1–152.
10. ———, *Orderings, valuations and quadratic forms*, Regional Conf. Series Math., vol. 52, A.M.S., 1982.
11. M. Marshall, *Classification of finite spaces of orderings*, Can. J. Math. **31** (1979), 320–330.
12. ———, *Quotients and inverse limits of spaces of orderings*, Can. J. Math. **31** (1979), 604–616.
13. ———, *Spaces of orderings* IV, Can. J. Math. **32** (1980), 603–627.
14. ———, *The Witt ring of a space of orderings*, Trans. Amer. Math. Soc. **258** (1980), 505–521.
15. J. L. Merzel, *Quadratic forms over fields with finitely many orderings, in*, Ordered Fields and Real Algebraic Geometry (D. W. Dubois and T. Recio, eds.), Contemporary Math., vol. 8, A.M.S., 1982, pp. 185–229.
16. A. Prestel and M. Ziegler, *Model-theoretic methods in the theory of topological fields*, J. reine angew. Math. **299/300** (1978), 318–341.
17. C. Scheiderer, *Stability index of real varieties*, Invent. Math. **97** (1989), 467–483.
18. D. J. A. Welsh, *Matroid theory*, Academic Press, London and New York, 1976.
19. N. White, N. (ed.), *Theory of matroids*, Encyclopedia Math. and Appl., Vol. 26, Cambridge Univ. Press, 1986.
20. ———, *Combinatorial geometries*, Encyclopedia Math. and Appl., Vol. 29, Cambridge Univ. Press, 1988.

Equipe de logique mathématique, UFR de Mathématiques, Université Paris VII, Tour 45–55, 5^{e}ét.—2, Place Jussieu, 75251 Paris Cedex 05, FRANCE

E-mail address: dickmann@logique.jussieu.fr

Contemporary Mathematics
Volume **155**, 1994

Formal Determination of Polynomial Consequences of Real Orthogonal Matrices

M. J. GONZÁLEZ-LOPEZ AND T. RECIO

1. Motivation

Geometric problems (as motion planning in robotics) dealing with position of rigid bodies in physical n-space (for instance in real two or three dimensional euclidean space) are often approached by attaching a reference system to the body and considering the algebraic set in $\mathbb{R}^{n^2+n}$ of all possible positions of this reference system with respect to a fixed external reference. The points of this algebraic set represent direct motions, i.e. pairs composed by a real $n \times n$ orthogonal matrix (rotation) and an n-array (translation), and in this way the algebraic statement (and algorithmic solution) of a variety of geometric problems involves systems of polynomial equations in the rotational and translational variables. A representative example of these problems is the computation of the inverse geometric model of a robot manipulator, which consists in determining the placement and orientation of the arms of the robot knowing the end effector placement and orientation. Buchberger [**B**] has studied this problem assuming as input a system of algebraic equalities that describes the relations satisfied by all the variables of the model (for instance, parameterizing rotational and translational variables for each body in the robot), so that finding a solution to the inverse model problem consists in obtaining the arm-variables of the system as a function of the end effector coordinates. Roughly, triangularization of the system by means of a Grobner basis with respect to some pure lexicographical order in which the end effector variables are smaller than the remaining ones is regarded by Buchberger as to facilitate the searching of the inverse model.

1991 *Mathematics Subject Classification.* Primary: 13P10 14P05 68Q40 Secondary: 14L35 13A50.

Partially supported by CICyT PB 89/0379/C02/01, TIC 88/0197 and Esprit/Bra 6846 (Posso).

This paper is submitted in final form and no version of it will be submitted for publication elsewhere

0271-4132/94 \$1.00 + \$.25 per page

EXAMPLE. Let's consider the very simple robot manipulator with two bodies and two rotational degrees of freedom moving in the plane as in the figure.

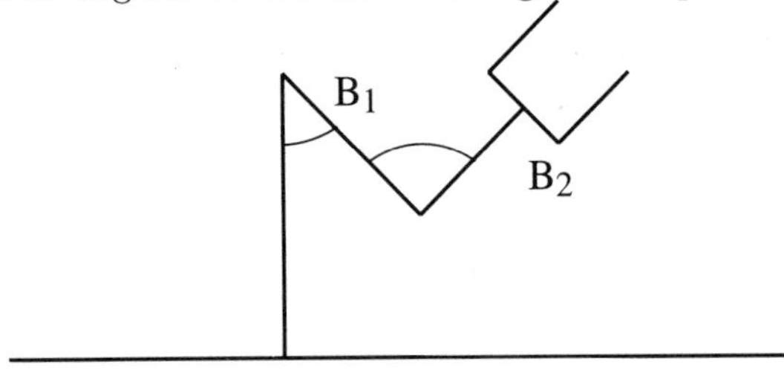

We denote by $P_i := (v_i, A_i)$ the position of the body B_i, $i = 1, 2$, where

$$v_i := (a_i, b_i) \in \mathbb{R}^2,\ A_i := \begin{pmatrix} x_i & y_i \\ z_i & t_i \end{pmatrix} \in SO(2).$$

The geometric constraints that express conditions on P_i's to represent the position of a body in the plane and relations between positions of the two bodies are given through the system of algebraic equalities:

$$A_i A_i^t = I,\ \det(A_i) = 1,\ i = 1, 2,$$

$$(1 \ \ 0 \ \ 0) \cdot \begin{pmatrix} 1 & a_1 & b_1 \\ 0 & x_1 & y_1 \\ 0 & z_1 & t_1 \end{pmatrix} = (1 \ \ 0 \ \ 0),$$

$$(1 \ \ 1 \ \ 0) \cdot \begin{pmatrix} 1 & a_1 & b_1 \\ 0 & x_1 & y_1 \\ 0 & z_1 & t_1 \end{pmatrix} = (1 \ \ 0 \ \ 0) \cdot \begin{pmatrix} 1 & a_2 & b_2 \\ 0 & x_2 & y_2 \\ 0 & z_2 & t_2 \end{pmatrix}$$

where $(1 \ \ 0 \ \ 0)$ and $(1 \ \ 1 \ \ 0)$ represent, in homogeneous coordinates, the extreme points of the bodies. A Grobner basis of the ideal generated by these polynomials, with respect to the pure lexicographical order in which $x_1 > y_1 > z_1 > t_1 > a_1 > b_1 > x_2 > y_2 > z_2 > t_2 > a_2 > b_2$ provides the equivalent (triangularized) system:

$$x_1 = a_2,\ y_1 = b_2,\ z_1 = -b_2,\ t_1 = a_2,\ a_1 = 0,\ b_1 = 0,$$
$$a_2^2 + b_2^2 = 1,\ z_2^2 + t_2^2 = 1,\ y_2 = -z_2,\ x_2 = t_2$$

where we obtain, in particular, the coordinates in P_1 as a function of the coordinates in P_2.

Let us remark that equations describing that a matrix is orthogonal ($AA^t = I$) or proper orthogonal ($AA^t = I$ and $\det(A) = 1$) appear for every body of the robot under consideration. More generally, an algebraic description of the orthogonal group $O(n)$ or of the special orthogonal group $SO(n)$ by equations describing these groups as subgroups of the set $\mathcal{M}(n, \mathbb{R})$ of matrices $n \times n$ with real coefficients appears in many algorithms in robotics other than the inverse geometric model: namely, in the computation of the direct geometric model, in the computation of the degrees of freedom of a robot system, in the checking of irredundancy, etc. Thus we quite naturally yield to the study of these algebraic descriptions, and, in particular, to the question of finding a "well behaved" polynomial description for the mentioned groups. For instance, we will like to obtain a description such that:

1) all and only all matrices in the group satisfy the polynomial equations;
2) any polynomial consequence of the matrix group (see definition below) is in the ideal generated by the polynomials in the description;
3) we can exhibit an algorithm to test whether a polynomial is a consequence or not of the matrix group;
4) there is a geometric or matricial interpretation of the polynomial equations in the description.

DEFINITION. Let G be a subgroup of $\mathcal{M}(n, \mathbb{R})$. A polynomial consequence for G is a polynomial $P(\underline{x}) \in \mathbb{R}[\underline{x}]$, $\underline{x} = (x_{1,1}, \ldots, x_{1,n}, \ldots, x_{n,1}, \ldots, x_{n,n})$, such that:

$$P(a_{1,1}, \ldots, a_{1,n}, \ldots, a_{n,1}, \ldots, a_{n,n}) = 0, \quad \text{for all} \begin{pmatrix} a_{1,1} & \ldots & a_{1,n} \\ \vdots & & \vdots \\ a_{n,1} & \ldots & a_{n,n} \end{pmatrix} \in G.$$

Examples of (almost trivial) polynomial consequences for $O(n)$ are:

- $P_i(\underline{x}) = x_{i,1}^2 + \cdots + x_{i,n}^2 - 1$, $i = 1, \ldots, n$;
- $Q_i(\underline{x}) = x_{1,i}^2 + \cdots + x_{n,i}^2 - 1$, $i = 1, \ldots, n$;
- $P_{i,j}(\underline{x}) = x_{i,1}x_{j,1} + \cdots + x_{i,n}x_{j,n}$, $i, j = 1, \ldots, n$;
- $Q_{i,j}(\underline{x}) = x_{1,i}x_{1,j} + \cdots + x_{n,i}x_{n,j}$, $i, j = 1, \ldots, n$;
- $\begin{vmatrix} x_{1,1} & \ldots & x_{1,n} \\ \vdots & & \vdots \\ x_{n,1} & \ldots & x_{n,n} \end{vmatrix}^2 - 1$;
- $H_{i,j}(\underline{x}) = x_{i,j}^2 - \tilde{x}_{i,j}^2$,

where

$$\tilde{x}_{i,j} := (-1)^{i+j} \begin{vmatrix} x_{1,1} & \ldots & x_{1,j-1} & x_{1,j+1} & \ldots & x_{1,n} \\ \vdots & & & & & \vdots \\ x_{i-1,1} & \ldots & x_{i-1,j-1} & x_{i-1,j+1} & \ldots & x_{i-1,n} \\ x_{i+1,1} & \ldots & x_{i+1,j-1} & x_{i+1,j+1} & \ldots & x_{i+1,n} \\ \vdots & & & & & \vdots \\ x_{n,1} & \ldots & x_{n,j-1} & x_{n,j+1} & \ldots & x_{n,n} \end{vmatrix}, i, j = 1, \ldots, n.$$

Again, a specific polynomial consequence for $SO(n)$ is:

$$H_{i,j}(\underline{x}) = x_{i,j} - \tilde{x}_{i,j}, \ i, j = 1, \ldots, n.$$

From the point of view of algebraic geometry, properties 1 and 2 above mean just that the polynomial description gives a basis of the ideal of the algebraic variety of all matrices in the group (identifying matrices and points in $\mathbb{R}^{n^2}$). Property 3, computationally oriented, says just that the sought polynomial basis of the ideal solves the ideal membership problem . As is well known, the Grobner basis of an ideal also has this property, but there could be other bases, not necessarily Grobner, for solving the same problem. Polynomial descriptions having these four properties could be of use in the algorithmic approach to some of the problems in robotics we have mentioned at the beginning, as it is clear that the lack of property 2 will imply, for instance in the example of the inverse model, the consideration of multiple-redundant solutions. On the other hand having property 3 could ease the computation of the Grobner basis or the triangularization of the given system (see §5 below). One could claim that it could be better to this purpose to have computed directly a Grobner basis of the ideal of matrices in the group, but it turns out that we have been unable to find a geometric or matricial interpretation of the elements of the Grobner basis of this ideal for many conceivable orders, while the basis we have found has a quite natural interpretation (property 4). In this way we can write explicitly a basis, depending on any n, with the four properties for any orthogonal or proper orthogonal group; while in the case of Grobner basis we have not been able to do so for general n. Moreover for the case $n = 2$, $n = 3$ the basis we have found happens to be also Grobner basis for the proper orthogonal ideal with respect to several orderings.

2. Main results

Coming to this point let us state formally the main results of this paper.

NOTATION. Let $\mathbb{K}$ be the field of real numbers $\mathbb{R}$ or the field of complex numbers $\mathbb{C}$. We identify a $n \times n$ matrix $A = (a_{i,j})$ with entries in $\mathbb{K}$, with the array $a = (a_{1,1}, \dots, a_{1,n}, \dots, a_{n,1}, \dots, a_{n,n})$ in $\mathbb{K}^{n^2}$. If $\mathbb{K}[\underline{x}] = \mathbb{K}[x_{1,1}, \dots, x_{1,n}, \dots, x_{n,1}, \dots, x_{n,n}]$ is the polynomial ring in n^2 variables with coefficients in $\mathbb{K}$ and $f(\underline{x}) \in \mathbb{K}[\underline{x}]$, we denote by $f(A) := f(a) \in \mathbb{K}$. Let $A(\underline{x}) = (x_{i,j})$ be a matrix whose entries are the variables in $\mathbb{K}[\underline{x}]$. We consider the polynomials:

(i) $\det(\underline{x}) := \det(A(\underline{x}))$.

(ii) $D^*_{i,j}(\underline{x}) := \sum_{k=1}^{n} x_{k,i} . x_{k,j} - \delta_{i,j} \; ; \quad i, j \in 1, \dots, n, \; i \leq j.$

(iii) $D_{i,j}(\underline{x}) := \sum_{k=1}^{n} x_{i,k} . x_{j,k} - \delta_{i,j} \; ; \quad i, j \in 1, \dots, n, \; i \leq j.$

(iv) Let $\pi = (i_1, \dots, i_a, I_1, \dots, I_b)$ and $\omega = (k_1, \dots, k_a, K_1, \dots, K_b)$ be even permutations of $(1, \dots, n)$, $b \leq a \leq n - 1$ and $a + b = n$; we denote:

$$D(\pi, \omega; a, b) := \begin{vmatrix} x_{i_1,k_1} & \dots & x_{i_1,k_a} \\ \vdots & & \vdots \\ x_{i_a,k_1} & \dots & x_{i_a,k_a} \end{vmatrix} - \begin{vmatrix} x_{I_1,K_1} & \dots & x_{I_1,K_b} \\ \vdots & & \vdots \\ x_{I_b,K_1} & \dots & x_{I_b,K_b} \end{vmatrix} .$$

And the ideals in $\mathbb{K}[\underline{x}]$:

$$\mathbf{so} := \langle \det(\underline{x}) - 1, \{D_{i,j}(\underline{x}) / i, j \in \{1, \dots, n\}, i \leq j\}\rangle,$$

$$\mathbf{o} := \langle \{D_{i,j}(\underline{x}) / i, j \in \{1, \dots, n\}, i \leq j\}\rangle.$$

THEOREM 1. (See Theorem 3.3 in §3.) *The ideal* $\mathbf{o}$ *in* $\mathbb{R}[\underline{x}]$ *is real (i.e. it is the ideal of the real algebraic set* $V(\mathbf{o}, \mathbb{R})$*; there is an equivalent purely algebraic description by the Dubois-Risler Nullstellensatz* **[D][R]**). *Analogoulsly the ideal* $\mathbf{so}$ *in* $\mathbb{R}[\underline{x}]$ *is real and prime.*

THEOREM 2. (See Theorems 4.4.1 and 4.4.2 in §4.) *Any real polynomial of degree d vanishing over all real orthogonal matrices is a combination of*

$$\{D_{i,j},\ D^*_{i,j} / i, j \in \{1, \dots, n\}, i \leq j\}$$

with polynomial coefficients of degree at most $d-2$. *Analogously, any real polynomial of degree d vanishing over all proper real orthogonal matrices is a combination of*

$$\{D_{i,j},\ D^*_{i,j} / i, j \in \{1, \dots, n\}, i \leq j\}$$
$$\cup$$
$$\{D(\pi, \omega; a, b) / \omega, \pi \text{ even permutations of } (1, \dots, n)\ , b \leq a \leq n-1, a+b=n\}$$

by means of coefficients which are polynomials of degree at most $d-2$, $d-2$ *and* $d-a$ *respectively.*

THEOREM 3. (See technical results 3.1.2 and 3.1.4 in §3.) *One can exhibit the polynomials in* $\{D^*_{i,j} / i, j \in \{1, \dots, n\}, i \leq j\}$ *as a polynomial combination of those in* $\{D_{i,j} / i, j \in \{1, \dots, n\}, i \leq j\}$. *Also* $\{D(\pi, \omega; a, b) / \omega, \pi \text{ even permutations of } (1, \dots, n)\ , b \leq a \leq n-1, a+b=n\}$ *as a polynomial combination of* $\{D_{i,j} / i, j \in \{1, \dots, n\}, i \leq j\}$ *and* $\det(\underline{x}) - 1$.

THEOREM 4. (see Corollary 4.4.3 in §4.) *One can determine algorithmically in a priori bounded number of steps whether any real polynomial p of degree d is a polynomial consequence of the group of real orthogonal (real proper orthogonal) matrices and if it is so, to construct the combination of polynomials that gives the membership to the ideal.*

It follows from the results above that $\{D_{i,j}, D^*_{i,j}\}$ (respectively $\{D_{i,j},\ D^*_{i,j},\ D(\pi, \omega; a, b)\}$) are the "well behaved" basis of $\mathbf{o}$ (respectively $\mathbf{so}$) that we were looking for, verifying properties 1 to 4 of §1. For reasons that will be clear in a moment we propose to name such basis the **Weyl basis** of the corresponding ideal. From the purely mathematical point of view, the interest of finding a basis with these properties is already present in the reputed invariant-theoretic book of H. Weyl, *The Classical Groups* **[W]**. In fact one can follow his fascinating winding way towards this basis during the first one hundred pages of the book. The modern approach to invariant theory seems to overpass the need for such a basis, as we have not found references to its existance or to its construction in more recent but also classical books ([**D-C**], [**F**], [**N**], [**S**], [**BO**]).

In fact we can say that our main results in the paper are just a reformulation of some theorems of Weyl (namely his Theorems 5.2.C, 5.3.B, 5.4.B, 5.4.C, 5.4.D). We have moreover also followed some of his proofs. The (if any) originality of the paper has to be therefore carefully explained. First, we have made a shorter and direct proof to the reality of the orthogonal or the proper orthogonal ideal, which is intrinsically tangled in Weyl's proof with the algorithmic property 3 of the basis. Second, we prove this algorithmic property without using (as it is done in his book) a cumbersome detour to Cayley's parameterization and to "formalized" main theorems for invariants (here "formalized" is used in a technical sense, as in Weyl's Chapter II, §11, meaning roughly invariants for generic orthogonal matrices). Third, we have formalized (in the sense of being less literary than the wonderful prose of Weyl) many of the concepts used in Weyl's proof and which are scattered throughout one hundred pages of his book, as he introduces them for a variety of purposes. We have also tried to make a self-contained proof, referring only to modern texts for some auxiliary results needed (in particular, a variation of Wedderburn's theorem from representation theory). Finally, our proof is a final check to Weyl's proof—as he does not detail the case of proper orthogonal matrices, leaving to the reader the task of adaptation from the non-proper orthogonal case proof—but in the case of non-proper orthogonal matrices Weyl's proof has a gap ([**W**] p. 143, Theorem 5.3.A, Supplement, Theorem 5.3.B and Corollary), as recognized by him in the Errata to the second edition, which includes a brief sketch about how to correct the mistake. We want to acknowledge our thanks to professor E. Becker who first mentioned the relation between our originally naively posed problem in robotics and the result of Weyl; to the late professor P. Menal, who helped us with the aspects in representation theory a few weeks before his sudden tragic death; and to professors T. Mora, C. Traverso and M. Coste for their help regarding the relations with Grobner basis.

3. Technical results and reality questions

In this paragraph we prove one of the main results (§3.3), namely that the ideals $\mathbf{o}.\mathbb{R}[\underline{x}]$ and $\mathbf{so}.\mathbb{R}[\underline{x}]$ are real and moreover that $\mathbf{so}.\mathbb{R}[\underline{x}]$ is prime (see notation in §2). In §3.1 we introduce some technical results, obtaining different basis for these ideals that will be used later. The proof of the reality and primality first considers the complex zeroes of the ideals, checking that the local rings of $\mathbf{o}.\mathbb{C}[\underline{x}]$ and $\mathbf{so}.\mathbb{C}[\underline{x}]$ at any complex zero are regular. This follows from the computation of Jacobians. Then it is a standard algebraic result to conclude that $\mathbf{o}.\mathbb{C}[\underline{x}]$ and $\mathbf{so}.\mathbb{C}[\underline{x}]$ are radical ideals. Next we prove (§3.2) that $V(\mathbf{so}, \mathbb{C})$ is connected by using a parameterization with skew-symmetric matrices via the exponential mapping. Thus, as $V(\mathbf{so}, \mathbb{C})$ is non-singular and connected, we conclude that $\mathbf{so}.\mathbb{C}[\underline{x}]$ is prime and therefore $\mathbf{so}.\mathbb{R}[\underline{x}]$ is prime. Again computation of local dimension allows us to prove that $\mathbf{so}.\mathbb{R}[\underline{x}]$ is real. For $\mathbf{o}.\mathbb{R}[\underline{x}]$ the result follows from its representation as intersection of two copies of $so.\mathbb{R}[\underline{x}]$.

3.1. Let us denote, following the notation in §2:

$$\begin{aligned}
\mathbf{so} &:= \langle \det(\underline{x}) - 1, \{D_{i,j}(\underline{x})/i,j \in \{1,\ldots,n\}, i \leq j\}\rangle,\\
\mathbf{so}' &:= \langle \det(\underline{x}) - 1, \{D^*_{i,j}(\underline{x})/i,j \in \{1,\ldots,n\}, i \leq j\}\rangle,\\
\mathbf{o} &:= \langle \{D_{i,j}(\underline{x})/i,j \in \{1,\ldots,n\}, i \leq j\}\rangle,\\
\mathbf{o} &:= \langle \{D^*_{i,j}(\underline{x})/i,j \in \{1,\ldots,n\}, i \leq j\}\rangle.
\end{aligned}$$

Let $I\!K$ be the field of real numbers $\mathbb{R}$ or the field of complex numbers $\mathbb{C}$. If $\mathbf{q}$ is an ideal in $I\!K[\underline{x}]$, let $V(\mathbf{q}, I\!K) := \{a \in I\!K^{n^2}/p(a) = 0, p(\underline{x}) \in \mathbf{q}\}$. If $B(\underline{x})$ and $C(\underline{x})$ are matrices $n \times n$ with entries in $K[\underline{x}]$ we write $B(\underline{x}) = C(\underline{x})(mod\,\mathbf{q})$ if all the entries of $B(\underline{x}) - C(\underline{x})$ belong to $\mathbf{q}$. We shall use the following properties:

- if $B(\underline{x}) = C(\underline{x})(mod\,\mathbf{q})$, then $\det(B(\underline{x})) = \det(C(\underline{x}))(mod\,\mathbf{q})$;
- if $H(\underline{x})$ is another matrix, and $B(\underline{x}) = C(\underline{x})(mod\,\mathbf{q})$, then $H(\underline{x})B(\underline{x}) = H(\underline{x})C(\underline{x})(mod\,\mathbf{q})$ and $B(\underline{x})H(\underline{x}) = C(\underline{x})H(\underline{x})(mod\,\mathbf{q})$.

The following results are clear from a geometric point of view (i.e. they are trivial to check for any orthogonal or proper orthogonal matrix, according to the case) but we are interested at the ideal (formal) level. The claims are valid over $I\!K$, real or complex.

3.1.1. $\det(\underline{x})^2 - 1 \in \mathbf{o} \cap \mathbf{o}'$.

Note that polynomials $\{D_{i,j}(\underline{x})/i,j \in \{1,\ldots,n\}, i \leq j\}$ are the entries of the (symmetric) matrix $A(\underline{x})A(\underline{x})^t - I$. Thus $A(\underline{x})A(\underline{x})^t = I(mod\,\mathbf{o})$, implies $\det(A(\underline{x})A(\underline{x})^t) = \det(I)(mod\,\mathbf{o})$, and $\det(\underline{x})^2 = \det(A(\underline{x})A(\underline{x})^t) = 1(mod\,\mathbf{o})$.

Likewise we obtain $\det(\underline{x})^2 = 1(mod\ \mathbf{o}')$ using instead the entries of $A(\underline{x})^t$ $A(\underline{x}) - I$.

3.1.2. $\mathbf{o} = \mathbf{o}'$ and $\mathbf{so} = \mathbf{so}'$.

Let $adj(A(\underline{x}))$ be the adjoint matrix of $A(\underline{x})$. We have the following list of formal implications:

$$\begin{aligned}
adj(A(\underline{x}))A(\underline{x})^t &= \det(\underline{x})I\\
A(\underline{x})^tA(\underline{x}) &= I\ (mod\,\mathbf{o}')\\
adj(A(\underline{x}))A(\underline{x})^tA(\underline{x}) &= adj(A(\underline{x}))\ (mod\,\mathbf{o}')\\
\det(\underline{x})A(\underline{x}) &= adj(A(\underline{x}))\ (mod\,\mathbf{o}')\\
\det(\underline{x})A(\underline{x})A(\underline{x})^t &= adj(A(\underline{x}))A(\underline{x})^t\ (mod\,\mathbf{o}')\\
\det(\underline{x})A(\underline{x})A(\underline{x})^t &= \det(\underline{x})I\ (mod\,\mathbf{o}')\\
\det(\underline{x})^2A(\underline{x})A(\underline{x})^t &= \det(\underline{x})^2I\ (mod\,\mathbf{o}')\\
A(\underline{x})A(\underline{x})^t &= I\ (mod\,\mathbf{o}'),
\end{aligned}$$

so that finally $\{D_{i,j}(\underline{x})/i,j \in \{1,\ldots,n\}, i \leq j\} \subset \mathbf{o}'$. Similarly $\{D^*_{i,j}(\underline{x})/i,j \in \{1,\ldots,n\}, i \leq j\} \subset \mathbf{o}$. So $\mathbf{o} = \mathbf{o}'$ and obviously $\mathbf{so} = \mathbf{so}'$.

3.1.3. Let $\Delta_{i,j}$ be the (i,j)-entry of the matrix $adj(A(\underline{x}))$. Then we have $x_{i,j} - \Delta_{i,j} \in \mathbf{so}$, $i,j \in \{1,\ldots,n\}$. In fact we have the following implications:

$$A(\underline{x})^t A(\underline{x}) = I(mod\,\mathbf{so})$$
$$adj(A(\underline{x}))A(\underline{x})^t A(\underline{x}) = adj(A(\underline{x}))(mod\,\mathbf{so})$$
$$\det(\underline{x})A(\underline{x}) = adj(A(\underline{x}))(mod\,\mathbf{so})$$
$$A(\underline{x}) = adj(A(\underline{x}))(mod\,\mathbf{so})$$
$$x_{i,j} - \Delta_{i,j} \in \mathbf{so},\; i,j \in \{1,\ldots,n\}.$$

3.1.4. With notation as in §2, let

$$W := \{D(\pi,\omega;a,b)/\omega, \pi \text{ even permutations of } (1,\ldots,n) \quad b \leq a \leq n-1, a+b=n\}.$$

Then $W \subset \mathbf{so}$.

PROOF. Let A be a $n \times n$ matrix with complex entries and let $C := adj(A)$. Then:

$$\begin{pmatrix} c_{i_1,k_1} & \cdots & c_{i_1,k_a} & c_{i_1,K_1} & \cdots & c_{i_1,K_b} \\ \vdots & & \vdots & \vdots & & \vdots \\ c_{i_a,k_1} & \cdots & c_{i_a,k_a} & c_{i_a,K_1} & \cdots & c_{i_a,K_b} \\ 0 & \cdots & 0 & 1 & \cdots & 0 \\ \vdots & & \vdots & \vdots & & \vdots \\ 0 & \cdots & 0 & 0 & \cdots & 1 \end{pmatrix} \cdot$$

$$\begin{pmatrix} a_{i_1,k_1} & \cdots & a_{i_a,k_1} & a_{I_1,k_1} & \cdots & a_{I_b,k_1} \\ \vdots & & \vdots & \vdots & & \vdots \\ a_{i_1,k_a} & \cdots & a_{i_a,k_a} & a_{I_1,k_a} & \cdots & a_{I_b,k_a} \\ a_{i_1,K_1} & \cdots & a_{i_a,K_1} & a_{I_1,K_1} & \cdots & a_{I_b,K_1} \\ \vdots & & \vdots & \vdots & & \vdots \\ a_{i_1,K_b} & \cdots & a_{i_a,K_b} & a_{I_1,K_b} & \cdots & a_{I_b,K_b} \end{pmatrix} =$$

$$\begin{pmatrix} \det(A) & \cdots & 0 & 0 & \cdots & 0 \\ \vdots & & \vdots & \vdots & & \vdots \\ 0 & \cdots & \det(A) & 0 & \cdots & 0 \\ a_{i_1,K_1} & \cdots & a_{i_a,K_1} & a_{I_1,K_1} & \cdots & a_{I_b,K_1} \\ \vdots & & \vdots & \vdots & & \vdots \\ a_{i_1,K_b} & \cdots & a_{i_a,K_b} & a_{I_1,K_b} & \cdots & a_{I_b,K_b} \end{pmatrix}.$$

Thus we have:

$$\begin{vmatrix} c_{i_1,k_1} & \cdots & c_{i_1,k_a} \\ \vdots & & \vdots \\ c_{i_a,k_1} & \cdots & c_{i_a,k_a} \end{vmatrix} \det(A) = \det(A)^a \begin{vmatrix} a_{I_1,K_1} & \cdots & a_{I_1,K_b} \\ \vdots & & \vdots \\ a_{I_b,K_1} & \cdots & a_{I_b,K_b} \end{vmatrix}.$$

As this identity is proved for all matrices we conclude that it is also formally true, which means that polynomials:

$$p(\pi,\omega;a,b) := \begin{vmatrix} adj(A(\underline{x}))_{i_1,k_1} & \cdots & adj(A(\underline{x}))_{i_1,k_a} \\ \vdots & & \vdots \\ adj(A(\underline{x}))_{i_a,k_1} & \cdots & adj(A(\underline{x}))_{i_a,k_a} \end{vmatrix}$$

$$\det(A(\underline{x})) - \det(A(\underline{x}))^a \begin{vmatrix} x_{I_1,K_1} & \cdots & x_{I_1,K_b} \\ \vdots & & \vdots \\ x_{I_b,K_1} & \cdots & x_{I_b,K_b} \end{vmatrix}$$

are identically null, thus they belong to **so**. Since $\det(A(\underline{x})) = 1 (mod\ \mathbf{so})$ and $adj(A(\underline{x})) = x_{i,j} (mod\ \mathbf{so})$ for all i, j, we conclude that polynomials $D(\pi,\omega;a,b)$ belong to **so**. $\square$

THEOREM 3.2. *$V(\mathbf{so},\mathbb{C})$ is a connected algebraic variety.*

PROOF. We will denote by $SS(\mathbb{R})$ the set of skew-symmetric $n \times n$ matrices $(A + A^t = 0)$ with entries in $\mathbb{R}$. This set can be identified with $\mathbb{R}^{\frac{n(n-1)}{2}}$ and it is therefore connected in the set $\mathcal{M}(n,\mathbb{R})$ of $n \times n$ matrices with entries in $\mathbb{R}$ endowed with topology given by the euclidean norm. Our theorem will follow from the construction of a continuous surjective map:

$$h: SS(\mathbb{R}) \times V(\mathbf{so},\mathbb{R}) \longrightarrow V(\mathbf{so},\mathbb{C})$$

as it is well known that $V(\mathbf{so},\mathbb{R})$ is also connected. We define this map by $h(S,R) := R\exp(iS)$. Then it is easy to check that h is well defined using some properties of the exp mapping with respect to the trace and the determinant. Moreover h is clearly continuous. To prove surjectivity let $B \in V(\mathbf{so},\mathbb{C})$; then B^*B (where B^* is the conjugate transpose matrix of B) is hermitian, positive-definite and orthogonal, and therefore using a result of Gantmacher (cf. [**G**, vol. II, Chap. 11, Lemma 1, §1]) there exists $S' \in SS(\mathbb{R})$ such that $B^*B = \exp(iS')$. Let $S := \frac{1}{2}S' \in SS(\mathbb{R})$ and $R := B\exp(-iS)$. It is clear that $B = R\exp(iS)$. To conclude it is enough to prove that $R \in V(\mathbf{so},\mathbb{R})$. Since $\exp(-iS) \in V(\mathbf{so},\mathbb{C})$, we have that $R \in V(\mathbf{so},\mathbb{C})$. Besides R has real entries because it is also unitary; in fact:

$$\begin{aligned} R^*R &= (B\exp(-iS))^*B\exp(-iS) = (\exp(-iS))^*B^*B\exp(-iS) = \\ &= \exp(-iS)\exp(iS')\exp(-iS) = \exp(-iS)\exp(2iS)\exp(-iS) = \\ &= \exp(-iS)\exp((2iS)+(-iS)) = \exp(-iS)\exp(iS) = I. \quad \square \end{aligned}$$

THEOREM 3.3. *The ideal* **so** *is real and prime in $\mathbb{R}[\underline{x}]$. The ideal* **o** *is real and radical in $\mathbb{R}[\underline{x}]$.*

PROOF. Let $a \in V(\mathbf{o}, I\!K)$ for $I\!K$ either the real or complex numbers, and denote $J(\mathbf{o})(a)$ the jacobian matrix evaluated in the point a. Then some easy computation yields $rank(J(\mathbf{o})(a)) = \frac{n(n+1)}{2}$. Analogously for $a \in V(\mathbf{so}, I\!K)$, we have $rank(J(\mathbf{so})(a)) = \frac{n(n+1)}{2}$. Next we remark that ideals **so** and **o** are equal in $I\!K[\underline{x}]_{\mathbf{m}_a}$, where $\mathbf{m}_a = \{p(\underline{x}) \in I\!K[\underline{x}]/p(a) = 0\}$, using 3.1.1. As **o** is generated by $\frac{n(n+1)}{2}$ polynomials it follows that $I\!K[\underline{x}]_{\mathbf{m}_a}/(\mathbf{o}.I\!K[\underline{x}]_{\mathbf{m}_a})$ is a regular local ring of dimension $\frac{n(n-1)}{2}$ and therefore also $I\!K[\underline{x}]_{\mathbf{m}_a}/(\mathbf{so}.I\!K[\underline{x}]_{\mathbf{m}_a})$ is a regular local ring of dimension $\frac{n(n-1)}{2}$. Now recall that if $\mathbf{b} = \langle h_1(\underline{x}), \ldots, h_r(\underline{x})\rangle$ is an ideal in $\mathbb{C}[\underline{x}]$, $a \in \mathbb{C}^n$; $\delta_a \colon \mathbb{C}[\underline{x}] \longrightarrow \mathbb{C}[\underline{x}]_{\mathbf{m}_a}$ the canonical inclusion, and $V(\mathbf{b}, \mathbb{C})$ the set of zeros in $\mathbb{C}^n$ of $\mathbf{b}$, then

$$\mathbf{b} = \bigcap_{a \in V(\mathbf{b},\mathbb{C})} \delta_a^{-1}(\mathbf{b}.\mathbb{C}[\underline{x}]_{\mathbf{m}_a}).$$

It follows that $\mathbf{so}.\mathbb{C}[\underline{x}]$ and $\mathbf{o}.\mathbb{C}[\underline{x}]$ are radical ideals and therefore also $\mathbf{so}.\mathbb{R}[\underline{x}]$ and $\mathbf{o}.\mathbb{R}[\underline{x}]$. We conclude that $V(\mathbf{so}, \mathbb{C})$ is an irreducible algebraic variety as it is non singular and connected (3.2). Therefore $\mathbf{so}.\mathbb{C}[\underline{x}]$ is prime and the same follows for $\mathbf{so}.\mathbb{R}[\underline{x}]$. As the dimension of $\mathbf{so}.\mathbb{R}[\underline{x}]$ is equal to $\frac{n(n-1)}{2}$ and agrees with the topological dimension of $V(\mathbf{so}, \mathbb{R})$ we conclude that $\mathbf{so}.\mathbb{R}[\underline{x}]$ is real. The reality of $\mathbf{o}.\mathbb{R}[\underline{x}]$ follows, again using 3.1.1, from $\mathbf{o}.\mathbb{R}[\underline{x}] = \mathbf{so}.\mathbb{R}[\underline{x}] \cap \mathbf{so}^-.\mathbb{R}[\underline{x}]$, where $\mathbf{so}^-$ is defined by

$$\mathbf{so}^- := \langle \det(\underline{x}) + 1, \{D_{i,j}(\underline{x})/i,j \in \{1,\ldots,n\}, i \leq j\}\rangle . \quad \square$$

4. Weyl basis

In this paragraph we obtain basis for the orthogonal and proper orthogonal ideals verifying the properties 1 to 4 of §1. First we introduce some notions from invariant and representation theory, namely the concept of enveloping algebra and the double centralizer property (4.1.1) for semi-simple rings. The idea of Weyl is, very roughly speaking, that using Kronecker products we are able to linearize polynomial consequences of orthogonal (or proper orthogonal) matrices. Next some linear conditions satisfied for all orthogonal (or proper orthogonal) matrices (again we recall that no precision is intended in this explanation) are introduced (4.3.1) so that the main point (4.3.2) is to prove that, conversely, any matrix satisfying such conditions is orthogonal (or proper orthogonal). Here we use the criterion given by the double centralizer property and therefore checking orthogonality is reduced to checking commutativity of such matrices with all commutators of orthogonal matrices. But this commutativity implies the invariancy of functions of the entries of the matrices with respect to the group considered (4.2.3). Therefore using the first main theorem (4.1.2 and 4.1.3) of the invariant theory for the group we have an easier way of checking the double commutativity. Finally the ideal membership problem is reduced to the same problem for an ideal generated by linear polynomials and we are done. Our proof, for the reasons explained in §2, is done for the special orthogonal group, but of course the simpler case for the orthogonal group follows along the same lines and it is therefore omitted. In what follows [**W**] will be the standard reference, but we have also included references to more modern texts for the basic concepts.

4.1. Let K be a field. If U is a set of $n \times n$ matrices with entries in K we define its linear closure $[U]$ in K as the set of all finite linear combinations

$$a_1 A_1 + \cdots + a_r A_r$$

of matrices A_i in U by means of coefficients a_i in K. If U is a (multiplicative) group, then addition of two matrices, multiplication of a matrix by an element in K and multiplication of two matrices are three operations closed in $[U]$, so that this set is an (matrix) algebra in K which is called the **enveloping algebra** of the group U (cf. also [**SH**, vol. I, 3.5.1]). The **commutator** of U is the set:

$$C(U) := \{B \in \mathcal{M}(n,K)/AB = BA \text{ for all } A \in U\}.$$

Clearly we see that $C(U)$ is a K-algebra and $C(U) = C([U])$. Let V be a n-dimensional K-vector space, $W \subset V$ a subspace and $U \subset \mathcal{M}(n,K)$ a group of matrices acting on V by means of:

$$\begin{array}{rccl} A\colon & V & \longrightarrow & V, \quad A \in U \\ & \underline{x} & \longmapsto & \underline{x}A\,. \end{array}$$

If $A \in \mathcal{M}(n,K)$ we say that W is A-**invariant** if $wA \in W$ for all $w \in W$; and W is $U-$invariant if it is A-invariant for all $A \in U$. Note that W is U-invariant if and only if W is $[U]$-invariant. The group U is **fully reducible** if we can write $V = V_1 \oplus \cdots \oplus V_k$, where V_i is an U-invariant subspace for all $i = 1, \ldots, k$. (cf. also [**C-R**, §10], as completely reducible).

The following theorem (**the double centralizer property**) is an extension of the theorem of Wedderburn on simple rings to semi-simple rings.

THEOREM 4.1.1. (Cf. [**C-R**, §59] or [**SH**, p. 124]). *The enveloping algebra of a fully reducible matrix set U is the commutator algebra of the commutator algebra of U, i.e. $C(C(U)) = [U]$.*

For the applications of this theorem let us remark that clearly any set of orthogonal transformations over a real field K is fully reducible and in particular the groups $O(n)$ and $SO(n)$ are fully reducible.

DEFINITION 4.1.2. Let $f(x^{(1)}, \ldots, x^{(k)})$ a function on V^k. We say that f is invariant on W associated to k vectors, with respect to U, if $f(x^{(1)}, \ldots, x^{(k)}) = f(x^{(1)}A, \ldots, x^{(k)}A)$ for all A in U and all x_i in W.

THEOREM 4.1.3. (First main theorem on invariants of the orthogonal group.) *If f is a function invariant on $\mathbb{R}^n$ associated to k vectors with respect to $O(n)$, then there exists a polynomial p (in k^2 variables) such that:*

$$f(x^{(1)}, \ldots, x^{(k)}) = p(\langle x^{(1)}, x^{(1)}\rangle, \langle x^{(1)}, x^{(2)}\rangle, \ldots, \langle x^{(1)}, x^{(k)}\rangle, \ldots, \langle x^{(k)}, x^{(1)}\rangle, \ldots, \langle x^{(k)}, x^{(k)}\rangle)$$

for all $x^{(i)} \in \mathbb{R}^n, i = 1, \dots, k$, *where* $\langle -, - \rangle$ *denotes the canonical scalar product in* $\mathbb{R}^n$.

THEOREM 4.1.4. (First main theorem on invariants of the special orthogonal group.) *If* f *is a function invariant on* $\mathbb{R}^n$ *associated to* k *vectors with respect to* $SO(n)$, *then* f *is in the* $\mathbb{R}$*-algebra generated by functions of the form:*

(i) $[x^{(i_1)}, \dots, x^{(i_n)}]$,
(ii) $g(x^{(1)}, \dots, x^{(k)})$,

where

$$[x^{(i_1)}, \dots, x^{(i_n)}] := \begin{vmatrix} x_1^{(i_1)} & \dots & x_n^{(i_1)} \\ \vdots & & \vdots \\ x_1^{(i_n)} & \dots & x_n^{(i_n)} \end{vmatrix},$$

$$i_j \in \{1, \dots, k\}, j = 1, \dots, n, \quad \text{(bracket factor)},$$

and g *is a polynomial combination of the scalar products* $\langle x^{(i)}, x^{(j)} \rangle$, $i, j \in \{1, \dots, k\}$.

DEFINITION 4.2. Let $A \in \mathcal{M}(n \times m, K)$, $B \in \mathcal{M}(n' \times m', K)$. The Kronecker product of A and B is the $nn' \times mm'$ matrix:

$$A \otimes B := \begin{pmatrix} a_{1,1}B & \dots & a_{1,n}B \\ \vdots & & \vdots \\ a_{m,1}B & \dots & a_{m,n}B \end{pmatrix}.$$

Thus if r is a natural number and $A \in \mathcal{M}(n, K)$ the r-Kronecker power of A is the $n^r \times n^r$ matrix:

$$\Pi_r(A) := A \otimes \dots^{(r)} \dots \otimes A =$$

$$\begin{pmatrix} a(1,1,\dots,1;1,1,\dots,1) & a(1,1,\dots,1;1,1,\dots,2) \cdots a(1,1,\dots,1;n,n,\dots,n) \\ a(1,1,\dots,2;1,1,\dots,1) & a(1,1,\dots,2;1,1,\dots,2) \cdots a(1,1,\dots,2;n,n,\dots,n) \\ \vdots & \vdots \qquad\qquad \vdots \\ a(n,n,\dots,n;1,1,\dots,1) & a(n,n,\dots,n;1,1,\dots,2) \cdots a(n,n,\dots,n;n,n,\dots,n) \end{pmatrix}$$

where $a(i_1, \dots, i_r; k_1, \dots, k_r) := a(i_1, k_1)a(i_2, k_2) \dots a(i_r, k_r)$.

4.2.1. Some basic properties of $\Pi_r(A)$ are:

(1) $\Pi_r(A)$ is bisymmetric, i.e., if η is a permutation of $(1, \dots, r)$, then

$$a(i_1, \dots, i_r; k_1, \dots, k_r) = a(i_{\eta(1)}, \dots, i_{\eta(r)}; k_{\eta(1)}, \dots, k_{\eta(r)})$$

for all indices i_j, k_m.

(2) $\Pi_r(AB) = \Pi_r(A)\Pi_r(B)$, thus if U is a matrix group then the set

$$\Pi_r(U) := \{\Pi_r(A)/A \in U\}$$

is also a matrix group.

(3) If A is an orthogonal matrix then $\Pi_r(A)$ is also orthogonal.

(4) If $\det(A) = 1$ then $\det(\Pi_r(A)) = 1$.

We shall proceed further considering the set $R^{(r)}$ of matrices of the form:

$$A^{(r)} := \begin{pmatrix} A_r & & & 0 \\ & A_{r-1} & & \\ & & \ddots & \\ 0 & & & A_0 \end{pmatrix}$$

where A_v is a $n^v \times n^v$ bisymmetric matrix and we denote its entries by $a(i_1, \ldots, i_v; k_1, \ldots, k_v) \in K$, $i_j, k_s \in \{1, \ldots, n\}$, $v = 1, \ldots, r$, and $A_0 := a(-,-) \in K$. Remark that $R^{(r)} \subset \mathcal{M}(m, K)$, where $m = 1 + n + n^2 + \cdots + n^r = \frac{n^{r+1}-1}{n-1}$.

4.2.2. Some properties of $R^{(r)}$ are:

(1) $R^{(r)}$ is a matrix algebra. In fact the product $A(r)B(r)$ is a bisymmetric matrix that has the form:

$$\begin{pmatrix} A_r B_r & & & 0 \\ & A_{r-1}B_{r-1} & & \\ & & \ddots & \\ 0 & & & A_0 B_0 \end{pmatrix}.$$

(2) A subset of $R^{(r)}$ is the set of matrices of the form:

$$\Pi^{(r)}(A) := \begin{pmatrix} \Pi_r(A) & & & 0 \\ & \Pi_{(r-1)}(A) & & \\ & & \ddots & \\ 0 & & & \Pi_0(A) \end{pmatrix},$$

where $A \in SO(n)$ and $\Pi_0(A) := 1$.

It is easy to verify that $\Pi^{(r)}(AB) = \Pi^{(r)}(A)\Pi^{(r)}(B)$, so the set $\Pi^{(r)}(SO(n)) := \{\Pi^{(r)}(A)/A \in SO(n)\}$ is a group; besides $\Pi^{(r)}(A)\Pi^{(r)}(A)^t = I$, so $\Pi^{(r)}(SO(n))$ is a set of orthogonal matrices and consequently fully reducible.

We are interested in describing the enveloping algebra of $\Pi^{(r)}(SO(n))$, which agrees with $C(C(\Pi^{(r)}(SO(n))))$, after Theorem 4.1.1. Note first that if we denote:

$$B = \begin{pmatrix} B_{r,r} & B_{r,r-1} & \ldots & B_{r,0} \\ \vdots & \vdots & & \vdots \\ B_{0,r} & B_{0,r-1} & \ldots & B_{0,0} \end{pmatrix}$$

where $B_{u,v}$ are matrices of dimension $n^u \times n^v$ and coefficients $b(i_1, \ldots, i_u; j_1, \ldots, j_v)$, with $i_k, j_l \in \{1, \ldots, n\}$, then:

$$C(\Pi^{(r)}(SO(n))) = \{B \in M(m,K)/B\Pi^{(r)}(A) = \Pi^{(r)}(A)B, \text{ for all } A \in SO(n)\}.$$

PROPOSITION 4.2.3. *If $B \in C(\Pi^{(r)}(SO(n)))$, and for each $u, v \in \{0, \dots, r\}$, we consider the multilinear form asociated to $u+v$ vectors:*

$$\begin{array}{cccc} f_{u,v}: & (\mathbb{R}^n)^{u+v} & \longrightarrow & \mathbb{R} \\ & \begin{array}{c}(x^{(1)}, \dots, x^{(u)}; \\ y^{(1)}, \dots, y^{(v)})\end{array} & \longmapsto & \displaystyle\sum_{i_j, k_l \in \{1,\dots,n\}} b(i_1, \dots, i_u; k_1, \dots, k_v) \\ & & & x^{(1)}_{i_1} \dots x^{(u)}_{i_u} y^{(1)}_{k_1} \dots y^{(v)}_{k_v} \end{array}$$

defined by means of the (u,v)-entry $B_{u,v}$ of B, then $f_{u,v}$ is invariant asociated to $u+v$ vectors with respect to the group $SO(n)$.

PROOF. First note that

$$f_{u,v}(x^{(1)}, \dots, x^{(u)}; y^{(1)}, \dots, y^{(v)}) = (x^{(1)} \otimes \dots \otimes x^{(u)}) B_{u,v}((y^{(1)})^t \otimes \dots \otimes (y^{(v)})^t).$$

Let $A \in SO(n)$; since $B \in C(\Pi^{(r)}(SO(n)))$ it is easy to check that $B_{u,v} = \Pi_u(A) B_{u,v}(\Pi_v(A))^t$ for all $u, v \in \{0, \dots, r\}$ and for all $A \in SO(n)$. Then we have the following identities:

$$\begin{aligned} & f_{u,v}(x^{(1)}A, \dots, x^{(u)}A, y^{(1)}A, \dots, y^{(v)}A) = \\ &= (x^{(1)}A \otimes \dots \otimes x^{(u)}A) B_{u,v}((y(1)A)^t \otimes \dots \otimes (y(v)A)^t) = \\ &= (x^{(1)} \otimes \dots \otimes x^{(u)}) \Pi_u(A) B_{u,v}(\Pi_v(A))^t ((y(1))^t \otimes \dots \otimes (y(v))^t) = \\ &= (x^{(1)} \otimes \dots \otimes x^{(u)}) B_{u,v}((y^{(1)})^t \otimes \dots \otimes (y^{(v)})^t) = \\ &= f_{u,v}(x^{(1)}, \dots, x^{(u)}; y^{(1)}, \dots, y^{(v)}). \quad \square \end{aligned}$$

COROLLARY 4.2.4. *With the notation above, if $B \in C(\Pi^{(r)}(SO(n)))$, then $f_{u,v}$ is in the $\mathbb{R}$-algebra generated by functions of the form:*

(1) $[z^{(i_1)}, \dots, z^{(i_n)}]$,

(2) $g(z^{(1)}, \dots, z^{(k)})$,

where $[z^{(i_1)}, \dots, z^{(i_n)}]$, is the bracket factor, g is a polynomial combination of the scalar products $\langle z^{(i)}, z^{(j)} \rangle$, $i, j \in \{1, \dots, k\}$, and the $z^{(i)}$ are choosen among the $x^{(l)}$ or $y^{(j)}$ variables. Moreover f can be written in such a way that there is no repeated variable in every monomial.

PROOF. After theorem 4.1.4, it is obvious that $f_{u,v}$ belongs to the $\mathbb{R}$-algebra generated by the functions stated in (i) and (ii). Besides $f_{u,v}$ has degree 1 in every variable $x^{(i)}_j$, so that functions generating it cannot have repeated variables. (Remark also that bracket factors with repeated variables $z^{(i)}$ are identically zero.) $\square$

Explaining a little more on the form of functions $f_{u,v}$ we remark that it is a linear combination of functions of the form:

$$\begin{aligned} (i)\ f^{\sigma\eta ab}_{u,v}(x^{(1)}, \dots, x^{(u)}; y^{(1)}, \dots, y^{(v)}) := \langle x^{(\sigma(1))}, x^{(\sigma(2))} \rangle \cdot \ldots \cdot \langle x^{(\sigma(2a-1))}, \\ x^{(\sigma(2a))} \rangle \cdot \langle y^{(\eta(1))}, y^{(\eta(2))} \rangle \cdot \ldots \cdot \langle y^{(\eta(2b-1))}, y^{(\eta(2b))} \rangle \cdot \\ \cdot \langle x^{(\sigma(2a+1))}, y^{(\eta(2b+1))} \rangle \cdot \ldots \cdot \langle x^{(\sigma(u))}, y^{(\eta(v))} \rangle. \end{aligned}$$

$$(ii)\ g^{\sigma\eta\omega\alpha}_{\substack{cda'b'\\u,v}}(x^{(1)},\dots,x^{(u)};y^{(1)},\dots,y^{(v)}) := [x^{(\sigma(1))},\dots,x^{(\sigma(c))},y^{(\eta(1))},\\ \dots,y^{(\eta(d))}]\cdot f^{\omega\alpha a'b'}_{u-c,v-d}(x^{(\sigma(c+1))},\dots,x^{(\sigma(u))};y^{(\eta(d+1))},\dots,y^{(\eta(v))})$$

where σ and η are permutations of $(1,\dots,u)$ and $(1,\dots,v)$ respectively, $u-2a = v-2b$, $u-c-2a' = v-d-2b'$, $c+d=n$ and, finally, ω and α are permutations of $(\sigma(c+1),\dots,\sigma(u))$ and $(\eta(d+1),\dots,\eta(v))$ respectively. Remark that powers of bracket factors do not appear because of the relation:

$$[x^{(1)},\dots,x^{(n)}]\cdot[y^{(1)},\dots,y^{(n)}] = \begin{vmatrix} \langle x^{(1)},y^{(1)}\rangle & \dots & \langle x^{(1)},y^{(n)}\rangle \\ \vdots & & \vdots \\ \langle x^{(n)},y^{(1)}\rangle & \dots & \langle x^{(n)},y^{(n)}\rangle \end{vmatrix}.$$

Let's denote by $\delta^{\sigma\eta ab}_{u,v}(i_{1,\dots,u};k_{1,\dots,v})$ the function:

$$\delta(i_{\sigma(1)},i_{\sigma(2)})\cdot\ldots\cdot\delta(i_{\sigma(2a-1)},i_{\sigma(2a)})\cdot\delta(k_{\eta(1)},k_{\eta(2)})\cdot\ldots\cdot\\ \delta(k_{\eta(2b-1)},k_{\eta(2b)})\cdot\delta(i_{\sigma(2a+1)},k_{\eta(2b+1)})\cdot\ldots\cdot\delta(i_{\sigma(u)},k_{\eta(v)})$$

and by $\gamma^{\sigma\eta cd}_{u,v}(i_{1,\dots,u};k_{1,\dots,v})$ the function:

$$\#\gamma(i_{\sigma(1)},i_{\sigma(2)})\cdot\ldots\cdot\gamma(i_{\sigma(1)},i_{\sigma(c)})\cdot\gamma(i_{\sigma(1)},k_{\eta(1)})\cdot\ldots\cdot\gamma(i_{\sigma(1)},k_{\eta(d)})\cdot\\ \cdot\gamma(i_{\sigma(2)},i_{\sigma(3)})\cdot\ldots\cdot\gamma(i_{\sigma(2)},i_{\sigma(c)})\cdot\gamma(i_{\sigma(2)},k_{\eta(1)})\cdot\ldots\cdot\gamma(i_{\sigma(2)},k_{\eta(d)})\cdot\\ \vdots\\ \cdot\gamma(i_{\sigma(c)},k_{\eta(1)})\cdot\ldots\cdot\gamma(i_{\sigma(c)},k_{\eta(d)})\cdot\\ \cdot\gamma(k_{\eta(1)},k_{\eta(2)})\cdot\ldots\cdot\gamma(k_{\eta(1)},k_{\eta(d)})\cdot\\ \dots\\ \cdot\gamma(k_{\eta(d-1)},k_{\eta(d)})$$

where $\gamma(i,j) := 1-\delta(i,j)$ and $\#$ takes value 1 or -1 according to the parity of the permutations σ and η ; in fact when the function $\gamma^{\sigma\eta cd}_{u,v}(i_{1,\dots,u};k_{1,\dots,v})$ is different from 0 the n-uple $(\sigma(1),\dots,\sigma(c),\eta(1),\dots,\eta(d))$ is a permutation of $(1,\dots,n)$; if it is even then $\# := 1$ and $\# := -1$ in other case. Remark that both $\delta_{u,v}$ and $\gamma_{u,v}$ are $\{0,1,-1\}$ valued functions.

COROLLARY 4.2.5. *With the notation above, if $B \in C(\Pi^{(r)}(SO(n)))$ then we have, for all indices i_j and k_l that:*

$$b(i_1,\dots,i_u;k_1,\dots,k_v) =$$

$$= \sum_{\sigma,\eta,a,b}\lambda_{\sigma\eta ab}\cdot\delta^{\sigma\eta ab}_{u,v}(i_{1,\dots,u};k_{1,\dots,v})+$$

$$+\sum_{\substack{\sigma,\eta,\omega,\alpha\\c,d,a,b}}\beta_{\substack{\sigma\eta\omega\alpha\\cdab}}\cdot\gamma^{\sigma\eta cd}_{u,v}(i_{1,\dots,u};k_{1,\dots,v})\cdot\delta^{\omega\alpha ab}_{u-c,v-d}(i_{\sigma(c+1),\dots,\sigma(u)};k_{\eta(d+1),\dots,\eta(v)})$$

where $\lambda_{\sigma\eta ab}$ *and* $\beta_{\substack{\sigma\eta\omega\alpha \\ cdab}}$ *are constants.*

PROOF. As we have seen, it is possible to write:

$$f_{u,v} = \sum_{\sigma,\eta,a,b} \lambda_{\sigma\eta ab} \cdot f_{u,v}^{\sigma\eta ab} + \sum_{\substack{\sigma,\eta,\omega,\alpha \\ c,d,a,b}} \beta_{\substack{\sigma\eta\omega\alpha \\ cdab}} \cdot g_{u,v}^{\substack{\sigma\eta\omega\alpha \\ cdab}} .$$

In particular we can write:

$$\sum_{\sigma,\eta,a,b} \lambda_{\sigma\eta ab} \cdot f_{u,v}^{\sigma\eta ab} = \sum_{i_j,k_l} \Big[\sum_{\sigma,\eta,a,b} \lambda_{\sigma\eta ab} \cdot \delta_{u,v}^{\sigma\eta ab}(i_{1,\dots,u}; k_{1,\dots,v}) \Big] x_{i_1}^{(1)} \dots x_{i_u}^{(u)} y_{k_1}^{(1)} \dots y_{k_v}^{(v)}$$

and analogously:

$$\sum_{\substack{\sigma,\eta,\omega,\alpha \\ c,d,a,b}} \beta_{\substack{\sigma\eta\omega\alpha \\ cdab}} \cdot g_{u,v}^{\substack{\sigma\eta\omega\alpha \\ cdab}} =$$

$$= \sum_{i_j,k_l} \Big[\sum_{\sigma,\eta,a,b} \beta_{\substack{\sigma\eta\omega\alpha \\ cdab}} \cdot \gamma_{u,v}^{\sigma\eta ab}(i_{1,\dots,u}; k_{1,\dots,v}) +$$

$$+ \sum_{\substack{\sigma,\eta,\omega,\alpha \\ c,d,a,b}} \beta_{\substack{\sigma\eta\omega\alpha \\ cdab}} \cdot \gamma_{u,v}^{\sigma\eta cd}(i_{1,\dots,u}; k_{1,\dots,v}) \cdot \delta_{u-c,v-d}^{\omega\alpha ab}(i_{\sigma(c+1),\dots,\sigma(u)}; k_{\eta(d+1),\dots,\eta(v)}) \Big]$$

$$x_{i_1}^{(1)} \dots x_{i_u}^{(u)} y_{k_1}^{(1)} \dots y_{k_v}^{(v)} .$$

So, identifying in the first expression the coefficients corresponding to the monomial $x_{i_1}^{(1)} \dots x_{i_u}^{(u)} y_{k_1}^{(1)} \dots y_{k_v}^{(v)}$, we obtain the desired equality. □

4.3. After this preparation we are ready to prove the main point, namely the description of the set of "linearized orthogonal matrices" (i.e. $[\Pi^{(r)}(SO(n))]$) by means of a set $U^{(r)}$ of linear equations in the entries of the matrices in $R^{(r)}$. As explained at the beginning of this paragraph, we shall use theorem 4.1.1 to test equality between $U^{(r)}$ and $[\Pi^{(r)}(SO(n))]$. Naturally we shall profit from the special form of writing the elements in $C(\Pi^{(r)}(SO(n)))$ given by corollary 4.2.5.

DEFINITION 4.3.1. Consider the sets of matrices:[1]

$$\begin{aligned} U^{(r)} := \Big\{ & A^{(r)} \in R^{(r)} / \\ & \sum_{k=1}^{n} a(i_1, \dots, i_v; k, k, k_3, \dots, k_v) = \delta(i_1, i_2) a(i_3, \dots, i_v; k_3, \dots, k_v), \\ & \sum_{i=1}^{n} a(i, i, i_3, \dots, i_v; k_1, \dots, k_v) = \delta(k_1, k_2) a(i_3, \dots, i_v; k_3, \dots, k_v), \\ & \qquad v \in \{2, \dots, r\}; i_j, k_j \in \{1, \dots, n\} \Big\}. \end{aligned}$$

[1]The set $U^{(r)}$ alone is the one needed for the proof of the orthogonal case.

We will denote by π_a a permutation of $(1,\dots,a)$, and $|\pi_a|$ takes value 1 if π_a is even and -1 if it is odd.

$$\begin{aligned} T^{(r)} := \Big\{ A^{(r)} \in R^{(r)} / & \\ & \sum_{\pi_a} |\pi_a| a(i_1,\dots,i_a,i_{a+1},\dots,i_v;k_{\pi_a(1)},\dots,k_{\pi_a(a)},k_{a+1},\dots,k_v) \\ & = \sum_{\omega_b} |\omega_b| a(I_1,\dots,I_b,i_{a+1},\dots,i_v;K_{\omega_b(1)},\dots,K_{\omega_b(b)},k_{a+1},\dots,k_v), \\ & \quad (i_1,\dots,i_a,I_1,\dots,I_b) \text{ and } (k_1,\dots,k_a,K_1,\dots,K_b) \\ & \quad \text{even permutations of } (1,\dots,n), i_{a+1},\dots,i_v,k_{a+1},\dots,k_v \\ & \qquad \in \{1,\dots,n\}, a+b=n, a \geq b, 1 \leq v \leq r \Big\}. \end{aligned}$$

Finally let us call $\underline{U}^{(r)} := U^{(r)} \cap T^{(r)}$.

It is an easy exercise to prove that all these sets are $I\!K$-algebras.

THEOREM 4.3.2. (Cf. [**W**].) *For any natural number* r,

$$\underline{U}^{(r)} := [\Pi^{(r)}(SO(n))].$$

PROOF. For the inclusion $[\Pi^{(r)}(SO(n))] \subset \underline{U}^{(r)}$, since $\underline{U}^{(r)}$ is a $I\!K$-algebra, it suffices to prove $\Pi^{(r)}(SO(n)) \subset \underline{U}^{(r)}$. Let $\Pi^{(r)}(A) \in \Pi^{(r)}(SO(n))$, $A \in SO(n)$; as we have seen, $\Pi^{(r)}(A) \in R^{(r)}$ and it verifies:

$$\begin{aligned} \sum_{k=1}^{n} a(i_1,\dots,i_v;k,k,k_3,\dots,k_v) &= \Big[\sum_{k=1}^{n} a(i_1,k)a(i_2,k)\Big] x(i_3,k_3)\dots x(i_v,k_v) \\ &= \delta(i_1,i_2) a(i_3,\dots,i_v;k_3,\dots,k_v). \end{aligned}$$

The same reasoning concludes the other condition needed in order to prove $\Pi^{(r)}(A) \in U^{(r)}$.

On the other hand, technical result 3.1.4 allows us to establish the central identity among the following, to obtain that $\Pi^{(r)}(A) \in T^{(r)}$.

$$\begin{aligned} & \sum_{\pi_a} |\pi_a| a(i_1,\dots,i_a,i_{a+1},\dots,i_v;k_{\pi_a(1)},\dots,k_{\pi_a(a)},k_{a+1},\dots,k_v) = \\ & = \Big[\sum_{\pi_a} |\pi_a| a(i_1,k_{\pi_a(1)})\dots,a(i_a,k_{\pi_a(a)})\Big] a(i_{a+1},k_{a+1})\dots,a(i_v,k_v) = \\ & = \begin{vmatrix} a(i_1,k_1) & \dots & a(i_1,k_a) \\ \vdots & & \vdots \\ a(i_a,k_1) & \dots & a(i_a,k_a) \end{vmatrix} a(i_{a+1},k_{a+1})\dots,a(i_v,k_v) = \\ & = \begin{vmatrix} a(I_1,K_1) & \dots & a(I_1,K_b) \\ \vdots & & \vdots \\ a(I_b,K_1) & \dots & a(I_b,K_b) \end{vmatrix} a(i_{a+1},k_{a+1})\dots,a(i_v,k_v) = \\ & = \Big[\sum_{\omega_b} |\omega_b| a(I_1,K_{\omega_b(1)})\dots,a(I_b,K_{\omega_b(b)})\Big] a(i_{a+1},k_{a+1})\dots,a(i_v,k_v) = \\ & = \sum_{\omega_b} |\omega_b| a(I_1,\dots,I_b,i_{a+1},\dots,i_v;K_{\omega_b(1)},\dots,K_{\omega_b(b)},k_{a+1},\dots,k_v). \end{aligned}$$

With respect to the other inclusion, after theorem 4.1.1., it suffices to prove $\underline{U}^{(r)} \in C(C(\Pi^{(r)}(SO(n))))$. Let $A^{(r)} \in \underline{U}^{(r)}$ and $B \in C(\Pi^{(r)}(SO(n)))$. We will prove, as needed, that $B_{u,v}A_v = A_u B_{u,v}$ for all $u, v \in \{0, \ldots, r\}$. Multiplying the $(i_1, \ldots, i_u)$-row of $B_{u,v}$ times the $(k_1, \ldots, k_v)$-column of A_v we obtain:

$$\sum_j b(i_1, \ldots, i_u; j_1, \ldots, j_v) a(j_1, \ldots, j_v; k_1, \ldots, k_v)$$

$$= \sum_j \Big[\sum_{\sigma,\eta,a,b} \lambda_{\sigma\eta ab} \delta^{\sigma\eta ab}_{u,v}(i_{1,\ldots,u}; k_{1,\ldots,v})$$

$$+ \sum_{\substack{\sigma,\eta,\omega,\alpha \\ c,d,a,b}} \beta_{\substack{\sigma\eta\omega\alpha \\ cdab}} \gamma^{\sigma\eta cd}_{u,v}(i_{1,\ldots,u}; k_{1,\ldots,v}) \delta^{\omega\alpha ab}_{u-c,v-d}(i_{\sigma(c+1),\ldots,\sigma(u)}; k_{\eta(d+1),\ldots,\eta(v)}) \Big]$$

$$a(j_1, \ldots, j_v; k_1, \ldots, k_v) = \sum_{\sigma,\eta,a,b} \lambda_{\sigma\eta ab} \beta_{\substack{\sigma\eta\omega\alpha \\ cdab}}$$

$$\Big[\sum_j \delta^{\sigma\eta ab}_{u,v}(i_{1,\ldots,u}; k_{1,\ldots,v}) a(j_1, \ldots, j_v; k_1, \ldots, k_v) \Big]$$

$$+ \sum_{\substack{\sigma,\eta,\omega,\alpha \\ c,d,a,b}} \Big[\sum_j \gamma^{\sigma\eta cd}_{u,v}(i_{1,\ldots,u}; k_{1,\ldots,v}) \delta^{\omega\alpha ab}_{u-c,v-d}$$

$$(i_{\sigma(c+1),\ldots,\sigma(u)}; k_{\eta(d+1),\ldots,\eta(v)}) a(j_1, \ldots, j_v; k_1, \ldots, k_v) \Big].$$

Making the equivalent computation with $A_u B_{u,v}$ we see that it suffices to prove, for $\sigma, \eta, \omega, \alpha, a, b, c, d$ fixed, the two identities:

(I_1)

$$\sum_j \delta^{\sigma\eta ab}_{u,v}(i_{1,\ldots,u}; j_{1,\ldots,v}) a(j_1, \ldots, j_v; k_1, \ldots, k_v)$$

$$= \sum_j \delta^{\sigma\eta ab}_{u,v}(j_{1,\ldots,u}; k_{1,\ldots,v}) a(i_1, \ldots, i_u; j_1, \ldots, j_u).$$

(I_2)

$$\sum_j \gamma^{\sigma\eta cd}_{u,v}(i_{1,\ldots,u}; j_{1,\ldots,v}) \delta^{\omega\alpha ab}_{u-c,v-d}(i_{\sigma(c+1),\ldots,\sigma(u)}; j_{\eta(d+1),\ldots,\eta(v)})$$

$$a(j_1, \ldots, j_v; k_1, \ldots, k_v) = \sum_j \gamma^{\sigma\eta cd}_{u,v}(j_{1,\ldots,u}; k_{1,\ldots,v}) \delta^{\omega\alpha ab}_{u-c,v-d}(j_{\sigma(c+1),\ldots,\sigma(u)};$$

$$k_{\eta(d+1),\ldots,\eta(v)}) a(i_1, \ldots, i_u; j_1, \ldots, j_u).$$

Playing with the expressions in (I_1) and since $A^{(r)} \in U^{(r)}$, we obtain that both sides of the equality are equal to:

$$\delta(i_{\sigma(1)}, i_{\sigma(2)}) \ldots \delta(i_{\sigma(2a-1)}, i_{\sigma(2a)}) \delta(k_{\eta(1)}, k_{\eta(2)}) \ldots \delta(k_{\eta(2b-1)}, k_{\eta(2b)})$$

$$a(i_{\sigma(2a+1)}, \ldots, i_{\sigma(u)}; k_{\eta(2b+1)}, \ldots, k_{\eta(v)}).$$

With respect to (I_2) in order to simplify [sic] the notation, we will suppose that the fixed permutations are identities. Again playing with the indices and after using the definition of $U^{(r)}$ we have that the identity in (I_2) is equivalent to:

$$\sum_j \#a(j_1, \dots, j_d, i_{c+2a+1}, \dots, i_u; k_1, \dots, k_d, k_{d+2b+1}, \dots, k_v)$$
$$= \sum_{j'} \#a(i_1, \dots, i_c, i_{c+2a+1}, \dots, i_u; j'_1, \dots, j'_c, k_{d+2b+1}, \dots, k_v)$$

where $\{i_1, \dots, i_c, j_1, \dots, j_d\}$ and $\{k_1, \dots, k_d, j'_1, \dots, j'_c\}$ are pairwise distinct; thus, and since $c + d = n$, the sums in j and j' are sums in the permutations of $\{1, \dots, n\} - \{i_1, \dots, i_c\}$ and $\{1, \dots, n\} - \{k_1, \dots, k_d\}$ respectively, so that the last identity is one of the stated in the definition of the set $T^{(r)}$. We remark that in this definition the conditions on the permutations $(i_1, \dots, i_a, I_1, \dots, I_b)$ and $(k_1, \dots, k_a, K_1, \dots, K_b)$ to be even is not essential in this proof. □

4.4. Finally we arrive at the description of the Weyl basis for $SO(n)$ and $O(n)$ in terms of the polynomials $\{\{D_{i,j}\}_{i,j}, \{D^*_{i,j}\}_{i,j}, \{D(\pi, \omega; a, b)\}_{\pi,\omega,a,b}\}$ introduced in §2.

THEOREM 4.4.1. *Let* $p(x_{1,1}, \dots, x_{1,n}, \dots, x_{n,1}, \dots, x_{n,n}) \in \mathbb{R}[\underline{x}]$ *of degree* r *such that it vanishes on all proper real orthogonal matrices. Then* p *can be written in the form:*

$$p = \sum_{i,j} L_{i,j} . D_{i,j} + \sum_{i,j} L^*_{i,j} . D^*_{i,j} + \sum_{\pi,\omega,a,b} H_{\pi,\omega,a,b} . D(\pi, \omega; a, b)$$

where $\deg(L_{i,j}) \leq r - 2, \deg(L^*_{i,j}) \leq r - 2$ *and* $\deg(H_{\pi,\omega,a,b}) \leq r - a$.

PROOF. We can write p in bisymmetric form, i.e.:

$$p(\underline{x}) = \sum_{v=0}^{r} \Gamma(i_1, \dots, i_v; k_1, \dots, k_v) x_{i_1,k_1} \cdots x_{i_v,k_v}$$

where $\Gamma(i_1, \dots, i_v; k_1, \dots, k_v) = \Gamma(i_{\sigma(1)}, \dots, i_{\sigma(v)}; k_{\sigma(1)}, \dots, k_{\sigma(v)})$ for all σ permutations of $(1, \dots, v)$. Consider the function:

$$g = \sum_{v=0}^{r} \sum_{i_j, k_l} \Gamma(i_1, \dots, i_v; k_1, \dots, k_v) x(i_1, \dots, i_v; k_1, \dots, k_v) .$$

Remark that g is a linear form in the variables

$$x(-, -),$$
$$x(1, 1), x(1, 2), \dots, x(n, n),$$
$$\dots,$$
$$x(1, \dots^{(r)} \dots, 1; 1, \dots^{(r)} \dots, 1), x(1, \dots, 2; 1, \dots, 1), \dots, x(n, \dots, n; n, \dots, n)$$

that vanishes on all matrices in $\Pi^{(r)}(SO(n))$, because $\Pi^{(r)}(A) = p(A) = 0$ for all $A \in SO(n)$, so it vanishes also in $[\Pi^{(r)}(SO(n))] = \underline{U}^{(r)}$. Since $\underline{U}^{(r)}$ is the set of zeroes of the linear forms:

$$H_v(i_{1,\ldots,v}; k_{3,\ldots,v}) := \sum_{k=1}^{n} x(i_1, \ldots, i_v; k, k, k_3, \ldots, k_v) - \delta(i_1, i_2)$$
$$x(i_3, \ldots, i_v; k_3, \ldots, k_v),$$
$$F_v(i_{3,\ldots,v}; k_{1,\ldots,v}) := \sum_{i=1}^{n} x(i, i, i_3, \ldots, i_v; k_1, \ldots, k_v) - \delta(k_1, k_2)$$
$$x(i_3, \ldots, i_v; k_3, \ldots, k_v),$$
$$T_v(i_{1,\ldots,v}; k_{1,\ldots,v}; \sigma) := x(i_1, \ldots, i_v; k_1, \ldots, k_v) - x(i_{\sigma(1)}, \ldots, i_{\sigma(v)};$$
$$k_{\sigma(1)}, \ldots, k_{\sigma(v)}), Q_v(i_{1,\ldots,v}; k_{1,\ldots,v}; I_{1,\ldots,v}; K_{1,\ldots,v}; a, b; \pi_a, \omega_b)$$
$$:= \sum_{\pi_a} |\pi_a| a(i_1, \ldots, i_a, i_{a+1}, \ldots, i_v; k_{\pi_a(1)}, \ldots, k_{\pi_a(a)},$$
$$k_{a+1}, \ldots, k_v)$$
$$- \sum_{\omega_b} |\omega_b| a(I_1, \ldots, I_b, i_{a+1}, \ldots, i_v; K_{\omega_b(1)}, \ldots, K_{\omega_b(b)},$$
$$k_{a+1}, \ldots, k_v)$$

indices and permutations running over above indicated ranges, then g is a linear combination of these forms, i.e.:

$$\sum_{v=0}^{r} \Gamma(i_1, \ldots, i_v; k_1, \ldots, k_v) x_{i_1,k_1} \ldots x_{i_v,k_v}$$
$$= \sum_{v=2}^{r} \sum_{i,k} h_v(i_{1,\ldots,v}; k_{3,\ldots,v}) H_v(i_{1,\ldots,v}; k_{3,\ldots,v})$$
$$+ \sum_{v=2}^{r} \sum_{i,k} f_v(i_{3,\ldots,v}; k_{1,\ldots,v}) F_v(i_{3,\ldots,v}; k_{1,\ldots,v})$$
$$+ \sum_{v=1}^{r} \sum_{i,k,\sigma} t_v(i_{1,\ldots,v}; k_{1,\ldots,v}; \sigma) T_v(i_{1,\ldots,v}; k_{1,\ldots,v}; \sigma)$$
$$+ \sum_{v=1}^{r} \sum_{i,k,I,K,\pi_a,\omega_b} q_v(i_{1,\ldots,v}; k_{1,\ldots,v}; I_{1,\ldots,v}; K_{1,\ldots,v}; a, b; \pi_a, \omega_b)$$
$$Q_v(i_{1,\ldots,v}; k_{1,\ldots,v}; I_{1,\ldots,v}; K_{1,\ldots,v}; a, b; \pi_a, \omega_b)$$

where h_v, f_v, t_v and q_v are constants in the field.

Putting, in particular, $x(i_1, \ldots, i_v; k_1, \ldots, k_v) = x(i_1, k_1) x(i_2, k_2) \ldots x(i_v, k_v)$

and $x(-,-)=1$, we obtain:

$$p(\underline{x}) =$$

$$= \sum_{v=2}^{r} \sum_{i,k} h_v(i_{1,\ldots,v}; k_{3,\ldots,v})(x(i_3,k_3)\ldots x(i_v,k_v) D_{i_1,i_2}(\underline{x}))+$$

$$+ \sum_{v=2}^{r} \sum_{i,k} f_v(i_{1,\ldots,v}; k_{3,\ldots,v})(x(i_3,k_3)\ldots x(i_v,k_v) D^*_{i_1,i_2}(\underline{x}))+$$

$$+ \sum_{v=1}^{r} \sum_{i,k,I,K,\pi_a,\omega_b} q_v(i_{1,\ldots,v}; k_{1,\ldots,v}; I_{1,\ldots,v}; K_{1,\ldots,v}; a,b; \pi_a, \omega_b)$$

$$\left(x(i_{a+1},k_{a+1})\ldots x(i_v,k_v) D_{\pi_a,\omega_b;a,b}(\underline{x}) \right),$$

which is the desired combination verifying the requirements on the degrees. □

Analogously we have for the orthogonal ideal the following basis:

THEOREM 4.4.2. *Let* $p(x_{1,1},\ldots,x_{1,n},\ldots,x_{n,1},\ldots,x_{n,n}) \in \mathbb{R}[\underline{x}]$ *of degree* r *such that it vanishes on all real orthogonal matrices. Then* p *can be written in the form:*

$$p = \sum_{i,j} L_{i,j}.D_{i,j} + \sum_{i,j} L^*_{i,j}.D^*_{i,j}$$

where $\deg(L_{i,j}) \le r-2$ *and* $\deg(L^*_{i,j}) \le r-2$, $i,j \in 1,\ldots,n$.

COROLLARY 4.4.3.

(i) $\mathbf{so}.\mathbb{R}[\underline{x}] = \langle \{D_{i,j}\}_{i,j}, \{D^*_{i,j}\}_{i,j}, \{D(\pi,\omega;a,b)\}_{\pi,\omega,a,b}\rangle.\mathbb{R}[\underline{x}]$, *and this basis solves the ideal membership problem for* $\mathbf{so}.\mathbb{R}[\underline{x}]$.

(ii) $\mathbf{o}.\mathbb{R}[\underline{x}] = \langle \{D_{i,j}\}_{i,j}, \{D^*_{i,j}\}_{i,j}\rangle.\mathbb{R}[\underline{x}]$, *and this basis solves the ideal membership problem for* $\mathbf{o}.\mathbb{R}[\underline{x}]$.

PROOF. The equality between ideals is obvious from the theorems above and the technical results in 3.1. Concerning the ideal membership problem let us remark that given an element $p(\underline{x})$ in $\mathbf{so}.\mathbb{R}[\underline{x}]$ of degree r we can formally express an identity:

$$p = \sum_{i,j} L_{i,j}.D_{i,j} + \sum_{i,j} L^*_{i,j}.D^*_{i,j} + \sum_{\pi,\omega,a,b} H_{\pi,\omega,a,b}.D(\pi,\omega;a,b)$$

where $L_{i,j} L^*_{i,j}$ and $H_{\pi,\omega,a,b}$ are given with indeterminated coefficients as their degrees are bounded by $\deg(L_{i,j}) \le r-2, \deg(L^*_{i,j}) \le r-2$ and $\deg(H_{\pi,\omega,a,b}) \le r-a$. This identity yields a linear system of equations when we identify the coefficients of $p(\underline{x})$ with the linear combinations in the indeterminated coefficients. Solving this linear system gives us either that $p(\underline{x})$ does not belong to $\mathbf{so}.\mathbb{R}[\underline{x}]$ (if no solution exists) or the coefficients of the representation of $p(\underline{x})$ in terms of the Weyl basis. The same applies for the case $\mathbf{o}.\mathbb{R}[\underline{x}]$. □

5. Some examples and computational remarks

5.1 Weyl basis for $O(2)$ and $SO(2)$.

Weyl basis for $O(2)$	Weyl basis for $SO(2)$
$x_{1,1}^2 + x_{2,1}^2 - 1,$	$x_{1,1}^2 + x_{2,1}^2 - 1,$
$x_{1,2}^2 + x_{2,2}^2 - 1,$	$x_{1,2}^2 + x_{2,2}^2 - 1,$
$x_{1,1}x_{1,2} + x_{2,1}x_{2,2},$	$x_{1,1}x_{1,2} + x_{2,1}x_{2,2},$
$x_{1,1}^2 + x_{1,2}^2 - 1,$	$x_{1,1}^2 + x_{1,2}^2 - 1,$
$x_{2,1}^2 + x_{2,2}^2 - 1,$	$x_{2,1}^2 + x_{2,2}^2 - 1,$
$x_{1,1}x_{2,1} + x_{1,2}x_{2,2}$	$x_{1,1}x_{2,1} + x_{1,2}x_{2,2},$
	$x_{1,1} - x_{2,2},$
	$x_{1,2} + x_{2,1}$

5.2 Weyl basis for $O(3)$ and $SO(3)$.

Weyl basis for $O(3)$	Weyl basis for $SO(3)$
$x_{1,1}^2 + x_{2,1}^2 + x_{3,1}^2 - 1,$	$x_{1,1}^2 + x_{2,1}^2 + x_{3,1}^2 - 1,$
$x_{1,2}^2 + x_{2,2}^2 + x_{3,2}^2 - 1,$	$x_{1,2}^2 + x_{2,2}^2 + x_{3,2}^2 - 1,$
$x_{1,3}^2 + x_{2,3}^2 + x_{3,3}^2 - 1,$	$x_{1,3}^2 + x_{2,3}^2 + x_{3,3}^2 - 1,$
$x_{1,1}x_{1,2} + x_{2,1}x_{2,2} + x_{3,1}x_{3,2},$	$x_{1,1}x_{1,2} + x_{2,1}x_{2,2} + x_{3,1}x_{3,2},$
$x_{1,1}x_{1,3} + x_{2,1}x_{2,3} + x_{3,1}x_{3,3},$	$x_{1,1}x_{1,3} + x_{2,1}x_{2,3} + x_{3,1}x_{3,3},$
$x_{1,2}x_{1,3} + x_{2,2}x_{2,3} + x_{3,2}x_{3,3},$	$x_{1,2}x_{1,3} + x_{2,2}x_{2,3} + x_{3,2}x_{3,3},$
$x_{1,1}^2 + x_{1,2}^2 + x_{1,3}^2 - 1,$	$x_{1,1}^2 + x_{1,2}^2 + x_{1,3}^2 - 1,$
$x_{2,1}^2 + x_{2,2}^2 + x_{2,3}^2 - 1,$	$x_{2,1}^2 + x_{2,2}^2 + x_{2,3}^2 - 1,$
$x_{3,1}^2 + x_{3,2}^2 + x_{3,3}^2 - 1,$	$x_{3,1}^2 + x_{3,2}^2 + x_{3,3}^2 - 1,$
$x_{1,1}x_{2,1} + x_{1,2}x_{2,2} + x_{1,3}x_{2,3},$	$x_{1,1}x_{2,1} + x_{1,2}x_{2,2} + x_{1,3}x_{2,3},$
$x_{1,1}x_{3,1} + x_{1,2}x_{3,2} + x_{1,3}x_{3,3},$	$x_{1,1}x_{3,1} + x_{1,2}x_{3,2} + x_{1,3}x_{3,3},$
$x_{2,1}x_{3,1} + x_{2,2}x_{3,2} + x_{2,3}x_{3,3}$	$x_{2,1}x_{3,1} + x_{2,2}x_{3,2} + x_{2,3}x_{3,3},$
	$x_{1,1} - (x_{2,2}x_{3,3} - x_{3,2}x_{2,3}),$
	$x_{1,2} - (-x_{2,1}x_{3,3} + x_{3,1}x_{2,3}),$
	$x_{1,3} - (x_{2,1}x_{3,2} - x_{3,1}x_{2,2}),$
	$x_{2,1} - (x_{3,2}x_{1,3} - x_{1,2}x_{3,3}),$
	$x_{2,2} - (x_{1,1}x_{3,3} - x_{3,1}x_{1,3}),$
	$x_{2,3} - (x_{3,1}x_{1,2} - x_{1,1}x_{3,2}),$
	$x_{3,1} - (x_{1,2}x_{2,3} - x_{2,2}x_{1,3}),$
	$x_{3,2} - (x_{2,1}x_{1,3} - x_{1,1}x_{2,3}),$
	$x_{3,3} - (x_{1,1}x_{2,2} - x_{2,1}x_{1,2})$

5.3. Weyl bases are a particular case of what we have denominated Macaulay basis according to the following definition:

DEFINITION. A finite basis B of an ideal $\mathbf{I}$ in $\mathbb{K}[\underline{x}]$ is a Macaulay basis of $\mathbf{I}$ if for every $f \in \mathbf{I}$ there exist $h_1, \ldots, h_k \in B$ and $l_1, \ldots, l_k \in \mathbb{K}[\underline{x}]$ such that:

(i) $f = \sum_{j=1}^{k} l_j.h_j$,
(ii) $\deg(l_i) \leq \deg(f) - \deg(h_i) \geq 0,\ i \in \{1, \ldots, k\}$.

It is easy to see the following equivalence:

PROPOSITION. *Let $B = \{g_1, \ldots, g_r\}$ be a basis of $\mathbf{I}$; we denote by $\mathbf{I}^h$ the homogeneous ideal asociated to $\mathbf{I}$ with respect to a new variable x_0 and we write f^h to denote the polynomial f homogeneized with x_0. The following sentences are equivalent:*

(i) *B is a Macaulay basis of $\mathbf{I}$.*
(ii) *$\{g_1^h, \ldots, g_r^h\}$ is a basis of $\mathbf{I}^h$.*

5.4 Remark. Macaulay basis and Grobner basis share properties (both can be used to test ideal membership and also to find a basis of the homogeneized ideal, for instance) but the following example shows that although Grobner bases for degree compatible ordering are also Macaulay bases, there is a strict inclusion between the two concepts:

For $B := \{x + y, xy\} \subset \mathbb{C}[x, y]$ and $\mathbf{I} := \langle B \rangle$, we have that B is not a Grobner basis with respect to any order degree compatible. In fact we have only two possibilities: $x > y$ and $x < y$; and Grobner basis are, respectively $\{x + y, y^2\}$ and $\{x + y, x^2\}$. However B is Macaulay basis because, since $\{x + y, x^2\}$ is a Grobner basis with respect to an order degree compatible and $x^2 = x.(x+y) - xy$, each $f \in \mathbf{I}$ can be written as $f = l_1.(x + y) + l_2.x^2 = (l_1 + l_2.x).(x + y) - l_2.xy$ with $\deg(l_1) \leq \deg(f) - 1 \geq 0$ and $\deg(l_2) \leq \deg(f) - 2 \geq 0$. Thus $\deg(-l_2) \leq \deg(f) - 2 \geq 0$ and $\deg(l_1 + l_2.x) \leq \deg(f) - 1 \geq 0$.

5.5. The searching for a general rule to describe Grobner basis with respect to a suitable order for the orthogonal and proper orthogonal ideals has produced several negative results that we consider could be of interest when compared with Weyl basis. In the following summary we collect information concerning, for $n = 2, 3, 4$, when the Weyl basis is also a Grobner basis with respect to a degree compatible order.

	$n=2$	$n=3$	$n=4$
$SO(n)$	Yes, for all possible orders	Yes, for the many orders checked	No, for the row-order(*)
$O(n)$	Yes, for circular orders(**) No, for the remaining orders	No, for the many orders checked	No, for the row-order(*)

(*) The row-order for n is the degree compatible order in which

$$x_{1,1} > x_{1,2} > \ldots x_{1,n} > x_{2,1} > x_{2,2} > \cdots > x_{2,n} > \cdots > x_{n,1} > x_{n,2} > \ldots x_{n,n}.$$

(**) A circular order for $n = 2$ is a degree compatible order in which the ordering over the variables is of the kind:

$$x_{1,1} > x_{2,1} > x_{2,2} > x_{1,2}, \text{ or}$$
$$x_{2,1} > x_{2,2} > x_{1,2} > x_{1,1}, \text{ or}$$
$$x_{2,2} > x_{1,2} > x_{1,1} > x_{2,1}, \text{ or}$$
$$x_{1,2} > x_{1,1} > x_{2,1} > x_{2,2}$$

or the corresponding ones replacing in these the symbol $>$ by $<$.

5.6. Finally we include here a Grobner basis for the group $O(3)$ with respect to the row-order to give the reader an idea of the difficult interpretation in geometric or matricial terms of the polynomials in the basis.

$$x_{1,1}^2 - x_{2,2}^2 - x_{2,3}^2 - x_{3,2}^2 - x_{3,3}^2 + 1,$$
$$x_{1,2}^2 + x_{2,2}^2 + x_{3,2}^2 - 1,$$
$$x_{1,3}^2 + x_{2,3}^2 + x_{3,3}^2 - 1,$$
$$x_{1,1}x_{1,2} + x_{2,1}x_{2,2} + x_{3,1}x_{3,2},$$
$$x_{1,1}x_{1,3} + x_{2,1}x_{2,3} + x_{3,1}x_{3,3},$$
$$x_{1,2}x_{1,3} + x_{2,2}x_{2,3} + x_{3,2}x_{3,3},$$
$$x_{1,2}x_{2,3}^2 + x_{1,2}x_{3,3}^2 - x_{1,3}x_{2,2}x_{2,3} - x_{1,3}x_{3,2}x_{3,3} - x_{1,2},$$
$$x_{1,1}x_{2,3}^2 + x_{1,1}x_{3,3}^2 - x_{1,3}x_{2,1}x_{2,3} - x_{1,3}x_{3,1}x_{3,3} - x_{1,1},$$
$$x_{1,2}x_{2,2}x_{2,3} + x_{1,2}x_{3,2}x_{3,3} - x_{1,3}x_{2,2}^2 - x_{1,3}x_{3,2}^2 + x_{1,3},$$
$$x_{1,2}x_{2,1}x_{2,3} + x_{1,2}x_{3,1}x_{3,3} - x_{1,3}x_{2,1}x_{2,2} - x_{1,3}x_{3,1}x_{3,2},$$
$$x_{1,1}x_{2,2}x_{2,3} + x_{1,1}x_{3,2}x_{3,3} - x_{1,3}x_{2,1}x_{2,2} - x_{1,3}x_{3,1}x_{3,2},$$
$$x_{1,1}x_{2,2}^2 - x_{1,1}x_{3,3}^2 - x_{1,2}x_{2,1}x_{2,2} + x_{1,3}x_{3,1}x_{3,3},$$

$$x_{1,2}x_{2,2}x_{3,3}^2 - x_{1,2}x_{2,3}x_{3,2}x_{3,3} - x_{1,3}x_{2,2}x_{3,2}x_{3,3} + x_{1,3}x_{2,3}x_{3,2}^2 - x_{1,2}x_{2,2} - x_{1,3}x_{2,3},$$
$$x_{1,1}x_{2,2}x_{3,3}^2 - x_{1,1}x_{2,3}x_{3,2}x_{3,3} - x_{1,3}x_{2,2}x_{3,1}x_{3,3} + x_{1,3}x_{2,3}x_{3,1}x_{3,2} - x_{1,1}x_{2,2},$$
$$x_{1,2}x_{2,1}x_{3,3}^2 - x_{1,2}x_{2,3}x_{3,1}x_{3,3} - x_{1,3}x_{2,1}x_{3,2}x_{3,3} + x_{1,3}x_{2,3}x_{3,1}x_{3,2} - x_{1,2}x_{2,1},$$
$$x_{2,2}^2x_{3,3}^2 - 2x_{2,2}x_{2,3}x_{3,2}x_{3,3} + x_{2,3}^2x_{3,2}^2 - x_{2,2}^2 - x_{2,3}^2 - x_{3,2}^2 - x_{3,3}^2 + 1,$$
$$x_{2,1}x_{2,2}x_{3,3}^2 - x_{2,1}x_{2,3}x_{3,2}x_{3,3} - x_{2,2}x_{2,3}x_{3,1}x_{3,3} + x_{2,3}^2x_{3,1}x_{3,2} - x_{2,1}x_{2,2} - x_{3,1}x_{3,2},$$
$$x_{2,1}x_{2,2}x_{3,2} + x_{2,1}x_{2,3}x_{3,3} - x_{2,2}^2x_{3,1} - x_{2,3}^2x_{3,1} + x_{3,1},$$
$$x_{1,1}x_{2,2}x_{3,2} + x_{1,1}x_{2,3}x_{3,3} - x_{1,2}x_{2,2}x_{3,1} - x_{1,3}x_{2,3}x_{3,1},$$
$$x_{1,2}x_{2,1}x_{3,2} - x_{1,2}x_{2,2}x_{3,1} + x_{1,3}x_{2,1}x_{3,3} - x_{1,3}x_{2,3}x_{3,1},$$
$$x_{2,1}x_{3,2}^2 + x_{2,1}x_{3,3}^2 - x_{2,2}x_{3,1}x_{3,2} - x_{2,3}x_{3,1}x_{3,3} - x_{2,1},$$
$$x_{1,1}x_{3,2}^2 + x_{1,1}x_{3,3}^2 - x_{1,2}x_{3,1}x_{3,2} - x_{1,3}x_{3,1}x_{3,3} - x_{1,1},$$
$$x_{2,1}x_{3,1} + x_{2,2}x_{3,2} + x_{2,3}x_{3,3},$$
$$x_{2,1}^2 + x_{2,2}^2 + x_{2,3}^2 - 1,$$
$$x_{3,1}^2 + x_{3,2}^2 + x_{3,3}^2 - 1,$$
$$x_{1,1}x_{2,1} + x_{1,2}x_{2,2} + x_{1,3}x_{2,3},$$
$$x_{1,1}x_{3,1} + x_{1,2}x_{3,2} + x_{1,3}x_{3,3}$$

References

[B] B. Buchberger., *Applications of Grobner basis in non-linear computational geometry. Trends in computer algebra*, Lecture Notes in Computer Sci., vol. 296 (R. Jansen, ed.), Springer-Verlag, Berlin and New York, 1989.

[BO] A. Borel, *Linear algebraic groups*, Math. Lecture Note Series, W. A. Benjamin, Inc., 1969.

[C-R] C. W. Curtis and I. Reiner., *Representation theory of finite groups and associative algebras*, Pure and Applied Math., vol. XI, Interscience Publishers, 1962.

[D] D. Dubois, *A Nullstellensatz for ordered fields*, Ark. Mat. **8** (1969), 111–114.

[D-C] J. A. Dieudonne and J. B. Carrell, *Invariant theory old and new*, Academic Press, 1971.

[F] J. Fogarty, *Invariant theory*, Math. Lecture Note Series, W. A. Benjamin, Inc., 1969.

[G] F. R. Gantmacher, *Théorie des matrices*, Collection Universitaire des Mathématiques, vol. 2, DUNOD, Paris, 1966.

[N] D. G. Northcott, *Affine sets and affine groups*, London Math. Soc., Lecture Note Series, vol. 39, Cambridge University Press, 1980.

[R] J. J. Risler, *Une caractérisation des variétés algébriques réelles*, C. R. Acad. Sci. Paris **271** (1970), 1171–1173.

[S] T. A. Springer, *Invariant theory*, Lecture Notes in Math., vol. 585 (A. Dold and B. Eckmann, eds.), Springer-Verlag, Berlin and New York, 1977.

[SH] R. Shaw, *Linear algebra and group representation*, Vol. I, Academic Press, 1982.

[W] H. Weyl, *The classical groups*, Second edition, Princeton University Press, 1946.

Dpto. Matemáticas, Estadística y Computación, Facultad de Ciencias, Universidad de Cantabria, Santander 39071, Spain

E-mail address: g_lopez@ccucvx.unican.es and recio@ccuvx.unican.es

Contemporary Mathematics
Volume **155**, 1994

On Valuation Spectra

ROLAND HUBER AND MANFRED KNEBUSCH

INTRODUCTION

We have seen in the last decade how important it is to switch from the consideration of particular orderings of fields to a study of the set of all orderings of all residue class fields of a commutative ring A, i.e., the real spectrum $\operatorname{Sper} A$.

Now why not do the same with valuations? This leads to the definition of "valuation spectra." In principle the points of the valuation spectrum $\operatorname{Spv} A$ should be pairs $(\mathfrak{p}, v)$ consisting of a prime ideal $\mathfrak{p}$ of A, i.e., a point of $\operatorname{Spec} A$, and a Krull valuation v of the residue class field $qf(A/\mathfrak{p})$. Different valuations of $qf(A/\mathfrak{p})$ which have the same valuation ring are identified.

M. J. de la Puente has written a thesis under the guidance of G. Brumfiel at Stanford about such a valuation spectrum $\operatorname{Spv} A$ (which she calls the "Riemann surface" of A [**Pu**]. Without being aware of the work of Puente (which had not yet appeared), one of us (R.H.) in 1987 started a thorough investigation of valuation spectra [**Hu**, Chap. I]. Puente and Huber both arrive at the same definition of $\operatorname{Spv} A$.

The motivations of Puente (and Brumfiel) and Huber are different. Puente and Brumfiel want to use valuation spectra for compactification of affine algebraic varieties. Here we should also mention a recent paper by N. Schwartz [**S**], where he uses a related "absolute value spectrum" (which he also calls the "valuation spectrum") for the same purpose. The authors of the present article have been driven by some striking analogies between semialgebraic geometry and rigid analytic geometry, a subject started by John Tate (cf. [**BGR**], [**FP**]). This led Huber to a new "abstract" approach to rigid analytic geometry by use of "analytic spectra," which are natural descendents of valuation spectra [**Hu**]. (Only recently have we become aware of the extensive work of V. Berkovich [**Be**],

---

1991 *Mathematics Subject Classification.* Primary 13A18.
This paper is submitted in final form and no version of it will be submitted for publication elsewhere.

who studies rank 1 valuations of Banack algebras and applies his theory to rigid analytic geometry. This is another "abstract" approach to rigid geometry.)

Since Huber's abstract rigid geometry is close in spirit to abstract real algebraic geometry, it is not surprising that these two theories can be "mixed." One result of such a mixture is Huber's recent paper on semirigid functions [**Hu**$_4$], which permits studies of real phenomena of rigid analytic varieties. As has been amply demonstrated by the Spanish school (Andradas, Ruiz, ... cf. also their article in this volume), semianalytic geometry is amenable to methods from abstract real algebraic geometry. We have high hopes that the same will turn out to be true of semirigid geometry.

The spaces on which the semirigid functions are defined are derivates of *real* valuation spectra. The *real valuation spectrum* $\operatorname{Sperv} A$ of a commutative ring is a refinement of the real spectrum $\operatorname{Sper} A$. Its points are the triples $(\mathfrak{p}, P, C)$ with $\mathfrak{p} \in \operatorname{Spec} A$, P an ordering on $qf(A/\mathfrak{p})$, and C a convex subring of $qf(A/\mathfrak{p})$ with respect to P. Notice that $(\mathfrak{p}, P)$ is a point of $\operatorname{Sper} A$ and $(\mathfrak{p}, C)$ is a point of $\operatorname{Spv} A$. In this way $\operatorname{Sperv} A$ may be viewed as a natural subspace of the fibre product of $\operatorname{Sper} A$ and $\operatorname{Spv} A$ over $\operatorname{Spec} A$.

Real valuation spectra are indispensable in real *rigid* geometry. They seem to be also a valuable tool in real *algebraic* geometry, as is indicated by the very frequent occurence of real valuation rings in arguments in this area. All this has motivated us to give several talks about valuation spectra in the Ragsquad seminar and also talks about semirigid functions, both in the Ragsquad seminar and at the AMS conference at San Francisco in January '91. This is also the motivation for the present article.

In this article we intend to give a comprehensive account of basic facts about valuation spectra, as defined in [**Hu**]. We also give some applications to algebraic geometry in order to demonstrate that valuation spectra are already useful there. We have decided not to go on to real valuation spectra and real geometry in this article, because we want to keep the picture as simple as possible. (A [very] brief treatment of real valuation spectra can be found in §1 of [**Hu**$_4$].) Once the reader has obtained a firm grasp of valuation spectra and a feeling about possible applications in algebraic geometry, he or she will have no difficulty understanding real valuation spectra, and will hopefully be able to explore applications in real algebraic geometry. The reader will also find the door open to abstract rigid geometry, which is a very extensive—but useful—enlargement of classical rigid geometry.

Thinking about applications of valuation spectra in algebraic or real algebraic geometry, we should remember that valuations played a central role in Zariski's approach from the late 1930's, building up algebraic geometry by algebraic means. Later this role was reduced by Grothendieck and others in favour of prime ideals. Valuations survived, for example, in various valuative criteria and the resolution of singularities, but lost their dominance in algebraic geometry. Recently, in the Ragsquad seminar and elsewhere, we experienced a revived interest in Zariski's work. This should not be surprising since valuations occur so frequently and in such a natural way in real algebraic geometry.

Valuation spectra may be viewed as a refinement of Zariski spectra. We hope that Chapter 4 of the present article will convince the reader that this refinement, which brings us closer to Zariski's work, can be useful for problems of very different type in algebraic geometry.

1. The valuation spectrum of a ring

1.1. Definition of the valuation spectrum. Let A be a ring. (All rings are tacitly assumed to be commutative with unit element.) We recall the definition of a valuation of A. Let Γ be a totally ordered commutative group written additively. We adjoin an element ∞ to Γ and extend the addition and the ordering of Γ to $\Gamma_\infty :=\Gamma \cup \{\infty\}$ by $\alpha + \infty = \infty + \alpha = \infty$ and $\alpha \leq \infty$ for every $\alpha \in \Gamma_\infty$.

DEFINITION. [**B**, VI.3.1]. A *valuation* of A with values in Γ_∞ is a mapping $v\colon A \to \Gamma_\infty$ such that

i) $v(x+y) \geq \min(v(x), v(y))$ for all $x, y \in A$,
ii) $v(x \cdot y) = v(x) + v(y)$ for all $x, y \in A$,
iii) $v(0) = \infty$ and $v(1) = 0$.

Let $v\colon A \to \Gamma_\infty$ be a valuation. The subgroup of Γ generated by $\{v(a) | a \in A,\ v(a) \neq \infty\}$ is called the *value group* of v and is denoted by Γ_v. The valuation is called *trivial* if $\Gamma_v = \{0\}$. The convex subgroup $c\Gamma$ of Γ generated by $\{v(a) | a \in A,\ v(a) \leq 0\}$ is called the *characteristic subgroup* of v. The set $\operatorname{supp}(v) := v^{-1}(\infty)$ is a prime ideal of A and is called the *support* of v. The valuation v factorizes uniquely in $A \xrightarrow{g} qf(A/\operatorname{supp}(v)) \xrightarrow{\bar{v}} \Gamma_\infty$ where g is the canonical mapping and $\bar{v}$ is a valuation of the quotient field $qf(A/\operatorname{supp}(v))$ of $A/\operatorname{supp}(v)$. The valuation ring of $\bar{v}$ is denoted by $A(v)$.

Two valuations v and w of A are called equivalent if the following equivalent conditions are satisfied

i) For all $a, b \in A$, $v(a) \geq v(b)$ iff $w(a) \geq w(b)$.
ii) There is an isomorphism $f\colon (\Gamma_v)_\infty \longrightarrow (\Gamma_w)_\infty$ with $w = f \circ v$.
iii) $\operatorname{supp}(v) = \operatorname{supp}(w)$ and $A(v) = A(w)$.

Remark. The model theoretic result that the theory of algebraically closed fields with non-trivial valuation-divisibility relation has elimination of quantifiers implies that the equivalence classes of valuations of A correspond bijectively to the elementary equivalence classes of ring homomorphisms from A to non-trivial valued algebraically closed fields.

DEFINITION. i) $S(A)$ denotes the set of all equivalence classes of valuations of A. (In the following we often do not distinguish between a valuation and its equivalence class.)

ii) $K(A)$ denotes the boolean algebra of subsets of $S(A)$ generated by the subsets of the form $\{v \in S(A) | v(a) \geq v(b)\}$ $(a, b \in A)$.

We equip $S(A)$ with the topology $\mathcal{T}$ generated by the sets of the form $\{v \in S(A) | v(a) \geq v(b) \neq \infty\}$ $(a, b \in A)$, and call the topological space $\operatorname{Spv} A := (S(A), \mathcal{T})$ the *valuation spectrum* of A. This notation is justified by the following proposition.

PROPOSITION (1.1.1). *Spv A is a spectral space. $K(A)$ is the boolean algebra of constructible subsets of Spv A.*

PROOF. Every valuation v of A defines a binary relation $|_v$ on A by

$$a|_v b :\iff v(a) \leq v(b).$$

Two valuations v and w of A are equivalent if and only if $|_v = |_w$. Therefore we have an injective mapping $\varphi: S(A) \to \mathcal{P}(A \times A)$, $v \mapsto |_v$. ($\mathcal{P}(A \times A)$ denotes the power set of $A \times A$.) We equip $\{0,1\}$ with the discrete topology and $\mathcal{P}(A \times A) = \{0,1\}^{A \times A}$ with the product topology. Then $\mathcal{P}(A \times A)$ is a compact Hausdorff space. The image $\operatorname{im}(\varphi)$ of φ is closed in $\mathcal{P}(A \times A)$ since $\operatorname{im}(\varphi)$ is the set of all binary relations following conditions

1) $a|b$ or $b|a$.
2) If $a|b$ and $b|c$ then $a|c$.
3) If $a|b$ and $a|c$ then $a|b+c$.
4) If $a|b$ then $ac|bc$.
5) If $ac|bc$ and $0 \nmid c$ then $a|b$.
6) $0 \nmid 1$. □

We equip $S(A)$ with the topology such that φ is a topological embedding. Then $S(A)$ is a compact Hausdorff space and $K(A)$ is the set of all subsets of $S(A)$ which are open and closed. Now Proposition (1.1.1) follows from Hochster's result [**H**, Prop. 7]. For convenience we recall this result in the following lemma.

LEMMA (1.1.2 [**H**]). *Let (X, T) be a quasi-compact topological space and $\mathcal{L}$ be the set of all subsets of X which are open and closed. Let $\mathcal{T}$ be a topology of X such that $\mathcal{T}$ is generated by elements of $\mathcal{L}$ and $(X, \mathcal{T})$ is a T_0-space. Then $(X, \mathcal{T})$ is a spectral space and $\mathcal{L}$ is the set of all constructible subsets of $(X, \mathcal{T})$.*

Let $f: A \to B$ be a ring homomorphism. Then f induces a mapping Spv(f): Spv $B \to$ Spv A. (We often write $v|A$ instead of Spv$(f)(v)$.) Spv(f) is continuous, even more, Spv(f) is spectral.

Remark (1.1.3). i) If K is a field then $\operatorname{Spv} K$ is the abstract Riemann surface of [**ZS**, VI.17] (with the difference that in [**ZS**] the trivial valuation is excluded).

ii) The set M of all trivial valuations of A is a pro-constructible subset of $\operatorname{Spv} A$.

iii) The support mapping $\operatorname{supp} : \operatorname{Spv} A \to \operatorname{Spec} A$, $v \mapsto \operatorname{supp}(v)$ is spectral. The restriction of supp to the set M of all trivial valuations of A is a homeomorphism from M to $\operatorname{Spec} A$.

1.2. Specializations in the valuation spectrum. Let A be a ring and $v\colon A \longrightarrow \Gamma_\infty$ a valuation of A. To every convex subgroup H of Γ we have the mappings

$$v/H\colon A \longrightarrow (\Gamma/H)_\infty, \quad a \mapsto \begin{cases} v(a) \bmod H & \text{if } v(a) \neq \infty \\ \infty & \text{if } v(a) = \infty, \end{cases}$$

$$v|H\colon A \longrightarrow H_\infty, \quad a \mapsto \begin{cases} v(a) & \text{if } v(a) \in H \\ \infty & \text{if } v(a) \notin H. \end{cases}$$

One can easily check:

LEMMA (1.2.1). i) *v/H is a valuation of A and v/H is a generalization of v in Spv A.*

ii) *$v|H$ is a valuation of A iff $c\Gamma \subseteq H$, and in this case $v|H$ is a specialization of v in Spv A.*

The generalizations of v in Spv A of the form $v|H$ are called the *secondary generalizations* of v, and the specializations of v in Spv A of the form $v|H$ are called the *primary specializations* of v. A valuation w of A is called a *generalized primary specialization* of v if w is a primary specialization of v or if $c\Gamma_v = \{0\}$, w is trivial and supp $(v|c\Gamma_v) \subseteq$ supp (w) (in the latter case we have by (1.1.3 iii) a chain of specializations $v \succ v|c\Gamma_v \succ w$).

Remark (1.2.2). Let $A \xrightarrow{g} qf(A/\mathrm{supp}\,(v)) \xrightarrow{\bar{v}} \Gamma_\infty$ be the canonical factorization of v and H a convex subgroup of Γ. Let $\mathfrak{p}$ be the prime ideal $\{x \in A(v) | \bar{v}(x) > H\}$ of $A(v)$. Then

i) $\mathrm{supp}\,(v/H) = \mathrm{supp}\,(v)$ and $A(v/H) = A(v)_\mathfrak{p}$.

ii) $g(A) \subseteq A(v)_\mathfrak{p}$ iff $c\Gamma \subseteq H$. Let us assume $c\Gamma \subseteq H$. Then g induces a mapping $A \to K\colon = A(v)_\mathfrak{p}/\mathfrak{p}$, and $Q\colon = A(v)/\mathfrak{p}$ is a valuation ring of the field K. This ring homomorphism of A into the valued field (K, Q) induces the valuation $v|H$ on A.

A subset T of A is called *v-convex* if for all $a, b, c \in A$ holds: $v(a) \geq v(c) \geq v(b)$, $a \in T$, $b \in T \Longrightarrow c \in T$. (If $0 \in T$ this means: $v(c) \geq v(b)$, $b \in T \Longrightarrow c \in T$.)

LEMMA (1.2.3). *The supports of the primary specializations of v are the v-convex prime ideals of A.*

PROOF. Let $\mathfrak{p}$ be a v-convex prime ideal of A. Then $v(A \backslash \mathfrak{p}) < v(\mathfrak{p})$, especially $v(A \setminus \mathfrak{p}) \subseteq \Gamma$. Let G be the subgroup of Γ generated by $v(A \setminus \mathfrak{p})$. Then $v(\mathfrak{p}) > G$. (Indeed, assume to the contrary $g \geq v(c)$ for some $g \in G$ and $c \in \mathfrak{p}$. Since $v(A \setminus \mathfrak{p})$ is additively closed, there exist $a, b \in A \setminus \mathfrak{p}$ with $v(a) - v(b) = g$. Then $v(a) \geq v(bc)$, in contradiction to $v(A \setminus \mathfrak{p}) < v(\mathfrak{p})$.) Let H be the convex hull of G in Γ. Then H is a convex subgroup of Γ with $v(\mathfrak{p}) > H$ and $v(A \setminus \mathfrak{p}) \subseteq H$, hence $c\Gamma \subseteq H$ and $\mathfrak{p} = \mathrm{supp}\,(v|H)$. □

Now we can describe all specializations of v in $\operatorname{Spv} A$.

PROPOSITION (1.2.4). *Every specialization of v is a secondary specialization of a generalized primary specialization of v, and also a primary specialization of a secondary specialization of v.*

PROOF. i) Let $w \in \operatorname{Spv} A$ be a specialization of v. We show that w is a secondary specialization of a generalized primary specialization of v. If $c\Gamma_v = \{0\}$ and $v(a) \leq 0$ for each $a \in A \setminus \operatorname{supp}(w)$ then the trivial valuation u of A with $\operatorname{supp}(u) = \operatorname{supp}(w)$ is a generalized primary specialization of v and w is a secondary specialization of u. It remains to consider the case that $c\Gamma_v \neq \{0\}$ or $v(a) > 0$ for some $a \in A \backslash \operatorname{supp}(w)$. We notice for arbitrary $a, b \in A$.

(1) If $v(a) \geq v(b)$, $w(a) \neq \infty$ and $w(b) = \infty$, then $v(a) = v(b) \neq \infty$. (Indeed, we have $w(b) \geq w(a) \neq \infty$ and hence $v(b) \geq v(a) \neq \infty$.) First we show that $\operatorname{supp}(w)$ is v- convex. Let $x, y \in A$ with $y \in \operatorname{supp}(w)$ and $v(x) \geq v(y)$. We have to show $x \in \operatorname{supp}(w)$. Assume to the contrary $x \notin \operatorname{supp}(w)$. Then

(2) $v(x) \geq v(y)$, $w(x) \neq \infty$, $w(y) = \infty$.
We deduce from (1) and (2)

(3) $v(x) = v(y) \neq \infty$.
By our supposition there exists a $a \in A$ with (I) $v(a) < 0$ or (II) $v(a) > 0$ and $w(a) \neq \infty$. In case (I) we have $v(x) \geq v(ay)$, $w(x) \neq \infty$, $w(ay) = \infty$ (by (2)) and hence $v(x) = v(ay)$ (by (1)), in contradiction to (3). In case (II) we have $v(ax) \geq v(y)$, $w(ax) \neq \infty$, $w(y) = \infty$ (by (2)) and hence $v(ax) = v(y)$ (by (1)), in contradiction to (3). Thus we have proved that $\operatorname{supp}(w)$ is v-convex. By (1.2.3) there exists a primary specialization u of v with $\operatorname{supp}(u) = \operatorname{supp}(w)$. We show that w is a secondary specialization of u. Since $\operatorname{supp}(u) = \operatorname{supp}(w)$, it suffices to show: If $a, b \in A$ with $w(a) \geq w(b) \neq \infty$ then $u(a) \geq u(b)$. Let a, b be elements of A with $w(a) \geq w(b) \neq \infty$. Since w is a specialization of v, we have $v(a) \geq v(b)$ and hence $u(a) \geq u(b)$ (since u is a primary specialization of v).

ii) Let w be a specialization of v. We show that w is a primary specialization of a secondary specialization of v. By i), w is the secondary specialization of a generalized primary specialization u of v. If u is a primary specialization of v, then the assertion follows from the subsequent Lemma (1.2.5 ii). Now assume that u is not a primary specialization of v. By the subsequent Lemma (1.2.6) there exists a primary generalization w' of w with $\operatorname{supp}(w') = \operatorname{supp}(v|c\Gamma_v)$. Then w' is a secondary specialization of $v|c\Gamma_v$. By (1.2.5 ii) there exists some $v' \in \operatorname{Spv} A$ such that v' is a secondary specialization of v and a primary generalization of w'. Then w is a primary specialization of v'. $\square$

LEMMA (1.2.5). *Let w be a primary specialization of v.*

i) *Let v' be a secondary specialization of v.*
Then there exists a unique secondary specialization w' of w such that w' is a primary specialization of v'.

ii) *Let w' be a secondary specialization of w.*
Then there exists a secondary specialization v' of v such that w' is a primary specialization of v'.

iii) *Let v' be a secondary generalization of v.*
Then there exists a unique secondary generalization w' of w such that w' is a generalized primary specialization of v'.

iv) *Let w' be a secondary generalization of w.*
Then there exists a secondary generalization v' of v such that w' is a primary specialization of v'.

PROOF. We prove only ii). By (1.2.2) there is a prime ideal $\mathfrak{p}$ of $A(v)$ such that $(qf(A/\mathrm{supp}\,(w)),\ A(w)) \subseteq (A(v)_\mathfrak{p}/\mathfrak{p},\ A(v)/\mathfrak{p})$ is an extension of valued fields. Let B be a valuation ring of $A(v)_\mathfrak{p}/\mathfrak{p}$ with $B \subseteq A(v)/\mathfrak{p}$ and $B \cap qf(A/\mathrm{supp}\,(w)) = A(w')$. Let v' be the valuation of A with $\mathrm{supp}\,(v') = \mathrm{supp}\,(v)$ and $A(v') = \lambda^{-1}(B)$ where λ is the canonical mapping $A(v) \longrightarrow A(v)/\mathfrak{p}$. Then v' is a secondary specialization of v and a primary generalization of w'. □

LEMMA (1.2.6). *Let $\mathfrak{p}$ be a prime ideal of A with $\mathfrak{p} \subseteq \mathrm{supp}\,(v)$. Then there exists a primary generalization w of v with $\mathfrak{p} = \mathrm{supp}\,(w)$.*

PROOF. Let $(B, \mathfrak{m})$ be a valuation ring of $qf(A/\mathfrak{p})$ which dominates the local ring $(A/\mathfrak{p})_{\mathrm{supp}\,(v)/\mathfrak{p}}$. Let C be a valuation ring of $B/\mathfrak{m}$ with $C \cap qf(A/\mathrm{supp}\,(v)) = A(v)$. Let w be the valuation of A with $\mathrm{supp}\,(w) = \mathfrak{p}$ and $A(w) = \lambda^{-1}(C)$ where $\lambda\colon B \to B/\mathfrak{m}$ is the canonical mapping. Then w is a primary generalization of v. □

For later use we remark:

LEMMA (1.2.7). *Assume that $\mathrm{supp}\,(v)$ is a maximal ideal of A. Then a valuation w of A is a primary generalization of v if and only if $w(a) \geq 0$ for all $a \in A$ with $v(a) \geq 0$ and $w(a) > 0$ for all $a \in A$ with $v(a) > 0$.*

PROOF. Put $G = \{w \in \mathrm{Spv}\,A | w(a) \geq 0$ for all $a \in A$ with $v(a) \geq 0$ and $w(a) > 0$ for all $a \in A$ with $v(a) > 0\}$. Then $v \in G$ and G is closed under primary specializations and primary generalizations. Let $w \in G$ be given. Then $w(a) > 0$ for all elements a of the maximal ideal $\mathfrak{m}$ of A. This implies $w(a) > c\Gamma_w$ for all $a \in \mathfrak{m}$. (Indeed, if $a \in \mathfrak{m}$ and $x \in A$, then $w(ax) > 0$.) Hence $w|c\Gamma_w \in \{g \in G | \mathfrak{m} = \mathrm{supp}\,(g)\} = \{v\}$. □

1.3. Some other topologies on $S(A)$. Let A be a ring. Beside the topology $\mathcal{T}$ from (1.1), there are other useful topologies on the set $S(A)$, for example the topologies $\mathcal{T}'$ and $\mathcal{T}''$ with

$\mathcal{T}'$: = topology generated by the sets $\{v \in S(A) | v(a) > v(b)\}$, $a, b \in A$,

$\mathcal{T}''$: = topology generated by $\mathcal{T} \cup \mathcal{T}'$.

We put $\mathrm{Spv}' A := (S(A), \mathcal{T}')$ and $\mathrm{Spv}'' A := (S(A), \mathcal{T}'')$.

Proposition (1.1.1) and (1.1.2) imply

PROPOSITION (1.3.1). *Spv′A and Spv″A are spectral spaces. $K(A)$ is the set of constructible subsets of both Spv′A and Spv″A.*

Let us study the specializations in $\mathrm{Spv}' A$ and $\mathrm{Spv}'' A$. First we consider $\mathrm{Spv}' A$. Obviously, for a valuation $v: A \to \Gamma_\infty$, the valuations v/H (H a convex subgroup of Γ) and $v|H$ (H a convex subgroup of Γ with $c\Gamma \subseteq H$) are specializations of v in $\mathrm{Spv}' A$. We call v/H a secondary specialization of v and $v|H$ a primary specialization of v. Similarly to (1.2.4) one can prove

PROPOSITION (1.3.2). *Every specialization of a point v in Spv′A is a secondary specialization of a primary specialization of v, and also a primary specialization of a secondary specialization of v.*

Propositions (1.2.4) and (1.3.2) imply

PROPOSITION (1.3.3). *Let v and w be points of Spv″A. Then w is a specialization of v in Spv″A if and only if there exists a convex subgroup H of Γ_v with $c\Gamma_v \subseteq H$ and $w = v|H$.*

Remark (1.3.4). i) If K is a field, then $\mathrm{Spv}' K$ is the inverse spectral space to $\mathrm{Spv}\, K$ in the sense of [**H**, Prop. 8].

ii) The support mapping supp : $\mathrm{Spv}' A \to \mathrm{Spec}\, A$ is spectral.

iii) Let M be the set of all trivial valuations of A. Then M is closed in $\mathrm{Spv}' A$ and $\mathrm{supp}\,|M: M \to \mathrm{spec}\, A$ is a homeomorphism if we equip $\mathrm{spec}\, A$ with the constructible topology of the spectral space $\mathrm{Spec}\, A$.

Let us motivate the topologies $\mathcal{T}, \mathcal{T}', \mathcal{T}''$. Let k be an algebraically closed field complete with respect to a rank 1 valuation $\alpha: k \to \Gamma_\infty$. In rigid analytic geometry one associates to every (affine) variety $X = \mathrm{Spec}\, E$ over k an analytic space whose underlying "topological space" is the set $X(k)$ of k-rational points of X equipped with a Grothendieck topology G [**BGR**], [**FP**]. The admissible open sets of G are sets of the form

$$(*) \quad \{x \in X(k) | \alpha(f_i(x)) \geq \alpha(g_i(x)) \neq \infty \quad \text{for } i = 1, \ldots, n\} \quad \text{with } f_i, g_i \in E.$$

(Notice that weak inequalities $\geq$ are used in order to define the admissible open sets.) The description $(*)$ of admissible open sets suggests to work with the topology $\mathcal{T}$. As is shown in [**Hu**], there is a strong relation between $(X(k), G)$ and $\mathrm{Spv}\, E$.

Concerning the topology $\mathcal{T}'$ there is, for example, the following application: In [**Be**], Berkovich constructs to $\operatorname{Spec} E$ an analytic space but instead of $(X(k), G)$ he uses the topological subspace $\{v \in \operatorname{Spv}' E | v$ has rank 1 and $v|k = \alpha\}$ of $\operatorname{Spv}' E$.

We are interested in $\mathcal{T}''$ since there are applications of Spv'' in algebraic geometry and analytic geometry (cf. (4.2) and [**Hu$_2$**]). The spectrum Spv'' has many properties in common with the real spectrum, for example:

a) If K is a field, then any constructible subset of $\operatorname{Spv}'' K$ is open.
b) The specializations of a point in $\operatorname{Spv}'' A$ form a chain and are uniquely determined by their supports.
c) Let k be an algebraically closed field, α a nontrivial valuation of k and E a finitely generated k-algebra. By $\operatorname{Spv}''(\alpha, E)$ we denote the pro-constructible subspace $\{v \in \operatorname{Spv}'' E \mid \alpha = v|k\}$ of $\operatorname{Spv}'' E$. Then a constructible subset L of $\operatorname{Spv}''(\alpha, E)$ is open if and only if $L \cap (\operatorname{Spec} E)(k)$ is open in the strong topology of $(\operatorname{Spec} E)(k)$ induced by α (cf. (3.2)).
d) If A is universally catenary, then we have a curve selection lemma for $\operatorname{Spv}'' A$ (cf. (2.3)).
e) If the topological space $\operatorname{Spec} A$ is noetherian, then the closure of a constructible subset of $\operatorname{Spv}'' A$ is constructible (cf. (2.2)).

But Spv'' has a big disadvantage in comparison with Spv and Spv'. Namely, $\operatorname{Spv}'' A$ is disconnected if $\dim A \geq 1$, whereas $\operatorname{Spv} A$ is connected iff $\operatorname{Spv}' A$ is connected iff $\operatorname{Spec} A$ is connected. Even in the geometric situation we have: Let α be a henselian valuation of a field k and A a finitely generated k-algebra. Then $\operatorname{Spv}''(\alpha, A)$ has infinitely many connected components if $\dim A \geq 1$, but $\operatorname{Spv}(\alpha, A)$ is connected iff $\operatorname{Spv}'(\alpha, A)$ is connected iff $\operatorname{Spec} A$ is connected. (Remark: Let Z be the set of closed points of $\operatorname{Spv}'' A$ (resp. $\operatorname{Spv}''(\alpha, A)$). For every $z \in Z$, let $G(z)$ be the set of generalizations of z in $\operatorname{Spv}'' A$ (resp. $\operatorname{Spv} \diamond''(\alpha, A)$). Then $(G(z) | z \in Z)$ is the family of connected components of $\operatorname{Spv}'' A$ (resp. $\operatorname{Spv}''(\alpha, A)$).)

Remark (1.3.5). Schwartz uses in [**S**] a modification of $\mathcal{T}'$, namely the topology T of $S(A)$ generated by the sets $\{v \in S(A) | \infty \neq v(a) > v(b)\}$, $\{v \in S(A) | \infty \neq v(a)\}$ $(a, b \in A)$. We have (M denotes the set of trivial valuations of A):

i) T is weaker than $\mathcal{T}'$, and $T | S(A) \setminus M = \mathcal{T}' | S(A) \setminus M$.
ii) M is closed in $(S(A), T)$ and $\operatorname{supp} : (M, T|M) \to \operatorname{Spec} A$ is a homeomorphism.
iii) $(S(A), T)$ is a spectral space and $K(A)$ is the set of constructible subsets of $(S(A), T)$.
iv) Let v and w be valuations of A. Then w is a specialization of v in $(S(A), T)$ if and only if w is a specialization of v in $\operatorname{Spv}' A$ or w is a trivial valuation with $\operatorname{supp}(v) \subseteq \operatorname{supp}(w)$.

PROOF. ii) is trivial, and iii) follows from (1.1.1) and (1.1.2).

i) The mapping $\mathrm{Spv}'A \to (S(A), T), v \mapsto v$ is spectral. By [**S**, Prop. 26], $\mathrm{Spv}'A$ and $(S(A), T)$ have the same specializations on $S(A) \setminus M$. Hence $T'|S(A) \setminus M = T|S(A) \setminus M$.

iv) If w is a specialization of v in $\mathrm{Spv}'A$, then w is a specialization of v in $(S(A), T)$ by i), and if w is trivial with $\mathrm{supp}\,(v) \subseteq \mathrm{supp}\,(w)$, then w is specialization of v in $(S(A), T)$ by definition of T. Conversely, assume that w is a specialization of v in $(S(A), T)$. If v is trivial, then w is trivial and $\mathrm{supp}\,(v) \subseteq \mathrm{supp}\,(w)$ by ii). Assume that v is not trivial. If w is not trivial, then w is a specialization of v in $\mathrm{Spv}'A$ by i), and if w is trivial, then $\mathrm{supp}\,(v) \subseteq \mathrm{supp}\,(w)$ since $\mathrm{supp} : (S(A), T) \to \mathrm{Spec}\, A$ is continuous. □

2. Some general results on the valuation spectrum

2.1. Morphisms. By (1.1.1) we know the constructible subsets of the valuation spectrum. Then the following proposition is an immediate consequence of the fact that the theory of algebraically closed fields with non-trivial valuation-divisibility relation has elimination of quantifiers [**P**, 4.17].

PROPOSITION (2.1.1). *Let $f\colon A \to B$ be a ring homomorphism of finite presentation and let L be a constructible subset of $\mathrm{Spv}\, B$. Then $\mathrm{Spv}(f)(L)$ is a constructible subset of $\mathrm{Spv}\, A$.*

Let $f\colon A \to B$ be a ring homomorphism. We want to study the relation between the specializations (resp. generalizations) of a point v in $\mathrm{Spv}\, B$ and the specializations (resp. generalizations) of $\mathrm{Spv}\,(f)(v)$ in $\mathrm{Spv}\, A$. By (1.2.4) it suffices to consider secondary specializations (resp. secondary generalizations) and primary specializations (resp. primary generalizations). Concerning the secondary specializations (resp. secondary generalizations), we have the following trivial remark.

Remark (2.1.2). Let v be a point of $\mathrm{Spv}\, B$ and $w\colon = \mathrm{Spv}\,(f)(v)$. Let $S(v)$ (resp. $G(v)$) be the set of all secondary specializations (resp. secondary generalizations) of v in $\mathrm{Spv}\, B$, analogously $S(w)$ (resp. $G(w)$). Then $\mathrm{Spv}\,(f)\colon \mathrm{Spv}\, B \to \mathrm{Spv}\, A$ induces surjective mappings $S(v) \to S(w)$ and $G(v) \to G(w)$.

Let v be a point of $\mathrm{Spv}\, B$. We call $\mathrm{Spv}\,(f)$ *primarily generalizing* at v if for every primary generalization y of $\mathrm{Spv}\,(f)(v)$ in $\mathrm{Spv}\, A$ there is a primary generalization x of v in $\mathrm{Spv}\, B$ with $y = \mathrm{Spv}\,(f)(x)$. We call $\mathrm{Spv}\,(f)$ *universally primarily generalizing* at v if, for every base extension $g\colon C \to C \otimes_A B$ of f and every point w of $\mathrm{Spv}\, C \otimes_A B$ lying over v, the mapping $\mathrm{Spv}\,(g)$ is primarily generalizing at w. Analogously we define (*universally*) *primarily specializing*. With this definition we have

PROPOSITION (2.1.3). *Let v be a valuation of B. Then the following conditions are equivalent.*

i) *$\mathrm{Spv}(f)$ is universally primarily generalizing at v.*

ii) *Spec*(f) *is universally generalizing at supp*(v).

PROOF. ii) follows from i) by (1.2.6). Let us assume ii). Let t be a primary generalization of $s := \mathrm{Spv}(f)(v)$ in $\mathrm{Spv}\, A$. We have to show that there exists a primary generalization w of v in $\mathrm{Spv}\, B$ with $t = \mathrm{Spv}(f)(w)$. By (1.2.2) there exist valuation rings A', C and a ring homomorphism $h: A \to A'$ such that $C \subseteq A'$, $qf(C) = qf(A')$ and $\mathrm{Spv}(h)(t') = t$ and $\mathrm{Spv}(h)(s') = s$ where t' and s' are the points of $\mathrm{Spv}\, A'$ given by the valuation rings C and $C/\mathfrak{m}_{A'}$. Let $f': A' \to A' \otimes_A B =: B'$ be the ring homomorphism induced by f. Let v' be a valuation of B' with $v'|B = v$ and $v'|A' = s'$. It suffices to show that there is a primary generalization w' of v' in $\mathrm{Spv}\, B'$ with $w'|A' = t'$. Let $\mathfrak{p}$ be a prime ideal of B' with $f'^{-1}(\mathfrak{p}) = \{0\}$ and $\mathfrak{p} \subseteq \mathrm{supp}(v')$. By (1.2.6) there exists a primary generalization w' of v' with $\mathfrak{p} = \mathrm{supp}(w')$. Then $w'|A' = t'$ since s' has only one primary generalization in $\mathrm{Spv}\, A'$ with support $\{0\}$. $\square$

COROLLARY (2.1.4). *If f is flat and finitely presented then the mappings $Spv(f)\colon Spv\,B \to Spv\,A$, $Spv'(f)\colon Spv'B \to Spv'A$ and $Spv''(f)\colon Spv''B \to Spv''A$ are open.*

PROPOSITION (2.1.5). *Let v be a valuation of B. If Spec(f) is universally specializing at supp(v), then Spv(f) is universally primarily specializing at v.*

PROOF. Let t be a primary specialization of $s := \mathrm{Spv}(f)(v)$ in $\mathrm{Spv}\, A$. We have to show that there exists a primary specialization w of v in $\mathrm{Spv}\, B$ with $t = \mathrm{Spv}(f)(w)$. By (1.2.2) there exists a valuation ring D of $K := qf(A/\mathrm{supp}(s))$ such that D contains $A(s)$ and the image of the mapping $A \to K$ and t is induced by the mapping of A into the valued field $(D/\mathfrak{m}_D,\ A(s)/\mathfrak{m}_D)$. Let E be a valuation ring of $F := qf(B/\mathrm{supp}(v))$ with $B(v) \subseteq E$ and $E \cap K = D$. Since $\mathrm{Spec}(f)$ is universally specializing at $\mathrm{supp}(v)$, E contains the image of the mapping $B \to F$. Hence we have a mapping of B to the valued field $(E/\mathfrak{m}_E,\ B(v)/\mathfrak{m}_E)$, which induces a valuation w of B. Then w is a primary generalization of v with $t = \mathrm{Spv}(f)(w)$. $\square$

COROLLARY (2.1.6). i) *If f is integral, then Spv(f) is universally primarily specializing at every point.*

ii) *If f is integral and injective, A integral and normal, and B integral, then Spv(f) is universally primarily generalizing at every point.*

PROOF. i) follows from (2.1.5). The assumptions of ii) imply that $\mathrm{Spec}(f)$ is universally generalizing at every point [**EGA**, IV, 14.4.2]. Hence ii) follows from (2.1.3). $\square$

COROLLARY (2.1.7). *We consider the mappings $Spv(f)\colon Spv\,B \to Spv\,A$, $Spv'(f)\colon Spv'B \to Spv'A$ and $Spv''(f)\colon Spv''B \to Spv''A$.*

i) *If f is integral, then there are no specializations in the fibres of $Spv(f)$, $Spv'(f)$, $Spv''(f)$.*

ii) *If f is integral, then the mappings $Spv(f)$, $Spv'(f)$, $Spv''(f)$ are closed.*

iii) *If f is integral, injective and finitely presented, A integral and normal, and B integral, then the mappings $Spv(f)$, $Spv'(f)$, $Spv''(f)$ are open.*

PROOF. i) is obvious, since there are no specializations in the fibres of f; ii) and iii) follow from (2.1.6). $\square$

2.2. *Closure of constructible subsets.*

PROPOSITION (2.2.1). *Let A be a ring such that the topological space $\operatorname{Spec} A$ is noetherian and let L be a constructible subset of $Spv''A$. Then the closure $\bar{L}$ of L in $Spv''A$ is constructible.*

The analogous statements for $\operatorname{Spv} A$ and $\operatorname{Spv}' A$ are not true. Examples:

i) The closure of $\{v \in \operatorname{Spv} \mathbb{C}[T] | v(2) > 0,\ v(T) \geq 0\}$ in $\operatorname{Spv} \mathbb{C}[T]$ is not constructible (by the results of (3.2)).

ii) The closure of $\{v \in \operatorname{Spv}' \mathbb{Z} | v(2) > 0\}$ in $\operatorname{Spv}' \mathbb{Z}$ is not constructible.

In order to prove (2.2.1) we need the following lemma.

LEMMA (2.2.2). *Let A be a local ring with maximal ideal $\mathfrak{m}$ and residue field κ. Let L be the set of all points of $Spv''A$ which have a specialization with support $\mathfrak{m}$ (i.e. $supp(v|c\Gamma_v) = \mathfrak{m}$). Let π be the mapping $L \to Spv''\kappa$, $v \mapsto v|c\Gamma_v$ (here we identify the subspace $\{v \in Spv''A | supp(v) = \mathfrak{m}\}$ of $Spv''A$ with $Spv''\kappa$). Then*

i) *$L = \{v \in Spv''A | v(a) > 0$ for all $a \in \mathfrak{m}\}$.*

ii) *π is spectral.*

iii) *If K is a constructible subset of L, then $\pi(K)$ is a constructible subset of $Spv''\kappa$.*

PROOF. i) If $\operatorname{supp}(v|c\Gamma_v) = \mathfrak{m}$, then $v(a) > c\Gamma_v$ for all $a \in \mathfrak{m}$, especially $v(a) > 0$. Conversely, let v be a valuation of A with $v(a) > 0$ for all $a \in \mathfrak{m}$. Then, for all $x \in A \setminus \operatorname{supp}(v)$ and $a \in \mathfrak{m}$, $v(xa) > 0$. Hence $v(a) > c\Gamma_v$ for all $a \in \mathfrak{m}$, which means $\operatorname{supp}(v|c\Gamma_v) = \mathfrak{m}$.

ii) Let f be an element of κ and a an element of A with $f = a \bmod \mathfrak{m}$. Then $\pi^{-1}(\{v \in \operatorname{Spv}''\kappa | v(f) \geq 0\}) = \{v \in L | v(a) \geq 0\}$. Hence π is spectral.

iii) Let K be a constructible subset of L. By ii), $\pi(K)$ is pro-constructible in $\operatorname{Spv}''\kappa$. Let v be an element of K. We have to show that there is a constructible subset W of $\operatorname{Spv}''\kappa$ with $\pi(v) \in W \subseteq \pi(K)$. Choose $a_i, b_i, c_i, d_i \in A$ $(i = 1, \dots, n)$ such that $v \in \{x \in L | x(a_i) \geq x(b_i),\ x(c_i) > x(d_i)$ for $i = 1, \dots, n\} \subseteq K$. Then $d_i \notin \operatorname{supp}(v)$ for $i = 1, \dots, n$. We may assume $b_i \notin \operatorname{supp}(v)$ for $i = 1, \dots, m$ and $b_i \in \operatorname{supp}(v)$ for $i = m+1, \dots, n$. Then $a_i \in \operatorname{supp}(v)$ for $i = m+1, \dots, n$. By (1.2.2) there exists a valuation ring B of $qf(A/\operatorname{supp}(v))$ which dominates $A/\operatorname{supp}(v)$ and contains $A(v)$. Let k be the residue field of B and $f\colon A \to qf(A/\operatorname{supp}(v))$, $g\colon B \to k$, $h\colon \kappa \to k$ the canonical mappings. We have $\lambda_i\colon = \frac{f(a_i)}{f(b_i)} \in A(v) \subseteq B$ for $i = 1, \dots, m$

and $\mu_i: = \frac{f(c_i)}{f(d_i)} \in A(v) \subseteq B$ for $i = 1, \dots, n$. Put $S: = \{x \in \operatorname{Spv} k | x(g(\lambda_i)) \geq 0$ for $i = 1, \dots, m$ and $x(g(\mu_i)) > 0$ for $i = 1, \dots, n\}$. By the subsequent lemma, $\operatorname{Spv}''(h)(S)$ is constructible in $\operatorname{Spv}''\kappa$. We have $\pi(v) \in \operatorname{Spv}''(h)(S) \subseteq \pi(K)$. □

LEMMA (2.2.3). *Let $E \hookrightarrow F$ be an extension of fields and L a constructible subset of $Spv F$. Then the image of L under the mapping $Spv F \to Spv E$ is constructible in $Spv E$.*

PROOF. We choose a field G and a constructible subset M of $\operatorname{Spv} G$ such that $E \hookrightarrow G \hookrightarrow F$, G is finitely generated over E and L is the preimage of M under the mapping $f: \operatorname{Spv} F \to \operatorname{Spv} G$. Since f is surjective, we have to show that $g(M)$ is constructible in $\operatorname{Spv} E$ where g is the mapping $\operatorname{Spv} G \to \operatorname{Spv} E$. Let A be a finitely generated E-subalgebra of G with $G = qf(A)$ and $h: \operatorname{Spv} A \to \operatorname{Spv} E$ be the canonical mapping. Let N be a constructible subset of $\operatorname{Spv} A$ such that N is closed under primary generalizations in $\operatorname{Spv} A$ and $M = N \cap \operatorname{Spv} G$. Then $g(M) = h(N)$. (Indeed, let $v \in N$ be given. By (1.2.6), there exists a primary generalization w of v in $\operatorname{Spv} A$ with $\{0\} = \operatorname{supp}(w)$. Then $w \in M$ and $g(w) = h(v)$.) Now (2.1.1) shows that $g(M)$ is constructible in $\operatorname{Spv} E$. □

Now we prove (2.2.1). We use ideas from [**Ru**]. $\bar{L}$ is pro-constructible in $\operatorname{Spv}''A$. Let v be an element of $\bar{L}$. We have to show that there exists a constructible subset M of $\operatorname{Spv}''A$ with $v \in M \subseteq \bar{L}$. Put $F = qf(A/\operatorname{supp}(v))$. Applying (2.2.2 iii) to the local ring $A_{\operatorname{supp}(v)}$, we obtain a constructible subset N of $\operatorname{Spv}''F$ with $v \in N \subseteq \bar{L}$. Let M be an open constructible subset of $\operatorname{Spv}''A/\operatorname{supp}(v)$ with $M \cap \operatorname{Spv}''F = N$. By (1.2.6), M is contained in the closure of N. Hence $v \in M \subseteq \bar{L}$. Since $\operatorname{Spec} A$ is noetherian, M is a constructible subset of $\operatorname{Spv}''A$.

2.3. Curve selection lemma. We have the following abstract version of the curve selection lemma. Concrete versions will be deduced from it in (3.2.6) and [**Hu$_2$**].

PROPOSITION (2.3.1). *Let A be a noetherian ring and v a point of $Spv''A$. We assume that A is universally catenary or that A is local and henselian with maximal ideal $supp(v)$. Put $T = \{w \in Spv''A | w$ specializes to $v\}$ and $T_0 = \{w \in T | ht(supp(v)/supp(w)) \leq 1\}$. Then T is the closure of T_0 in the constructible topology of $Spv''A$.*

PROOF. We may assume that A is local with maximal ideal $\mathfrak{m} = \operatorname{supp}(v)$. Let L be a non-empty constructible subset of T. We have to show $L \cap T_0 \neq \emptyset$. We may assume $L = \{x \in T | x(a_i) \geq x(b_i)$ and $x(c_i) > x(d_i)$ for $i = 1, \dots, n\}$ with $a_i, b_i, c_i, d_i \in A$. Let w be an element of L. Assume $ht(\mathfrak{m}/\operatorname{supp}(w)) \geq 2$. Then we will show that there exists a $u \in \operatorname{Spv}''A$ with

a) $\operatorname{supp}(w) \subsetneqq \operatorname{supp}(u)$
b) $u \in L$.

Then we are done, since $\dim A$ is finite. □

Without loss of generality we can assume that A is an integral domain and $\mathrm{supp}\,(w) = \{0\}$. Furthermore we may assume $b_i \neq 0$ for $i = 1, \dots, m$ and $b_i = 0$ for $i = m+1, \dots, n$ (which implies $a_i = 0$ for $i = m+1, \dots, n$). We have $d_i \neq 0$ for $i = 1, \dots, n$.

Let B be the subring $A\big[\frac{a_i}{b_i},\ i = 1, \dots, m; \frac{c_i}{d_i}, i = 1, \dots, n\big]$ of $qf(A)$, and let $f\colon \mathrm{Spec}\, B \to \mathrm{Spec}\, A$ be the morphism of schemes induced by the inclusion $A \subseteq B$. Then we have

(1) There exists a valuation $\bar{v}$ of B with the following properties

i) $\bar{v}\big(\frac{a_i}{b_i}\big) \geq 0$ for $i = 1, \dots, m$

ii) $\bar{v}\big(\frac{c_i}{d_i}\big) > 0$ for $i = 1, \dots, n$

iii) $v = \bar{v}|A$

iv) $\mathrm{supp}\,(\bar{v})$ is a closed point of the fibre $f^{-1}(\mathfrak{m})$.

PROOF. The valuation ring $A(w)$ of $qf(A) = qf(B)$ defines a valuation $\bar{w}$ of B with $w = \bar{w}|A$ and $\bar{w}(x) \geq 0$ for every $x \in \{\frac{a_i}{b_i} | i = 1, \dots, m\} \cup \{\frac{c_i}{d_i} | i = 1, \dots, n\}$. Hence the characteristic subgroup $c\Gamma_{\bar{w}}$ is the convex hull of $c\Gamma_w$ in $\Gamma_{\bar{w}} \leq$.

Since v is a specialization of w in $\mathrm{Spv}''A$, there exists a smallest convex subgroup H of Γ_w with $c\Gamma_w \subseteq H$ and $v = w|H$. Let $\bar{H}$ be the convex hull of H in $\Gamma_{\bar{w}}$. Then $c\Gamma_{\bar{w}} \subseteq \bar{H}$. Hence we have a specialization $s\colon = \bar{w}|\bar{H}$ of $\bar{w}$ in $\mathrm{Spv}''B$ with $v = s|A$. □

Now we distinguish two cases.

First case: v is trivial. Let $\bar{v}$ be a trivial valuation of B such that $\mathrm{supp}\,(s) \subseteq \mathrm{supp}\,(\bar{v})$ and $\mathrm{supp}\,(\bar{v})$ is closed in $f^{-1}(\mathfrak{m})$. Then, clearly i), iii), iv) are satisfied. Since $s\big(\frac{c_i}{d_i}\big) > 0$ for $i = 1, \dots, n$ and s is trivial, we have $\frac{c_i}{d_i} \in \mathrm{supp}\,(s) \subseteq \mathrm{supp}\,(\bar{v})$ for $i = 1, \dots, n$. Hence ii) is fulfilled.

Second case: v is non-trivial. Then the existence of a valuation $\bar{v}$ of B satisfying i)–iv) follows from the fact that s fulfills i), ii), iii) and the result that the theory of algebraically closed fields with non-trivial valuation-divisibility relation has elimination of quantifiers ([**P**, 4.17], cf. (3.2.3)).

Put $h = \big(\prod_{i=1}^{m} b_i\big) \cdot \big(\prod_{i=1}^{n} d_i\big) \in A$. Then we have

(2) There exists a $t \in \mathrm{Spv}''B$ with

i) t is a generalization of $\bar{v}$,

ii) $h \notin \mathrm{supp}\,(t)$,

iii) $\{0\} \neq \mathrm{supp}\,(t)$.

PROOF. First we observe that $ht(\mathrm{supp}\,(\bar{v})) \geq 2$. Indeed, if A is universally catenary, then the dimension formula $ht(\mathfrak{m}) + \mathrm{trdeg}\,(B|A) = ht(\mathrm{supp}\,(\bar{v})) + \mathrm{trdeg}\,(B/\mathrm{supp}\,(\bar{v})|A/\mathfrak{m})$ [**EGA**, IV.5.6.1] gives $ht(\mathrm{supp}\,(\bar{v})) = ht(\mathfrak{m}) \geq 2$, since $qf(A) = qf(B)$ and $B/\mathrm{supp}\,(\bar{v})$ is algebraic over $A/\mathfrak{m}$ (the latter by (1 iv)). Now assume that A is henselian. If $\mathrm{supp}\,(\bar{v})$ has a proper generalization $\mathfrak{q}$ in $f^{-1}(\mathfrak{m})$, then $\{0\} \subsetneqq \mathfrak{q} \subsetneqq \mathrm{supp}\,(v)$ and hence $ht(\mathrm{supp}\,(\bar{v})) \geq 2$. If $\mathrm{supp}\,(\bar{v})$ has no proper generalization in $f^{-1}(\mathfrak{m})$, then $\operatorname{supp}(\bar{v})$ is isolated in $f^{-1}(\mathfrak{m})$ (by (1 iv)) and hence $B_{\operatorname{supp}(\bar{v}}$ is finite over A, which implies $ht(\operatorname{supp}(\bar{v})) = ht(\mathfrak{m}) \geq 2$.

Now, as $ht(\operatorname{supp}(\bar{v})) \geq 2$, the equivalence of a) and f) in [**EGA**, IV.10.5.1] shows that the localization $(B_{\operatorname{supp}(\bar{v})})_h$ is not a field. This means that there exists a prime ideal $\mathfrak{p}$ of B with $\{0\} \neq \mathfrak{p}$, $h \notin \mathfrak{p}$ and $\mathfrak{p} \subseteq \operatorname{supp}(\bar{v})$. By (1.2.6), there exists a generalization of $\bar{v}$ in $\operatorname{Spv}''B$ with support $\mathfrak{p}$. This shows (2). □

We claim that the conditions a) and b) are satisfied with $u\colon = t|A$. Since $A_h = B_h$, (2 ii) and (2 iii) imply $\operatorname{supp}(w) = \{0\} \subsetneqq \operatorname{supp}(u)$. Since d_i is a divisor of h, we have $d_i \notin \operatorname{supp}(t)$ by (2 ii). Then (1 ii) and (2 i) give $u(c_i) > u(d_i)$ for $i = 1, \ldots, n$. (1 i) and (2 i) imply $u(a_i) \geq u(b_i)$ for $i = 1, \ldots, m$. Note that $u(a_i) \geq u(b_i)$ for $i = m+1, \ldots, n$, since $a_i = b_i = 0$ for $i = m+1, \ldots, n$. According to (1 iii) and (2 i), u is a generalization of v in $\operatorname{Spv}''A$. Hence $u \in L$.

2.4. Connected components. In this paragraph we study the connected components of pro-constructible subsets of valuation spectra $\operatorname{Spv} A$.

We begin with a general remark on connected components of spectral spaces.

LEMMA (2.4.1). *Let $(X_i|i \in I)$ be a cofiltered system of spectral spaces such that all transition maps $X_i \to X_j$ are spectral. Let X be the projective limit of $(X_i|i \in I)$ in the category of topological spaces. Then*

i) *X is spectral. Each clopen (= closed and open) subset of X is the preimage of a clopen subset of some X_i. In particular, if each X_i is connected, then X is connected.*

ii) *Let Z be a connected component of X. For each $i \in I$, let Z_i be the connected component of X_i containing the image of Z. Then $Z = \varprojlim_{i\in I} Z_i \subseteq X$.*

iii) *For each $i \in I$, let Z_i be a connected component of X_i such that $\varphi(Z_i) \subseteq Z_j$ for each transition map $\varphi\colon X_i \to X_j$. Then $\varprojlim_{i\in I} Z_i \subseteq X$ is a connected component of X.*

iv) *Let Y be a spectral space and Z a connected component of Y. Then Z is the intersection of the clopen subsets of Y containing Z.*

PROOF. i) follows from (1.1.2) and [$\mathbf{B}_1$, I.9.6].

ii) We have $Z \subseteq \varprojlim Z_i \subseteq X$. By i), $\varprojlim Z_i$ is connected. Hence $Z = \varprojlim Z_i$.

iii) By i), $T\colon = \varprojlim Z_i$ is connected. Let Z be the connected component of X containing T. Then the image of Z in X_i is contained in Z_i. Hence $T = Z$.

iv) Let A be a ring with $Y \cong \operatorname{Spec} A$. We have $\operatorname{Spec} A \cong \varprojlim_j \operatorname{Spec} A_j$, where each A_j is a finitely generated $\mathbb{Z}$-algebra. Now the assertion follows from ii). □

PROPOSITION (2.4.2). *Let K be a field and D, E subsets of K. We consider the pro-constructible subset $L = \{v \in Spv\,K | v(d) \geq 0$ for all $d \in D$ and $v(e) > 0$ for all $e \in E\}$ of $Spv\,K$. Let A be the integral closure in K of the subring generated by $D \cup E$. Then there is a canonical bijection from the set of clopen subsets of $Spec\,A/E \cdot A$ to the set of clopen subsets of L. In particular, the connected components of $Spec\,A/E \cdot A$ correspond to the connected components of L.*

In order to prove (2.4.2), we first recall Zariski's representation of the valuation spectrum of a field as a projective limit of schemes [**ZS**, VI.17]:

LEMMA (2.4.3). *Let A be a ring, K a field and $s\colon A \to K$ a ring homomorphism. Let I be the following category. The objects are the triples (X, f, g) with X an integral scheme, $f\colon X \to Spec\,A$ a projective morphism and $g\colon Spec\,K \to X$ a dominant morphism such that $Spec(s) = f \circ g$. The morphisms $(X, f, g) \to (X', f', g')$ are the morphisms of schemes $h\colon X \to X'$ with $g' = h \circ g$ (and hence $f = f' \circ h$). Let c be the functor from I to the category of topological spaces which assigns to $(X, f, g) \in I$ the topological space $|X|$ underlying X. Put $Y = \{v \in Spv\,K | s(A) \subseteq K(v)\}$. For every $i = (X, f, g) \in I$, we have a continuous mapping $\varphi_i\colon Y \to |X|$. Namely, if v is an element of Y and if $t\colon Spec\,K(v) \to Spec\,A$ and $\bar{g}\colon Spec\,K(v) \to X$ are the extensions of $Spec(s)$ and g with $t = f \circ \bar{g}$, then $\varphi_i(v)$ is defined to be the image of the closed point of $Spec\,K(v)$ under $\bar{g}$.*

With these arrangements we have: $(Y, (\varphi_i | i \in I))$ is the projective limit of c.

Now we come to the proof of (2.4.2). Let V denote the subspace $\{\mathfrak{p} \in \operatorname{Spec} A | E \subseteq \mathfrak{p}\}$ of $\operatorname{Spec} A$. We have $v(a) \geq 0$ for every $v \in L$, $a \in A$. Let φ be the mapping $L \to V$, $v \mapsto \operatorname{supp}(w|c\Gamma_w)$ with $w\colon = v|A \in \operatorname{Spv} A$. The following two properties, i) and ii) of φ, show that $U \mapsto \varphi^{-1}(U)$ gives a bijection from the set of clopen subsets of V to the set of clopen subsets of L.

i) φ is spectral, specializing and surjective.

ii) Each fibre of φ is connected.

To i): For every $f \in A$, $\varphi^{-1}(D(f)) = \{v \in L | v(f) \leq 0\}$. Hence φ is spectral. Let v be an element of L and $\mathfrak{q}$ a specialization of $\varphi(v)$. Then the trivial valuation of A with support $\mathfrak{q}$ is a generalized primary specialization of $v|A$. Hence by (1.2.4) and (2.1.2), there exist a specialization w of v in L with $\varphi(v) = \mathfrak{q}$. The surjectivity of φ follows from (1.2.6).

To ii): We have to show that, for every local subring B of K which is integrally closed in K, the subset $\{v \in \operatorname{Spv} K | K(v)$ dominates $B\} \subseteq \operatorname{Spv} K$ is connected. By (2.4.1 i) we may assume that B is noetherian. Then the assertion follows from (2.4.1 i), (2.4.3) and [**EGA**, III.4.3.5].

PROPOSITION (2.4.4). *Let A be a ring, A_0 a subring of A and I an ideal of A_0 such that A_0 is henselian along I [**EGA**, IV.18.5.5]. We consider the subspace $L = \{v \in Spv\,A | v(a) \geq 0$ for all $a \in A_0$ and $v(a) > 0$ for all $a \in I\} \subseteq Spv\,A$. Let*

$\lambda: L \to \operatorname{Spec} A$ be the support mapping. Then $U \mapsto \lambda^{-1}(U)$ gives a bijection from the set of clopen subsets of Spec A to the set of clopen subsets of L. In particular, the connected components of Spec A correspond to the connected components of L.

PROOF. The assertion follows from the following two properties of λ.

i) λ is spectral, generalizing and surjective.

ii) Each fibre of λ is connected.

To i): Since L is closed under primary generalizations, λ is generalizing by (1.2.6). In order to show the surjectivity of λ, let $\mathfrak{p} \in \operatorname{Spec} A$ be given. Since A_0 is henselian along I, and thus I lies in the Jacobson radical of A_0, $\mathfrak{p} \cap A_0$ specializes to a prime ideal $\mathfrak{q} \in \operatorname{Spec} A_0$ with $I \subseteq \mathfrak{q}$. By (1.2.6) there exists a $w \in \operatorname{Spec} A_0$ such that $\mathfrak{p} \cap A_0 = \operatorname{supp}(w)$ and the trivial valuation of A_0 with support $\mathfrak{q}$ is a primary specialization of w. Let v be a valuation of A with $\mathfrak{p} = \operatorname{supp}(v)$ and $w = v|A_0$. Then $v \in L$.

To ii): Let $\mathfrak{p}$ be a prime ideal of A. Let B be the integral closure of A_0 in $qf(A/\mathfrak{p})$. Then $\operatorname{Spec} B/I \cdot B$ is connected, since A_0 is henselian along I and Spec B is connected. We conclude from (2.4.2) that $\lambda^{-1}(\mathfrak{p})$ is connected. □

Remark (2.4.5). Let A be a ring, A_0 a subring of A and I an ideal of A_0. But now we do not assume that A_0 is henselian along I. So we cannot apply (2.4.4) directly. But it is obvious what we have to do. Let $(\bar{A}_0, \bar{I})$ be a henselization of (A_0, I) [**R**, XI.2]. We consider the tensor product

$$\begin{array}{ccc} \bar{A}: = \bar{A}_0 \otimes_{A_0} A & \overset{f}{\longleftarrow} & A \\ \bar{i} \uparrow & & \uparrow i \\ \bar{A}_0 & \underset{f_0}{\longleftarrow} & A_0 . \end{array}$$

Then $\bar{A}_0$ is henselian along $\bar{I}$, and $\bar{i}$ is injective. Put $L = \{v \in \operatorname{Spv} A | v(a) \geq 0$ for all $a \in A_0$ and $v(a) > 0$ for all $a \in I\}$ and $\bar{L} = \{v \in \operatorname{Spv} \bar{A} | v(a) \geq 0$ for all $a \in \bar{A}_0$ and $v(a) > 0$ for all $a \in \bar{I}\}$. Then we have: The mapping Spv (f) induces a homeomorphism $g: \bar{L} \to L$. (Application: If A_0 is noetherian and A of finite type over A_0, then L has finitely many connected components.)

PROOF. We show that g is bijective and generalizing. Since $\bar{I} = I \cdot A_0$, $\bar{L}$ is closed under generalizations in $\operatorname{Spv}(f)^{-1}(L)$. Then (2.1.3) and (2.1.2) imply that $g\ g(v_1) = g(v_2)$. Let K_i be an algebraic closure of $qf(\bar{A}/\operatorname{supp}(v_i))$ and A_i a valuation ring of K_i extending $\bar{A}(v_i)$ $(i = 1, 2)$. Let $h_i: \bar{A} \to K_i$ be the canonical ring homomorphism $(i = 1, 2)$. Since $g(v_1) = g(v_2)$, there exists an isomorphism $h: K_1 \to K_2$ with $h(A_1) = A_2$ and $h \circ h_1 \circ f = h_2 \circ f$. We consider the ring homomorphisms $g_i: = h_i \circ \bar{i}: \bar{A}_0 \to A_i$ $(i = 1, 2)$. Then $(h \circ g_1) \circ f_0 = g_2 \circ f_0$. Since A_2 is henselian and $f_0: (A_0, I) \to (\bar{A}_0, \bar{I})$ a henselization of (A_0, I), we conclude $h \circ g_1 = g_2$. Now $h \circ h_1 \circ f = h_2 \circ f$ and $h \circ h_1 \circ \bar{i} = h_2 \circ \bar{i}$ imply

$h \circ h_1 = h_2$, and therefore $v_1 = v_2$. A similar argument (representing a $v \in L$ by a henselian valuation ring ...) shows that g is surjective. $\square$

If X is an irreducible normal complex analytic space, L a connected open subset of X, and M a closed complex analytic subspace of X with $\dim M < \dim X$, then $L \setminus M$ is connected. In the next proposition we prove an analogous result for the valuation spectrum.

PROPOSITION (2.4.6). *Let A be a normal integral domain, L a connected pro-constructible subset of $\operatorname{Spv} A$ which is closed under primary generalizations, and M a subset of $\operatorname{Spv} A$ such that there is a $a \in A \setminus \{0\}$ with $a \in \operatorname{supp}(v)$ for all $v \in M$. Then $L \setminus M$ is connected, too.*

PROOF. Put $T = \{v \in L | a \in \operatorname{supp}(v)\}$. By (1.2.6), $L \setminus T$ is dense in L. In particular, $L\setminus T$ is dense in $L\setminus M$. Hence it suffices to show that $L\setminus T$ is connected. Assume to the contrary that $L \setminus T$ is not connected. Let $L \setminus T = U_1 \cup U_2$ be a partition of $L \setminus T$ into non-empty closed subsets. Since L is connected and $L \setminus T$ dense in L, there exists a $t \in T$ having generalizations in U_1 and U_2. Then by (1.2.4), t has primary generalizations in U_1 and U_2. Let $f\colon A \to B$ be the strict henselization of A at $\operatorname{supp}(t)$ [**R**, VIII.2]. We consider the mapping $g = \operatorname{Spv}(f)\colon \operatorname{Spv} B \to \operatorname{Spv} A$. Let s be a valuation of B with $t = g(s)$. Put $C = \{b \in B | s(b) \geq 0\}$ and $I = \{b \in B | s(b) > 0\}$. Then C is a subring of B and I is an ideal of C. More precisely, C is a local ring with maximal ideal I. Since B and $B(s)$ are henselian and $B(s)$ is integrally closed in the residue field of B, C is henselian. Put $G = \{v \in \operatorname{Spv} B | v(c) \geq 0$ for all $c \in C$ and $v(i) > 0$ for all $i \in I\}$. B is integral, since A is normal. Hence $\{\mathfrak{p} \in \operatorname{Spec} B | f(a) \notin \mathfrak{p}\}$ is connected. Now we know by (2.4.4) that $H\colon = \{v \in G | f(a) \notin \operatorname{supp}(v)\}$ is connected. According to (1.2.7) and (2.1.3), $g(G)$ is the set of primary specializations of t in $\operatorname{Spv} A$. Hence $H \subseteq g^{-1}(U_1) \cup g^{-1}(U_2)$, $g^{-1}(U_1) \cap g^{-1}(U_2) = \emptyset$, $g^{-1}(U_1) \cap H \neq \emptyset$, $g^{-1}(U_2) \cap H \neq \emptyset$, in contradiction to the connectedness of H. $\square$

In the rest of this paragraph and in §3.4 we will investigate the following question: Let $f\colon A \to B$ be a ring homomorphism of finite presentation and let L be a pro-constructible subset of $\operatorname{Spv} A$ such that every constructible subset of L has finitely many connected components. Under what conditions has every constructible subset of $\operatorname{Spv}(f)^{-1}(L) \subseteq \operatorname{Spv} B$ finitely many connected components, too? For example, we will show that every constructible subset of $\operatorname{Spv}(f)^{-1}(L)$ has finitely many connected components if A is a Nagata ring [**M**, Ch. 12] and L is closed under primary generalizations or if every valuation $v \in L$ is non-trivial. But in general, not every constructible subset of $\operatorname{Spv}(f)^{-1}(L)$ has finitely many connected components, as the following example shows: Let A be a noetherian ring, $B = A[T]$ the polynomial ring in one variable over A, $f\colon A \to B$ the canonical ring homomorphism and L the set of all trivial valuations of A. Then L is pro-constructible in $\operatorname{Spv} A$ and every constructible subset of L has finitely many connected components (1.1.3). But in $M\colon = \{v \in \operatorname{Spv}(f)^{-1}(L) | v(T) > 0$ and

$v(T) \neq \infty\}$ there are no proper specializations, and hence M is totally disconnected. (M is homeomorphic to L equipped with the constructible topology.)

LEMMA (2.4.7). *Let A be a ring such that the topological space $\operatorname{Spec} A$ is noetherian. Let L be a pro-constructible subset of $\operatorname{Spv} A$ which is closed under primary generalizations. We consider the following two conditions:*

i) *For every residue field K of A, every constructible subset of $L \cap \operatorname{Spv} K$ has finitely many connected components.*

ii) *Every constructible subset of L has finitely many connected components.*

Then i) *implies* ii). *If, for every prime ideal $\mathfrak{p}$ of A, the set $\{x \in \operatorname{Spec} A/\mathfrak{p} | (A/\mathfrak{p})_x$ is normal$\}$ contains a nonempty open subset of $\operatorname{Spec} A/\mathfrak{p}$, then* ii) *implies* i).

PROOF. Assume i). Let M be a constructible subset of L. In order to prove that M has finitely many connected components we show that, for every $x \in M$, there is a connected constructible subset of M containing x. Let $x \in M$ be given. Put $K = qf(A/\operatorname{supp}(x))$. By assumption there exists a connected constructible subset T of $M \cap \operatorname{Spv} K$ containing x. Let Z be a constructible subset of $\operatorname{Spv}(A/\operatorname{supp}(x))$ such that $Z \cap L \cap \operatorname{Spv} K = T$ and Z is closed under primary generalizations in $\operatorname{Spv}(A/\operatorname{supp}(x))$. Let $\lambda\colon \operatorname{Spv}(A/\operatorname{supp}(x)) \to \operatorname{Spec}(A/\operatorname{supp}(x))$ be the support mapping. Since $Z \cap L \cap \operatorname{Spv} K \subseteq M$, there exists a non-empty open subset U of $\operatorname{Spec}(A/\operatorname{supp}(x))$ with $S := Z \cap L \cap \lambda^{-1}(U) \subseteq M$. Since S is closed under primary generalizations in $\operatorname{Spv}(A/\operatorname{supp}(x))$ and $T = S \cap \operatorname{Spv} K$ is connected, we conclude by (1.2.6) that S is connected. Since $\operatorname{Spec} A$ is noetherian, S is constructible in L. We have $x \in S$ by construction of S.

Now assume ii). Let $\mathfrak{p}$ be a prime ideal of A such that the set $\{x \in \operatorname{Spec} A/\mathfrak{p} | (A/\mathfrak{p})_x$ is normal$\}$ contains a non-empty open affine subset U of $\operatorname{Spec} A/\mathfrak{p}$. Put $K = qf(A/\operatorname{supp}(x))$. Let M be a constructible subset of $L \cap \operatorname{Spv} K$. Let $\lambda\colon \operatorname{Spv} A/\mathfrak{p} \to \operatorname{Spec} A/\mathfrak{p}$ be the support mapping. Choose a constructible subset Z of $\lambda^{-1}(U)$ such that $Z \cap L \cap \operatorname{Spv} K = M$ and Z is closed under primary generalizations in $\lambda^{-1}(U)$. Since $\operatorname{Spec} A$ is noetherian, $Z \cap L$ is constructible in L. So by assumption, $Z \cap L$ has finitely many connected components $L_1, \dots, L_n$. Each L_1 is pro-constructible in $\lambda^{-1}(U)$ and closed under primary generalizations in $\lambda^{-1}(U)$. Hence by (2.4.6), $L_i \cap \lambda^{-1}(V)$ is connected for every open subset V of U. Put

$$M_i = \bigcap_{\substack{V \subseteq U \text{ open},\\ V \neq \emptyset}} L_i \cap \lambda^{-1}(V).$$

Then M_i is connected (by (2.4.1 i)) and $M = \bigcup_{i=1}^{n} M_i$. Hence M has finitely many connected components. $\square$

In §3.4 we will prove

LEMMA (2.4.8). *Let $E \hookrightarrow F$ be a finitely generated extension of fields. We consider the mapping $g\colon \mathit{Spv}\, F \to \mathit{Spv}\, E$. Let L be a pro-constructible subset of $\mathit{Spv}\, E$.*

i) *If L has finitely many connected components, then $g^{-1}(L)$ has finitely many connected components, too. More precisely: Assume that L is connected. Then if F is purely transcendental over E, then $g^{-1}(L)$ is connected, and if F is finite over E, then the number of connected components of $g^{-1}(L)$ is at most $[F\colon E]_s$ (the separable degree of F over E).*

ii) *If every constructible subset of L has finitely many connected components, then every constructible subset of $g^{-1}(L)$ has finitely many connected components.*

Lemmata (2.4.7) and (2.4.8) imply

COROLLARY (2.4.9). *Let A be a Nagata ring and $f\colon A \to B$ a ring homomorphism of finite type. Let L be a pro- constructible subset of $\mathit{Spv}\, A$ such that L is closed under primary generalizations and every constructible subset of L has finitely many connected components. Then every constructible subset of $\mathit{Spv}(f)^{-1}(L) \subseteq \mathit{Spv}\, B$ has finitely many connected components.*

EXAMPLE. Let $A = S^{-1}B$ be the localization of a finitely generated $\mathbb{Z}$-algebra B by a multiplicative system $S \subseteq B$. Then every constructible subset of $\mathrm{Spv}\, A$ has finitely many connected components.

In §3.4 we will also prove

PROPOSITION (2.4.10). *Let $f\colon A \to B$ be a ring homomorphism of finite presentation and let L be a pro-constructible subset of $\mathit{Spv}\, A$ such that every constructible subset of L has finitely many connected components and every valuation $v \in L$ is non-trivial. Then every constructible subset of $\mathit{Spv}(f)^{-1}(L)$ has finitely many connected components, too.*

Remark (2.4.11). If K is a field, then $\mathrm{Spv}'K$ is the dual spectral space to $\mathrm{Spv}\, K$. Hence (2.4.2) and (2.4.8) remain true if we write Spv' instead of Spv. Then the proofs of (2.4.4), (2.4.5), (2.4.6) and (2.4.7) show that these results are also true for Spv'. One can show that (2.4.10) is true for Spv', even without the assumption that every $v \in L$ is non-trivial.

3. Valuation spectrum of rings over fields

3.1. Affine schemes over fields. Let k be a field, α a valuation of k and A a k-algebra. We put $S(\alpha, A) = \{v \in S(A) |\ v|k = \alpha\}$. Then $S(\alpha, A) = \{v \in S(A) | v(a) \geq 0$ for all $a \in k$ with $\alpha(a) \geq 0$ and $v(a) > 0$ for all $a \in k$ with $\alpha(a) > 0\}$. We equip $S(\alpha, A)$ with the subspace topologies of $\mathrm{Spv}\, A$, $\mathrm{Spv}'A$, $\mathrm{Spv}''A$, and denote the resulting topological spaces by $\mathrm{Spv}(\alpha, A)$, $\mathrm{Spv}'(\alpha, A)$, $\mathrm{Spv}''(\alpha, A)$,

$$\mathrm{Spv}\,(\alpha, A) = (S(\alpha, A), \mathcal{T}|S(\alpha, A))$$
$$\mathrm{Spv}\,'(\alpha, A) = (S(\alpha, A), \mathcal{T}'|S(\alpha, A))$$
$$\mathrm{Spv}\,''(\alpha, A) = (S(\alpha, A), \mathcal{T}''|S(\alpha, A)).$$

Then $\mathrm{Spv}\,(\alpha, A)$, $\mathrm{Spv}\,'(\alpha, A)$, $\mathrm{Spv}\,''(\alpha, A)$ are convex pro-constructible subspaces of $\mathrm{Spv}\,A$, $\mathrm{Spv}\,'A$, $\mathrm{Spv}\,''A$, are closed under primary specializations and primary generalizations, and their constructible topologies coincide.

LEMMA (3.1.1). *Let (K, β) be a valued field extending (k, α). We consider the canonical mappings $Spv(\beta, A \otimes_k K) \to Spv(\alpha, A)$, $Spv'(\beta, A \otimes_k K) \to Spv'(\alpha, A)$, $Spv''(\beta, A \otimes_k K) \to Spv''(\alpha, A)$.*

i) *They are surjective and spectral.*

ii) *If K is algebraic over k, then they are open and closed and map constructible subsets to constructible subsets.*

iii) *If (K, β) is a henselization of (k, α), then they are homeomorphisms.*

PROOF. i) is obvious.

ii) Let L be a constructible subset of $\mathrm{Spv}\,(\beta, A \otimes_k K)$. We choose a finite extension F of k with $k \hookrightarrow F \hookrightarrow K$ and a constructible subset M of $\mathrm{Spv}\,(\beta|F, A \otimes_k F)$ with $L = p^{-1}(M)$ where $p\colon \mathrm{Spv}\,(\beta, A \otimes_k K) \to \mathrm{Spv}\,(\beta|F, A \otimes_k F)$ is the canonical mapping. Since p is surjective, we have $r(L) = q(M)$ with $r\colon \mathrm{Spv}\,(\beta, A \otimes_k K) \to \mathrm{Spv}\,(\alpha, A)$ and $q\colon \mathrm{Spv}\,A \otimes_k F \to \mathrm{Spv}\,A$. $\mathrm{Spv}\,(\beta|F, A \otimes_k F)$ is open, closed and constructible in $q^{-1}(\mathrm{Spv}\,(\alpha, A))$. Hence by (2.1.1), $r(L)$ is constructible in $\mathrm{Spv}\,(\alpha, A)$. If L is open (resp. closed), then we choose M open (resp. closed) in $\mathrm{Spv}\,(\beta|F, A \otimes_k F)$. By (2.1.4) (resp. (2.1.7 ii)), we obtain that $r(L)$ is open (resp. closed) in $\mathrm{Spv}\,(\alpha, A)$.

iii) follows from (2.4.5) and (2.4.11). □

PROPOSITION (3.1.2). *Let (K, β) be a henselization of (k, α). Then*

i) *There is a canonical bijection between the set of clopen subsets of $Spv(\alpha, A)$ and the set of clopen subsets of $Spec\,A \otimes_k K$. Especially, the connected components of $Spv(\alpha, A)$ correspond to the connected components of $Spec\,A \otimes_k K$.*

ii) *There is a canonical bijection between the set of clopen subsets of $Spv'(\alpha, A)$ and the set of clopen subsets of $Spec\,A \otimes_k K$. Especially, the connected components of $Spv'(\alpha, A)$ correspond to the connected components of $Spec\,A \otimes_k K$.*

PROOF. By (3.1.1 iii), $\mathrm{Spv}\,(\beta, A \otimes_k K) \xrightarrow{\sim} \mathrm{Spv}\,(\alpha, A)$ and $\mathrm{Spv}\,'(\beta, A \otimes_k K) \xrightarrow{\sim} \mathrm{Spv}\,(\alpha, A)$. Hence we may assume $(K, \beta) = (k, \alpha)$, i.e. α is henselian. Then the assertion follows from (2.4.4) and (2.4.11). □

PROPOSITION (3.1.3). *If A is finitely generated over k, then every constructible subset of $Spv(\alpha, A)$ or $Spv'(\alpha, A)$ has finitely many connected components.*

PROOF. (2.4.9) and (2.4.11). □

PROPOSITION (3.1.4). *If the topological space Spec A is noetherian, then the closure of a constructible subset of Spv*$''(\alpha, A)$ *is constructible.*

PROOF. Let L be a constructible subset of $\mathrm{Spv}''(\alpha, A)$. We choose a constructible subset M of $\mathrm{Spv}''A$ with $L = M \cap \mathrm{Spv}''(\alpha, A)$. Let $\bar{L}$ (resp. $\bar{M}$) be the closure of L (resp. M) in $\mathrm{Spv}''(\alpha, A)$ (resp. $\mathrm{Spv}''A$). Then $\bar{L} = \bar{M} \cap \mathrm{Spv}''(\alpha, A)$, since $\mathrm{Spv}''(\alpha, A)$ is closed under generalizations in $\mathrm{Spv}''A$. Now (2.2.1) implies that $\bar{L}$ is constructible in $\mathrm{Spv}''(\alpha, A)$. □

The (combinatorial) dimension $\dim X$ of a spectral space X is the supremum of lengths of chains of specializations in X.

PROPOSITION (3.1.5). i) $\dim \mathit{Spv}''A = \dim \mathit{Spv}''(\alpha, A) = \dim \mathit{Spec}\, A$.

ii) *Assume that* α *is non-trivial. Then* $\dim \mathit{Spv}(\alpha, A) = \dim \mathit{Spv}'(\alpha, A) = \sup\, \{\mathit{trdeg}(K|k)\ |K$ *residue field of* $A\}$. *In particular, if* A *is finitely generated over* k, *then* $\dim \mathit{Spv}(\alpha, A) = \dim \mathit{Spv}'(\alpha, A) = \dim \mathit{Spec}\, A$.

PROOF. i) follows from (1.2.6).

ii) Put $s = \sup\{\mathrm{trdeg}\,(K|k)|K$ residue field of $A\}$. Obviously, $\dim \mathrm{Spv}\,(\alpha, A) \geq s$ and $\dim \mathrm{Spv}'(\alpha, A) \geq s$. We show $\dim \mathrm{Spv}\,(\alpha, A) \leq s$ and $\dim \mathrm{Spv}'(\alpha, A) \leq s$. Let g be the length of a chain of specializations in $\mathrm{Spv}\,(\alpha, A)$. Then by (1.2.4) and (1.2.5 ii), there exist a chain of specializations $v_0 \succ v_1 \succ \cdots \succ v_m = w_0 \succ w_1 \succ \cdots \succ w_n$ in $\mathrm{Spv}\,(\alpha, A)$ such that $v_0 \succ v_1 \succ \cdots \succ v_m$ are secondary specializations, $w_0 \succ w_1 \succ \cdots \succ w_n$ are primary specializations and $g = m + n$. Let $\sum$ be the value group of v_m, Γ the value group of α and K the residue field of A at $\mathrm{supp}\,(v_m)$. There exist convex subgroups $G_0, \dots, G_m$ and $H_0, \dots, H_n$ of $\sum$ with $G_0 \underset{\neq}{\supset} G_1 \underset{\neq}{\supset} \cdots \underset{\neq}{\supset} G_m = \{0\}$, $G_i \cap \Gamma = \{0\}$ for $i = 0, \dots, m$ and $\sum = H_0 \underset{\neq}{\supset} H_1 \underset{\neq}{\supset} \cdots \underset{\neq}{\supset} H_n \supseteq \Gamma$. Then $g = m + n \leq \dim_{\mathbb{Q}}(\sum/\Gamma) \otimes_{\mathbb{Z}} \mathbb{Q} \leq \mathrm{trdeg}\,(K|k) \leq s$. Hence $\dim \mathrm{Spv}\,(\alpha, A) \leq s$. Similarly, one can show $\dim \mathrm{Spv}'(\alpha, A) \leq s$. □

Remark (3.1.6). Assume that k is algebraically closed, α non-trivial and A finitely generated over k. Let $f\colon A \to B$ be an etale ring homomorphism. We consider the mappings $g\colon \mathrm{Spv}\,(\alpha, B) \to \mathrm{Spv}\,(\alpha, A)$, $g'\colon \mathrm{Spv}'(\alpha, B) \to \mathrm{Spv}'(\alpha, A)$ and $g''\colon \mathrm{Spv}''(\alpha, B) \to \mathrm{Spv}''(\alpha, A)$ induced by f. Let x be a point of $S(\alpha, B)$. If $\mathrm{supp}\,(x)$ is a maximal ideal of B, then g, g', g'' are local homeomorphisms at x. But if $\mathrm{supp}\,(x)$ is not a maximal ideal of B, then in general g, g', g'' are not local homeomorphisms at x. Example: Let n be a natural number with $n \geq 2$ and $\mathrm{char}\,(k) \nmid n$. We consider the mapping $g\colon \mathrm{Spv}\,(\alpha, k[T]) \to \mathrm{Spv}\,(\alpha, k[T])$ induced by the k-algebra homomorphism $f\colon k[T] \to k[T]$, $T \mapsto T^n$. Let v be the Gauss valuation of $k[T]$ extending α, i.e. $v(a_0 + a_1 T + \cdots + a_m T^m) = \min\{\alpha(a_i)|i = 0, \dots, m\}$. Then there exists no constructible subset L of $\mathrm{Spv}\,(\alpha, k[T])$ such that $v \in L$ and $g|L$ is injective. Indeed, let L be a constructible subset of $\mathrm{Spv}\,(\alpha, k[T])$ with $v \in L$. The extension of fields $k(T) \hookrightarrow k(T)$ induced by f is galois, and

$v \circ \mu = v$ for every $\mu \in$ Gal $(k(T)|k(T))$. Hence $g^{-1}(g(v)) = \{v\} \subseteq L$. This implies that there is a constructible subset M of Spv $(\alpha, k[T])$ with $g(v) \in M$ and $g^{-1}(M) \subseteq L$. By the subsequent Proposition (3.2.3), there exists a $w \in M$ such that supp (w) is a maximal ideal of $k[T]$ and supp $(w) \neq T \cdot k[T]$. Then $g^{-1}(w) \subseteq L$ and $\#g^{-1}(w) = n$.

3.2. Semialgebraic sets. Let k be a field, $\alpha\colon k \twoheadrightarrow \Gamma_\infty$ a non-trivial valuation of k and A a k-algebra.

Let $\tilde{\Gamma}$ be the divisible hull of Γ. We put

$$\begin{aligned}\operatorname{Max}(\alpha, A) = \{v\colon A \to \tilde{\Gamma}_\infty | \, & v \text{ is a valuation of A,} \\ & A/\operatorname{supp}(v) \text{ is algebraic over } k, \\ & v \text{ extends } \alpha\colon k \to \Gamma_\infty\end{aligned}$$

and equip Max (α, A) with the weakest topology such that, for every $a \in A$, the mapping Max $(\alpha, A) \to \tilde{\Gamma}_\infty$, $v \mapsto v(a)$ is continuous where $\tilde{\Gamma}_\infty$ carries the order-induced topology.

LEMMA (3.2.1). i) *Assume that k is algebraically closed and A is generated over k by $a_1, \ldots, a_n \in A$. Equip k with the valuation topology of α and k^n with the product topology. Then Max$(\alpha, A) \to k^n$, $v \mapsto (a_1 \operatorname{mod} \operatorname{supp}(v), \ldots, a_n \operatorname{mod} \operatorname{supp}(v))$ is a topological embedding.*

ii) *The canonical mappings Max$(\alpha, A) \to$ Spv$A$, Max$(\alpha, A) \to$ Spv$'A$, Max$(\alpha, A) \to$ Spv$''A$ are topological embeddings.*

PROOF. i) is obvious.

ii) We show that the canonical mapping φ: Max $(\alpha, A) \to$ Spv A is a topological embedding. Obviously φ is injective. Let $a \in A$ and $\gamma \in \tilde{\Gamma}_\infty$ be given. Put $U = \{v \in \operatorname{Max}(\alpha, A) | v(a) < \gamma\}$ and $V = \{v \in \operatorname{Max}(\alpha, A) | v(a) > \gamma\}$. We have to show that there exist open subsets U' and V' of Spv A with $U = U' \cap \operatorname{Max}(\alpha, A)$ and $V = V' \cap \operatorname{Max}(\alpha, A)$. We may assume $\gamma \neq \infty$. For every $\delta \in \tilde{\Gamma}$ we choose $n(\delta) \in \mathbb{N}$ and $k(\delta) \in k^*$ with $n(\delta) \cdot \delta = \alpha(k(\delta))$ and put $U(\delta) = \{v \in \operatorname{Spv} A | v(a^{n(\delta)}) \leq v(k(\delta))\}$ and $V(\delta) = \{v \in \operatorname{Spv} A | v(a^{n(\delta)}) \geq v(k(\delta))\}$. Then $U(\delta)$, $V(\delta)$ are open in Spv A and we have

$$U = \operatorname{Max}(\alpha, A) \cap \bigcup_{\substack{\delta \in \tilde{\Gamma} \\ \delta < \gamma}} U(\delta),$$

$$V = \operatorname{Max}(\alpha, A) \cap \bigcup_{\substack{\delta \in \tilde{\Gamma} \\ \delta > \gamma}} V(\delta).$$

Let $a, b \in A$ be given and put $U = \{v \in \operatorname{Spv} A | v(a) \geq v(b) \neq \infty\}$. We have to show that $\varphi^{-1}(U)$ is open in Max (α, A). Let $x \in \varphi^{-1}(U)$ be given. We consider the element $c\colon = \frac{a}{b} \in A(x) \subseteq A/\operatorname{supp}(x)$. Since the residue field of $A(x)$ is algebraic over the residue field of $k(\alpha)$, there exists a monic polynomial

$p(T) = T^n + e_1 T^{n-1} + \cdots + e_n \in k(\alpha)[T]$ such that $p(c)$ is contained in the maximal ideal of $A(x)$. Put $d := a^n + e_1 ba^{n-1} + e_2 b^2a^{n-2} + \cdots + e_nb^n \in A$. Then $x(d) > x(b^n)$, and hence we can choose an element $\gamma \in \tilde{\Gamma}$ with $x(d) > \gamma > x(b^n)$. Put $V := \{v \in \operatorname{Max}(\alpha, A) | v(d) > \gamma > v(b^n)\}$. Then V is open in $\operatorname{Max}(\alpha, A)$ and $x \in V \subseteq \varphi^{-1}(U)$. Thus we have proved that $\operatorname{Max}(\alpha, A) \to \operatorname{Spv} A$ is a topological embedding. Obviously, $\operatorname{Max}(\alpha, A) \to \operatorname{Spv}' A$ is a topological embedding. Hence $\operatorname{Max}(\alpha, A) \to \operatorname{Spv}'' A$ is a topological embedding, too. □

Remark (3.2.2). Since $\operatorname{Max}(\alpha, A) \to \operatorname{Spv} A$ is continuous, the topology of $\operatorname{Max}(\alpha, A)$ is the weakest topology on $\operatorname{Max}(\alpha, A)$ such that, for all $a \in A$, the mapping $\operatorname{Max}(\alpha, A) \to \tilde{\Gamma}_\infty$, $v \mapsto v(a)$ is continuous where we now equip $\tilde{\Gamma}_\infty$ with the topology generated by the sets $\{\gamma\}$, $\{x \in \tilde{\Gamma}_\infty | x > \gamma\}$ with $\gamma \in \tilde{\Gamma}$.

The theory of algebraically closed fields with non-trivial valuation-divisibility relation has elimination of quantifiers [**P**, 4.17]. This implies

PROPOSITION (3.2.3). *Assume that A is finitely generated over k. Then $Spv''(\alpha, A)$ is the closure of $Max(\alpha, A)$ in the constructible topology of $Spv'' A$.*

COROLLARY (3.2.4). *If A is finitely generated over k, then the closure of $Max(\alpha, A)$ in the constructible topology of $Spv'' A$ is closed under generalization and specialization in $Spv'' A$.*

DEFINITION. If A is finitely generated over k, we call a subset S of $\operatorname{Max}(\alpha, A)$ *semialgebraic* if S is a finite boolean combination of sets of the type $\{v \in \operatorname{Max}(\alpha, A) | v(a) > v(b)\}$ $(a, b \in A)$.

By (3.2.3) there is a canonical bijection $S \mapsto \tilde{S}$ from the set of semialgebraic subsets of $\operatorname{Max}(\alpha, A)$ onto the set of constructible subsets of $\operatorname{Spv}''(\alpha, A)$, namely $\tilde{S}$ is the unique constructible subset of $\operatorname{Spv}'' A$ with $S = \tilde{S} \cap \operatorname{Max}(\alpha, A)$.

PROPOSITION (3.2.5). *Assume that A is finitely generated over k. Let L be a constructible subset of $Spv''(\alpha, A)$ and let $\bar{L}$ be the closure of L in $Spv''(\alpha, A)$. Then*

i) *$\bar{L}$ is constructible in $Spv''(\alpha, A)$, and $\bar{L} \cap Max(\alpha, A)$ is the closure of $L \cap Max(\alpha, A)$ in $Max(\alpha, A)$.*

ii) *L is closed (resp. open) in $Spv''(\alpha, A)$ if and only if $L \cap Max(\alpha, A)$ is closed (resp. open) in $Max(\alpha, A)$.*

PROOF. i) By (3.1.4), $\bar{L}$ is constructible in $\operatorname{Spv}''(\alpha, A)$. By (3.2.1 ii) and (3.2.3), $\bar{L} \cap \operatorname{Max}(\alpha, A)$ is the closure of $L \cap \operatorname{Max}(\alpha, A)$ in $\operatorname{Max}(\alpha, A)$.

ii) follows from i) and (3.2.3). □

Proposition (3.2.5) means that the operation $\sim$ commutes with the closure operations in $\operatorname{Max}(\alpha, A)$ and $\operatorname{Spv}''(\alpha, A)$.

PROPOSITION (3.2.6). *We assume that A is finitely generated over k. Let S be a semialgebraic subset of $Max(\alpha, A)$ and let $x \in Max(\alpha, A)$ be a point of the closure of S in $Max(\alpha, A)$. Then there exist a finitely generated, 1-dimensional,* *regular k-algebra B, a k-algebra homomorphism $f\colon A \to B$, an open subset U of $Max(\alpha, B)$ and a point $x_0 \in U$ such that $g(x_0) = x$ and $g(U \setminus \{x_0\}) \subseteq S$ where $g\colon Max(\alpha, B) \to Max(\alpha, A)$ is the mapping induced by f.*

PROOF. We distinguish the cases $x \in S$ and $x \notin S$. First assume $x \in S$. Let K be the residue field at $\operatorname{supp}(x)$ and β the valuation of K induced by x. Let f be the k-algebra homomorphism $A \to K[T]$, $a \mapsto a \bmod \operatorname{supp}(x) \in K \subseteq K[T]$. Put $U = \operatorname{Max}(\beta, K[T]) \subseteq \operatorname{Max}(\alpha, K[T])$. Then U is open in $\operatorname{Max}(\alpha, K[T])$ and $g(U) = \{x\}$. Now assume $x \notin S$. Let L be a constructible subset of $\operatorname{Spv}''A$ with $L \cap \operatorname{Max}(\alpha, A) = S$. Then x lies in the closure of L in $\operatorname{Spv}''A$. By (2.3.1) there exists a generalization v of x in $\operatorname{Spv}''A$ with $v \in L$ and $ht(\operatorname{supp}(x)/\operatorname{supp}(v)) = 1$. Let B be the normalization of $A/\operatorname{supp}(v)$ and $f\colon A \to B$ the canonical ring homomorphism. Then B is a finitely generated, 1–dimensional, regular k-algebra. Put $h\colon = \operatorname{Spv}''(f)$. There exist a $x_0 \in \operatorname{Max}(\alpha, B)$ and a generalization v_0 of x_0 in $\operatorname{Spv}''B$ with $x = h(x_0)$ and $v = h(v_0)$. Then $\{x_0\}$ is constructible in $\operatorname{Spv}''(\alpha, B)$ and v_0 is the unique proper generalization of x_0 in $\operatorname{Spv}''(\alpha, B)$. Since $v_0 \in h^{-1}(L)$, there exists an open constructible subset V of $\operatorname{Spv}''(\alpha, B)$ with $x_0 \in V$ and $V \setminus \{x_0\} \subseteq h^{-1}(L)$. Put $U = V \cap \operatorname{Max}(\alpha, B)$. Then U is a neighbourhood of x_0 in $\operatorname{Max}(\alpha, B)$ with $h(U \setminus \{x_0\}) \subseteq S$. □

3.3. An example. Let k be an algebraically closed field and $\alpha\colon k \twoheadrightarrow \Gamma_\infty$ a non-trivial valuation of k. In this section, we study $\operatorname{Spv}(\alpha, k[T])$ where $k[T]$ is the polynomial ring in one variable over k. (Analogous results hold for $\operatorname{Spv}'(\alpha, k[T])$.)

We consider $\operatorname{Max}(\alpha, k[T])$ as a subspace of $\operatorname{Spv}(\alpha, k[T])$ (by (3.2.1 ii)) and identify $\operatorname{Max}(\alpha, k[T])$ with k (by (3.2.1 i)).

By (3.2.3), the points of $\operatorname{Spv}(\alpha, k[T])$ correspond to the ultrafilters of semialgebraic subsets of k. Our first aim is to define a subset C of the set of semialgebraic subsets of k and to show that the points of $\operatorname{Spv}(\alpha, k[T])$ correspond to the filters of C.

DEFINITION. i) Let S be a subset of k. S is called an *o-disk* if there exist $a \in k$ and $\gamma \in \Gamma$ with $S = \mathbb{B}^+(a, \gamma)\colon = \{x \in k | \alpha(x - a) \geq \gamma\}$. S is called a *c-disk* if $|S| = 1$, or if there exist $a \in k$ and $\gamma \in \Gamma$ with $S = \mathbb{B}^-(a, \gamma)\colon = \{x \in k | \alpha(x - a) > \gamma\}$. S is called a *disk* if S is an o-disk or a c-disk.

ii) Let S be a subset of $\operatorname{Spv}(\alpha, k[T])$. S is called an *o-disk* if there exist $a \in k$, $b \in k^*$ with $S = \{v \in \operatorname{Spv}(\alpha, k[T]) | v(T - a) \geq v(b)\}$. S is called a *c-disk* if there exist $a, b \in k$ with $S = \operatorname{Spv}(\alpha, k[T]) \setminus \{v \in \operatorname{Spv}(\alpha, k[T]) | v(b) \geq v(T - a) \neq \infty\}$. S is called a *disk* if S is an o-disk or a c-disk.

By $S \mapsto \tilde{S}$ we have a bijection from the set of o-disks (resp. c-disks) of k to the set of o-disks (resp. c-disks) of $\operatorname{Spv}(\alpha, k[T])$). If A, B are two disks of k (resp. $\operatorname{Spv}(\alpha, k[T])$), then $A \cap B = \emptyset$ or $A \subseteq B$ or $B \subseteq A$.

DEFINITION. Let C be the set of disks of k. A *filter* of C is a subset F of C such that

a) If $A \in F$ and $B \in C$ with $A \subseteq B$, then $B \in F$.
b) If $A, B \in F$, then $A \cap B \in F$.

Remark (3.3.1). i) To $a \in k$, $b \in k^*$ there exists a unique valuation $v = v(a, \alpha(b))$ of $k(T)$ extending α such that $v(\frac{T-a}{b}) \geq 0$ and the image t of $\frac{T-a}{b}$ in the residue field of v is transcendental over the residue field $\bar{k}$ of α [**B**, VI.10.1, Prop. 2]. We have $v\colon k(T) \to \Gamma_\infty$,

$$v\Big(\sum_{i=0}^{n} a_i(T-a)^i\Big) = \min\{\alpha(a_i) + i \cdot \alpha(b) | i = 0, \dots, n\}.$$

The residue field of v is $\bar{k}(t)$. Conversely, if v is an extension of α to $k(T)$ such that the residue field of v is a proper extension of $\bar{k}$, then $v = v(a, \gamma)$ for some $a \in k$, $\gamma \in \Gamma$.

ii) Let M be a minor subset of Γ (i.e. if $x \in M$, $y \in \Gamma$ with $y \leq x$ then $y \in M$) and a an element of k. Then there exists, up to equivalence, a unique valuation $v = v(a, M)$ of $k(T)$ extending α such that $v(T-a) \notin \Gamma$ and $M = \{\gamma \in \Gamma | \gamma < v(T-a)\}$ (cf. [**B**, VI.10.1, Prop. 1]). v can be constructed as follows. We consider Γ as a subgroup of $\Gamma \oplus \mathbb{Z}$ by $\gamma = (\gamma, 0)$, and extend the ordering of Γ to the group ordering of $\Gamma \oplus \mathbb{Z}$ such that $M = \{\gamma \in \Gamma | \gamma < (0,1)\}$. Then $v\colon k(T) \to (\Gamma \oplus \mathbb{Z})_\infty$,

$$v\Big(\sum_{i=0}^{n} a_i(T-a)^i\Big) = \min\{\alpha(a_i) + (0, i) | i = 0, \dots, n\}.$$

The residue field of v is equal to that of α. Conversely, if v is an extension of α to $k(T)$ such that $\Gamma_v \supsetneqq \Gamma$, then $v = v(a, M)$ for some $a \in k$ and minor subset M of Γ.

PROPOSITION (3.3.2). *Let F be a filter of C. Then there exists an unique point $\varphi(F) \in Spv(\alpha, k[T])$ with $F = \{S \in C | \varphi(F) \in \tilde{S}\}$. Distinguishing four cases, we can give a precise description of $\varphi(F)$.*

I) *If there exists a $a \in k$ with $F = \{S \in C | a \in S\}$, then $\varphi(F)$ is the point of $Spv(\alpha, k[T])$ with support $(T-a) \cdot k[T]$, i.e. $\varphi(F) = a$.*
II) *If there exist $a \in k$, $\gamma \in \Gamma$ with $F = \{S \in C | \mathbb{B}^+(a, \gamma) \subseteq S\}$, then $\varphi(F) = v(a, \gamma)$.*
III) *If $\bigcap_{S \in F} S = \emptyset$, then $\varphi(F)$ is an immediate extension of α to $k(T)$ and can be constructed as follows. Let $\frac{p(T)}{q(T)} \in k(T)^*$ be given. Choose $S \in F$, which is disjoint to the zero set of $p(T) \cdot q(T)$. Then there exists a $\gamma \in \Gamma$ with $\alpha\Big(\frac{p(x)}{q(x)}\Big) = \gamma$ for every $x \in S$, and we have $\varphi(F)\Big(\frac{p(T)}{q(T)}\Big) = \gamma$.*

IV) *Assume that F is not of type* I *or* II *or* III. *Choose* $\alpha \in \bigcap_{S \in F} S$ *and put* $M = \{\gamma \in \Gamma | \mathbb{B}^-(a, \gamma) \in F\}$. *Then* $\varphi(F) = v(a, M)$.

The mapping $F \mapsto \varphi(F)$ *is a bijection from the set of filters of* C *to the set* $Spv(\alpha, k[T])$.

PROOF. 1) For every $a \in k$, there exists obviously an unique $x \in$ Spv $(\alpha, k[T])$ with $\{S \in C | a \in S\} = \{S \in C | x \in \tilde{S}\}$. Namely, x is the point of Spv (α, A) with support $(T - a) \cdot k[T]$. Hence, in the following steps 2) and 3) we deal only with filters F such that $|S| > 1$ for every $S \in F$ and with points x of Spv $(\alpha, k[T])$ such that $\{0\} = \operatorname{supp}(x)$.

2) Let F be a filter of C. Then there exists at most one $v \in$ Spv $(\alpha, k[T])$ with $F = \{S \in C | v \in \tilde{S}\}$. Indeed, let such a v be given. For every $a \in k$ put $M_a = \{\gamma \in \Gamma | v(T - a) \geq \gamma\}$ and $N_a = \{\gamma \in \Gamma | v(T - a) > \gamma\}$. Then M_a and N_a are uniquely determined since $M_a = \{\gamma \in \Gamma | v \in \mathbb{B}^+(a, \gamma)^\sim\}$ and $N_a = \{\gamma \in \Gamma | v \in \mathbb{B}^-(a, \gamma)^\sim\}$. M_a and N_a are minor subsets of Γ with $N_a \subseteq M_a$. If $M_a = N_a$ for some $a \in k$, then $v(T - a) \notin \Gamma$, and hence v is uniquely determined by (3.3.1 ii). So assume $N_a \subsetneqq M_a$ for all $a \in k$. Then $v(T - a) \in \Gamma$ and $v(T - a) = \max M_a$ for every $a \in k$. Hence v is uniquely determined on the set $\{T - a | a \in k\} \subseteq k[T]$, and therefore v is uniquely determined.

3) Let F be a filter of type II, III or IV, and let $\varphi(F)$ be the valuation as defined in II, III or IV. One can easily check that $F = \{S \in C | \varphi(F) \in \tilde{S}\}$.

4) By 1), 2), 3) we have at every filter F of C an unique point $\varphi(F) \in$ Spv $(\alpha, k[T])$ with $F = \{S \in C | \varphi(F) \in \tilde{S}\}$. Hence we have a mapping φ from the set of filters of C to the set Spv $(\alpha, k[T])$. Obviously, φ is injective. To show the surjectivity of φ, let $x \in$ Spv $(\alpha, k[T])$ be given. Then $F = \{S \in C | x \in \tilde{S}\}$ is a filter of C with $x = \varphi(F)$.

5) Let v be a valuation as in III. It remains to show that v is an immediate extension of α. We have $\Gamma_v = \Gamma$ by construction of v. We deduce from (3.3.1 i) the injectivity of φ and II that the residue field of v is equal to that of α. □

COROLLARY (3.3.3). *The boolean algebra of constructible subsets of Spv* $(\alpha, k[T])$ *is generated by the disks of Spv* $(\alpha, k[T])$.

PROOF. Let B be the boolean algebra generated by the disks of Spv $(\alpha, k[T])$. Let L be a constructible subset of Spv $(\alpha, k[T])$ and let x be a point in the complement of L. By (3.3.2) there exists, for every $y \in L$, an element M of B with $y \in M$ and $x \notin M$. Hence there exists a $K \in B$ with $L \subseteq K$ and $x \notin K$. That means

$$L = \bigcap_{\substack{K \in B \\ L \subseteq K}} K$$

which implies $L \in B$.

A subset S of Spv $(\alpha, k[T])$ is called a *generalized disk* if we can write

$$(*) \qquad S = B \setminus \bigcup_{i=1}^{n} B_i$$

where $n \in \mathbb{N}_0$ and $B, B_1, \ldots, B_n$ are disks or Spv $(\alpha, k[T])$. S is called a *generalized o-disk* if there exists a representation $(*)$ where B is an o-disk or Spv $(\alpha, k[T])$ and $B_1, \ldots, B_n$ are c-disks, and S is called a *generalized c-disk* if there exists a representation $(*)$ where B is a c-disk or Spv $(\alpha, k[T])$ and $B_1, \ldots, B_n$ are o-disks. (Note that we can assume $B_i \subseteq B$ for $i = 1, \ldots, n$ and $B_i \cap B_j = \emptyset$ for $i \neq j$.) $\square$

The following corollary is a reformulation of (3.3.3).

COROLLARY (3.3.4). *Every constructible subset of Spv $(\alpha, k[T])$ is a finite and disjoint union of generalized disks.*

Next we consider the specializations in Spv $(\alpha, k[T])$. We call $\mathbb{B}^+(a, \lambda)$ the associated o-disk to the c-disk $\mathbb{B}^-(a, \lambda)$. A c-disk U of k is called associated to an o-disk V of k if V is associated to U. Every o-disk of k is the disjoint union of its associated c-disks.

PROPOSITION (3.3.5). i) *A point x of Spv$(\alpha, k[T])$ has a proper generalization in Spv$(\alpha, k[T])$ if and only if there exists a c-disk B of k with $x = \varphi(\{S \in C | B \subseteq S\})$ or an o-disk B of k with $x = \varphi(\{S \in C | B \subsetneqq S\})$. In the following points* ii), iii), iv), *we describe these generalizations.*

ii) *Let a be an element of k. Then $a = \varphi(\{S \in C | a \in S\})$ has a unique proper generalization in Spv$(\alpha, k[T])$, namely the point $\varphi(\{S \in C | \{a\} \subsetneqq S\})$, and this is a primary generalization.*

iii) *Let B be a c-disk of k with $|B| > 1$ and let D be the associated o-disk. Then $\varphi(\{S \in C | B \subseteq S\})$ has a unique proper generalization in Spv$(\alpha, k[T])$, namely the point $\varphi(\{S \in C | D \subseteq S\})$, and this is a secondary generalization.*

iv) *Let B be an o-disk of k. Then $\varphi(\{S \in C | B \subsetneqq S\})$ has a unique proper generalization in Spv$(\alpha, k[T])$, namely the point $\varphi(\{S \in C | B \subseteq S\})$, and this is a secondary generalization.*

PROOF. Let x be a point of Spv $(\alpha, k[T])$. We consider the generalizations of x.

1) Assume that supp (x) is the maximal ideal $(T - a) \cdot k[T]$ of $k[T]$. Then x has no proper secondary generalization in Spv $(\alpha, k[T])$, and hence according to (1.2.4) every generalization of x in Spv $(\alpha, k[T])$ is primary. Since the local ring $k[T]_{\text{supp}\,(x)}$ is a valuation ring of dimension 1, x has a unique proper primary generalization v (cf. (1.2.2 ii)). One easily checks that v corresponds to the filter $\{S \in C | \{a\} \subsetneqq S\}$.

2) Assume $\operatorname{supp}(x) = \{0\}$. Then every generalization v of x in $\operatorname{Spv}(\alpha, k[T])$ is a secondary generalization, i.e. $v = x/H$ where H is a convex subgroup of Γ_x with $H \cap \Gamma = \{0\}$. If x is of type II or III in (3.3.2), then $\Gamma_x = \Gamma$ and hence $v = x$. So we assume that x is of type IV in (3.3.2), $x = \varphi(F) = v(a, M)$. There exists a non-trivial convex subgroup H of Γ_x with $H \cap \Gamma = \{0\}$ if and only there exists a $h \in \Gamma_x$ such that $0 < h < \gamma$ for every non-negative $\gamma \in \Gamma$, and that holds true if and only if M has a greatest element or $\Gamma \setminus M$ has a smallest element. In both cases there is only one non-trivial convex subgroup H of Γ_x with $H \cap \Gamma = \{0\}$. M has a greatest element if and only if there exists a c-disk B of k such that $|B| > 1$ and $F = \{S \in C | B \subseteq S\}$. $\Gamma \setminus M$ has a smallest element if and only if there exists an o-disk B of k with $F = \{S \in C | B \subsetneqq S\}$. Let γ be the greatest element of M (resp. smallest element of $\Gamma \setminus M$). Then $v = x/H = v(a, M)/H = v(a, \gamma)$. Hence by (3.3.2 II), v is the point given in iii) and iv). □

The following proposition is a reformulation of (3.3.5).

PROPOSITION (3.3.6). i) *A point x of $Spv(\alpha, k[T])$ has a proper specialization in $Spv(\alpha, k[T])$ if and only if there exists a $a \in k$ with $x = \varphi(\{S \in C | \{a\} \subsetneqq S\})$ or an o-disk B of k with $x = \varphi(\{S \in C | B \subseteq S\})$. These specializations are described in the following points* ii) *and* iii).

ii) *Let a be an element of k. Then $\varphi(\{S \in C | \{a\} \subsetneqq S\})$ has an unique specialization in $Spv(\alpha, k[T])$, namely the point $a \in Spv(\alpha, k[T])$, and this specialization is primary.*

iii) *Let B be an o-disk of k. We consider the point $x = \varphi(\{S \in C | B \subseteq S\})$. Let $B_i, i \in I$ be the c-disks associated to B. Then $\varphi(\{S \in C | B_i \subseteq S\})$, $i \in I$ and $\varphi(\{S \in C | B \subsetneqq S\})$ are the specializations of x in $Spv(\alpha, k[T])$. All these specializations are secondary.*

The following two corollaries are consequences of (3.3.5) and (3.3.6).

COROLLARY (3.3.7). i) *Let B be a c-disk of $Spv(\alpha, k[T])$. Then there exist unique points $x \in B$ and $y \in Spv(\alpha, k[T]) \setminus B$ such that y is a generalization of x.*

ii) *Let B be an o-disk of $Spv(\alpha, k[T])$. Then there exist unique points $x \in B$ and $y \in Spv(\alpha, k[T]) \setminus B$ such that y is a specialization of x.*

COROLLARY (3.3.8). i) *A generalized disk is closed in $Spv(\alpha, k[T])$ if and only if it is a generalized c- disk.*

ii) *A generalized disk is open in $Spv(\alpha, k[T])$ if and only if it is a generalized o-disk.*

PROPOSITION (3.3.9).

i) *Every open constructible subset of $Spv(\alpha, k[T])$ is a finite and disjoint union of generalized o-disks.*

ii) *Every closed constructible subset of Spv*$(\alpha, k[T])$ *is a finite and disjoint union of generalized c-disks.*

PROOF. Let L be an open constructible subset of Spv$(\alpha, k[T])$. By (3.3.4), $L = L_1 \cup \cdots \cup L_n$ where $L_1, \ldots, L_n$ are generalized disks. Since L is closed under generalizations in Spv$(\alpha, k[T])$, we conclude from (3.3.4) and (3.3.5) that, for every $i = 1, \ldots, n$, there exists a generalized o-disk M_i with $L_i \subseteq M_i \subseteq L$. Assertion ii) can be proved analogously. □

One can conclude from (3.3.4) and (3.3.5):

PROPOSITION (3.3.10). *A constructible subset of Spv*$(\alpha, k[T])$ *is connected if and only if it is a generalized disk.*

Then (3.3.4) and (3.3.10) imply

COROLLARY (3.3.11). *Every constructible subset of Spv*$(\alpha, k[T])$ *has finitely many connected components.*

3.4. Supplement to (2.4). In this paragraph we want to give a proof of (2.4.8) and (2.4.10).

LEMMA (3.4.1). *Let* $f\colon X \to Y$ *be a spectral mapping between spectral spaces. Let* Z *be a clopen subset of a fibre* $F = f^{-1}(y)$ *of* f*, and let* U *be a constructible subset of* X *containing* Z*. Then there exist a constructible subset* V *of* U *and a constructible subset* W *of* Y *such that* $Z = F \cap V$ *and* V *is a clopen subset of* $f^{-1}(W)$.

PROOF. One can prove (3.4.1) by use of (2.4.1 i). But we give here another proof. Let P (resp. Q) be the specializations (resp. generalizations) of points of $F \setminus Z$ in X. Then $P \cup Q$ is a pro-constructible subset of X with $Z \cap (P \cup Q) = \emptyset$. Let R be a constructible subset of X with $Z \subseteq R \subseteq U \setminus (P \cup Q)$. Let S (resp. T) be the set of specializations (resp. generalizations) of points of R. Then $H\colon = (S \cup T) \setminus R$ is a pro-constructible subset of X with $H \cap F = \emptyset$. Hence $f(H)$ is a pro-constructible subset of Y with $y \notin f(H)$. Let W be a constructible subset of Y with $y \in W$ and $f(H) \cap W = \emptyset$. Put $V = R \cap f^{-1}(W)$. Then $V \subseteq U \cap f^{-1}(W)$, $Z = F \cap V$ and V is closed under specializations and generalizations in $f^{-1}(W)$, hence V is clopen in $f^{-1}(W)$. □

COROLLARY (3.4.2). *Let* $f\colon A \to B$ *be a flat, quasi-finite and finitely presented ring homomorphism. Let* L *be a pro-constructible subset of Spv*A *such that every constructible subset of* L *has finitely many connected components. Then every constructible subset of Spv*$(f)^{-1}(L) \subseteq$ *Spv*B *has finitely many connected components.*

PROOF. By [**EGA**, IV.18.12.13] there exist ring homomorphisms $g\colon C \to B$ and $h\colon A \to C$ such that $f = g \circ h$, h is finite and $\operatorname{Spec}(g)\colon \operatorname{Spec} B \to \operatorname{Spec} C$ is an open embedding of schemes. We consider $\operatorname{Spv} B$ as an open subspace of $\operatorname{Spv} C$ via $\operatorname{Spv}(g)$. Let U be a constructible subset of $\operatorname{Spv}(f)^{-1}(L)$ and x an element of U. We have to show that there exists a constructible subset V of U which is connected and contains x. Put $p = \operatorname{Spv}(h)\colon \operatorname{Spv} C \to \operatorname{Spv} A$. By (3.4.1) there exist a constructible subset W of L and a constructible subset V of U such that $\{x\} = p^{-1}(p(x)) \cap V$ and V is a clopen subset of $p^{-1}(W)$. Since W has finitely many connected components, we may assume that W is connected. We claim that then V is connected, too. Assume by way of contradiction that there exists a decomposition $V = V_1 \cup V_2$ into nonempty clopen subsets. Let $q\colon V \to W$ be the restriction of p. By (2.1.7 ii), q is closed and by (2.1.4) q is open. Hence $q(V_1)$ and $q(V_2)$ are clopen subsets of W. So $q(V_1) = q(V_2)$, in contradiction to $\#q^{-1}(q(x)) = 1$ and $V_1 \cap V_2 = \emptyset$. $\square$

LEMMA (3.4.3). *Let k be an algebraically closed field and α the trivial valuation of k. We consider $Spv(\alpha, k[T])$, where $k[T]$ is the polynomial ring in one variable over k. Let v_0 be the trivial valuation of $k(T)$, v_1 the valuation of $k(T)$ with valuation ring $k[T^{-1}]_{T^{-1}\cdot k[T^{-1}]}$, and, for every $a \in k$, $t(a)$ the valuation of $k(T)$ with valuation ring $k[T]_{(T-a)\cdot k[T]}$ and $s(a)$ the trivial valuation of $k[T]$ with support $(T-a)\cdot k[T]$. A subset L of $Spv(\alpha, k[T])$ is called a generalized disk if $|L| \leq 1$ and $L \neq \{v_0\}$ or if $L = \{s(a), t(a)\}$ for some $a \in k$ or if $L = Spv(\alpha, k[T]) \setminus M$ where M is a finite subset of $Spv(\alpha, k[T]) \setminus \{v_0\}$. Then*

i) *$Spv(\alpha, k[T]) = \{v_0, v_1\} \cup \{s(a), t(a) | a \in k\}$.*
ii) *Every subset L of $Spv(\alpha, k[T]) \setminus \{v_0\}$ with $|L| \leq 1$ is constructible, namely $\{v_1\} = \{v \in Spv(\alpha, k[T]) | v(T) < 0\}$, $\{s(a)\} = \{v \in Spv(\alpha, k[T]) | v(T - a) = \infty\}$ and $\{t(a)\} = \{v \in Spv(\alpha, k[T]) | v(T-a) > 0$ and $v(T-a) \neq \infty\}$.*
iii) *A subset L of $Spv(\alpha, k[T])$ is constructible if and only if there exists a finite subset M of $Spv(\alpha, k[T]) \setminus \{v_0\}$ with $L = M$ or $L = Spv(\alpha, k[T]) \setminus M$.*
iv) *The proper specializations in $Spv(\alpha, k[T])$ are $v_0 \succ t(a) \succ s(a)$ $(a \in k)$ and $v_0 \succ v_1$.*
v) *A constructible subset of $Spv(\alpha, k[T])$ is connected if and only if it is a generalized disk.*

PROOF. iii) Let L be a constructible subset of $\operatorname{Spv}(\alpha, k[T])$. If $v_0 \notin L$, then L is finite by ii), and if $v_0 \in L$, then $\operatorname{Spv}(\alpha, k[T]) \setminus L$ is finite by ii).

v) Every constructible subset of $\operatorname{Spv}(\alpha, k[T])$ containing v_0 is connected since $v_0 \succ x$ for every $x \in \operatorname{Spv}(\alpha, k[T])$. $\square$

In the following we use the following notation: Let k be a field and α a valuation of k. We call a subset L of $\operatorname{Spv}(\alpha, k[T])$ a generalized disk if $\lambda^{-1}(L)$

is a generalized disk of $\operatorname{Spv}(\bar{\alpha}, \bar{k}[T])$ where $\bar{k}$ is an algebraic closure of k, $\bar{\alpha}$ an extension of α to $\bar{k}$ and $\lambda\colon \operatorname{Spv}(\bar{\alpha}, \bar{k}[T]) \to \operatorname{Spv}(\alpha, k[T])$ the canonical mapping. Since λ is surjective, (3.3.10) and (3.4.3 v) imply that generalized disks are connected.

PROOF OF (2.4.8). i) Let L be a connected pro-constructible subset of $\operatorname{Spv} E$. First assume that $F = E(T)$ is a transcendental extension of E. By (3.1.2), $\operatorname{Spv}(\alpha, E[T])$ is connected for every $\alpha \in L$. Then by (2.4.6) and (2.4.1 i), $\operatorname{Spv}(\alpha, E(T)) = g^{-1}(\alpha)$ is connected. Since g is open (by (2.2.3) and (2.1.2)), we conclude that $g^{-1}(L)$ is connected. Now consider the case that F is finite over E. Assume that the number of connected components of $g^{-1}(L)$ is greater than $[F\colon E]_s$. Then there exists a decomposition $g^{-1}(L) = M_1 \cup \cdots \cup M_n$ of $g^{-1}(L)$ into non-empty clopen subsets $M_1, \ldots, M_n$ with $n > [F\colon E]_s$. Since g is open and closed, we have $L = g(M_1) = \cdots = g(M_n)$. Hence $\#g^{-1}(x) \geq n$ for every $x \in L$, contradiction.

ii) Let L be a pro-constructible subset of $\operatorname{Spv} E$ such that every constructible subset of L has finitely many connected components. We will show that every constructible subset of $g^{-1}(L)$ has finitely many connected components. By (3.4.2) we may assume that F is purely transcendental over E, and then by induction we may assume that $F = E(T)$ has transcendence degree 1 over E. We consider the canonical mapping $f\colon \operatorname{Spv} E[T] \to \operatorname{Spv} E$. By (2.4.7), it suffices to show that every constructible subset of $f^{-1}(L)$ has finitely many connected components. Let M be a constructible subset of $f^{-1}(L)$ and x an element of M. We will show that there exists a connected constructible subset of M containing x. Let $\bar{E}$ be the algebraic closure of E. We consider the commutative diagram (with canonical morphisms)

$$\begin{array}{ccc} \operatorname{Spv} \bar{E}[T] & \xrightarrow{h_T} & \operatorname{Spv} E[T] \\ {\scriptstyle \bar{f}}\downarrow & & \downarrow{\scriptstyle f} \\ \operatorname{Spv} \bar{E} & \xrightarrow[h]{} & \operatorname{Spv} E\,. \end{array}$$

Let $\bar{x}$ be a point of $\operatorname{Spv} \bar{E}[T]$ lying over x. Put $\bar{y} = \bar{f}(\bar{x}) \in \operatorname{Spv} \bar{E}$.

We distinguish five cases.

First case: $\bar{y}$ is non-trivial. By (3.3.4) there exists a generalized disk B of $\operatorname{Spv}(\bar{y}, \bar{E}[T]) = \bar{f}^{-1}(\bar{y})$ with $\bar{x} \in B \subseteq h_T^{-1}(M)$. Let $\bar{e}$ be an element of $\bar{E}$ such that the point of $\operatorname{Spv}(\bar{y}, \bar{E}[T])$ with support $(T - \bar{e}) \cdot \bar{E}[T]$ is contained in B. Choose a description of $B \subseteq \operatorname{Spv}(\bar{y}, \bar{E}[T])$ by polynomials $p_1, \ldots, p_n \in \bar{E}[T]$ of the form $p_i = T - a$ or $p_i = a$ with $a \in \bar{E}$. Let K be the subfield of $\bar{E}$ generated by $E, \bar{e}$ and the coefficients of $p_1, \ldots, p_n$. We consider the commutative diagram

$$(*) \qquad \begin{array}{ccccc} \operatorname{Spv} \bar{E}[T] & \xrightarrow{j_T} & \operatorname{Spv} K[T] & \xrightarrow{i_T} & \operatorname{Spv} E[T] \\ \bar{f}\downarrow & & \downarrow p & & \downarrow f \\ \operatorname{Spv} \bar{E} & \xrightarrow[j]{} & \operatorname{Spv} K & \xrightarrow[i]{} & \operatorname{Spv} E\,. \end{array}$$

Put $y = j(\bar{y})$. There exists a constructible subset U of $\operatorname{Spv} K[T]$ such that, for every $z \in \operatorname{Spv} K$, $p^{-1}(z) \cap U \subseteq p^{-1}(z) = \operatorname{Spv}(z, K[T])$ is a generalized disk of $\operatorname{Spv}(z, K[T])$ and $j_T(B) = p^{-1}(y) \cap U$. Then $p^{-1}(y) \cap U \subseteq i_T^{-1}(M)$. Hence there exists a constructible subset V of $i^{-1}(L)$ with $y \in V$ and $p^{-1}(V) \cap U \subseteq i_T^{-1}(M)$. Let $s\colon \operatorname{Spv} K \to \operatorname{Spv} K[T]$ be the mapping induced by the K-algebra homomorphism $K[T] \to K$, $T \mapsto \bar{e}$. Then s is a section of p with $s(y) \in U$. Making V smaller, we can assume $s(V) \subseteq U$. By (3.4.2), V has finitely many connected components. Hence we may assume that V is connected. Then since $p^{-1}(z) \cap U$ is connected for every $z \in V$ and $s(V) \subseteq p^{-1}(V) \cap U$ is connected, we obtain that $H\colon = p^{-1}(V) \cap U$ is connected. Now $i_T(H)$ is a connected constructible subset of M which contains x.

Second case: $\bar{y}$ is trivial and $\bar{x}$ is the trivial valuation of $\bar{E}(T)$. Then x is the trivial valuation of $E(T)$. Hence $x \succ z$ for every $z \in \operatorname{Spv} E[T]$, which implies that M is connected.

Third case: $\bar{y}$ is trivial and $\bar{x}(T) < 0$. Let y be the trivial valuation of E. Then x is the unique point v of $\operatorname{Spv}(y, E[T]) = f^{-1}(y)$ with $v(T) < 0$. Put $U = \{v \in \operatorname{Spv} E[T] | v(T) < 0\}$. Since $f^{-1}(y) \cap U \subseteq M$, there exists a constructible subset V of L with $y \in V$ and $f^{-1}(V) \cap U \subseteq M$. For every $z \in V$, $f^{-1}(z) \cap U$ is connected and contains a point which is a specialization of x. Hence $f^{-1}(V) \cap U$ is connected.

Fourth case: $\bar{y}$ is trivial and there exists a $\bar{e} \in \bar{E}$ with $\bar{x}(T - \bar{e}) > 0$ and $\bar{x}(T - \bar{e}) \neq \infty$. We consider the diagram $(*)$ with $K = E(\bar{e})$. Put $y = j(\bar{y})$ and $U = \{v \in \operatorname{Spv} K[T] | v(T - \bar{e}) > 0 \text{ and } v(T - \bar{e}) \neq \infty\}$. Then $p^{-1}(y) \cap U = \{j_T(\bar{x})\} \subseteq i_T^{-1}(M)$. Hence there exists a constructible subset V of $i^{-1}(L)$ with $y \in V$ and $p^{-1}(V) \cap U \subseteq i_T^{-1}(M)$. For every $z \in V$, $f^{-1}(z) \cap U$ is connected and contains a point which is a specialization of $j_T(x)$. Hence $H\colon = p^{-1}(V) \cap U$ is connected. Then $i_T(H)$ is a connected constructible subset of M with $x \in i_T(H)$.

Fifth case: $\bar{y}$ is trivial and there exists a $\bar{e} \in \bar{E}$ with $\bar{x}(T - \bar{e}) = \infty$. We consider the diagram $(*)$ with $K = E(\bar{e})$. Put $y = j(\bar{y})$ and $U = \{v \in \operatorname{Spv} K[T] | v(T - \bar{e}) = \infty\}$. Then $p^{-1}(y) \cap U = \{j_T(\bar{x})\} \subseteq i_T^{-1}(M)$. Hence there exists a constructible subset V of $i^{-1}(L)$ with $y \in V$ and $p^{-1}(V) \cap U \subseteq i_T^{-1}(M)$. Since every point of $H\colon = p^{-1}(V) \cap U$ is a specialization of $j_T(\bar{x}) \in H$, H is connected. Hence $i_T(H)$ is a connected constructible subset of M which contains x. □

PROOF OF (2.4.10). We may assume that B is a polynomial ring over A and then by induction we may assume that $B = A[T]$ is the polynomial ring in one variable over A and $f\colon A \to B$ is the canonical ring homomorphism. Let M be a constructible subset of $\operatorname{Spv}(f)^{-1}(L)$ and x an element of M. We show that there exists a connected constructible subset of M which contains x. Let $(C_i | i \in I)$ be a filtered inductive system of flat, quasi-finite and finitely presented A-algebras such that $C := \varinjlim_{i \in I} C_i$ is a local ring with maximal ideal $\mathfrak{m}$ such that $\mathfrak{m}$ lies over $\operatorname{supp}(x) \cap A$ and $C/\mathfrak{m}$ is algebraically closed. We consider the commutative diagram

$$\begin{array}{ccc} \operatorname{Spv} C[T] & \xrightarrow{h_T} & \operatorname{Spv} A[T] \\ {\scriptstyle f}\downarrow & & \downarrow \\ \operatorname{Spv} C & \xrightarrow[h]{} & \operatorname{Spv} A. \end{array}$$

Let y be a point of $\operatorname{Spv} C[T]$ with $x = h_T(y)$ and $\operatorname{supp}(y) \cap C = \mathfrak{m}$. Put $z = f(y)$. By (3.3.4), there exists a generalized disk D of $f^{-1}(z)$ with $y \in D \subseteq h_T^{-1}(M)$. Choose a representation of D by polynomials $p_1, \ldots, p_n \in C/\mathfrak{m}[T]$ such that every p_i is of the form $p_i = T - \bar{a}_i$ or $p_i = \bar{a}_i$ with $\bar{a}_i \in C/\mathfrak{m}$, and choose a $\bar{b} \in C/\mathfrak{m}$ such that the point of $f^{-1}(z)$ with support $(T - \bar{b}) \cdot C/\mathfrak{m}[T]$ is contained in D. Let $b, a_1, \ldots, a_n \in C$ be representatives of $\bar{b}, \bar{a}_1, \ldots, \bar{a}_n$, and choose a $k \in I$ with $b, a_1, \ldots, a_n \in C_k$. Now, using the commutative diagram

$$\begin{array}{ccccc} \operatorname{Spv} C[T] & \xrightarrow{j_T} & \operatorname{Spv} C_k[T] & \xrightarrow{i_T} & \operatorname{Spv} A[T] \\ {\scriptstyle f}\downarrow & & \downarrow{\scriptstyle p} & & \downarrow \\ \operatorname{Spv} C & \xrightarrow[j]{} & \operatorname{Spv} C_k & \xrightarrow[i]{} & \operatorname{Spv} A, \end{array}$$

we can continue with the arguments of the first case in the proof of (2.4.8 ii). □

4. Applications of the valuation spectrum to algebraic geometry

4.1. Etale cohomology. In this short section we pursue two aims. First we indicate that a conjecture of Artin stated in [**SGA**, XII.6.13] is equivalent to the vanishing of the cohomology of constant torsion sheaves on certain pro-constructible subsets of the valuation spectrum of algebraically closed fields, and secondly we sketch that this equivalence allows us to prove both these statements. We are interested in the results of this paragraph also because they are very useful in order to study etale cohomology of rigid analytic varieties (cf. [**Hu**$_3$]). Precise proofs for all that is described in this section are contained in [**Hu**$_1$].

Let X be a topological space, Y a closed subspace of X and F an abelian sheaf on X such that $H^i(U, F) = 0$ for all $i \in \mathbb{N}$ and all open subsets U of X. We assume X is paracompact or X is a normal spectral space (in the sense of [**CC**]). Then $H^i(Y, F|Y) = 0$ for all $i \in \mathbb{N}$. In [**SGA**, XII.6.13], Artin conjectured that also the analogous statement for etale cohomology of affine schemes is true, more precisely:

(*) Let X be an affine scheme, Y a closed subscheme of X and F an abelian sheaf on X_{et} which is flasque (i.e. $H^i(U,F)_{et} = 0$ for every $i \in \mathbb{N}$ and every $U \in X_{et}$) and torsion. Then $H^i(Y, F|Y)_{et} = 0$ for every $i \in \mathbb{N}$.

Let us make two statements about the cohomology of the valuation spectrum of algebraically closed fields.

(**) Let K be an algebraically closed field and D, E subsets of K. Put $X = \{v \in \operatorname{Spv} K | v(d) \geq 0$ for all $d \in D$ and $v(e) > 0$ for all $e \in E\}$ and equip X with the subspace topology of $\operatorname{Spv} K$. Then $H^i(X, F) = 0$ for every $i \in \mathbb{N}$ and every abelian torsion group F.

(**)′ Let K be an algebraically closed field and D, E subsets of K. Put $X = \{v \in \operatorname{Spv}' K | v(d) \geq 0$ for all $d \in D$ and $v(e) > 0$ for all $e \in E\}$ and equip X with the subspace topology of $\operatorname{Spv}' K$. Then $H^i(X, F) = 0$ for every $i \in \mathbb{N}$ and every abelian torsion group F.

Then we have

LEMMA (4.1.1). *The following conditions are equivalent:*
 i) (*) *holds,*
 ii) (**) *holds,*
 iii) (**)′ *holds.*

To prove the equivalence of i) and ii), use Zariski's representation of Spv K as a projective limit of schemes (2.4.3) and the proper base change theorem for etale cohomology [**SGA**, XII.5.1]. The equivalence of ii) and iii) follows from a general result of Schwartz [$\mathbf{S}_1$], which says that for every normal spectral space X, every abelian group G and every $n \in \mathbb{N}$, there is a canonical isomorphism $H^n(X, G) \cong H^n(X^*, G)$, where X^* is the inverse spectral space to X (in the sense of [**H**, Prop. 8]. In our situation, $X = \{v \operatorname{Spv}' K \mid v(d) \geq 0$ for all $d \in D$ and $v(e) > 0$ for all $e \in E\}$ is a normal spectral space and $\{v \in \operatorname{Spv}\ K \mid v(d) \geq 0$ for all $d \in D$ and $v(e) > 0$ for all $e \in E\}$ is the inverse spectral space to X.

Then (4.1.1) and the following lemma show that (*), (**) and (**)′ are true.

LEMMA (4.1.2). (**)′ *holds.*

In order to prove (4.1.2) we first show (**)′ for $i = 1$ by using the equivalence of i) and iii) in (4.1.1). Then we proceed by induction on i. For that we use that the cohomological dimension of a normal spectral space X is bounded from above by the combinatorial dimension of X ([**CC**]) and that, if $f: X \to Y$ is a spectral and specializing map between spectral spaces and the specializations of every point of X form a chain, then $(R^n f_* F)_y \cong H^n(f^{-1}(y), F|f^{-1}(y))$ for every $n \in \mathbb{N}_0$, $y \in Y$ and abelian sheaf F on X.

4.2. Open morphisms. Let $f: X \to Y$ be a morphism of schemes where f is called universally open at a point $x \in X$ if, for every base change $f_{(Y')}: X' = X \times_Y Y' \to Y'$ of f and every point $x' \in X\diamond'$ lying over x, the mapping $f_{(Y')}$ is open at x'. If f is of finite presentation at $x \in X$, then f is universally open at x

if and only if f is universally generalizing at x. In the following two propositions we study points at which f is universally generalizing.

PROPOSITION (4.2.1). *Let k be a field, A and B finitely generated k-algebras and $f\colon A \to B$ a k-algebra homomorphism. Then the set of points $x \in \operatorname{Spec} B$ at which $\operatorname{Spec}(f)\colon \operatorname{Spec} B \to \operatorname{Spec} A$ is universally generalizing is constructible in $\operatorname{Spec} B$.*

Let $f\colon X \to Y$ be a morphism of schemes, and let L be the set of points of X at which f is universally open. In [**EGA**, IV.14.3.9], *Grothendieck asked whether L is constructible in X.*

Lemma (4.2.1) says that L is constructible in X if Y is locally of finite type over a field and f is locally of finite type. Parusinski proved a similar result in complex analytic geometry. Namely in [**Pa**] he showed that if $f\colon X \to Y$ is a morphism of complex analytic spaces, then the set of points of X at which f is (universally) open is constructible in X (in the complex analytic sense). It is obvious that one can apply Parusinski's ideas to the algebraic situation, and then one obtains, more general than (4.2.1), that L is constructible in X if Y is locally noetherian and f locally of finite type. The main tool in Parusinski's proof is the flattening technique ([**Hi**], [**RG**]), whereas in our proof of (4.2.1) we use simpler methods, namely the valuation spectrum and some model theory. Also for the proof of the following proposition we use valuations.

PROPOSITION (4.2.2). *Let A, B be noetherian adic rings and $A \to B$ a continuous ring homomorphism. Let $\hat{A}$ and $\hat{B}$ be the completions of A and B. Let $\mathfrak{p}$ be an open prime ideal of B and $\hat{\mathfrak{p}}$ the corresponding open prime ideal of $\hat{B}$. Then the following conditions are equivalent.*

i) *$\operatorname{Spec} B \to \operatorname{Spec} A$ is universally generalizing at $\mathfrak{p}$.*
ii) *$\operatorname{Spec} \hat{B} \to \operatorname{Spec} \hat{A}$ is universally generalizing at $\hat{\mathfrak{p}}$.*

PROOF OF (4.2.1). We need three lemmata.

LEMMA (4.2.3). *Let $f\colon A \to B$ be a ring homomorphism.*

i) *Let x be a point of $\operatorname{Spec} B$. If for all $n \in \mathbb{N}_0$ the morphism $\operatorname{Spec} B[T_1, \ldots, T_n] \to \operatorname{Spec} A[T_1, \ldots, T_n]$ induced by f is generalizing at every point $x' \in \operatorname{Spec} B[T_1, \ldots, T_n]$ lying over x, then $\operatorname{Spec}(f)$ is universally generalizing at x.*
ii) *Let x be a point of $\operatorname{Spv}'' B$. If for all $n \in \mathbb{N}_0$ the mapping $\operatorname{Spv}'' B[T_1, \ldots, T_n] \to \operatorname{Spv}'' A[T_1, \ldots, T_n]$ induced by f is generalizing at every point $x' \in \operatorname{Spv}'' B[T_1, \ldots, T_n]$ lying over x, then $\operatorname{Spv}''(f)$ is universally generalizing at x.*

PROOF. We show ii), whereas i) can be proved analogously. Let $A \to C$ be a ring homomorphism and let y be a point of $\operatorname{Spv}'' C \otimes_A B$ lying over x. We show that $g\colon \operatorname{Spv}'' C \otimes_A B \to \operatorname{Spv}'' C$ is generalizing at y. We represent C

as the inductive limit of finitely generated A-algebras, $C = \varinjlim_{i\in I} C_i$. Let $y_i \in \operatorname{Spv}''C_i\otimes_A B$ be the image of y under the mapping $\operatorname{Spv}''C\otimes_A B \to \operatorname{Spv}''C_i\otimes_A B$. We denote the mapping $\operatorname{Spv}''C_i \otimes_A B \to \operatorname{Spv}''C_i$ by g_i. The assumption implies that, for every $i \in I$, the set $G(y_i)$ of generalizations of y_i in $\operatorname{Spv}''C_i \otimes_A B$ is mapped onto the set $G(g_i(y_i))$ of generalizations of $g(y_i)$ in $\operatorname{Spv}''C_i$. Since the set $G(y)$ of generalizations of y in $\operatorname{Spv}''C \otimes_A B$ is the projective limit of $(G(y_i)|i \in I)$ and the set $G(g(y))$ of generalizations of $g(y)$ in $\operatorname{Spv}''C$ is the projective limit of $(G(g_i(y_i))|i \in I)$, the mapping $G(y) \to G(g(y))$ is surjective [$\mathbf{B}_1$, I.9.6]. □

LEMMA (4.2.4). *Let k be a field with a non-trivial valuation α. Let A, B be finitely generated k-algebras and $f\colon A \to B$ a k-algebra homomorphism. Let $g\colon \mathit{Spv}''B \to \mathit{Spv}''A$ and $\bar{g}\colon \mathit{Max}(\alpha, B) \to \mathit{Max}(\alpha, A)$ be the mappings induced by f. Let L be a constructible subset of $\mathit{Spv}''(\alpha, B)$ such that $\bar{g}$ is open at every point of $L \cap \mathit{Max}(\alpha, B)$. Then g is generalizing at every point of L.*

PROOF. Let $x \in L$ be given. Assume that g is not generalizing at x. Then there is a generalization y of $g(x)$ in $\operatorname{Spv}''(\alpha, A)$ such that no point of $g^{-1}(y)$ specializes to x. Hence there exists an open constructible subset U of $\operatorname{Spv}''(\alpha, B)$ with $x \in U$ and $g^{-1}(y) \cap U = \emptyset$. By assumption $\bar{g}(L \cap U \cap \operatorname{Max}(\alpha, B))$ is contained in the interior of $\bar{g}(U \cap \operatorname{Max}(\alpha, B))$ in $\operatorname{Max}(\alpha, A)$. By (3.2.3) we have $\bar{g}(L \cap U \cap \operatorname{Max}(\alpha, B)) = g(L \cap U) \cap \operatorname{Max}(\alpha, A)$ and $\bar{g}(U \cap \operatorname{Max}(\alpha, B)) = g(U) \cap \operatorname{Max}(\alpha, A)$. Therefore, $g(L \cap U)\operatorname{Max}(\alpha, A)$ is contained in the interior of $g(U) \cap \operatorname{Max}(\alpha, A)$ in $\operatorname{Max}(\alpha, A)$. By (2.1.1), $g(L \cap U)$ and $g(U)$ are constructible subsets of $\operatorname{Spv}''(\alpha, A)$. Now (3.2.5) implies that $g(L \cap U)$ lies in the interior of $g(U)$ in $\operatorname{Spv}''(\alpha, A)$. Since $x \in U \cap L$ and y specializes to $g(x)$, we have $y \in g(U)$, in contradiction to $g^{-1}(y) \cap U = \emptyset$. □

LEMMA (4.2.5). *Let*

$$(*) \qquad \begin{array}{ccc} U & \xrightarrow{\ i\ } & X \\ {\scriptstyle f}\big\downarrow & & \big\downarrow{\scriptstyle h} \\ V & \xrightarrow[\ g\ ]{} & Y \end{array}$$

be a commutative diagram of schemes. Let u be a point of U.

i) *If f is universally generalizing at u and g is universally generalizing at $f(u)$, then h is universally generalizing at $i(u)$.*
ii) *Assume that $(*)$ is cartesian and g is universally generalizing at $f(u)$. Then f is universally generalizing at u if and only if h is universally generalizing at $i(u)$.*

PROOF. i) is obvious and ii) follows from i). □

Now we prove (4.2.1). First we study the case that k is algebraically closed and has a non-trivial valuation α. We consider the mappings $g\colon \mathrm{Spv}''B \to \mathrm{Spv}''A$ and $\bar{g}\colon \mathrm{Max}(\alpha, B) \to \mathrm{Max}(\alpha, A)$ induced by f. Since the statement "$\bar{g}$ is open at x" can be expressed in the formal language of valued fields by a formula Φ, and since the theory of algebraically closed fields with non-trivial valuation has elimination of quantifiers [**P**, 4.17], there exists a constructible subset L of $\mathrm{Spv}''(\alpha, B)$ such that $L \cap \mathrm{Max}(\alpha, B)$ is the set of points of $\mathrm{Max}(\alpha, B)$ at which $\bar{g}$ is open. We will show that L is the set of points of $\mathrm{Spv}''(\alpha, B)$ at which g is universally generalizing. For every $n \in \mathbb{N}$ we consider the mappings $g_n\colon \mathrm{Spv}''B[T_1, \dots, T_n] \to \mathrm{Spv}''A[T_1, \dots, T_n]$ and $p_n\colon \mathrm{Spv}''(\alpha, B[T_1, \dots, T_n]) \to \mathrm{Spv}''(\alpha, B)$. The mapping $\bar{g}_n\colon \mathrm{Max}(\alpha, B[T_1, \dots, T_n]) \to \mathrm{Max}(\alpha, A[T_1, \dots, T_n])$ is open at a point $x \in \mathrm{Max}(\alpha, B[T_1, \dots, T_n])$ if $\bar{g}$ is open at $p_n(x)$. Hence $p_n^{-1}(L)$ is a constructible subset of $\mathrm{Spv}''(\alpha, B[T_1, \dots, T_n])$ such that $\bar{g}_n$ is open at every point of $p_n^{-1}(L) \cap \mathrm{Max}(\alpha, B[T_1, \dots, T_n])$. Then according to (4.2.4), g_n is generalizing at every point of $p_n^{-1}(L)$. Now (4.2.3 ii) implies that g is universally generalizing at every point of L. Conversely, let x be a point of $\mathrm{Spv}''(\alpha, B)$ at which g is universally generalizing. Let $c\colon B \to s$ be a ring homomorphism of B to an algebraically closed field s and β a valuation of s with $\beta | B = x$. We consider the mappings $g'\colon \mathrm{Spv}''B \otimes_k s \to \mathrm{Spv}''A \otimes_k s$ and $\bar{g}'\colon \mathrm{Max}(\beta, B \otimes_k s) \to \mathrm{Max}(\beta, A \otimes_k s)$. Let x' be the point $\mathrm{Spv}''(d)(\beta) \in \mathrm{Spv}''B \otimes_k s$ where d is the ring homomorphism $B \otimes_k s \to s$ with $d(b \otimes e) = c(b) \cdot e$. Then x' is an element of $\mathrm{Max}(\beta, B \otimes_k s)$ and is mapped to x under the projection $p\colon \mathrm{Spv}''(\beta, B \otimes_k s) \to \mathrm{Spv}''(\alpha, B)$. Since g is universally generalizing at x, g' is generalizing at x'. Hence $\bar{g}'$ is open at x'. Since the set of points of $\mathrm{Max}(\beta, B \otimes_k s)$ at which $\bar{g}'$ is open can be described by Φ, we have $x' \in p^{-1}(L)$, i.e. $x \in L$. Thus we have proved that L is the set of points of $\mathrm{Spv}''(\alpha, B)$ at which g is universally generalizing.

Let M be the set of points of $\mathrm{Spec}\, B$ at which $\mathrm{Spec}(f)$ is universally generalizing. Let $t\colon \mathrm{Spv}''(\alpha, B) \to \mathrm{Spec}\, B$ be the support mapping. By (2.1.3), $L = t^{-1}(M)$. Since t is surjective and spectral, we deduce that M is constructible in $\mathrm{Spec}\, B$.

Now we prove (4.2.1) in general. Let K be an extension field of k such that K is algebraically closed and carries a non-trivial valuation. We consider the cartesian square

$$\begin{array}{ccc}
\mathrm{Spec}\, B \otimes_k K & \xrightarrow{\ q\ } & \mathrm{Spec}\, B \\
{\scriptstyle h}\downarrow & & \downarrow{\scriptstyle g} \\
\mathrm{Spec}\, A \otimes_k K & \xrightarrow[\ p\ ]{} & \mathrm{Spec}\, A,
\end{array}$$

where g and h are induced by $f\colon A \to B$ and p, q are the canonical morphisms. Let S (bzw. T) be set of points of $\mathrm{Spec}\, B$ (resp. $\mathrm{Spec}\, B \otimes_k K$) at which g (resp. h) is universally generalizing. By (4.2.5 ii), we obtain $T = q^{-1}(S)$. We know already that T is constructible in $\operatorname{Spec} B \otimes_k K$. Since q is surjective and spectral, we deduce that S is constructible in $\operatorname{Spec} B$. □

PROOF OF (4.2.2). We have a commutative diagram

$$\begin{array}{ccc} \operatorname{Spec}\hat{B} & \xrightarrow{i} & \operatorname{Spec} B \\ \downarrow & & \downarrow \\ \operatorname{Spec}\hat{A} & \xrightarrow[g]{} & \operatorname{Spec} A. \end{array}$$

Since g is flat, (4.2.5 i) shows that i) follows from ii). Now we prove that i) implies ii). For that we use continuous valuations: A valuation v of a topological ring E is called continuous if for every $\gamma \in \Gamma_v$ there exists a neighbourhood U of $0 \in E$ with $v(u) > \gamma$ for every $u \in U$. Let I be an ideal of definition of A. Let C be the ring B equipped with the $I \cdot B$-adic topology. Then $\operatorname{Spec}\hat{B} \to \operatorname{Spec}\hat{A}$ factorizes in $\operatorname{Spec}\hat{B} \xrightarrow{j} \operatorname{Spec}\hat{C} \xrightarrow{k} \operatorname{Spec}\hat{A}$. Since j is flat, it suffices to prove that k is universally generalizing at $\hat{\mathfrak{p}} \cap \hat{C}$. This means we may assume that $I \cdot B$ is an ideal of definition of B. In the following we equip every A-algebra E with the $I \cdot E$-adic topology. Let $\mathfrak{q}$ be a prime ideal of $\hat{B}[X]\colon = \hat{B}[X_1, \ldots, X_n$ with $\mathfrak{q} \cap B = \hat{\mathfrak{p}}$. By (4.2.3 i), it suffices to show that $\operatorname{Spec}\hat{B}[X] \to \operatorname{Spec}\hat{A}[X]$ is generalizing at $\mathfrak{q}$. Let $\mathfrak{r} \in \operatorname{Spec}\hat{A}[X]$ be a proper generalization of $\mathfrak{q} \cap \hat{A}[X]$. We have to show that there exists a prime ideal $\mathfrak{s}$ of $\hat{B}[X]$ with $\mathfrak{s} \subseteq \mathfrak{q}$ and $\mathfrak{s} \cap \hat{A}[X] = \mathfrak{r}$. By [**EGA***, 0.6.5.8], there exists a rank 1 valuation $\hat{v}$ of $\hat{A}[X]$ such that $\operatorname{supp}(\hat{v}) = \mathfrak{r}$ and $\hat{A}[X](\hat{v})$ dominates the localization of $\hat{A}[X]/\mathfrak{r}$ at the prime ideal $(\mathfrak{q} \cap \hat{A}[X])/\mathfrak{r}$. Then $\hat{v}$ is continuous and the trivial valuation of $A[X]$ with support $\mathfrak{q} \cap A[X]$ is a primary specialization of $v\colon = \hat{v}|A[X]$ in $\operatorname{Spv} A[X]$. By (2.1.3) there exists a valuation w of $B[X]$ such that the trivial valuation of $B[X]$ with support $\mathfrak{q} \cap B[X]$ is a primary specialization of w in $\operatorname{Spv} B[X]$ and $w|A[X] = v$. Let H be the convex hull of Γ_v in Γ_w. Since $c\Gamma_w = \{0\}$, we have the valuation $u\colon = w|H$ of $B[X]$. Then u is continuous, $u|A[X] = v$ and $u|c\Gamma_u$ is the trivial valuation of $B[X]$ with support $\mathfrak{q} \cap B[X]$. Since the ring homomorphism $\varphi\colon B[X] \to \hat{B}[X]$ is continuous and $\operatorname{im}(\varphi)$ is dense in $\hat{B}[X]$, u extends to a continuous valuation $\hat{u}$ of $\hat{B}[X]$. Then $\hat{u}|c\Gamma_{\hat{u}}$ lies over $u|c\Gamma_u$. Since $\mathfrak{q}$ is the unique prime ideal of $\hat{B}[X]$ lying over $\mathfrak{q} \cap B[X]$, we have $\operatorname{supp}(\hat{u}) \subseteq \operatorname{supp}(\hat{u}|c\Gamma_{\hat{u}}) = \mathfrak{q}$. Since $\hat{v}$ and $\hat{u}|\hat{A}[X]$ are continuous valuations of $\hat{A}[X]$ and $\hat{v}|A[X] = (\hat{u}|\hat{A}[X])|A[X]$, we have $\hat{v} = \hat{u}|\hat{A}[X]$, especially $\mathfrak{r} = \operatorname{supp}(\hat{v}) = \operatorname{supp}(\hat{u}|\hat{A}[X]) = \operatorname{supp}(\hat{u}) \cap \hat{A}[X]$. Hence $\operatorname{supp}(\hat{u})$ is a prime ideal as desired. □

REFERENCES

[B] N. Bourbaki, *Commutative Algebra*, Hermann, Paris, 1972.

[B1] ———, *General Topology*, Hermann, Paris, 1967.

[Be] V. G. Berkovich, *Spectral theory and analytic geometry over non-archimedean fields*, Math. Surveys and Monographs, vol. 33, Amer. Math. Soc., Providence, 1990.

[BGR] S. Bosch, U. Güntzer, and R. Remmert, *Non-archimedean analysis*, Grundlehren 261, Springer, Berlin and Heidelberg, 1984.

[CC] M. Carral and M. Coste, *Normal spectral spaces and their dimensions*, J. Pure Appl. Algebra **30** (1983), 227–235.

[EGA] A. Grothendieck and J. Dieudonne, *Éléments de Géométrie Algébrique*, Pub. Math. Inst. Hautes Etudes Sci. **11** (1961), **20** (1964), **24** (1965), **28** (1966), **32** (1967).

[EGA*] ———, *Éléments de Géométrie Algébrique I*, Grundlehren 166, Springer, Berlin and Heidelberg, 1971.

[FP] J. Fresnel and M. van der Put, *Géométrie analytique rigide et applications*, Progress in Mathematics, vol. 18, Birkhäuser, Boston and Basel, 1981.

[H] M. Hochster, *Prime ideal structure in commutative rings*, Trans. Amer. Math. Soc. **142** (1969), 43–60.

[Hi] H. Hironaka, *Flattening theorem in complex-analytic geometry*, Amer. J. Math. **97** (1975), 503–547.

[Hu] R. Huber, *Bewertungsspektrum und rigide Geometrie, Habilitationsschrift*, Universität Regensburg, 1990.

[Hu1] ———, *Etale cohomology of henselian rings and cohomology of abstract Riemann surfaces of fields*, Universität Regensburg, 1991 (preprint).

[Hu2] ———, *On semianalytic sets in rigid analytic geometry*, Universität Regensburg, 1991 (preprint).

[Hu3] ———, *On the etale cohomology of rigid analytic varieties* (to appear).

[Hu4] ———, *Semirigide Funktionen*, Universität Regensburg, 1990 (preprint).

[M] H. Matsumura, *Commutative algebra*, second edition, Benjamin Publishing Company, London, 1980.

[P] A. Prestel, *Einführung in die mathematische Logik und Modelltheorie*, Aufbaukurs Mathematik, Vieweg, Braunschweig, 1986.

[Pa] A. Parusinski, *Constructibility of set of points where complex analytic morphism is open*, 1990 (preprint).

[Pu] M. de la Puente, *Riemann surfaces of a ring and compactifications of semi-algebraic sets*, Ph.D. Thesis, Stanford University, 1988.

[R] M. Raynaud, *Anneaux Locaux Henséliens*, Lecture Notes in Math., vol. 169, Springer, Berlin and Heidelberg, 1970.

[RG] M. Raynaud and L. Gruson, *Critères de platitude et de projectivité*, Inventiones math. **13** (1971), 1–89.

[Ru] J. Ruiz, *Constructibility of closures in real spectra*, Universidad Complutense, Madrid (preprint).

[S] N. Schwartz, *Compactifications of varieties*, Arkiv för matematik **28** (1990), 323–370.

[S1] ———, *Topology of real closed spaces*, (1990) (manuscript).

[SGA] M. Artin, A. Grothendieck, and J. L. Verdier, *Théorie des topos et cohomologie étale des schémas*, Lecture Notes in Math., vol. 305, Springer, Berlin and Heidelberg, 1973.

[ZS] O. Zariski and P. Samuel, *Commutative Algebra*, Van Nostrand, Princeton, 1960.

Fachbereich Mathematik, Universität Regensburg, Universitätsstr. 31, 8400 Regensburg, F.R.G.

E-mail address: knebuschvax1.rz.uni-regensburg.dbp.de

Contemporary Mathematics
Volume **155**, 1994

Minimal Generation of Basic Sets in the Real Spectrum of a Commutative Ring

M. MARSHALL

Let A be a commutative ring with 1 and let $\operatorname{Sper} A$ denote the real spectrum of A [**2**], [**3**], [**8**], [**9**]. In case A is Noetherian, Bröcker [**4**] characterizes basic open sets (resp., basic closed sets) in $\operatorname{Sper} A$ and determines bounds for the number of strict inequalities (resp., inequalities) required to describe such sets. This generalizes earlier work of Bröcker and Scheiderer [**5**], [**14**] in case A is the coordinate ring of a real variety. Actually, Scheiderer's result on basic open sets [**14**, Th. 2] itself applies to any Noetherian ring whose real singularities behave in a reasonable way.

In the present paper, it is explained how much of this theory extends to an arbitrary commutative ring. At the same time, the theory is developed here, not just for basic sets in $\operatorname{Sper} A$, but for (relative) basic sets $S \subseteq X$ where $X \subseteq \operatorname{Sper} A$ is an arbitrary saturated set. A subset $X \subseteq \operatorname{Sper} A$ is called *saturated* (or pro–basic) if it is describable by some conjunction in inequalities ($\geq$) and strict inequalities ($>$) , possibly infinitely many of each type. This generalization to saturated sets is fairly straightforward but, at the same time, is quite natural in view of the corresponding theory over fields. Saturated sets in $\operatorname{Sper} F$, F a field, are just sets of the form $X = W(T) = \{P \in \operatorname{Sper} F : P \supseteq T\}$ where $T \subseteq F$ is some preordering [**10**].

The most important result in the paper is the proof that the main theorem on basic open sets [**4**], [**14**] extends to basic open sets in an arbitrary saturated set in the real spectrum on an arbitrary commutative ring (see Th. 4.1 and Cor. 5.1). The proof given is quite similar to Mahé's proof in [**11**]. In particular, it is more elementary than the proofs in [**4**], [**14**] in that it avoids all reference to the theory of spaces of orderings of semi-local rings. For this reason, this proof is of interest even in the Noetherian case.

1991 *Mathematics Subject Classification.* Primary:14P10 Secondary: 12J15, 11E10.

This paper is submitted in final form and no version of it will be submitted for publication elsewhere.

The reader only interested in Th. 4.1 and Cor. 5.1 can proceed directly to §4 since the proof of Th. 4.1 and Cor. 5.1 is essentially independent of the material in §§1, 2, 3.

An important ingredient in many of the proofs (although *not* in the proof of Th. 4.1 and Cor. 5.1) is a certain version of the Hörmander-Łojasiewicz inequality [**1**], [**4**] (see Ths. 1.1 and 1.3). This is used in the description of the saturation of a closed set $X \subseteq \operatorname{Sper} A$ (see Cor. 1.5 and Th. 2.1). In turn, Cor. 1.5 and Th. 2.1 are used to generalize the characterization of basic sets given in [**4**], [**5**] (see Ths. 3.2 and 3.4) and in the description of the local s-invariant (see Th. 3.6). Th. 1.3 is also used in a modification of Bröcker's "pasting" argument [**1**], [**4**], [**5**] to obtain a bound for the $\overline{s}$-invariant (Th. 5.3).

We use the following notation (for $I \subseteq A$ any set):

$$\begin{aligned} U(I) &= \{P \in \operatorname{Sper} A\colon a >_P 0 \quad \forall \quad a \in I\} \\ W(I) &= \{P \in \operatorname{Sper} A\colon a \geq_P 0 \quad \forall \quad a \in I\} \\ Z(I) &= \{P \in \operatorname{Sper} A\colon a =_P 0 \quad \forall \quad a \in I\} \ . \end{aligned}$$

In particular,

$$\begin{aligned} U(I^2) &= \left\{P \in \operatorname{Sper} A\colon a^2 >_P 0 \quad \forall \quad a \in I\right\} \\ &= \{P \in \operatorname{Sper} A\colon a \neq_P 0 \quad \forall \quad a \in I\} \ . \end{aligned}$$

1. The Hörmander-Łojasiewicz inequality

The original version of this inequality is for polynomial functions defined on closed semi-algebraic sets in $\mathbb{R}^n$ [**3**]. Later, it was observed by several people that an abstract version holds for closed constructible sets in $\operatorname{Sper} A$, A an arbitrary commutative ring [**1**], [**4**]. The version given here is just a slight generalization of the version in [**1**], [**4**]. We drop the requirement that the set be constructible.

THEOREM (1.1). *Suppose $X \subseteq \operatorname{Sper} A$ is any closed set and $f, g \in A$ are such that $g = 0$ on $X \cap Z(f)$. Then $\exists\, p \in A$, $p > 0$ on $\operatorname{Sper} A$ and $m \geq 0$ such that $|g|^{2m+1} \leq p|f|$ on X.*

PROOF. Since X is closed, $X = \cup_i X_i$ where each X_i is defined by a conjunction of inequalities ($\geq$). That is, $X_i = W(T_i)$ for some preordering $T_i \subseteq A$. Thus $X_i \cap Z(f) \subseteq Z(g)$ so, by the Nullstellensatz [**9**, Th. 7.4], $\exists\, m_i \geq 0$, $a_i \in A$, $s_i \in T_i$ such that $-g^{2m_i} = s_i - a_i f$. Since $s_i \geq 0$ on X_i, $X = \cup_i X_i \subseteq \cup_i W(s_i)$ so, by compactness of X in the Tychonoff topology [**9**, Th. 4.1], we have some finite set of indices $i_1, \ldots, i_v$ such that $X \subseteq \cup_{k=1}^{v} W(s_{i_k})$. Multiplying by a suitable even power of g, we can assume $m_{i_1} = \cdots = m_{i_v} = m$ say. Thus, on $W(s_{i_k})$, $g^{2m} = -s_{i_k} + a_{i_k} f \leq a_{i_k} f = |a_{i_k}||f|$ so $|g|^{2m+1} \leq |g a_{i_k}||f| \leq p|f|$ where $p = 1 + g^2\left(a_{i_1}^2 + \cdots + a_{i_v}^2\right)$. Since $X \subseteq \cup_{k=1}^{v} W(s_{i_k})$, this implies $|g|^{2m+1} \leq p|f|$ on X.

NOTE (1.2). It is easy to arrange things so that $|g|^{2m+1} < p|f|$ on $X\backslash Z(f)$. (E.g., just replace p by $p+f^2$.) Once this is done, $f_1 := pf + g^{2m+1}$ and f have the same sign on X.

We use (1.1) several times, but always in the following special form:

THEOREM (1.3). *Suppose $X \subseteq \operatorname{Sper} A$ is closed and $f \geq 0$ on $X \cap Z(\mathfrak{a}) \cap U(\Sigma^2)$ for some sets $\mathfrak{a}$, $\Sigma \subseteq A$. Then $\exists\ f_1 \in A$ such that $f_1 \geq 0$ on X and f_1, f have the same sign on $Z(\mathfrak{a}) \cap U(\Sigma^2)$.*

PROOF. We can assume $\mathfrak{a}$ is an ideal of A and Σ is a multiplicative set. Let $X' = X \cap W(-f)$. Thus $X' \cap Z(\mathfrak{a}) \cap U(\Sigma^2) \subseteq Z(f)$ so, by compactness of $U(f^2) = \operatorname{Sper} A\backslash Z(f)$, $\exists\ g \in \mathfrak{a}$, $\quad h \in \Sigma$ such that $X' \cap Z(g) \cap U(h^2) \subseteq Z(f)$. $\Big($Note, in this regard that $Z(g_1) \cap Z(g_2) = Z\left(g_1^2 + g_2^2\right)$, $\quad U\left(h_1^2\right) \cap U\left(h_2^2\right) = U\left((h_1h_2)^2\right)\Big)$. Thus $fh^2 = 0$ on $X' \cap Z(g)$. Of course, $Z(g^2) = Z(g)$. Thus, by (1.1) and (1.2) $\quad \exists\ p > 0$ on $\operatorname{Sper} A$ and $m \geq 0$ such that $f_1 := pg^2 + (fh^2)^{2m+1}$ has the same sign as g^2 on X'. Now $f_1 \geq 0$ on X is clear. Also, f_1 and f have the same sign on $Z(g) \cap U(h^2)$. Since $Z(\mathfrak{a}) \cap U(\Sigma^2) \subseteq Z(g) \cap U(h^2)$, this completes the proof.

NOTE (1.4). For $\mathfrak{a} \subseteq A$ an ideal and $\Sigma \subseteq A$ a multiplicative set, we can form the ring $\overline{\Sigma}^{-1}\overline{A}$:= the localization of $\overline{A} := A/\mathfrak{a}$ at the multiplicative set $\overline{\Sigma} := (\Sigma + \mathfrak{a})/\mathfrak{a}$. $Z(\mathfrak{a}) \cap U\left(\Sigma^2\right)$ is exactly the image of $\operatorname{Sper} \overline{\Sigma}^{-1}\overline{A}$ in $\operatorname{Sper} A$ via the natural embedding.

For any (real) prime $\mathfrak{p} \subseteq A$, denote by $F(\mathfrak{p})$ the residue field of A at $\mathfrak{p}$, i.e., $F(\mathfrak{p})$ is the field of quotients of the domain $A/\mathfrak{p}$. Identify $\operatorname{Sper} F(\mathfrak{p}) \subseteq \operatorname{Sper} A/\mathfrak{p} \subseteq \operatorname{Sper} A$ in the natural way. Thus

$$\operatorname{Sper} A = \cup_{\mathfrak{p}} \operatorname{Sper} F(\mathfrak{p}) ,$$

$\mathfrak{p}$ running through the set of real primes of A. Also, if $X \subseteq \operatorname{Sper} A$,

$$X = \cup_{\mathfrak{p}} X(\mathfrak{p}) \text{ where } X(\mathfrak{p}) := X \cap \operatorname{Sper} F(\mathfrak{p}) .$$

We have the following application of (1.3).

COROLLARY (1.5). *Suppose $X \subseteq \operatorname{Sper} A$ is closed, $\mathfrak{p} \subseteq A$ is a real prime, and $f \in A$ satisfies $f \geq 0$ on $X(\mathfrak{p})$. Then $\exists\ g \in A$ such that $g \geq 0$ on X and such that f, g have the same sign on* $\operatorname{Sper} F(\mathfrak{p})$.

PROOF. Apply (1.3) taking $\mathfrak{a} = \mathfrak{p}$, $\quad \Sigma = A\backslash\mathfrak{p}$, so $Z(\mathfrak{a}) \cap U(\Sigma^2) = \operatorname{Sper} F(\mathfrak{p})$.

2. Saturated sets

Let us say $X \subseteq \operatorname{Sper} A$ is *saturated* (or pro-basic) if X is described by some conjunction of inequalities ($\geq$) and strict inequalities ($>$), possibly infinitely many of each type. Define the *saturation* of a set $X \subseteq \operatorname{Sper} A$ (denoted $\operatorname{Sat} X$) to be the smallest saturated set in $\operatorname{Sper} A$ containing X. Thus, if we set

$$T := \{a \in A\colon\ a \geq 0 \text{ on } X\}, \quad \Sigma := \{a \in A\colon\ a > 0 \text{ on } X\} ,$$

then $\operatorname{Sat} X = W(T) \cap U(\Sigma)$. Of course,

$$(a > 0) \Leftrightarrow (a \geq 0 \quad \wedge \quad a \neq 0) \Leftrightarrow \left(a \geq 0 \quad \wedge \quad a^2 > 0\right)$$

so $\operatorname{Sat} X$ is also described by the formula $\operatorname{Sat} X = W(T) \cap U(\Sigma^2)$.

The main case to be considered is when X is closed in $\operatorname{Sper} A$. In this case we have the following immediate consequence of (1.5):

THEOREM (2.1). *Suppose $X \subseteq \operatorname{Sper} A$ is closed and $T = \{a \in A\colon\ a \geq 0$ on $X\}$. Then $\operatorname{Sat} X = W(T) = \cup_{\mathfrak{p}} \operatorname{Sat} X(\mathfrak{p})$, $\mathfrak{p}$ running through the set of all real primes of A.*

PROOF. The inclusions $\cup_{\mathfrak{p}} \operatorname{Sat} X(\mathfrak{p}) \subseteq \operatorname{Sat} X \subseteq W(T)$ are clear. Thus it remains to prove $W(T) \subseteq \cup_{\mathfrak{p}} \operatorname{Sat} X(\mathfrak{p})$. Suppose $P \in \operatorname{Sper} F(\mathfrak{p})$, $P \notin \operatorname{Sat} X(\mathfrak{p})$. Thus $\exists\ f \in A$, $f \geq 0$ on $X(\mathfrak{p})$, $f <_P 0$. By (1.5), $\exists\ g \in T$ having the same sign as f on $\operatorname{Sper} F(\mathfrak{p})$. In particular, $g <_P 0$. This proves $P \notin W(T)$, and completes the proof.

NOTES (2.2).

(1) $\operatorname{Sat} \emptyset = \emptyset$ so we can always exclude the primes $\mathfrak{p}$ with $X(\mathfrak{p}) = \emptyset$.

(2) Th. 2.1 reduces the study of the saturation of a closed set to the field case. For fields, some results are known, e.g., if the stability index is finite; see [**12**], [**15**], but in general, the situation is pretty complicated.

(3) For $X \subseteq \operatorname{Sper} A$ arbitrary, $\operatorname{Sat} X = W(T) \cap U(\Sigma^2)$ where $T = \{a \in A\colon\ a \geq 0 \text{ on } X\}$, $\Sigma = \{a \in A\colon\ a > 0 \text{ on } X\}$. Let $\overline{X} \subseteq \operatorname{Sper} A$ denote the closure of X. Thus $a \geq 0$ on $X \Leftrightarrow a \geq 0$ on $\overline{X}$ so, by (2.1), $W(T) = \operatorname{Sat} \overline{X} = \cup_{\mathfrak{p}} \operatorname{Sat} \overline{X}(\mathfrak{p})$. Also, $U(\Sigma^2) = \{P \in \operatorname{Sper} A\colon\ \operatorname{Supp} P \cap \Sigma = \emptyset\} = \operatorname{Sper} B$ where $B = \Sigma^{-1}A$. Thus $\operatorname{Sat} X = W(T) \cap U(\Sigma^2) = \cup\left\{\operatorname{Sat} \overline{X}(\mathfrak{p})\colon \mathfrak{p} \cap \Sigma = \emptyset\right\}$. This is the same as the saturation of $\overline{X} \cap \operatorname{Sper} B$ in $\operatorname{Sper} B$.

For the rest of this section, and all of the next, we weaken the hypothesis that X is closed in $\operatorname{Sper} A$, assuming only that

$$(*) \qquad X \text{ is closed in } U\left(\Sigma^2\right) \text{ where } \Sigma := \{a \in A\colon\ a > 0 \text{ on } X\}\ .$$

For example, any saturated set $X \subseteq \operatorname{Sper} A$ satisfies $(*)$. For any set $S \subseteq X$, we say *S is saturated in X* if S is describable in X by a conjunction of inequalities and strict inequalities, i.e., if $S = X \cap \operatorname{Sat} S$. Of course, if X itself is saturated, then S is saturated in $X \Leftrightarrow S$ is saturated.

COROLLARY (2.3). *For any set $S \subseteq X$, the following are equivalent:*

(1) *$S = W(T) \cap X$ for some preordering $T \subseteq A$.*

(2) *S is closed and saturated in X.*

(3) *S is closed in X and $S(\mathfrak{p})$ is saturated in $X(\mathfrak{p})$ for all real primes $\mathfrak{p} \subseteq A$.*

PROOF. Going to the ring $B = \Sigma^{-1}A$ where $\Sigma = \{a \in A\colon\ a > 0 \text{ on } X\}$, we can assume X (and hence S) is closed in $\operatorname{Sper} A$. In this case, the result is immediate from (2.1).

Recall: If $P, Q \in \operatorname{Sper} A$, we say P *generalizes* Q (or Q *specializes* P) if $P \subseteq Q$. If $S \subseteq X$ is saturated and closed under specialization in X, then S is closed in X [**2**, Prop. 2.11] so, by (2.3), $S = W(T) \cap X$ for some preordering $T \subseteq A$. Analogously to this, one has:

THEOREM (2.4). *For any set $S \subseteq X$, the following are equivalent:*

(1) *$S = U(\Sigma) \cap X$ for some multiplicative set $\Sigma \subseteq A$.*

(2) *S is saturated and closed under generalization in X.*

PROOF. (1) $\Rightarrow$ (2) is clear. (2) $\Rightarrow$ (1): Let $B = \Sigma^{-1}A$ (the localization of A at Σ) where $\Sigma := \{a \in A\colon\ a > 0 \text{ on } S\}$. $\operatorname{Sper} B$ is naturally identified with $U(\Sigma^2) \subseteq \operatorname{Sper} A$. To prove (1), we show that if $P \in X \backslash S$, then $\exists\, a \in \Sigma,\quad a \leq_P 0$. Since $U(\Sigma^2) \supseteq U(\Sigma)$, this is clear if $P \notin U(\Sigma^2)$, so we can assume $P \in U(\Sigma^2)$. Thus, replacing A by B and X by $X \cap \operatorname{Sper} B$, we are reduced to the case where S, X are closed in $\operatorname{Sper} A$ so, by (2.3), $S = W(T) \cap X$ for some preordering $T \subseteq A$. Let $Q \in X$ be the unique maximal ordering such that $Q \supseteq P$. Since S is closed under generalization and $P \notin S$, it follows that $Q \notin S$. Since $S = W(T) \cap X,\quad \exists\, b \in T$ such that $b <_Q 0$. Thus $b < 0$ on $W(Q) = \{Q\}$ so, by the Positivstellensatz [**9**, Th. 7.4], $-b(1+s) = 1 + t$ for some $s, t \in Q$. But then

$$a := 1 + 2b(1+s)^2 = -\big(1 + 2(s + t + st)\big)$$

satisfies $a > 0$ on S and $a <_Q 0$ (so $a <_P 0$). This completes the proof.

3. Basics sets

In this section, the characterizations of *basic open* and *basic closed* given in [**4**] in the Noetherian case are shown to generalize to an arbitrary commutative ring. A characterization of arbitrary basic sets is also given. Also, at the end of the section, we apply Cor. 1.5 to obtain a characterization of the *local* stability index.

Again, we fix a set $X \subseteq \operatorname{Sper} A$ satisifying $(*)$. For example, take $X = \operatorname{Sper} A$ or any saturated set. If $S \subseteq X$ is any set which is saturated and constructible in X then, using the compactness of $X \backslash S$ (in the Tychonoff topology), S has the form

$$(**) \qquad S = U(a_1, \ldots, a_m) \cap W(b_1, \ldots, b_n) \cap X$$

for some $a_1, \ldots, a_m,\ b_1, \ldots, b_n \in A$. In this situation we will say that S is *basic* in X. As usual, we say S is *basic open* (resp., *basic closed*) in X if S is expressible as in $(**)$, but with $n = 0$ (resp., with $m = 0$).

NOTE (3.1).

$$(a_1 > 0 \wedge \cdots \wedge a_m > 0) \Leftrightarrow (a_1 \geq 0 \wedge \cdots \wedge a_m \geq 0 \wedge a_1 \cdots a_m \neq 0)\ .$$

Thus, any basic set $S \subseteq X$ has the form

$$S = U\left(a^2\right) \cap W\left(c_1, \ldots, c_k\right) \cap X$$

for some $a, c_1, \ldots, c_k \in A$.

THEOREM (3.2). *For any constructible $S \subseteq X$, the following are equivalent:*

(1) *S is basic closed in X.*
(2) *S is basic and closed in X.*
(3) *S is closed in X and $S(\mathfrak{p})$ is basic in $X(\mathfrak{p})$ for each real prime $\mathfrak{p} \subseteq A$.*

PROOF. (1) $\Rightarrow$ (2) and (2) $\Rightarrow$ (3) are trivial. To prove (3) $\Rightarrow$ (1), apply (2.3) to obtain $S = W(T) \cap X$ for some preordering $T \subseteq A$. Thus, by compactness of $X \backslash S$, $S = W(a_1, \ldots, a_n) \cap X$ for some $a_1, \ldots, a_n \in T$.

THEOREM (3.3). *For any constructible $S \subseteq X$, the following are equivalent:*

(2) *S is basic in X,*
(3) *$S \cap \text{z-cl}\left(\overline{S}_X \backslash S\right) = \emptyset$ and $S(\mathfrak{p})$ is basic in $X(\mathfrak{p})$ for each real prime $\mathfrak{p} \subseteq A$.*

Here, $\overline{S}_X$ denotes the closure of S in X and z–cl(Y) denotes the Zariski-closure of any $Y \subseteq \operatorname{Sper} A$, i.e., z–cl$(Y) = Z(\mathfrak{a})$ where $\mathfrak{a} := \{a \in A \colon a = 0 \text{ on } Y\}$. Of course, if S is closed in X, then the condition $S \cap \text{z-cl}\left(\overline{S}_X \backslash S\right) = \emptyset$ is vacuous, so condition (3) of (3.3) is just condition (3) of (3.2).

PROOF. (2) $\Rightarrow$ (3): Suppose $S = U\left(a^2\right) \cap W\left(b_1, \ldots, b_n\right) \cap X$. Thus, if $P \in \overline{S}_X \backslash S$, then $a =_P 0$. Thus $\overline{S}_X \backslash S \subseteq Z(a)$, and clearly $S \cap Z(a) = \emptyset$.

(3) $\Rightarrow$ (2): By compactness of S, $\exists\ a \in A$ such that $\overline{S}_X \backslash S \subseteq Z(a)$ and $S \cap Z(a) = \emptyset$. Consider the set $X' = U\left(a^2\right) \cap X$. Thus $S \subseteq X'$ and S is closed in X' by choice of a. Of course, X' satisfies $(*)$. Thus, by (3.2)

$$S = W\left(b_1, \ldots, b_n\right) \cap X' = U(a^2) \cap W\left(b_1, \ldots, b_n\right) \cap X$$

for some $b_1, \ldots, b_n \in A$.

THEOREM (3.4). *For any constructible $S \subseteq X$, the following are equivalent:*

(1) *S is basic open in X.*
(2) *S is basic and open in X.*
(3) *$S \cap \text{z-cl}(\delta_X S) = \emptyset$ and $S(\mathfrak{p})$ is basic in $X(\mathfrak{p})$ for each real prime $\mathfrak{p} \subseteq A$.*

Here, $\delta_X S$ denotes the boundary of S in X.

PROOF. (1) $\Leftrightarrow$ (2) is clear from (2.4), using compactness of $X \backslash S$. Also, since

$$S \cap \text{z-cl}(\delta_X S) = \emptyset \Leftrightarrow S \cap \text{z-cl}\left(\overline{S}_X \backslash S\right) = \emptyset \wedge S \text{ is open in } X\ ,$$

(2) $\Leftrightarrow$ (3) follows from (3.3).

NOTE (3.5). The examples in [**5**] show that the closure (resp. interior) of a basic set $S \subseteq X$ need not be basic.

For $P \in X$, we define the *local s-invariant* (or *local stability index*) $s_P(X)$ as in the field case [**12**], [**15**]. Namely, $s_P(X)$ is defined to be the least integer $k \geq 1$ such that P has a basis of neighbourhoods in X of the form $U\left(a_1, \ldots, a_k\right) \cap X$, $a_1, \ldots, a_k \in A$ (or $s_P(X) = \infty$ if no such finite k exists).

THEOREM (3.6). *Suppose $P \in X$ has support $\mathfrak{p}$. Then $s_P(X) = s_P(X(\mathfrak{p}))$.*

PROOF. ($\geq$) is clear. To prove ($\leq$), suppose $C \subseteq X$ is closed in X and $P \notin C$ and suppose $a_1, \dots, a_k \in A$ are such that $P \in U(a_1, \dots, a_k)$ and $U(a_1, \dots, a_k)$ has empty intersection with $C(\mathfrak{p})$. Set $X' = X \cap U(a_2, \dots, a_k)$ and $C' = C \cap X'$. Thus $a_1 >_P 0$ and $a_1 \leq 0$ on $C'(\mathfrak{p})$. Applying (1.5), we obtain $a_1' \in A$ such that $a_1' >_P 0$ and $a_1' \leq 0$ on C'. (Actually, to be able to apply (1.5), we must first go to the localization $B = \Sigma^{-1}A$ where $\Sigma = \{a \in A : a > 0 \text{ on } X'\}$ to insure that X', and hence C', is closed in Sper A.) Anyway, we now have $P \in U(a_1', a_2, \dots, a_k)$ and $U(a_1', a_2, \dots, a_k) \cap C = \emptyset$, so the proof is complete.

The relationship between the local stability indices $s_P(X)$, $P \in X$, and the (global) stability index $s(X)$ will be clarified in §5 after proving Th. 4.1.

4. Minimal generation of basic open sets

The following result is proved in [**4**] in case A is Noetherian and $X = \operatorname{Sper} A$. The result in [**4**] generalizes still earlier results of Bröcker and Scheiderer in [**5**] and [**14**].

THEOREM (4.1). *Suppose $X \subseteq \operatorname{Sper} A$ is saturated and $S \subseteq X$ is basic open in X. Suppose $\exists\ k \geq 0$ such that, for each real prime $\mathfrak{p} \subseteq A$ $\exists\ b_1, \dots, b_k \in A$ (depending on $\mathfrak{p}$) such that $S(\mathfrak{p}) = U(b_1, \dots, b_k) \cap X(\mathfrak{p})$. Then $\exists\ a_1, \dots, a_k \in A$ such that $S = U(a_1, \dots, a_k) \cap X$.*

REMARK (4.2). The main ingredients in the proof are:

(i) Reduction to the case where A has "enough" units.
(ii) A careful analysis of Pfister forms.
(iii) A weak local-global principle for T-modules, where $T \subseteq A$ is a preordering.

Actually, after this was written up initially, it was discovered that ingredients (i) and (ii) are (essentially) just those used by Mahé in [**11**] and, moreover, that ingredient (iii) has been available for some time; see [**6**]. Thus, the proof given here is essentially just (an easy generalization of) Mahé's proof in [**11**], but with [**6**, Satz 1.8] replacing [**11**, Th. 4.2].

If $T \subseteq A$ is preordering, a *T-module* is a subset $M \subseteq A$ satisfying $M + M \subseteq M$, $T \subseteq M$, and $TM \subset M$. M is called *proper* if $-1 \notin M$. If $\mathfrak{p} \subseteq A$ is a prime, $T(\mathfrak{p})$ (resp. $M(\mathfrak{p})$) denotes the preordering (resp. $T(\mathfrak{p})$-module) in $F(\mathfrak{p})$ induced by T (resp. M).

THEOREM (4.3). (*Weak local-global principle for T-modules*). *For any preordering $T \subseteq A$ and any proper T-module $M \subseteq A$, there exists a (real) prime $\mathfrak{p} \subseteq A$ such that $-1 \notin M(\mathfrak{p})$ (so $M(\mathfrak{p})$ is a proper $T(\mathfrak{p})$-module).*

For the proof of (4.3) see [**6**], [**13**], or [**16**]. The proof in [**13**] has the advantage that it also holds for higher level preorderings. Of course, we only need the level 1 case here.

We also use a little notation from the reduced theory of quadratic forms: A^* := the unit group of A. If $\phi = \langle a_1, \dots, a_n \rangle$, $a_1, \dots, a_n \in A^*$ (or A), and $T \subseteq A$ is a preordering, the *T-value set of ϕ* is

$$D_T(\phi) := \{x \in A\colon x = t_1a_1 + \cdots + t_na_n, \quad t_1, \dots, t_n \in T\} \ .$$

The operations $\oplus$, $\otimes$ on quadratic forms are defined in the standard way. As usual, $\langle\langle a_1, \dots, a_n \rangle\rangle$ denotes the n-fold Pfister form $\langle 1, a_1 \rangle \otimes \cdots \otimes \langle 1, a_n \rangle$.

PROOF OF (4.1). The case $k = 0$ is trivial. (If $S(\mathfrak{p}) = X(\mathfrak{p})$ for all real primes $\mathfrak{p} \subseteq A$, then $S = X$.) Assume $k \geq 1$. By hypothesis $\exists\ a_1, \dots, a_n \in A$ such that $S = U(a_1, \dots, a_n) \cap X$. If $n \leq k$, we are done. Thus we assume $n > k$ and try to reduce from n to $n-1$.

CLAIM. We can assume each $a \in A$ satisfying $a > 0$ on S is in A^*. For, consider the localization $B = \Sigma^{-1}A$ where $\Sigma = \{a \in A\colon a > 0 \text{ on } S\}$. $\operatorname{Sper} B$ is identified with $U(\Sigma^2) \subseteq \operatorname{Sper} A$. Thus $S \subseteq \operatorname{Sper} B$ and it suffices to prove the result for $S \subseteq X' \subseteq \operatorname{Sper} B$ where X' denotes the saturated set in $\operatorname{Sper} B$ defined by $X' = X \cap \operatorname{Sper} B$. For suppose we have proved $S = U(b_2, \dots, b_n) \cap X'$ for some $b_2, \dots b_n \in B$. Clearing fractions, we can assume $b_2, \dots, b_n \in A$ so, in $\operatorname{Sper} A$,

$$S = U(b_2, \dots, b_n) \cap X' = U(b_2, \dots, b_n) \cap X \cap U(\Sigma^2) \ .$$

Thus, by compactness of $X \backslash S$ in the Tychonoff topology,

$$S = U(b_2, \dots, b_n) \cap X \cap U(a^2)$$

for some $a \in \Sigma$. But then $S = U(a^2b_2, b_3, \dots, b_n) \cap X$. Thus, in this way, replacing A by B, we can assume $\Sigma \subseteq A^*$. The proves the claim.

Thus, for example, since $2 > 0$ and $a_i > 0$ on S, $i = 1, \dots, n$, we have $2, a_1, \dots, a_n \in A^*$. Also, any $a \in A$ satisfying $a > 0$ on X is in A^* (since $S \subseteq X$) so our saturated set X is of the form $X = W(T)$ for some preordering $T \subseteq A$. Thus $X(\mathfrak{p}) = W(T(\mathfrak{p}))$ for each real prime $\mathfrak{p} \subseteq A$. Now let $\phi = \langle\langle a_1, \dots, a_n \rangle\rangle$ and consider the T-module $M := D_T(\langle 1 \rangle \oplus -\phi')$ where ϕ' denotes the quadratic form defined by $\phi = \langle 1 \rangle \oplus \phi'$. If $-1 \notin M$, then M is a proper T-module so, by (4.3),

$$-1 \notin M(\mathfrak{p}) = D_{T(\mathfrak{p})}(\langle 1 \rangle \oplus -\phi')$$

for some real prime $\mathfrak{p} \subseteq A$. Since $k < n$, this contradicts our hypothesis. After all, $\phi \cong_{T(\mathfrak{p})} 2^{n-k} \times \langle\langle b_1, \dots, b_k \rangle\rangle$ for some $b_1, \dots, b_k \in A \backslash \mathfrak{p}$, so $\langle 1 \rangle \oplus -\phi'$ is $T(\mathfrak{p})$-isotropic; e.g. see [**10**, Cor. 1.20]. Thus $-1 \in M$ so $-1 = t - b$, i.e. $b = 1 + t$, for some $b \in D_T(\phi')$, $t \in T$. We now prove the following general fact by induction on n:

LEMMA (4.4). *Suppose $X = W(T)$ for some preordering $T \subseteq A$ and*

$$S = U(a_1, \dots, a_n) \cap X, \quad \phi = \langle\langle a_1, \dots, a_n \rangle\rangle$$

for some $a_1, \dots, a_n \in A$. *Suppose* $b \in D_T(\phi')$ *satisfies* $b > 0$ *on* S. *Then* $\exists\, b_2, \dots, b_n \in A$ *such that* $S = U(b, b_2, \dots, b_n) \cap X$.

Applying (4.4) with $b = 1 + t$ as above gives the desired reduction:

$$S = U(b, b_2, \dots, b_n) \cap X = U(b_2, \dots, b_n) \cap X$$

(since $b > 0$ on X). Thus, it only remains to prove (4.4).

PROOF OF (4.4). This is clear if $n = 1$, so we assume $n \geq 2$. If $B = \Sigma^{-1}A$ where $\Sigma = \{a \in A \colon a > 0 \text{ on } S\}$ and $S = U(b, b_2, \dots, b_n) \cap X'$ where $X' = X \cap \operatorname{Sper} B$, then, by compactness,

$$S = U(b, b_2, \dots, b_n) \cap X \cap U(a^2) = U(b, a^2 b_2, b_3, \dots, b_n) \cap X$$

for some $a \in \Sigma$. Thus, replacing A by B, we may as well assume to begin with, that $\Sigma \subseteq A^*$. Write $\phi = \psi \oplus a_1\psi$ where $\psi = \langle\langle a_2, \dots, a_n \rangle\rangle$, so $\phi' = \psi' \oplus a_1\psi$. Thus $b = c + da_1$ with $c \in D_T(\psi')$, $d \in D_T(\psi)$. We can assume $c, d \in A^*$. To prove this, use the identity

$$x = \left(\frac{x+1}{2}\right)^2 - \left(\frac{x-1}{2}\right)^2$$

to write

$$\frac{a_1 + a_2}{b} = r^2 - s^2, \quad r, s, \in A.$$

Thus $r^2 b = s^2 b + a_1 + a_2$ so $(1 + r^2) b = (1 + s^2) b + a_1 + a_2 = (1 + s^2)(c + da_1) + a_1 + a_2 = (1 + s^2) c + a_2 + \big((1 + s^2) d + 1\big) a_1$. Thus $b = c' + d' a_1$ where

$$c' = \frac{(1 + s^2) c + a_2}{1 + r^2}, \quad d' = \frac{(1 + s^2) d + 1}{1 + r^2}.$$

Thus $c' \in D_T(\psi')$, $d' \in D_T(\psi)$. The point is, c', d' are strictly positive on S so $c', d' \in A^*$. Thus, replacing c, d by c', d', we can assume $c, d \in A^*$. Thus, by induction on n, $\exists\, b_3, \dots, b_n \in A$ such that

$$(*) \qquad U(a_2, \dots, a_n) \cap X = U(c, b_3, \dots, b_n) \cap X.$$

We *claim* this implies $S = U(b, a_1cd, b_3, \dots, b_n) \cap X$, so we are done, taking $b_2 = a_1cd$. The proof of this claim is a standard argument from the theory of quadratic forms over fields but, in any case, can be verified directly as follows: Suppose $P \in S = U(a_1, \dots, a_n) \cap X$. Then $b, c, d >_P 0$ (since $b, c, d \in D_T(\phi)$) so, using (*), $P \in U(b, a_1cd, b_3, \dots, b_n) \cap X$. Conversely, suppose $P \in U(b, a_1cd, b_3, \dots, b_n) \cap X$. Then, using $b = c + a_1 d$ (i.e., $bc = c^2 + a_1cd$) one deduces $bc >_P 0$ so $c >_P 0$. Thus $a_2, \dots, a_n >_P 0$ using (*), so $d >_P 0$ (since $d \in D_T(\psi)$). Finally, from $a_1cd >_P 0$, it follows that $a_1 >_P 0$ too.

REMARKS (4.5).

(1) As stated, Ths. 4.1 and 3.4 do not exploit possible connections between the residue fields. This can be rectified as in [**5**], [**14**]. Suppose there exists a set of real primes $\mathcal{D}$ satisfying:

(i) For each real prime $\mathfrak{p} \subseteq A$, there exists $\mathfrak{q} \in \mathcal{D}$ such that $\mathfrak{q} \subseteq \mathfrak{p}$ and such that the natural homomorphism $\lambda\colon A/\mathfrak{q} \longrightarrow A/\mathfrak{p}$ extends to a place $\lambda\colon F(\mathfrak{q}) \longrightarrow F(\mathfrak{p}) \cup \{\infty\}$ and

(ii) The saturated set $X \subseteq \operatorname{Sper} A$ is closed under generalization.

Condition (ii) is required to insure that any pull–back of $P \in X(\mathfrak{p})$ along the place λ lies in $X(\mathfrak{q})$, e.g., (ii) holds if $X = \operatorname{Sper} A$. Then the hypothesis of Th. 4.1 can be weakened: One does not need to assume that $S(\mathfrak{p}) = U(b_1, \ldots, b_k) \cap X(\mathfrak{p})$ for all real primes $\mathfrak{p}$, but only for all $\mathfrak{p} \in \mathcal{D}$. The proof is a simple application of the Krull–Baer theorem [**10**, Th. 3.10]. Similarly, in Th. 3.4(3), one does not need to assume $S(\mathfrak{p})$ is basic in $X(\mathfrak{p})$ for all real primes $\mathfrak{p}$, but only for all $\mathfrak{p} \in \mathcal{D}$.

(2) Scheiderer shows in [**14**] that if A is Noetherian and the real singular set of A is reasonably well–behaved, then there is a natural finite choice for $\mathcal{D}$.

5. The invariants $s, \overline{s}$

Fix a saturated set $X \subseteq \operatorname{Sper} A$. $s(X)$ (resp. $\overline{s}(X)$) is defined to be the smallest integer $k \geq 1$ such that each basic open (resp., basic closed) set $S \subseteq X$ is expressible as

$$S = U(a_1, \ldots, a_k) \cap X \quad \left(\text{resp., } S = W(a_1, \ldots, a_k) \cap X\right),$$

$a_1, \ldots, a_k \in A$. If no such finite k exists, then $s(X) := \infty$ (resp., $\overline{s}(X) := \infty$). $s(X)$ is referred to as the (global) *s–invariant* or the (global) *stability index.* $\overline{s}(X)$ is simply called the *$\overline{s}$–invariant.*

COROLLARY (5.1). (*compare to* [**4**], [**5**], [**11**], [**14**]).

$$s(X) = \sup\left\{s(X(\mathfrak{p}))\colon\ \mathfrak{p} \subseteq A \text{ is a real prime, } X(\mathfrak{p}) \neq \emptyset\right\}.$$

PROOF. Immediate from (4.1).

COROLLARY (5.2). $s(X) = \sup\left\{s_P(X)\colon\ P \in X\right\}$.

PROOF. By (5.1) and (3.6), we are reduced to the field case. In the field case the result follows from results in [**12**] or [**15**].

In order to get a bound for $\overline{s}(X)$ it is necessary to assume some extra finiteness condition on X. If $X \subseteq \operatorname{Sper} A$ is Tychonoff closed, $X \neq \emptyset$, define the *d–invariant* $d(X)$ to be the supremum of all integers $k \geq 0$ such that $\exists\ P_i, Q_i \in X,\ \ i = 1, \ldots, k$ with $P_i \subset Q_i,\ \ i = 1, \ldots, k$ and $\operatorname{Supp} Q_i \subseteq \operatorname{Supp} P_{i+1},\ \ i = 1, \ldots, k-1$. (Convention: $d(\emptyset) = -1$.)

THEOREM (5.3). *Suppose $X \subseteq \operatorname{Sper} A$ is saturated and $d(X) = d < \infty$. For $i = 0, \ldots, d$ define*

$$s_i = \sup\left\{s(X(\mathfrak{p}))\colon\ \mathfrak{p} \subseteq A \text{ is a real prime s.t. } d(X \cap Z(\mathfrak{p})) \leq i\right\}.$$

Then $\overline{s}(X) \leq s_0 + s_1 + \cdots + s_d$. *In particular,* $s(X) \leq \overline{s}(X) \leq (d+1)s(X)$.

REMARKS (5.4).

(1) $d(X)$ could be called the *dimension* of X, but this would be a bit confusing since there are other definitions of dimension in the literature: Let $d_1(X)$ denote the Krull dimension of the ring $A/\cap\{\operatorname{Supp} P\colon\ P \in X\}$ [**1**], [**4**]. Also, let $d_0(X)$ denote supremum of all integers $k \geq 0$ such that $\exists\ P_i \in X,\quad i = 0, \ldots, k$ with $P_0 \subset \cdots \subset P_k$ [**3**].

(2) Clearly $d_0(X) \leq d(X) \leq d_1(X)$. Also, examples exist where both inequalities are strict. On the other hand, if $X = \operatorname{Sper} A$ where A is the coordinate ring of an algebraic set $V \subseteq R^N$, R real closed, then $d_0(X) = d(X) = d_1(X) = \dim(V)$ [**3**, Prop. 7.5.6].

(3) (5.3) is some improvement of the results in [**4**]. Mainly, this is because there is no requirement in (5.3) that A be Noetherian. But also, since $d(X)$ can be strictly less than $d_1(X)$, the bound for $\overline{s}(X)$ given in (5.3) is in general better than that obtained in [**4**, Prop. 5.3].

In the course of the proof of (5.3) we use (1.3), (5.1), and the following two results:

LEMMA (5.5). *Suppose* $X \neq \emptyset$ *is Tychonoff closed,* $S \subseteq X$ *is constructible in* X, *and* $X' := X \cap \text{z-cl}(\delta_X S)$. *Then* $d(X') < d(X)$.

PROOF. This is clear if $X' = \emptyset$ so we assume $X' \neq \emptyset$. Thus $\delta_X S \neq \emptyset$ so $\exists\ Q \in \delta_X S = \overline{S} \cap \overline{X \backslash S} \cap X$. By [**2**, Prop. 2.11], $\exists\ P \in S,\quad P' \in X\backslash S,\quad P, P' \subseteq Q$, and clearly one of P, P' is properly contained in Q. Thus $d(X) \geq 1$ so we are done if $d(X') = 0$. Now suppose $d(X') = k \geq 1$ so $\exists\ P_i, Q_i \in X',\quad i = 1, \ldots, k$ with $P_i \subset Q_i,\quad \operatorname{Supp} Q_i \subseteq \operatorname{Supp} P_{i+1}$. Let $\mathfrak{a} = \cap\{\operatorname{Supp} Q\colon Q \in \delta_X S\}$. Thus z-cl$(\delta_X S) = Z(\mathfrak{a})$ so $\operatorname{Supp} P_1 \supseteq \mathfrak{a}$. Thus, if $a \in A\backslash \operatorname{Supp} P_1$, then $a \notin \mathfrak{a}$ so $\exists\ Q \in \delta_X S$ with $a \notin \operatorname{Supp} Q$. Thus, by compactness,

$$\cap\left\{\delta_X S \cap U\left(a^2\right) : a \in A\backslash \operatorname{Supp} P_1\right\} \neq \emptyset\ .$$

Any Q_0 in this intersection satisfies $\operatorname{Supp} Q_0 \subseteq \operatorname{Supp} P_1$. Also, $Q_0 \in \delta_X S$ so, as above, $\exists\ P_0 \in X,\quad P_0 \subset Q_0$. Thus $d(X) \geq k+1$.

THEOREM (5.6). *Suppose* $S \subseteq X$ *is basic and clopen and suppose* $\exists\ a_1, \ldots, a_s \in A$ *such that* $S = U(a_1, \ldots, a_s) \cap X$. *Then* $\exists\ b_1, \ldots, b_s \in A$ (*same* s) *such that* $S = U(b_1, \ldots, b_s) \cap X = W(b_1, \ldots, b_s) \cap X$.

PROOF. We can assume X, and hence S, is closed in $\operatorname{Sper} A$. Thus, by (2.1), $S = W(T)$ for some preordering $T \subseteq A$. Now $a_i > 0$ on $S = W(T)$ so, by the Positivstellensatz [**9**, Th. 7.4], $a_i(1+s_i) = 1 + t_i$ for some $s_i, t_i \in T$. Thus $b_i := -1 + 2a_i(1+s_i)^2 = 1 + 2(s_i + t_i + s_i t_i)$ satisfies $b_i > 0$ on S and $b_i < 0$ on $W(-a_i),\quad i = 1, \ldots, s$.

PROOF OF (5.3). Since $s\big(X(\mathfrak{p})\big) = \overline{s}\big(X(\mathfrak{p})\big) \leq \overline{s}(X)$, the "in particular" follows from (5.1) once we have established $\overline{s}(X) \leq s_0 + \cdots + s_d$. We proceed by induction on d. Let $S \subseteq X$ be basic closed. By (5.5), $d(X \cap Z(\mathfrak{a})) < d$ where $\mathfrak{a} := \{a \in A\colon a = 0 \text{ on } \delta_X S\}$. By induction on d, $\exists\ n \leq s_0 + \cdots + s_{d-1}$ and $b_1, \ldots, b_n \in A$ such that

$$S \cap Z(\mathfrak{a}) = W(b_1, \ldots, b_n) \cap X \cap Z(\mathfrak{a})\ .$$

By compactness, $\exists\ b \in \mathfrak{a}$ such that

$$S \cap Z(b) = W(b_1, \ldots, b_n) \cap X \cap Z(b)\ .$$

The point is, b vanishes on $\delta_X S$ so $S \cap U(b^2)$ is clopen in $X \cap U(b^2)$. Thus, by (5.1) and (5.6) applied to $S \cap U(b^2) \subseteq X \cap U(b^2)$, $\exists\ a_1, \ldots a_m \in A$, $m \leq s_d$ such that

$$S \cap U(b^2) = W(a_1, \ldots, a_m) \cap X \cap U(b^2)\ .$$

By (1.3), for $i = 1, \ldots n$, $\exists\ c_i \in A$, $c_i \geq 0$ on S, such that b_i, c_i have the same sign on $Z(b)$. Thus

$$S = W(b^2 a_1, \ldots, b^2 a_m, c_1, \ldots, c_n) \cap X\ .$$

Since $m + n \leq s_0 + \cdots + s_d$, this completes the proof.

Remarks (5.7).

(1) If we are in the special case where $d \geq 1$ and $|X \cap Z(\mathfrak{p})| \leq 1$ holds for all real primes $\mathfrak{p} \subseteq A$ satisfying $d(X \cap Z(\mathfrak{p})) = 0$, then $\overline{s}(X) \leq s_1 + \cdots + s_d$ (i.e., we can take $s_0 := 0$ in this case). This applies, for example, if A is the coordinate ring of a real algebraic set. The proof is the same as the proof of (5.3) but with one modification: If $d(X \cap Z(\mathfrak{a})) = 0$, then, by our special hypothesis, $S \cap Z(\mathfrak{a}) = X \cap Z(\mathfrak{a})$. (If $P \in X \cap Z(\mathfrak{a})$ then, as in the proof of (5.5), $\exists\ Q \in \delta_X S$ such that $\operatorname{Supp} Q \subseteq \operatorname{Supp} P$. But then, by our hypothesis, $P = Q$ so $P \in S$.) Thus, we can take $n = 0$ in this case.

(2) Suppose $X = \operatorname{Sper} A$ where A is the coordinate ring of a real algebraic set $V \subseteq R^N$, R real closed. Then $d := d(X) = \dim(V)$ by (5.4)(2). Also, combining (5.1) (resp., (5.3) and Remark (1) above) with the stability formula for field extensions given in [**7**] and the correspondence between semi-algebraic sets in V and constructible sets in X [**9**, Th. 8.12] yields $s(V) = s(X) = d$ (resp., $\overline{s}(V) = \overline{s}(X) \leq d(d+1)/2$) if $d \geq 1$; see [**5**], [**14**]. Scheiderer [**14**] has shown that $\overline{s}(V) = \overline{s}(X) = d(d+1)/2$, i.e., the bound for $\overline{s}(X)$ is best possible in this situation.

(3) The reader may wonder if the material on the t–invariant in [**4**] also extends to saturated sets. Certainly, there is no problem if A is Noetherian. On the other hand, it is *not clear* that [**4**, Cor. 5.7] continues to hold if one drops the hypothesis that A is Noetherian.

REFERENCES

1. C. Andradas, L. Bröcker and J. M. Ruiz, *Minimal generation of basic open semi-analytic sets*, Invent. Math., **92** (1988), 409–430.
2. E. Becker, *On the real spectrum of a ring and its application to semi–algebraic geometry*, Bull. Amer. Math. Soc. **15** (1986), 19–61.
3. J. Bocknak, M. Coste, and M.-F. Roy, *Géometrie Algébrique Réele*, Ergeb. Math., Springer, Berlin, Heidelberg, New York, 1987.
4. L. Bröcker, *On the stability index of Noetherian rings* (to appear).
5. ______, *On basic semi-algebraic sets*, Expositiones Math. **9** (1991), 289–334.
6. ______, *Positivbereiche in kommutativen Ringen*, Abh. Math. Sem. Univ. Hamburg **52** (1982), 170–178.
7. ______, *Zur Theorie der quadratischen Formen über reelen Körpern*, Math. Ann. **210** (1974), 233–256.
8. M. Knebusch and C. Scheiderer, *Einführung in die reelle Algebra*, Friedr. Vieweg & Sohn, Braunschweig/Wiesbaden, 1989.
9. T.-Y. Lam, *An introduction to real algebra*, Rocky Mountain J. Math. **14** (1984), 767–814.
10. ______, *Orderings, valuations, and quadratic forms*, vol. 52, Conf. Ser. in Math. AMS, 1983.
11. L. Mahé, *Une démonstration élementaire du théorème de Bröcker–Scheiderer* (to appear).
12. M. Marshall, *Spaces of orderings: systems of quadratic forms, local structure, and saturation*, Comm. in Algebra **12** (1984), 723–743.
13. M. Marshall and L. Walter, *Signatures of higher level on rings with many units*, Math. Zeit. **204** (1990), 129–143.
14. C. Scheiderer, *Stability index of real varieties*, Invent. Math. **97** (1989), 467–483.
15. N. Schwartz, *Local stability and saturation in spaces or orderings*, Canad. J. of Math. **35** (1983), 454–477.
16. L. Walter, *Quadratic forms, orderings, and quaternion algebras over rings with many units*, M.Sc. thesis, University of Saskatchewan, 1988.

DEPARTMENT OF MATHEMATICS & STATISTICS, UNIVERSITY OF SASKATCHEWAN, SASKATOON, SK, CANADA, S7N 0W0

E-mail address: marshall@snoopy.usask.ca

Contemporary Mathematics
Volume **155**, 1994

A New Proof of the Homogeneous Nullstellensatz for p-Fields, and Applications to Topology

ALBRECHT PFISTER

In this talk I describe an elementary and reasonably short proof of the theorem in the title (Theorem 2 below) which has recently been found by my colleague H.-J. Fendrich. It uses properties of the Hilbert polynomial but not the theorem of Bezout or results from intersection theory. In the second part of the talk I derive the Borsuk-Ulam and Brouwer fixpoint theorem from the nullstellensatz for 2–fields.

DEFINITION 1. *A polynomial $P(z) \in \mathbb{Q}[z]$ of degree $d \geq 0$ is called numerical polynomial if*

$$P(z) = c_0 \binom{z}{d} + c_1 \binom{z}{d-1} + \cdots + c_d$$

where $c_i \in \mathbb{Z}, c_0 \neq 0$, $\binom{z}{d} = \frac{1}{d!} z(z-1)\cdots(z-d+1)$.
For $d < 0$ put $P(z) = 0$. Clearly: $i \in \mathbb{Z} \Rightarrow P(i) \in \mathbb{Z}$.

LEMMA 1. *Suppose $f(z) \in \mathbb{Q}[z], f(i) \in \mathbb{Z}$ for $i \gg 0$ (i.e. for all $i \geq i_0$ and some $i_0 \in \mathbb{Z}$) and*

$$(\Delta f)(i) := f(i+1) - f(i) = Q(i) \text{ for } i \gg 0$$

where Q is a numerical polynomial of degree $d \geq 0$.
Then there is a (unique) numerical polynomial P of degree $d+1$ with $f(i) = P(i)$ for $i \gg 0$.

1991 *Mathematics Subject Classification.* Primary 14A05 14P05 55MZ0 11E76.

This paper is submitted in final form and no version of it will be submitted for publication elsewhere.

PROOF. Let

$Q(z) = c_0\binom{z}{d} + \cdots + c_d$, put

$P(z) = c_0\binom{z}{d+1} + \cdots + c_d\binom{z}{1} + c_{d+1}$

where $c_{d+1} \in \mathbb{Z}$ has to be determined.

$$\binom{z+1}{r+1} - \binom{z}{r+1} = \binom{z}{r} \text{ for } r \in \mathbb{N}_0 \quad \Rightarrow (\Delta P)(z) = Q(z)$$

$\Rightarrow (\Delta(f-P))(i) = 0$ for $i \gg 0 \Rightarrow f(i) - P(i)$ is constant for big i

$\Rightarrow f(i) - P(i) = 0$ for $i \gg 0$ and suitable choice of $c_{d+1} \in \mathbb{Z}$. □

REMARK: Compare [**Ha**, p. 49].

Let K be a field, let $R = K[X_0, \ldots, X_n]$ be the polynomial ring over K in $n+1$ indeterminates ($n \geq 0$, for $n = -1$ put $R = K$). R is graded: $R = \bigoplus_{d \geq 0} R_d$ where R_d consists of the homogeneous polynomials of degree d (including 0). R_d is a K-vectorspace of dimension $\binom{d+n}{n}$. An R-module M is called graded if $M = \bigoplus_{i \geq 0} M_i$ with K-vectorspaces M_i such that $R_d M_i \subseteq M_{d+i}$. For $f \in R_d$ multiplication with f gives rise to exact sequences:

($\star$) $\quad 0 \to {}_fM_i \to M_i \xrightarrow{f} M_{i+d} \to M_{i+d}/fM_i \to 0$

Here ${}_fM_i := \{m \in M_i \mid fm = 0\}, i \geq 0$.

THEOREM 1. *(Hilbert-Serre) If M is a finitely generated graded R-module the M_i are finite-dimensional K-vectorspaces and there is a unique numerical polynomial $P = P_M$ such that*

$$\dim M_i = P(i) \textit{ for } i \gg 0.$$

PROOF. By induction on n

1) $n = -1 \Rightarrow R = K, M = M_0$ with $\dim M_0 < \infty, M_i = 0$ for $i > 0, P = 0$.

2) $n - 1 \to n$(for $n \geq 0$):

Put $X = X_n$. The R-modules ${}_XM$ and M/XM are annihilated by (multiplication with) X, hence can be considered as finitely generated R'-modules where $R' = K[X_0, \ldots, X_{n-1}]$. By induction:

$$\dim({}_XM_i) = P_{{}_XM}(i)$$

$$\dim(M_{i+1}/XM_i) = \dim(M/XM)_{i+1} = P_{M/XM}(i+1)$$

for $i \gg 0$. Now ($\star$) implies by induction on i (starting with $M_{-1} = 0$): $\dim M_i < \infty$ for all i,

$$\dim M_{i+1} - \dim M_i = P_{M/XM}(i+1) - P_{{}_XM}(i) =: Q(i)$$

for $i \gg 0$. Here Q is a numerical polynomial which is determined by M and $X = X_n$ but depends only on M.

Lemma 1 implies: $\dim M_i = P(i)$ for $i \gg 0$ where $P = P_M$ is a numerical polynomial, uniquely determined by M. Obviously $P = 0 \Leftrightarrow \dim_K M < \infty$. P is called "Hilbert polynomial" of M. $\square$

DEFINITION 2. *For a finitely generated graded R-module M with*

$$\begin{aligned} P_M &= c_0\binom{z}{d} + \cdots + c_d, \quad c_0 \neq 0 \\ \delta_M &:= \deg P_M = d \quad \textit{is called the degree of } M. \\ \mu_M &:= c_0 \quad \textit{is called the multiplicity of } M. \end{aligned}$$

$$\textit{Then } P_M(z) = \frac{\mu_M}{d!} z^d + \textit{ lower terms.}$$

EXAMPLE 1. $M = R = K[X_0, \ldots, X_n] \Rightarrow P_M(z) = \binom{z+n}{n} \Rightarrow \delta_M = n, \mu_M = 1$

DEFINITION 3. *M integral: $\Leftrightarrow$ for any $f \in R$ either $fM = 0$ or ${}_fM = 0$*

REMARK: Let P be a homogeneous ideal in R. Then we have: $M = R/P$ integral $\Leftrightarrow$ ${}_fM = 0$ for $f \notin P$ $\Leftrightarrow$ P is a prime ideal $\Leftrightarrow$ M is an integral domain.

LEMMA 2. *Suppose M integral, $d > 0$, $0 \neq f \in R_d$. Then $fM \neq 0 \Rightarrow \delta_{M/fM} = \delta_M - 1$, $\mu_{M/fM} = d\mu_M$*

PROOF. $(\star)$ and ${}_fM = 0$ imply

$$\begin{aligned} P_{M/fM}(i) &= P_M(i) - P_M(i-d) \\ &= \frac{\mu}{\delta!}(i^\delta - (i-d)^\delta) + \ldots \\ &= \frac{\mu d}{(\delta-1)!} i^{\delta-1} + \ldots \qquad \text{for } i \gg 0 \qquad \square \end{aligned}$$

LEMMA 3. *Let M be a (fin.-gen. graded) R-module with $\delta_M \geq 0$, let p be a prime number with $p \nmid \mu_M$.*

Then M has a homogeneous submodule S such that $N = M/S$ satisfies the conditions:

$$(1) \ \ \delta_N = \delta_M, \qquad (2) \ \ p \nmid \mu_N, \qquad (3) \ \ N \textit{ integral.}$$

PROOF. By the assumptions on the noetherian R-module M there exists a maximal homogeneous submodule S such that (1), (2) hold for N. We have to show that (3) follows.

Suppose $f \in R, fN \neq 0$, and w.l.o.g. $f \in R_d (d \geq 0)$. Then by $(\star)$

$$P_N(i) = P_{N/fN}(i) + P_{N/{}_fN}(i-d) \text{ for } i \gg 0.$$

By the maximality of S the module $N/fN \cong M/(S+fM)$ does not satisfy both conditions (1), (2). Hence either $\delta_{N/fN} < \delta_N$ or $\delta_{N/fN} = \delta_N$ and $p/\mu_{N/fN}$. In both cases we get:

$$\delta_{N/_fN} = \delta_N \text{ and } p\nmid \mu_{N/_fN}, \text{ i.e. (1) and (2) for } N/_fN$$

This implies $_fN = 0$ and N integral by the "minimality" of N. □

LEMMA 4. *Let $P < R$ be a homogeneous prime ideal such that the integral R-module $M = R/P$ satisfies:*

$$\delta_M = 0, \quad \mu_M = c \neq 0.$$

Then:
(1) $XM \neq 0$ *for some* $X = X_j \in \{X_0, \dots, X_n\}$
(2) $L = M/(X-1)M$ *is a field,* $[L : K] = c$.

PROOF.
1) $X_iM = 0$ for all $i = 0, \dots, n$ would imply $X_i \in P$ for all i, i.e. $M = K$, $P_M = 0$.
2) If $X = X_j$ and $XM \neq 0$ then $M \xrightarrow{X} M$ is injective since M is integral. Therefore $M_i \xrightarrow{X} M_{i+1}$ is bijective for $i \gg 0$ (where $\dim M_i = c$).

Multiplication by X induces the identity on $L = \pi M = \pi(\underset{i}{\oplus} M_i)$, hence $L = \pi M_k = \pi M_{k+1} = \dots$ for $k \geq k_0$ (π denotes the natural projection map). Now $\pi|_{M_k}$ is injective since $m = (X-1)m'$ implies $m = 0$ for $m \in M_k$. Therefore $\dim L = \dim M_k = c$. Furthermore L is an integral domain:

Suppose $l, l' \in L, ll' = 0$. Then $l = \pi(m), l' = \pi(m')$ with $m, m' \in M_k$ and $0 = ll' = \pi(mm')$ with $mm' \in M_{2k}$. This implies $mm' = 0$ since $\pi|_{M_{2k}}$ is injective. Therefore $m = 0$ or $m' = 0$ since M is integral. This gives $l = 0$ or $l' = 0$.

Finally any finite-dimensional integral domain over K is a field. □

THEOREM 2. (*Nullstellensatz for p-fields*)

Assumptions:
p is a prime number,
K a p-field, i.e. $[L : K]$ is a power of p for any finite field extension L/K,
$n \geq 1, \quad R = K[X_0, \dots, X_n]$,
$f_1, \dots, f_n$ homogeneous elements of R, $(p, \deg f_i) = 1$ for all i.

Claim: *There exists a common zero $0 \neq a = (a_0, \dots, a_n) \in K^{n+1}$ of the system of equations $f_1 = \dots = f_n = 0$.*

PROOF. (After Fendrich.)

Since $\delta_R = n, \mu_R = 1$ we can apply Lemmas 2 and 3 alternatively n times to find a maximal homogeneous prime ideal $P \subset R$ containing $f_1, \dots, f_n$ (and possibly some of the indeterminates X_i) such that

$$\delta_{R/P} = 0, \quad p \nmid \mu_{R/P} = c.$$

Apply Lemma 4 to $M = R/P$. Then $L = M/(X_j - 1)M$ is a field with $[L : K] = c$. This implies $c = 1$ since K is a p-field and $p \nmid c$. Let a_i be the image of X_i in $L = K$ $(i = 0, \dots, n)$. Then $a_j = 1$ and $f_i(a) = 0$ for $i = 1, \dots, n$. □

THEOREM 3. *Let k be a p-field, let K be a field of transcendence degree $i \geq 0$ over k. Let $f_1, \dots, f_r \in K[X_0, \dots, X_n]$ be homogeneous of degrees $d_1, \dots, d_r$ with $(p, d_j) = 1$ for $j = 1, \dots, r$. Suppose $n \geq d_1^i + \cdots + d_r^i$. Then the system $f_1 = \cdots = f_r = 0$ has a nontrivial common zero in K^{n+1}.*

PROOF. This follows by induction on i from Theorem 2. The proofs for C_i-fields (where k is algebraically closed) in [**L1**] or [**G**] can be taken over verbatim since all forms which appear in the induction process have degrees prime to p.

The most important case of Theorems 2 and 3 is the case where the groundfield K (resp. k) is real closed and $p = 2$. S. Lang then calls the field K "oddly C_i". □

OPEN PROBLEM: [**L2**]. Let $k = R$ be real closed, let $\mathrm{tr}(K/k) = i$ as in Theorem 3. Suppose in addition that K is non-real. Is then K a C_i-field, i.e. does the implication of Theorem 3 hold without the assumption $2 \nmid d_j$?

As far as I know this is solved only in the case $i = 1, r = 1$ and $d_1 = 2$. Namely, by an old result of Witt [**W**] we have $u(K) \leq 2$ for any non-real field of transcendence degree 1 over R.

On the other hand, no counter-examples to Lang's conjecture are known.

THEOREM 4. *(Borsuk-Ulam for polynomial maps)*
Let R be real closed, let $n \geq 1$, and let $q_1, \dots, q_n$ be "odd" polynomials in $R[X_1, \dots, X_{n+1}]$, i.e. we have

$$q_i(-X) = -q_i(X) \text{ for } i = 1, \dots, n$$

where $X = (X_1, \dots, X_{n+1})$, $-X = (-X_1, \dots, -X_{n+1})$.

Then there exists

$$a = (a_1, \dots, a_{n+1}) \in R^{n+1} \text{ with } \sum_1^{n+1} a_i^2 = 1 \text{ and}$$

$$q_1(a) = \cdots = q_n(a) = 0.$$

PROOF. We introduce an additional indeterminate X_0.

Let $\tilde{q}_i = \tilde{q}_i(X_0, \dots, X_{n+1})$ be the homogenized polynomial belonging to q_i, with $\deg \tilde{q}_i = \deg q_i = d_i$. Since the q_i contain only monomials of odd degree d_i is odd and X_0 occurs in $\tilde{q}_i$ in even powers only. Thus it is possible to replace X_0^2 by $X_1^2 + \cdots + X_{n+1}^2$ in $\tilde{q}_i$. This gives homogeneous polynomials $f_i(X_1, \dots, X_{n+1})$ of odd degrees $d_i (i = 1, \dots, n)$. By Theorem 2 the f_i have a common nontrivial zero $a = (a_1, \dots, a_{n+1}) \in R^{n+1}$. W.l.o.g. we can assume $\sum_1^{n+1} a_i^2 = 1$ since the f_i are homogeneous. Going back to the original polynomials q_i we get

$$q_i(a) = \tilde{q}_i(1, a) = f_i(a) = 0 \quad (i = 1, \dots, n) \qquad \square$$

THEOREM 5. (*Borsuk-Ulam for continuous maps*)
Let $\mathbb{R}$ be the field of real numbers, let $S^n \subset \mathbb{R}^{n+1}$ denote the n-dimensional unit-sphere.

Let $f : S^n \to \mathbb{R}^n$ with $f = (f_1, \dots, f_n)$ be an "odd" continuous map, i.e. $f(-x) = -f(x)$ for all $x \in S^n$. Then f vanishes for at least one point $a \in S^n$.

COROLLARY 1. *There exists no continuous map $g : S^n \to S^{n-1}$ such that $g(-x) = -g(x)$ for all $x \in S^n$.*

PROOF.

1) Since S^n is compact we can apply the approximation theorem of Weierstraß: For any real $\varepsilon > 0$ there exist polynomials $p_i \in \mathbb{R}[X_0, \dots, X_n]$ such that

$$| f_i(x) - p_i(x) | < \varepsilon \text{ for all } x \in S^n \text{ and } i = 1, \dots, n.$$

Putting $q_i(X) = \frac{1}{2}(p_i(X) - p_i(-X))$ one immediately derives that the p_i may be replaced by the "odd" polynomials q_i.
2) Assume by contradiction that the given map $f : S^n \to \mathbb{R}^n$ has no zero. Then, since S^n is compact, there exists a $\delta > 0$ such that

$$\text{Max}\{| f_1(x) |, \dots, | f_n(x) |\} \geq \delta$$

for all $x \in S^n$. Choose $0 < \varepsilon < \delta$ in part 1).

Then $\text{Max}\{| q_1(x) |, \dots, | q_n(x) |\} \geq \delta - \varepsilon > 0$ for all $x \in S^n$. This contradicts Theorem 4. □

REMARK: The Weierstraß approximation theorem does not hold for real closed fields R which contain infinitely small elements (relative to $\mathbb{Q} \subset R$).

Example: There is no polynomial $g(x) \in R[x]$ which approximates $f(x) = | x |$ better than ε on the interval $[-1, 1] \subset R$ if $0 < \varepsilon \in R, \varepsilon < q$ for all $q \in \mathbb{Q}_+$.

We shall now derive the Brouwer fixpoint-theorem from Theorem 5. Denote by
$E^n = \{(x_0, \dots, x_{n-1}) \in \mathbb{R}^n \mid \sum x_i^2 \leq 1\}$ the n-ball,
$E^n_{+-} = \{(x_0, \dots, x_n) \in S^n \subset \mathbb{R}^{n+1} \mid \pm x_n \geq 0\}$ the upper resp. lower hemisphere of S^n.
Then $S^{n-1} = \partial E^n \cong E^n_+ \cap E^n_-$.

Let π denote the projection $\mathbb{R}^{n+1} \to \mathbb{R}_n : \quad (x_0, \dots x_n) \to (x_0, \dots, x_{n-1})$. Then we have:

PROPOSITION 1. *There is no continuous retract*

$$r : E^n \to S^{n-1}, \quad r |_{S^{n-1}} = id$$

PROOF. Suppose that r exists. Define $f : S^n \to S^{n-1}$ by

$$f(x) = \begin{cases} r(-\pi(x)) & for \quad x \in E^n_+ \\ -r(\pi(x)) & for \quad x \in E^n_- \end{cases}$$

f is well defined: For $x \in E^n_+ \cap E^n_-$ we have $x_n = 0, \pi(x) = x$ and $r(-x) = -x = -r(x)$.
f is clearly continuous. Finally f is an antipodal map: For $x \in E^n_+$ we have

$$f(-x) = -r(\pi(-x)) = -r(-\pi(x)) = -f(x).$$

A similar equation holds for $x \in E^n_-$. Thus f contradicts the Corollary of Theorem 5. □

THEOREM 6. *(Fixpoint-theorem of Brouwer)*
A continuous map $f : E^n \to E^n$ always has a fixpoint $a \in E^n$ with $f(a) = a$.

PROOF.

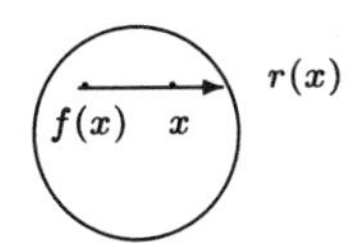

Suppose $f(x) - x \neq 0$ for all $x \in E^n$.
Then it is possible to define a continuous retract $r : E^n \to S^{n-1}$ as given by the picture: Intersect the arrow $\overrightarrow{f(x), x}$ with the border of E^n. □

QUESTION: What happens to Theorems 5 and 6 if $\mathbb{R}$ is replaced by an arbitrary real closed field R?
This question can be (partly) answered as follows:
(i) The Tarski principle (see [**BCR**]) implies that Theorems 5 and 6 hold for continuous semi-algebraic maps with the corresponding properties.
(ii) Theorems 5 and 6 are no longer true for all continuous maps unless R is order-isomorphic to $\mathbb{R}$: There are two types of real closed fields R:
(a) R is archimedean over $\mathbb{Q}$, i.e. to any $r \in R \; \exists \; n \in \mathbb{N}$ such that

$$-n < r < n$$

Then the map $r \to \sup\{q \in \mathbb{Q} \mid q < r\}$ defines an orderisomorphic imbedding of R in $\mathbb{R}$. Therefore we may assume $R \subsetneq \mathbb{R}$ in the archimedean case.
(b) R contains infinitely large elements $r \in R, r > n$ for all $n \in \mathbb{N}$, and therefore infinitely small elements $\varepsilon = \frac{1}{r} > 0$.

Consider now the following subsets of R:

$$\begin{aligned}
E^1 &= \{r \in R : -1 \le r \le 1\} \\
U_0 &= \{r \in R : \mid r \mid < r_0\} \text{ in case a), where } r_0 \in \mathbb{R}\backslash R, 0 < r_0 < 1 \\
U_0 &= \{\varepsilon \in R : \varepsilon \text{ infinitely small } \} \text{ in case b)} \\
U_1 &= \{r \in E^1 : r > 0, r \notin U_0\} \\
U_{-1} &= \{r \in E^1 : r < 0, r \notin U_0\} = -U_1.
\end{aligned}$$

Then it is easily seen that U_{-1}, U_0, U_1 are open in the order topology (also called interval or euclidean or strong topology) of E^1 and that E^1 is the disjoint union of U_{-1}, U_0 and U_1. Therefore E^1 is not connected (actually totally disconnected since the same reasoning applies to any closed subinterval of E^1) and not compact in its topology.

This immediately gives the following continuous maps f, g, h contradicting the analoga of Proposition 1, Theorem 6 and Theorem 5 respectively:

$$\begin{aligned}
f &: \quad E^1 \to S^0 = \{-1, 1\}, f(U_0 \cup U_1) = 1, f(U_{-1}) = -1 \\
g &: \quad E^1 \to E^1, g(U_0 \cup U_1) = -1, g(U_{-1}) = 1 \\
h &: \quad S^1 \to S^0, h(V_1) = 1, h(V_{-1}) = -1
\end{aligned}$$

Here $S^1 = \{(x, y) \in R^2 : x^2 + y^2 = 1\}$ is the disjoint union of the open subsets

$$\begin{aligned}
V_1 &:= \{(x, y) \in S^1 : \text{ either } x \in U_1, \text{ or } x \in U_0 \text{ and } y > 0\} \\
V_{-1} &:= S^1 \backslash V_1 = -V_1
\end{aligned}$$

HISTORICAL REMARKS:

(i) The Brouwer fixpoint theorem was proved in 1910 [**Br**].

(ii) The Borsuk-Ulam theorem was conjectured by Ulam and proved by Borsuk in 1932 [**Bo**].

(iii) The first proofs of Theorem 2 for $p = 2$ and $K = \mathbb{R}$ are from Borsuk, Hopf and Stiefel and are of topological nature (see [**Ho**]). The first purely algebraic proof is due to Behrend (1940) [**Be**]. Another proof (for $K = R$ real-closed) has been given by S. Lang (1953) [**L2**].

(iv) The first proof of Theorem 2 and Theorem 3 for arbitrary p-fields is due to Terjanian [**T**]. It uses a lot of local algebra, multiplicities, Bezout theorem etc. Another proof using henselizations and Bezout is due to Arason (see [**P**]). A proof of a more general theorem, based on intersection theory, is given in Chapter 13 of Fulton's book [**Fu**].

(v) The first algebraic proof of Theorem 4 is probably the one by Knebusch [**K**]. The proof given here is in [**AP**]. The idea of these proofs came from the wonderful paper of Dai-Lam-Peng on the existence of integral domains with prescribed level.

REFERENCES

[AP] J. K. Arason and A. Pfister, *Quadratische Formen über affinen Algebren und ein algebraischer Beweis des Satzes von Borsuk-Ulam*, J.r.a. Math. **331** (1982).

[Be] F. Behrend, *Über Systeme reeller algebraischer Gleichungen*, Compos. Math. **7** (1940).

[BCR] J. Bochnak, M. Coste, and M.-F. Roy, *Géométrie algébrique réelle*, Springer-Verlag 1987.

[Bo] K. Borsuk, *Drei Sätze über die n-dimensionale euklidische Sphäre*, Fund. Math. **20** (1933).

[Br] L. E. J. Brouwer, *Über Abbildung von Mannigfaltigkeiten*, Math. Ann. **71** (1912).

[Fe] H. J. Fendrich, *Eine Bemerkung zum homogenen Nullstellensatz für p-Körper* (to appear).

[Fu] W. Fulton, *Intersection theory*, Springer-Verlag, 1984.

[G] M. Greenberg, *Lectures on forms in many variables*, Benjamin, New York, 1969.

[Ha] R. Hartshorne, *Algebraic geometry*, Springer-Verlag, 1977.

[Ho] H. Hopf, *Ein topologischer Beitrag zur reellen Algebra*, Comment. Math. **13** (1940/41).

[K] M. Knebusch, *An algebraic proof of the Borsuk-Ulam theorem for polynomial mappings*, Proc. Amer. Math. Soc. **84** (1982).

[L1] S. Lang, *On quasi-algebraic closure*, Ann. of Math. **55** (1952).

[L2] ——— *The theory of real places*, Ann. of Math. **57** (1953).

[P] A. Pfister, *Systems of quadratic forms*, Bull Soc. Math. France, Mémoire **59** (1979).

[T] G. Terjanian, *Dimension arithmétique d'un corps*, J. of Alg. **22** (1972).

[W] E. Witt, *Zerlegung reeller algebraischer Funktionen in Quadrate, Schiefkörper über reellem Funktionenkörper*, J.r.a. Math. **171** (1934), 4–11.

FACHBEREICH MATHEMATIK, SAARSTRASSE 21, 6500 MAINZ
E-mail address: pfister@mzdmza.zdv.uni-mainz.de

Contemporary Mathematics
Volume **155**, 1994

The Compatible Valuation Rings of the Coordinate Ring of the Real Plane

M. J. DE LA PUENTE

ABSTRACT. Let A be the coordinate ring of the real plane, i.e., $A = \mathbf{R}[X, Y]$, where X, Y are algebraically independent over the field of real numbers $\mathbf{R}$. The purpose of this paper is to describe all the valuation rings of A which are compatible with some prime cone of A, i.e., to give a description of all the points of the real Riemann surface $S_r(A/\mathbf{R})$. Furthermore, we explain the geometric meaning of these points and indicate the specialization relations among them.

The relationship between the orders and the convex valuation rings of an orderable field already arises at the very beginning of the general theory of valuations, in the works of Baer [**2**] and Krull [**6**]. This has been used in the study of some aspects of real algebraic sets. With the aim of having an adequate framework for the systematic study of such a relationship, the real Riemann surface of a ring A has been defined in [**9**] as a certain topological space.

Particularizing the ring A, let A be the coordinate ring of the real plane, i.e., $A = \mathbf{R}[X, Y]$, where X, Y are algebraically independent over the field of real numbers $\mathbf{R}$. On one hand, in [**11**], Zariski discusses the various types of valuations on algebraic function fields of two variables over an algebraically closed field. On the other hand, in [**1**], Alonso et al. give a constructive description of the orders of $\mathbf{R}[[X, Y]]$, by means of formal power series. Using also compatible valuation rings, some geometric examples of orders of $\mathbf{R}(X, Y)$ have been given in [**4**, §8.12] and [**3**, Rem. 10.3.5]. Bringing together [**11**] and [**1**], the purpose of this paper is to describe all the valuation rings of A which are compatible with some prime cone of A, i.e., to give a description of all the points of the real

1991 *Mathematics Subject Classification.* 13F10, 13F25, 12D15.

Partially supported by the Consejería de Educaci6n de la Comunidad de Madrid and by CICYT PB860062.

This paper is submitted in final form and no version of it will be submitted for publication elsewhere.

Riemann surface $S_r(A/\mathbf{R})$. This requires the knowledge of the real spectrum $\mathrm{Spec}_r A$, for $A = \mathbf{R}[X,Y]$. Furthermore, we explain the geometric meaning of these points and indicate the specialization relations among them.

The paper is divided into five sections. In the first one, we gather some material and prove a pull-back result, for points of real Riemann surfaces. In the second one, the points of $S_r(A/\mathbf{R})$ near $(0,0)$ are described in detail, as well as the specialization relations for points on the same fiber. The third section is devoted to giving a geometric meaning to those points, in terms of half-branches of plane curves. We complete the description of specialization relations for points $S_r(A/\mathbf{R})$ near $(0,0)$, on possibly different fibers, in the fourth section. Finally, we deal with the rest of the points in $S_r(A/\mathbf{R})$, using the results of the previous sections.

We are grateful to T. Y. Lam for his invitation to participate in the Special Year on Real Algebraic Geometry and Quadratic Forms, at U.C. Berkeley, where this work was completed.

1. Notations and terminology

Recall that in any topological space, a point a *specializes* a point b if a lies in the closure of b. In this case, we write $a \leftarrow b$.

Let k be an ordered field and K be a field containing k. There is a bijective correspondence between the real Riemann surface of K over k (see the definition below), denoted $S_r(K/k)$, and the collection of pairs (β, B), where $\beta \in \mathrm{Spec}_r K/k$ and B is a convex valuation ring of K/k, with respect to β, see [**3**, §10.1]. B might equal K, the *trivial valuation ring* of K. This is, for the case of real fields, in the same spirit of the classical definition of [**12**, VI, §17]. Notice that $S_r(K/k) \neq \emptyset$ if and only if the order of k can be extended to K.

Now, turning to a more general situation, let A be a commutative real k–algebra. In these notes, $\mathrm{Spec}_r A/k$ will denote the set of prime cones of A containing the positive cone of k, see [**3**, Def. 4.3.1]. For every $\beta \in \mathrm{Spec}_r A/k$ let $\mathrm{supp}\,\beta$ denote the prime ideal $\beta \cap -\beta$, $\kappa(\beta)$ the quotient field of $A/\mathrm{supp}\,\beta$ and $\bar{\beta}$ the induced prime cone (or order) on $\kappa(\beta)$. Intuitively, the union

$$\cup_{p \in \mathrm{Spec} A} S_r(\kappa_p/k)$$

would be the real Riemann surface of A/k, where κ_p is the quotient field of A/p. More formally, see [**9**], a prime ideal $p \subset A$ defines an equivalence relation in $A \times (A \backslash p)$: $(x,y) \sim (x',y')$ if and only if $xy' - x'y \in p$. Rather than looking at an actual valuation ring of κ_p, we will consider its preimage in $A \times (A \setminus p) \subset A \times A$ under the map

$$\phi_p : A \times (A \setminus p) \to \kappa_p, \quad \phi_p(x,y) = \bar{x}/\bar{y},$$

where $\bar{x} = x + p$. The interpretation of (x,y) as $\bar{x}/\bar{y}$ is what we have in mind when we define a *valuation ring of A over k* as a subset $B \subset A \times A$ satisfying the following properties:

(1) $\pi(B) := \{y \in A : (y, y) \notin B\}$ is a proper ideal of A, possibly zero,
(2) $(x, y) \in B \Rightarrow y \notin \pi(B)$,
(3) $(x, y) \in B$, $y' \notin \pi(B)$, $x' \in A$ and $xy' - x'y \in \pi(B) \Rightarrow (x', y') \in B$,
(4) $(x, y) \in B$ and $(x', y') \in B \Rightarrow (xy' - x'y, yy') \in B$ and $(xx', yy') \in B$,
(5) $(x, y) \notin B$, $y \notin \pi(B) \Rightarrow (y, x) \in B$,
(6) $(c, 1) \in B$, for all $c \in k$.

It follows from (2) and (4) that the ideal $p = \pi(B)$ is prime and from (3) that B is saturated under $\pi(B)$–equivalence relation. It turns out that $B \subseteq A \times (A \backslash p)$ and that the field κ_p contains a valuation ring, $\mathcal{B}$, giving rise to B. Indeed, if B verifies (1)–(6) and $p = \pi(B)$, then $\phi_p(B)$ is a valuation ring of κ_p. Conversely, for every prime ideal $p \subset A$ and every valuation ring $k \subseteq \mathcal{B} \subseteq \kappa_p$, there exists exactly one B,

$$B = \{(x, y) \in A \times (A \setminus p) : \bar{x}/\bar{y} \in \mathcal{B}\}$$

verifying (1)–(6) such that $\phi_p(B) = \mathcal{B}$. In view of this, we see that (4) and (5) express, in terms of B, that $\mathcal{B}$ is a valuation ring of κ_p.

Let $\beta \in \operatorname{Spec}_r A/k$ and B be a valuation ring of A over k. β and B are said to be *compatible* if the two following conditions hold:

(7) $\operatorname{supp} \beta = \pi(B)$,
(8) $(x, y) \in B$, $(y', y') \in B$, $x' \in A$, $x'y' \in \beta$ and $xy' - x'y \in \beta$ imply $(x', y') \in B$.

Notice that the meaning of (8) is that $\mathcal{B}$ is convex with respect to $\bar{\beta}$.

If v is a valuation associated to $\mathcal{B}$, we say that β and v are *compatible* if β and B are. The rank, rational rank, dimension and value group of (β, B) (or of B) is the rank, rational rank, dimension and value group of $\mathcal{B}$, see [**12**, VI]. We say that (β, B) (or B) is discrete if $\mathcal{B}$ is.

The *real Riemann surface of A over k* is the collection $S_r(A/k) \subset 2^A \times 2^{A \times A}$ of compatible pairs (β, B). The set $S_r(A/k)$ is given the topology generated by the following sets

$$\{(\beta, B) : t \notin \beta\ ,\ (x, y) \in B\}$$

where t, x, y vary in A.

The subset of $S_r(A/k)$ of all the points with β as first entry is called the *fiber over β*.

REMARK 1.1. In [**10**, Th. 13 and 14] we see that given $(\beta, B) \in S_r(A/k)$, then the points (β', C) in the closure of (β, B) are precisely all the points meeting one of the following conditions:

(a) $\beta = \beta'$ and $\mathcal{C} \subseteq \mathcal{B}$,
(b) $\beta \subset \beta'$ and there exists a valuation ring $\mathcal{D}$ of $\kappa(\beta)$, convex with respect to $\bar{\beta}$, such that $\mathcal{D} \subseteq \mathcal{B}$ and $\mathcal{C} = \mathcal{D}/m_{\mathcal{B}} \cap \kappa(\beta') \subseteq \Delta_{\mathcal{B}}$, where $m_{\mathcal{B}}$ is the maximal ideal of $\mathcal{B}$, and $\Delta_{\mathcal{B}}$ is the residue field $\mathcal{B}/m_{\mathcal{B}}$, with $C = \phi^{-1}_{\operatorname{supp}\beta'}(\mathcal{C})$.

Notice that in both cases, $\beta \to \beta'$ in $\mathrm{Spec}_r AA/k$.

Let $k = \mathbf{R}$ and $A = \mathbf{R}[X,Y]$, where X, Y are algebraically independent over $\mathbf{R}$. Our objective is to describe all the pairs (β, B) in $S_r(A/\mathbf{R})$ for which we need, first of all, a description of $\mathrm{Spec}_r A = \mathrm{Spec}_r A/\mathbf{R}$. We will see that each point $\beta \in \mathrm{Spec}_r A$ is given by an $\mathbf{R}$-homomorphism

$$\psi : A \to K$$

where K is $\mathbf{R}$or an ordered field of formal power series, containing $\mathbf{R}$. The next result will lead us, latter on, to search for $S_r(K/\mathbf{R})$, for all the fields K involved.

PROPOSITION 1.2. *Let k be an ordered field, A a commutative real k-algebra and K a real field containing k and suppose that a k-homomorphism*

$$\psi : A \to K$$

is given. Then, a point $(\bar{\beta}, \mathcal{B}) \in S_r(K/k)$ induces a point $(\beta, \psi^*(\mathcal{B})) \in S_r(A/k)$, via ψ. The value group of $(\beta, \psi^*(\mathcal{B}))$ is a subgroup of the value group of $(\bar{\beta}, \mathcal{B})$.

PROOF. First of all, suppose A is a field. Then, it is well known that $\psi^{-1}(\bar{\beta}) \in \mathrm{Spec}_r A/k$ and that $\psi^{-1}(\mathcal{B})$ is a valuation ring of A/k convex with respect to the order $\psi^{-1}(\bar{\beta})$. In addition, the map ψ induces an embedding

$$\psi^{-1}(\mathcal{B})/\psi^{-1}(U_{\mathcal{B}}) \to \mathcal{B}/U_{\mathcal{B}}$$

from the value group of $\psi^{-1}(\mathcal{B})$ into the value group of $\mathcal{B}$, where $m_{\mathcal{B}}$ is the maximal ideal of $\mathcal{B}$ and $U_{\mathcal{B}} = \mathcal{B} \setminus m_{\mathcal{B}}$ is the set of units of $\mathcal{B}$.

In the general case, $\ker \psi$ is a prime ideal p of A, and ψ induces a k-algebra homomorphism

$$\psi_p : \kappa_p \to K.$$

Thus $(\psi_p^{-1}(\bar{\beta}), \psi_p^{-1}(\mathcal{B})) \in S_r(\kappa_p/k)$ and there exists a unique point in $S_r(A/k)$ associated to $(\psi_p^{-1}(\bar{\beta}), \psi_p^{-1}(\mathcal{B}))$. Let $(\beta, \psi^*(\mathcal{B}))$ denote such a point in $S_r(A/k)$. □

We proceed recalling some more definitions and results from [**5**] and [**8**]. Let k be a field, Γ a subgroup of the additive group of $\mathbf{R}$such that $1 \in \mathbf{Z}$ belongs to Γ. In particular, Γ is a totally ordered abelian additive group. Let

$$k((U_\Gamma)) = \{s : \Gamma \to k; \quad \mathrm{ex}\,(s) \text{ is well-ordered}\}$$

where $\mathrm{ex}\,(s) = \{\gamma \in \Gamma : s(\gamma) \neq 0\}$ is the *set of exponents of* s, (often called the *support of* s). Notice that $\mathrm{ex}\,(s)$ is countable, since $\Gamma \subseteq \mathbf{R}$. As $1 \in \mathbf{Z}$ belongs to Γ, then U^1, the *characteristic function of* 1, belongs to $k((U_\Gamma))$,

$$U^1 : \Gamma \to k, \quad U^1(1) = 1, \quad U^1(\gamma) = 0, \quad \forall \gamma \neq 1.$$

It is usually denoted U or U_Γ and interpreted as a transcendental element over k.

Associated to the constant group homomorphism mapping every element in Γ onto the identity automorphism of k, a field structure is defined on $k((U_\Gamma))$, so that it agrees with the usual "power series notation" for the elements of $k((U_\Gamma))$:

$$s = \sum_{\gamma \in \Gamma} s_\gamma U^\gamma,$$

where $s_\gamma = s(\gamma)$. In particular, the γ-th power of U coincides with the characteristic function of γ.

For instance, $\mathbf{R}((U_\mathbf{Z}))$ is the well-known field of *formal Laurent series* with real coefficients.

If $\operatorname{ex}(s)$ is an infinite set, let $\gamma^* \in \mathbf{R} \cup \{+\infty\}$ be the smallest cluster point of $\operatorname{ex}(s)$, ($\gamma^* = \lim a_i$, where a_i is the strictly increasing sequence defined by $a_0 = \min \operatorname{ex}(s)$ and $a_i = \min \operatorname{ex}(s) \setminus \{a_0, a_1, \dots, a_{i-1}\}$). Then the formal power series

$$s_1 = \sum_{\gamma < \gamma^*} s_\gamma U^\gamma$$

is called the *initial part of s of type ω*. If $\operatorname{ex}(s)$ is a finite set, let $s_1 = s$. If $\operatorname{ex}(s)$ is finite or has exactly one cluster point in $\mathbf{R} \cup \{+\infty\}$, then we say that s is *of type ω*. For example, every Puiseux series is of type ω.

There is a *natural valuation* v on $k((U_\Gamma))$, whose residue field is k and whose value group is Γ:

$$v(s) = \min \operatorname{ex}(s).$$

Let k be an ordered field and $K = k((U_\Gamma))$ a field of formal power series. Let 0_+ denote the natural order of K considered in [**5**, VIII, 5, Cor. 11]. It is defined by the condition:

$$s \in 0_+ \Leftrightarrow c \geq 0 \text{ in } k,$$

where $s = cU^{\gamma_0}(1+t)$, $c \in k$, $\gamma_0 \in \Gamma$ and $t \in K$ with $v(t) \geq 1$. The order 0_+ makes U positive and K non-archimedean over k. Indeed, U is smaller than any positive $c \in k$. The valuation ring of K/k associated to this natural valuation is K_k, the *convex hull of k in K*, i.e.,

$$K_k = \{s \in K : \exists a \in k, -a < s < a\}.$$

In the sequel, every field of formal power series will be tacitly endowed with the order 0_+, unless otherwise stated. For example, if $K = \mathbf{R}((U_\Gamma))((U_{\Gamma'}))$, then we consider in $\mathbf{R}((U_\Gamma))$ the order 0_+ and then in K the order 0_+ again.

PROPOSITION 1.3. *Let k be an ordered field, Γ a subgroup of the additive group of $\mathbf{R}$ such that $1 \in \Gamma$ and K the field of formal power series $k((U_\Gamma))$. Then the fiber of $S_r(K/k)$ over 0_+ consists of $(0_+, K_k)$ and $(0_+, K)$.*

PROOF. If $\mathcal{B}$ is a valuation ring of K/k compatible with 0_+, then $K_k \subseteq \mathcal{B} \subseteq K$. Since the rank of K_k is 1 and the rank of K is 0, then either $\mathcal{B} = K_k$ or $\mathcal{B} = K$. □

PROPOSITION 1.4. *Let k be a field of characteristic 0, Γ a subgroup of the additive group of* $\mathbf{R}$ *with finite rational rank and $1 \in \Gamma$, $U = U_\Gamma$ and*

$$\xi(U) = a_0 U^{\gamma_0} + a_1 U^{\gamma_1} + \cdots, \quad a_0 \neq 0, \quad \gamma_0 < \gamma_1 < \cdots, \quad \gamma_i \in \Gamma,$$

a formal power series in $K = k((U))$ of type ω. Suppose that there exist $1, t_1, t_2, \ldots, t_r \in \mathbf{R}$ such that each $\gamma_i \in ex(\xi)$ is a linear combination over $\mathbf{Q}$ of $1, t_1, t_2, \ldots, t_r$. Then the natural valuation on the field $k(U, U^{t_1}, U^{t_2}, \ldots, U^{t_r}, \xi)$ has a valuation group generated over $\mathbf{Q}$ by $\{1, t_1, t_2, \ldots, t_r\} \cup ex(\xi)$.

PROOF. The natural valuation on $k(U, U^{t_1}, U^{t_2}, \ldots, U^{t_r}, \xi)$ is given by $v(U) = 1$, $v(U^{t_i}) = t_i, \forall i$, $v(\xi) = \gamma_0$, $v(f) = \min \mathrm{ex}(f)$, for $f \in k[U, U^{t_1}, U^{t_2}, \ldots, U^{t_r}, \xi]$ and $v(f/g) = v(f) - v(g)$ for $f, g \in k[U, U^{t_1}, U^{t_2}, \ldots, U^{t_r}, \xi]$. The proofs of [**7**, Th. 3 and 4] carry out for any series of type ω. □

2. Description of the points of $S_r(A/\mathbf{R})$ near $(0,0)$, for $A = \mathbf{R}[X,Y]$

Let $k = \mathbf{R}$ and $A = \mathbf{R}[X,Y]$, where X, Y are algebraically independent over $\mathbf{R}$. Each point in $\mathrm{Spec}_r A$ is given by an $\mathbf{R}$-homomorphism

$$\psi : A \to K$$

where K is $\mathbf{R}$ or an ordered field of formal power series extending $\mathbf{R}$. Three different fields of formal power series K are involved: $F = \mathbf{R}((U))$, $H = \mathbf{R}((U))((T))$ and $E = \mathbf{R}((T))((U))$, where $U = U_{\mathbf{R}}$, $T = U_{\mathbf{Z}}$. All of them are endowed with the order 0_+.

In this section, we study the points of $S_r(A/\mathbf{R})$ near $(0,0)$, i.e., those points satisfying

(i) $\exists d \in \mathbf{R}, \quad d > 0, \quad |Y| < d|X|$,
(ii) $|X| < c, \quad \forall c \in \mathbf{R}, \quad c > 0$.

In addition, we can reduce to the case $|X| = X$, by considering the $\mathbf{R}[Y]$-automorphism of A mapping X to $-X$.

Let us pick a point in $\mathrm{Spec}_r A$ verifying (i)–(ii) and investigate its fiber in $S_r(A/\mathbf{R})$. The points of $\mathrm{Spec}_r A$ are classified according to the height of their support which, obviously is at most 2. Several cases arise:

CASE 2. Let $\delta \in \mathrm{Spec}_r A$, $\mathrm{ht}(\mathrm{supp}\,\delta) = 2$. Then $\mathrm{supp}\,\delta = (X,Y)$ and $\kappa(\delta) \simeq \mathbf{R}$. Composing it with the canonical map

$$\pi : A \to A/\mathrm{supp}\,\delta \to \kappa(\delta),$$

provides an $\mathbf{R}$-homomorphism

$$\psi : A \to \mathbf{R}.$$

Then $(\delta, \psi^*(\mathbf{R}))$—the trivial point over δ— is the only point in $S_r(A/\mathbf{R})$ over δ.

CASE 1. Let $\gamma \in \mathrm{Spec}_r A$, $\mathrm{ht}(\mathrm{supp}\,\gamma) = 1$. Then the real closure of $\kappa(\gamma)$ can be embedded into the above-named field F (each element of $\kappa(\gamma)$ is mapped to a Puiseux series on U, with coefficients in $\mathbf{R}$, algebraic over $\mathbf{R}(U)$). We get an $\mathbf{R}$-homomorphism

$$\psi : A \to F$$

by composing the mentioned embedding with the canonical map

$$\pi : A \to A/\operatorname{supp}\gamma \to \kappa(\gamma),$$

such that $\psi(X) = U$ and $\gamma = \psi^{-1}(0_+)$. By propositions 1.2 and 1.3, the points of $S_r(A/\mathbf{R})$ over γ are the following and they specialize as indicated by the arrow:

$$(\gamma, \psi^*(F))$$
$$\downarrow$$
$$(\gamma, \psi^*(F_{\mathbf{R}}))$$

The point at the top is the trivial point over γ and the point at the bottom is discrete of rank 1 and dimension 0. Indeed, $\psi(Y) = \xi(U)$ is some root of a polynomial generating the ideal $\operatorname{supp}\gamma$, so that it is a Puiseux series on U, with coefficients in $\mathbf{R}$. Therefore, $\exists n \in \mathbf{N}$ such that $\operatorname{ex}(\xi) \subseteq \{m/n : m \in \mathbf{Z}\} \subseteq \mathbf{Q}$. Thus, $\{1\} \cup \operatorname{ex}(\xi)$ generates a discrete subgroup of $\mathbf{R}$; then apply proposition 1.4.

CASE 0. Let $\beta \in \operatorname{Spec}_r A$ have $\operatorname{ht}(\operatorname{supp}\beta) = 0$. Such a β corresponds to an order $\bar{\beta}$ of the field $\mathbf{R}(X,Y)$. Following [**1**], $\beta = \psi^{-1}(0_+)$ where

$$\psi : A \to K$$

is an $\mathbf{R}$-homomorphism with

(1) $K = F, H$ or E,
(2) $\psi(X) = U$,
(3) $\psi(Y)$ is transcendental over $\mathbf{R}(U)$ and possibly depends on
  (a) a formal power series $\xi(U) \in F$ with rational exponents and order ≥ 1,
  (b) numbers $\sigma, \rho \in \{1, -1\}$,
  (c) an upper bound $\theta \in \mathbf{R} \cup \{+\infty\}$ for the set $\operatorname{ex}(\xi_1)$.

The $\mathbf{R}$-homomorphisms ψ, and consequently the prime cones β, fit one of the following subcases:

CASE 0 A. $\psi : A \to F$ and $\psi(Y) = \xi(U)$ is a formal power series, transcendental over $\mathbf{R}(U)$. By propositions 1.2 and 1.3, the points of $S_r(A/\mathbf{R})$ on the fiber over β are the following, and they specialize as indicated by the arrow:

$$(\beta, \psi^*(F))$$
$$\downarrow$$
$$(\beta, \psi^*(F_{\mathbf{R}}))$$

The point at the top is the trivial point over β and the point at the bottom has rank 1 and dimension 0. The value group of $\psi^*(F_{\mathbf{R}})$ is the subgroup of $\mathbf{R}$ generated by $\{1\} \cup \operatorname{ex}(\xi_1)$. It is discrete if and only if $\operatorname{ex}(\xi_1)$ is discrete.

CASE 0 B. $\psi : A \to H$ and $\psi(Y) = \xi(U) + \sigma T$, where $\xi(U)$ is a formal power series, algebraic over $\mathbf{R}(U)$. By propositions 1.2 and 1.3, the points of $S_r(A/\mathbf{R})$ on the fiber over β are the following, and they specialize as indicated by the arrows:

$$(\beta, \psi^*(H))$$
$$\downarrow$$
$$(\beta, \psi^*(H_F))$$
$$\downarrow$$
$$(\beta, \psi^*(H_{\mathbf{R}}))$$

The point at the top is the trivial point over β. The point in the middle is a prime divisor of the first kind (see §3), and the point at the bottom has rank 2, dimension 0 and is discrete. The value group of $\psi^*(H_F)$ is indeed isomorphic to $\mathbf{Z}$, since it is generated by the set of exponents of $\psi(Y)$, as a function of T. The value group of $\psi^*(H_{\mathbf{R}})$ is isomorphic to $\mathbf{Z} \times \Gamma$, where Γ is the subgroup of $\mathbf{R}$ generated by $\{1\} \cup \operatorname{ex}(\xi) \subset \mathbf{Q}$. Since $\xi(U)$ is a Puiseux series on U algebraic over $R(U)$, then arguing as in case Case 1, $\Gamma \simeq \mathbf{Z}$.

CASE 0 C. $\psi : A \to F$ and $\psi(Y) = \xi(U) + \sigma U^{\theta}$, where $\operatorname{ex}(\xi)$ is finite and $\theta \in \mathbf{R} \setminus \mathbf{Q}$. By propositions 1.2 and 1.3, the points of $S_r(A/\mathbf{R})$ on the fiber over β are the following, and they specialize as indicated by the arrow:

$$(\beta, \psi^*(F))$$
$$\downarrow$$
$$(\beta, \psi^*(F_{\mathbf{R}}))$$

The point at the top is the trivial point over β and the point at the bottom has rank 1, rational rank 2, dimension 0 and is indiscrete. The value group of $\psi^*(F_{\mathbf{R}})$ is the subgroup of $\mathbf{R}$ generated by $\{1, \theta\} \cup \operatorname{ex}(\xi)$.

CASE 0 D. $\psi : A \to E$ and $\psi(Y) = \xi(U) + \sigma T^{\rho} U^{\theta}$, where $\operatorname{ex}(\xi)$ is finite and $\theta \in \mathbf{Q}$. By propositions 1.2 and 1.3, the points of $S_r(A/\mathbf{R})$ on the fiber over β are the following, and they specialize as indicated by the arrows:

$$(\beta, \psi^*(E))$$
$$\downarrow$$
$$(\beta, \psi^*(E_{\mathbf{R}((T))}))$$
$$\downarrow$$
$$(\beta, \psi^*(E_{\mathbf{R}}))$$

The point at the top is the trivial point over β. The point in the middle is a prime divisor of the second kind (see §3), and the point at the bottom has rank 2, dimension 0 and is discrete. The value group of $\psi^*(E_{\mathbf{R}((T))})$ is the subgroup Γ of $\mathbf{R}$ generated by $\{1, \theta\} \cup \operatorname{ex}(\xi) \subset \mathbf{Q}$, and being $\operatorname{ex}(\xi)$ finite, then $\Gamma \simeq \mathbf{Z}$.

The value group of $\psi^*(E_{\mathbf{R}})$ is isomorphic to $\mathbf{Z} \times \mathbf{Z}$, as the second factor is the group generated by ρ.

3. Geometric interpretation of the points in $S_r(A/\mathbf{R})$ near $(0,0)$, for $A = \mathbf{R}[X,Y]$

Now, we show that the points in $S_r(A/\mathbf{R})$ near $(0,0)$ have natural interpretations, in terms of half-branches of plane curves.

CASE 2. $\delta \in \mathrm{Spec}_r A$ is such that $\mathrm{supp}\,\delta = (X,Y)$ and corresponds to the point $(0,0)$ in $\mathbf{R}^2$. Then for $f \in A$,

$$f \in \delta \Leftrightarrow f(0,0) \geq 0 \quad \text{in } \mathbf{R}.$$

$\kappa(\delta) \simeq \mathbf{R}$ and this is the only valuation ring of $\kappa(\delta)$ which is compatible with $\bar{\delta}$.

CASE 1. $\gamma \in \mathrm{Spec}_r A$ has as support the prime ideal corresponding to an algebraic curve $\mathcal{V} \subseteq \mathbf{R}^2$, defined and irreducible over $\mathbf{R}$, and γ corresponds to a half-branch $\mathcal{H}$ of $\mathcal{V}$ at the point $(0,0) \in \mathcal{V}$, [**3**, Prop. 10.3.3]. The pair $(U, \xi(U))$ gives a parametrization of $\mathcal{H}$. Then for $f \in A$,

$$f \in \gamma \Leftrightarrow \exists \epsilon \in \mathbf{R} \quad \epsilon > 0 \quad f(U, \xi(U)) \geq 0 \quad \forall 0 \leq U \leq \epsilon.$$

The valuations of $\kappa(\gamma)$ compatible with $\bar{\gamma}$ are the trivial one and the one measuring the orders of the power series in U given by substitution of x by U and of y by $\xi(U)$ in the elements of $\kappa(\gamma) = \mathbf{R}(x,y)$. In other words, the latter measures proximity to the algebraic half-branch $\mathcal{H}$.

CASES 0 A AND 0 C. $\beta \in \mathrm{Spec}_r A$ is such that the point $(U, \psi(Y))$ describes a half-branch $\mathcal{H}$ at $(0,0)$ of a non-algebraic curve $\mathcal{V} \subseteq \mathbf{R}^2$, defined over $\mathbf{R}$. Then for $f \in A$,

$$f \in \beta \Leftrightarrow f \geq 0 \quad \text{along } \mathcal{H}.$$

In addition to the trivial valuation of $\kappa(\beta)$, there is the $\bar{\beta}$–compatible valuation which measures the orders of the power series in U given by substitution in the rational functions of $\kappa(\beta) = \mathbf{R}(X,Y)$ of X by U and of Y by $\xi(U)$. This valuation measures proximity to the non-algebraic half-branch $\mathcal{H}$.

CASE 0 B. $\beta \in \mathrm{Spec}_r A$ is such that the point $(U, \xi(U))$ describes a half-branch $\mathcal{H}$ at $(0,0)$ of an algebraic curve $\mathcal{V} \subseteq \mathbf{R}^2$, defined over $\mathbf{R}$. The number σ chooses one of the two sides of $\mathcal{H}$. Then

$$f \in \beta \Leftrightarrow f \geq 0 \quad \text{on the chosen side of } \mathcal{H}.$$

Apart from the trivial one, there are two valuations in $\kappa(\beta)$ compatible with $\bar{\beta}$:

(a) one measures the vanishing order of a rational function at the half-branch $\mathcal{H}$; it is a prime divisor, i.e., it has dimension 1, with center $\mathcal{V}$ in $\mathbf{R}^2$ and, being $\mathcal{V}$ of dimension 1, it is of the first kind,

(b) the other one measures the vanishing order at $\mathcal{H}$ and, in addition, proximity to the point $(0,0)$.

CASE 0 D. $\beta \in \operatorname{Spec}_r A$ is such that the point $(U, \xi(U) + \sigma T^\rho U^\theta)$ describes a half-branch $\mathcal{H}$ at $(0,0)$ of an algebraic curve $\mathcal{V} \subseteq \mathbf{R}(T)^2$, defined and irreducible over $\mathbf{R}(T)$. The description of the $\bar{\beta}$–compatible valuations of $\kappa(\beta)$ is just like that of Case 0 B, with the exception that here the prime divisor is of the second kind, as its center in $\mathbf{R}^2$ is the point $(0,0)$, of dimension 0.

4. Further specialization relations in $S_r(A/\mathbf{R})$ near $(0,0)$

In §2, we have explained the specializations among points of $S_r(A/\mathbf{R})$ near $(0,0)$ which are on the same fiber, i.e., specializations of type (a) in remark 1.1. Now, let us consider those of type (b), i.e., specializations among points of $S_r(A/\mathbf{R})$ near $(0,0)$ which are on *different* fibers.

Specialization between trivial points. If $\beta \to \beta'$, $\beta \neq \beta'$, and we take $\mathcal{D} = \mathcal{B} = \kappa(\beta)$ in 1.1 (b), then $\mathcal{C} = \kappa(\beta')$. Thus, if $\beta \to \beta'$ in $\operatorname{Spec}_r A$ then the trivial points over β and β' specialize accordingly.

Specialization between convex hulls. If $\beta \to \beta'$, $\beta \neq \beta'$, let $\mathcal{B} = \kappa(\beta)$. The field $\mathbf{R}$ is a common subfield of $\kappa(\beta)$ and $\kappa(\beta')$ and if, in remark 1.1 (b), we consider $\mathcal{D} = \kappa(\beta)_{\mathbf{R}}$, then $\mathcal{C} = \kappa(\beta')_{\mathbf{R}}$. Thus, if $\beta \to \beta'$ in $\operatorname{Spec}_r A$, then the points associated to the convex hull of $\mathbf{R}$ on the fiber over β and on the fiber over β' specialize accordingly.

We apply these remarks to the various specialization chains $\delta \leftarrow \beta$, $\delta \leftarrow \gamma \leftarrow \beta$ in $\operatorname{Spec}_r A$, where $A = \mathbf{R}[X,Y]$, thus finding more specializations among the points in $S_r(A/\mathbf{R})$ near $(0,0)$. The specializations between trivial points or between convex hulls of $\mathbf{R}$ are indicated by starred arrows; all the other arrows have already been explained or follow by transitivity.

Let $\delta \in \operatorname{Spec}_r A$ be as in Case 2 and $\beta \in \operatorname{Spec}_r A$ as in Cases 0 A or 0 C. Then $\delta \leftarrow \beta$ in $\operatorname{Spec}_r A$ and so, in $S_r(A/\mathbf{R})$ we have

$$\begin{array}{ccc} & & (\beta, \psi_\beta^*(F)) \\ & \overset{\star}{\swarrow} & \downarrow \\ (\delta, \psi_\delta^*(\mathbf{R})) & \overset{\star}{\leftarrow} & (\beta, \psi_\beta^*(F_{\mathbf{R}})) \end{array}$$

Let $\delta \in \operatorname{Spec}_r A$ be as in Case 2, $\gamma \in \operatorname{Spec}_r A$ as in Case 1 and $\beta \in \operatorname{Spec}_r A$ as in Case 0 B. Suppose that $\psi_\beta(Y) = \xi(U) + \sigma T$, where $\xi(U) = \psi_\gamma(y)$, $y = Y + \operatorname{supp}\gamma$. Then $\delta \leftarrow \gamma \leftarrow \beta$ in $\operatorname{Spec}_r A$ and so, in $S_r(A/\mathbf{R})$ we have

$$\begin{array}{ccccc} & & & & (\beta, \psi_\beta^*(H)) \\ & & & \overset{\star}{\swarrow} & \downarrow \\ & & (\gamma, \psi_\gamma^*(F)) & \leftarrow & (\beta, \psi_\beta^*(H_F)) \\ & \overset{\star}{\swarrow} & \downarrow & \swarrow & \downarrow \\ (\delta, \psi_\delta^*(\mathbf{R})) & \overset{\star}{\leftarrow} & (\gamma, \psi_\gamma^*(F_{\mathbf{R}})) & \overset{\star}{\leftarrow} & (\beta, \psi_\beta^*(H_{\mathbf{R}})) \end{array}$$

Let $\delta \in \operatorname{Spec}_r A$ be as in Case 2 and $\beta \in \operatorname{Spec}_r A$ as in Case 0 D. Then $\delta \leftarrow \beta$ in $\operatorname{Spec}_r A$ and so, in $S_r(A/\mathbf{R})$ we have

$$\begin{array}{ccc} & & (\beta, \psi^*_\beta(E)) \\ & \overset{\star}{\swarrow} & \downarrow \\ & & (\beta, \psi^*_\beta(E_{\mathbf{R}((T))})) \\ & \swarrow & \downarrow \\ (\delta, \psi^*_\delta(\mathbf{R})) & \overset{\star}{\leftarrow} & (\beta, \psi^*_\beta(E_{\mathbf{R}})) \end{array}$$

5. The rest of the points of $S_r(A/\mathbf{R})$

By means of certain automorphisms of A or of $\mathbf{R}(X, Y)$, we reduce the study of the points of $S_r(A/\mathbf{R})$ which are *not* near $(0,0)$ to the analysis in §2–§4. Let us explain the changes to be done, when conditions (i)–(ii) in §2 are eliminated.

(1) If $\forall d \in \mathbf{R}$ is $d|X| < |Y|$, then consider the $\mathbf{R}$-automorphism of A mapping X to Y and Y to X. Interchanging the roles of the two variables, the analysis can be carried through as before.

(2) If $\exists a \in \mathbf{R}$ such that $|X - a| < c$, $\forall c \in \mathbf{R}$, $c > 0$, then the $\mathbf{R}[Y]$-automorphism of A mapping X to $X - a$, reduces the study to §2–§4. The ideal (X, Y) must be changed into $(X - a, Y)$ and the point $(0,0)$ into $(a, 0)$.

(3) If $|X| > c$, $\forall c \in \mathbf{R}$, then the $\mathbf{R}(Y)$-automorphism of $\mathbf{R}(X, Y)$ mapping X to X^{-1} transforms A into $A' = \mathbf{R}[X^{-1}, Y]$. Notice that any point of $S_r(A'/\mathbf{R})$ with support of height 2 will provide no point in $S_r(A/\mathbf{R})$. Thus, from §2–§4 we see that the points of $S_r(A/\mathbf{R})$ which are in this situation and the specializations among them are as follows: if β is in Cases 0 A or 0 C, then

$$\begin{array}{c} (\beta, \psi^*_\beta(F)) \\ \downarrow \\ (\beta, \psi^*_\beta(F_{\mathbf{R}})) \end{array}$$

if γ is in Case 1 and β is in Cases 0 B, then

$$\begin{array}{ccc} & & (\beta, \psi^*_\beta(H)) \\ & \swarrow & \downarrow \\ (\gamma, \psi^*_\gamma(F)) & \leftarrow & (\beta, \psi^*_\beta(H_F)) \\ \downarrow & \swarrow & \downarrow \\ (\gamma, \psi^*_\gamma(F_{\mathbf{R}})) & \leftarrow & (\beta, \psi^*_\beta(H_{\mathbf{R}})) \end{array}$$

and finally, if β is in Case 0 D, then

$$\begin{array}{c}(\beta, \psi_\beta^*(E)) \\ \downarrow \\ (\beta, \psi_\beta^*(E_{\mathbf{R}((T))})) \\ \downarrow \\ (\beta, \psi_\beta^*(E_{\mathbf{R}}))\end{array}$$

The geometric interpretation of these points is now explained with half-branches $\mathcal{H}$ going to infinity.

References

1. M. E. Alonso, J. M. Gamboa, and J. M. Ruiz, *On orderings in real surfaces*, J. Pure and Appl. Math. **36** (1985), 1–14.
2. R. Baer, *Über nicht-Archimedisch geordnete Körper, S. B.*, Heidelberger Akad. Wiss. Math. Natur. **8** (1927), (Beiträge zur Algebra 1).
3. J. Bochnak, M. Coste, and M.-F. Roy, *Géométrie algébrique réelle*, Ergebnisse der Math. 3, Folge Band 12, Springer-Verlag, Berlin-Heidelberg-New York, 1987.
4. G. Brumfiel, *Partially ordered rings and semi-algebraic geometry*, LMS Lecture Note Series, vol. 37, Cambridge University Press, 1979.
5. L. Fuchs, *Partially ordered algebraic systems*, Internat. Series Monog. Pure and Appl. Math., vol. 28, Pergamon Press, Oxford-London-New York-Paris, 1963.
6. W. Krull, *Allgemeine Bewertungstheorie*, J. Reine Angew. Math. **167** (1931), 160–196.
7. S. MacLane and O. F. G. Schilling, *Zero-dimensional branches of rank one on algebraic varieties*, Ann. of Math. **40** (1939), 507–520.
8. A. Prestel, *Lectures on formally real fields*, Lecture Notes in Math., vol. 1093, Springer-Verlag, Berlin-Heidelberg-New York, 1984.
9. M. J. de la Puente, *Riemann surfaces of a ring and compactifications of semi-algebraic sets,*, Ph.D. thesis, Stanford Univ., 1988.
10. ———, *The real Riemann surface of a ring*, prépublications de Équipe de Logique Mathématique, Séminaire de Structures Algébriques Ordonnées, 1988–1989 (F. Delon, M. Dickmann et D. Gondard, eds.), No. 2, Univ. Paris VII, 1990.
11. O. Zariski, *The reduction of the singularities of an algebraic surface*, Ann. of Math. **40** (1939), 639–689.
12. O. Zariski and P. Samuel, *Commutative algebra, vol. II*, GTM, vol. 29, Springer-Verlag, Berlin-Heidelberg-New York, 1960.

Departamento de Algebra, Fac. de Matemáticas, Universidad Complutense, 28040–Madrid, Spain

Contemporary Mathematics
Volume **155**, 1994

Estimates for Parametric Nonuniformity in Representations of a Definite Polynomial as a Sum of Fourth Powers

GILBERT STENGLE

Eberhard Becker [**1**] has developed an extended Artin-Schreier theory of fields which generalizes the relation between sums of squares and orderings of a real field to sums of $2n$-th powers. In the context of this theory the simple polynomial $P_0 = y^4 + ty^2 + 1$, viewed as a one-parameter family of elements of $\mathbb{R}[y]$, presents some interesting features. Becker's theory shows that for each fixed positive real t this is an element of $\mathbb{R}[y]$ which is a sum of fourth powers of rational functions (or, more briefly for purposes of this note, "has a representation"). However Prestel [**4, Theorem 2**] shows that such representation of P_0 must depend badly on t in the sense that as t increases without bound so must either the number of summands or the degrees of the polynomials appearing in the representation. He proves by model-theoretic reasoning (also by an unpublished alternate compactness argument) that otherwise the polynomial $z^4 + z^2$, which partially characterizes behavior of P_0 if t and y are both large, would also have a representation. But the double real zero at $z = 0$ precludes this since multiplicities divisible by four are a more or less evident necessary condition. He then concludes that in representations with a fixed number of summands, which are known to exist, the degrees necessarily tend to infinity. This adverse dependence on t contrasts strongly with known properties of representations of real polynomials as sums of squares. Such representability is equivalent to semidefiniteness (by the affirmative answer to Hilbert's seventeenth problem) and consequently is definable by the statement

$$\{\forall x_1, ..., x_n \exists a \text{ such that } P(x_1, ..., x_n) = a^2\}$$

1991 *Mathematics Subject Classification.* Primary:11E10 Secondary: 12J15,12Y05.

Research at MSRI is supported in part by NSF grant DMS-8505550. This paper is submitted in final form and no version of it will be submitted for publication elsewhere.

which is elementary in the coefficients of a represented polynomial P. The behavior of P_0 shows that there cannot be a corresponding elementary statement characterizing representability by sums of fourth powers over an arbitrary real closed field. This note gives some bounds for the nonuniformity in t of rational functions in representations of P_0. We show, by combining Prestel's results with an argument of Hilbert, that the degrees must tend to infinity in any family of representations parameterized by t. We also bound this growth for a certain special algorithmically computable family of represention.

THEOREM 1. *Let* $y^4 + ty^2 + 1 = \sum\{f_k/f_0\}^4, t \geq 0$ *where* $f_i \in \mathbb{R}[y,t]$. *Let* $D(t)$ *be the minimum degree of* f_0 *in any such representation. Then:*

a) $D(t)$ *tends to infinity with* t,

b) there is a constant C *for which the degrees of the* f_k *and the number of summands are bounded by* $C \log t$ *for all large* t.

PROOF. a) By Theorem 2 of [**4**] either the degree of f_0 or the number of summands must become large with t. It thus suffices to show that the degree of f_0 cannot remain bounded or, equivalently, that the degrees of the numerator cannot remain bounded. This follows by an argument of Hilbert [**3**]. If the latter degrees were bounded, then for each fixed t the fourth powers of polynomials appearing in the numerator would be elements of a finite dimensional vector space. If the number of summands exceeded this dimension, then a nontrivial linear combination of them would represent 0. This combination could be normalized to contain at least one negative coefficient and to make its most negative coefficient be -1. Adding this form of 0 to the numerator would then give a representation of P_0 with fewer summands and degrees satisfying the same bounds. In this fashion a family of representations with bounded degrees could be modified to yield a family with degrees and number of summands both bounded, which is impossible.

b) The rest of the proof will depend on a recursive algorithm, given t, for actually calculating a representation. It is convenient to associate to the polynomial with positive coefficients $P = ay^4 + by^2 + c$ the parameter $\mu(P) = 36ac/b^2$. The change of variables $P \to 36ab^{-2}F, y \to (b/6a)^{1/2}y$ then transforms P into the form $F(m) = y^4 + 6y^2 + m$ where $m = \mu(P)$. Obviously two polynomials of the form P are equivalent as far as representability if they share a common value of μ. We indicate this equivalence by $\sim$. The parameter m will serve as a numerical measure of the difficulty of representing F in this sense: the smaller the value of m, the higher the degree of the polynomials needed to represent F. An extreme case occurs if $m \geq 1$. Then the identity $F(m) = \{(y+1)^4 + (y-1)^4\}/2 + m - 1$ shows that $F(m)$ is a sum of fourth powers of polynomials. Thus it suffices to restrict attention to the range $0 < m < 1$. We next show that by introducing factors equivalent to $F(1)$ as denominators we can effectively magnify the salient value of m. Consider

$$F(1)F(m) = y^8 + 12y^6 + (37+m)y^4 + 6(1+m)y^2 + m = Q + R$$

where $Q = y^4(y^4 + 12y^2 + 4)$ and $R = (33 + m)y^4 + 6(1 + m)y^2 + m$. Since

$$\mu(y^4 + 12y^2 + 4) = 1,$$

Q is a sum of fourth powers of polynomials. Also

$$\mu(R) = m'(m) = (33 + m)m/(1 + m)^2.$$

Simple calculation shows that $m' \geq 1$ if and only if $m \geq 1/31$. For $m < 1/31$ the function $(33 + m)/(1 + m)^2$ is monotonically decreasing. Hence, for this latter range of m,

$$m' \geq m(33 + 1/31)/(1 + 1/31)^2 = 31m.$$

We conclude that $F(1)F(m) = Q + R$ where Q is a sum of fourth powers of polynomials and $R \sim F(m')$ where $m' \geq min(1, 31m)$.

We next establish by induction that if $31^{-k-1} \leq m < 31^{-k}$ then $F(m)$ has a representation with degree $(f_0) = 4(k + 1)$. For $k = 0$ the preceding estimate implies

$$F(m) = (Q + R)/F(1) = F(1)^3(Q + R)/F(1)^4.$$

Since $R \sim F(m')$ with $m' \geq 1, R$ is a sum of fourth powers of polynomials. So are $F(1)$ and Q. Since sums of fourth powers form a semiring, this is a representation of the required form. Now assume $31^{-k-2} \leq m < 31^{-k-1}$. Then $F(1)F(m) = Q+R$ where $R \sim F(M')$ with $31^{-k-1} \leq m' < 31^{-k}$. By the induction hypothesis $F(M')$ has a representation and therefor so does the equivalent R, say $R = \sum\{g_k/g_0\}^4$ with degree $(g_0) = 4(k + 1)$. Then

$$F(m) = F(1)^3\{Q(g_0)^4 + \sum(g_k)^4\}/\{F(1)g_0\}^4.$$

This is a representation with $f_0 = F(1)g_0$ of degree $4(k + 2)$. To obtain the conclusion of the theorem use $\mu(P_0) = 36/t^2$. Then P_0 has a representation with degree $(f_0) = 4(k+1)$ provided $31^{k-1} \leq 36/t^2 < 31^{-k}$. This gives an upper bound for k if $36/t^2 < 31^{-k}$ of the form

$$\text{degree } (f_0) = 4(k + 1) < 4 + 4[2\,log^+(t/6)/log\,31].$$

This in turn implies the less explicit estimate $C\,log\,t$ of the theorem for f_0 and therefore for the other f_k as well. Finally, since the dimension of the space of polynomials containing the fourth powers in the numerator is linear in their degree, Hilbert's argument above shows that the number of summands can be economized to $C\,log\,t$. □

It is clear that our reasoning depends on the archimedean character of the reals but applies to more general fields provided t is not infinitely large. By keeping more careful track of our algorithm it is also easy to estimate numbers of summands in our representations. These also tend to infinity with t. It is known [**2**] that no more than 24 summands are needed and recently Choi, Lam and Reznick have announced that 6 suffice. Presumably in such more constrained representations the degrees will grow more rapidly. We ask on one hand if there are more general analogous algorithmic processes for calculating representations and on the other hand if the general form of our estimate can be inferred without the use of such example-specific calculation, say by model-theoretic methods.

REFERENCES

1. E. Becker, *Summen n-ter Potenzen in Körpern,*, J. Reine angew. Math. **307/308** (1979), 8-30.
2. E. Becker, *The real holomorphy ring and sums of* $2n$*-th powers*, Lecture notes in Math., no. 959 (1982), Springer-Verlag, 139-181.
3. D. Hilbert, *Über die Darstellung definiter Formen als Summe von Formenquadraten*, Math. Ann. Bd. **33** (1889), 342-350.
4. A. Prestel, *Model theory of fields: An application to semidefinite polynomials*, Soc. Math. France, 2° serié, memorie **16** (1984), 53-66.

Contemporary Mathematics
Volume 155, 1994

On Generators for the Witt Ring

JÓN KR. ARASON, RICHARD ELMAN, AND BILL JACOB

Symmetric bilinear spaces over a scheme are, in general, far from being well understood. In the case of a proper algebraic scheme over a field the Krull-Schmidt theorem holds for vector bundles on the scheme and there might be some hope of understanding.

A major problem in using the Krull-Schmidt theorem is that an indecomposable vector bundle on a proper algebraic scheme over a field may not be indecomposable after an extension of the base field. In the case of curves, Agnes Tillmann began the study of this phenomenon in her unpublished thesis [**T**]. In our paper [**AEJ3**] her results are generalized and also extended to arbitrary proper algebraic schemes. In the first section of this paper we start by reviewing some results of [**AEJ3**] and then deduce results on generators of the Witt ring of symmetric bilinear forms over such schemes.

In the second section we shall show how our results give a reasonable set of generators for the Witt ring of an elliptic curve over a field (of characteristic different from 2). It turns out that although the Witt ring is, in general, not diagonalizable, i.e., not generated by symmetric bilinear spaces of rank 1, it is (additively) generated by traces of such spaces. In a subsequent paper [**AEJ4**] we shall attempt to describe the relations between these generators. We shall describe "elementary" relations but also show that these do not, in general, generate all relations.

In [AEJ4] we shall use a connection between the Witt ring of symmetric bilinear spaces over a curve and the étale cohomology of the curve. This connection is the theme of the third section of this paper. Besides results on the Witt rings of curves over elementary fields, a noteworthy result of this section is that the Witt ring of an elliptic curve does not have to be finitely generated over the Witt ring of the base field.

1991 *Mathematics Subject Classification.* 11E81 11G20 14F05 14F20 14H52.

Richard Elman and Bill Jacob were supported by the National Science Foundation.

This paper is submitted in final form and no version of it will be submitted for publication elsewhere.

In what follows K is always a field of characteristic different from 2.

1. Proper schemes

In this section X is an algebraic scheme over K. We shall assume that X is proper over K. (Actually, we only need that the endomorphism rings of vector bundles on X are finite dimensional over K.) We shall also assume that K is perfect.

Recall that a *symmetric bilinear space* over X is a pair $(\mathcal{M}, \beta)$, where $\mathcal{M}$ is a vector bundle on X and $\beta : \mathcal{M} \to \mathcal{M}^\vee$ is a self-dual isomorphism of vector bundles from $\mathcal{M}$ to its dual bundle $\mathcal{M}^\vee$. A symmetric bilinear space $(\mathcal{M}, \beta)$ over X is said to be *metabolic* if there is a subbundle $\mathcal{U}$ of $\mathcal{M}$ such that the orthogonal subbundle $\mathcal{U}^\perp$ of $\mathcal{U}$ equals $\mathcal{U}$. It is said to be *hyperbolic* if it is isomorphic to a symmetric bilinear space of the type $(\mathcal{U} \oplus \mathcal{U}^\vee, \left[\begin{smallmatrix} 0 & 1 \\ 1 & 0 \end{smallmatrix}\right])$. Clearly, every hyperbolic symmetric bilinear space is metabolic. The *Witt ring* $W(X)$ of X is, by definition, the Grothendieck ring of symmetric bilinear spaces over X modulo the ideal of the classes of metabolic spaces over X. (For details see [**K1**].) We shall often simply say *space* instead of symmetric bilinear space.

Of course, the study of symmetric bilinear spaces over X involves the study of vector bundles on X. We shall now collect the main results we need on vector bundles on X.

First we recall that the Krull-Schmidt theorem holds for vector bundles on X. This means that every vector bundle on X has a decomposition as a direct sum of indecomposable vector bundles, unique up to isomorphisms and order of summands. Furthermore, a vector bundle $\mathcal{M}$ on X is indecomposable if and only its reduced endomorphism algebra $D(\mathcal{M})$, i.e., the endomorphism algebra $\mathrm{End}(\mathcal{M})$ of $\mathcal{M}$ modulo its Jacobson radical, is a division algebra.

Next we describe the behavior of indecomposable vector bundles on X under base field extensions. For details see [**AEJ3**].

Let L be an algebraic field extension of K and let $Y = L \times_K X$ be the scheme obtained from X by extending the base from K to L (i.e., $\mathrm{Spec}(L) \times_{\mathrm{Spec}(K)} X$). For a vector bundle $\mathcal{M}$ on X we denote by $L \otimes_K \mathcal{M}$ the extension of $\mathcal{M}$ to Y, i.e., the inverse image of $\mathcal{M}$ under the natural morphism $Y \to X$. In the case $L = \overline{K}$, the algebraic closure of K, we write $\overline{X}$ instead of $\overline{K} \times_K X$ and $\overline{\mathcal{M}}$ instead of $\overline{K} \otimes_K \mathcal{M}$. The vector bundle $\mathcal{M}$ on X is said to be *absolutely indecomposable* if $\overline{\mathcal{M}}$ is indecomposable. As $D(\overline{\mathcal{M}}) = \overline{K} \otimes_K D(\mathcal{M})$, it is clear that $\mathcal{M}$ is absolutely indecomposable if and only if $D(\mathcal{M}) = K$.

If L is finite over K and $\mathcal{N}$ is a vector bundle on Y then we denote by $\mathrm{tr}_{L/K}(\mathcal{N})$ the direct image of $\mathcal{N}$ under the natural morphism $Y \to X$. It is a vector bundle on X called the *trace* of $\mathcal{N}$. The following proposition describes the indecomposable vector bundles on X in terms of the absolutely indecomposable ones. (Cf. [**AEJ3**, Theorem 1.8].)

PROPOSITION 1.1. *Let $\mathcal{M}$ be an indecomposable vector bundle on X and let L be a maximal subfield of $D(\mathcal{M})$. Then there is an absolutely indecomposable vector bundle $\mathcal{N}$ on Y such that $\mathcal{M} = tr_{L/K}(\mathcal{N})$.*

In the case that X is a curve, this is one of the main results in Tillmann's thesis [**T**].

As we are studying symmetric bilinear spaces over X, we are particularly interested in self-dual vector bundles on X.

Every self-dual indecomposable vector bundle $\mathcal{M}$ on X carries a regular symmetric or skew-symmetric bilinear form, i.e., there is an isomorphism $\beta : \mathcal{M} \to \mathcal{M}^\vee$ with $\beta^\vee = \pm\beta$. (Cf. [**S**, Ch. 7, Theorem 4.5(i)].) Such a form β induces an involution on $\mathrm{End}(\mathcal{M})$, namely by $\varphi \mapsto \beta^{-1} \circ \varphi^\vee \circ \beta$, and hence an involution * on $D(\mathcal{M})$. We say that $\mathcal{M}$ is of the *first* or *second kind* depending on whether * is of the first or second kind. It is easy to see that the kind of $\mathcal{M}$ does not depend on the choice of β.

For self-dual vector bundles on X we then have the following extension of Proposition 1.1. (Cf. [**AEJ3**, Proposition 5.4].)

PROPOSITION 1.2. *Let $\mathcal{M}$ be a self-dual indecomposable vector bundle on X. Write $\mathcal{M} = tr_{L/K}(\mathcal{N})$ with an absolutely indecomposable vector bundle $\mathcal{N}$. Then $\mathcal{N}$ is self-dual if and only if $\mathcal{M}$ is of the first kind.*

The following alternative version is also useful. (Cf. [**AEJ3**, Remark 5.5].)

PROPOSITION 1.3. *Let $\mathcal{M}$ be an indecomposable vector bundle on X. Then $\mathcal{M}$ is self-dual of the first kind if and only if all the indecomposable summands of $\overline{\mathcal{M}}$ are self-dual.*

Let $\mathcal{M}$ still be a self-dual indecomposable vector bundle on X, let β be a regular symmetric or skew-symmetric bilinear form on $\mathcal{M}$, and let * be the involution on $D(\mathcal{M})$ induced by β. If $\mathcal{M}$ is of the second kind then * is of the second kind. In this case we also say that $\mathcal{M}$ is of *unitary type*.

Now assume that $\mathcal{M}$ is of the first kind. Then * is of the first kind. If β is symmetric then we say that $\mathcal{M}$ is of *orthogonal type* if * is of orthogonal type and of *symplectic type* if * is of symplectic type. If β is skew-symmetric then we say that $\mathcal{M}$ is of *orthogonal type* if * is of symplectic type and of *symplectic type* if * is of orthogonal type. (Cf. [**S**, Ch. 8, Definition 7.6].) One easily sees that the type of $\mathcal{M}$ is independent of the choice of β. (Cf. [**S**, Ch. 8, Remark 7.7(i)].)

With this terminology we have the following companion to Proposition 1.2. (Cf. [**AEJ3**, Proposition 5.8].)

PROPOSITION 1.4. *Let $\mathcal{M}$ be a self-dual indecomposable vector bundle of the first kind on X. Write $\mathcal{M} = tr_{L/K}(\mathcal{N})$ with an absolutely indecomposable vector bundle $\mathcal{N}$. Then there is a regular symmetric bilinear form on $\mathcal{N}$ if and only if $\mathcal{M}$ is of orthogonal type and a regular skew-symmetric bilinear form on $\mathcal{N}$ if and only if $\mathcal{M}$ is of symplectic type.*

The following alternative version is also useful. (Cf. [**AEJ3**, Remark 5.9(i)].)

PROPOSITION 1.5. *Let $\mathcal{M}$ be an indecomposable vector bundle on X. Then $\mathcal{M}$ is self-dual of orthogonal type if and only if all the indecomposable summands of $\overline{\mathcal{M}}$ carry regular symmetric bilinear forms and $\mathcal{M}$ is self-dual of symplectic type if and only if all the indecomposable summands of $\overline{\mathcal{M}}$ carry regular skew-symmetric bilinear forms.*

We now come to some general results on symmetric bilinear spaces over X. We shall say that a symmetric bilinear space is *(linearly) indecomposable* if its underlying vector bundle is indecomposable.

First we note that there is a kind of a Krull-Schmidt theorem for symmetric bilinear spaces over X. Let us say that a space over X is *isotypical* of type $\mathcal{P}$, where $\mathcal{P}$ is an indecomposable vector bundle on X, if every summand in the Krull-Schmidt decomposition of the underlying vector bundle is isomorphic to $\mathcal{P}$ or $\mathcal{P}^\vee$. Then the theorem says that every symmetric bilinear space $(\mathcal{M}, \beta)$ over X is the orthogonal sum of isotypical subspaces $(\mathcal{M}_i, \beta_i)$ of, say, type $\mathcal{P}_i$, such that $\mathcal{P}_i$ is not isomorphic to $\mathcal{P}_j$ or ${\mathcal{P}_j}^\vee$ if $i \neq j$, and that this decomposition is unique up to isomorphisms and order. (Cf. [**S**, Ch. 7, Theorems 10.8 and 10.9(i)].)

Obviously, there are two classes of isotypical spaces over X. If $(\mathcal{M}, \beta)$ is isotypical of type $\mathcal{P}$, where $\mathcal{P}$ is not self-dual, then $(\mathcal{M}, \beta)$ is isomorphic to an orthogonal sum of the hyperbolic spaces $(\mathcal{P} \oplus \mathcal{P}^\vee, \left[\begin{smallmatrix} 0 & 1 \\ 1 & 0 \end{smallmatrix}\right])$, hence is hyperbolic. (Cf. [**S**, Ch. 7, Theorem 10.9(iii)].) So, let us assume that $(\mathcal{M}, \beta)$ is isotypical of type $\mathcal{P}$, where $\mathcal{P}$ is self-dual. Assume first that $\mathcal{P}$ carries no regular symmetric bilinear form. Then $D(\mathcal{P})$ must be a field and the involution on $D(\mathcal{P})$, induced by any regular skew-symmetric bilinear form on $\mathcal{P}$, must be trivial. It follows that $(\mathcal{M}, \beta)$ is hyperbolic. (Cf. [**S**, Ch. 7, Theorems 10.9(ii) and 8.1].) Now assume that $\mathcal{P}$ carries a regular symmetric bilinear form. Then $(\mathcal{M}, \beta)$ is an orthogonal sum of indecomposable symmetric bilinear spaces each with underlying vector bundle $\mathcal{P}$. (Cf. [**S**, Ch. 7, Theorems 10.9(ii) and 6.3].)

From all this we see that every space over X is the orthogonal sum of two subspaces such that the first one has no indecomposable orthogonal summand but the second one is the orthogonal sum of indecomposable subspaces. Furthermore, this decomposition is uniquely determined up to isomorphisms. Also, the first summand is hyperbolic. Although this decomposition is well-behaved under base extension, it turns out that it does not behave well under the trace map considered below. We shall, therefore, use another decomposition.

Since every subspace is an orthogonal summand, we see that any space can be written as an orthogonal sum of a hyperbolic space and a subspace containing no hyperbolic subspace. Since Witt cancellation holds (cf. [**S**, Ch. 7, Theorem 10.9(iv)]), we get (by the additivity of the hyperbolic functor) the following stronger result.

THEOREM 1.6. *Every symmetric bilinear space over X is the orthogonal sum of a hyperbolic space and a subspace containing no hyperbolic subspace. Furthermore, this decomposition is uniquely determined up to isomorphisms. Also,* *the second summand can be written as an orthogonal sum of indecomposable symmetric bilinear spaces.*

If $(\mathcal{M}, \beta)$ is a space over X then the unique (up to isomorphism) subspace without hyperbolic subspaces given by the theorem is called the *Witt kernel* of $(\mathcal{M}, \beta)$.

In a sense, the above reduces the study of spaces over X to the study of the indecomposable ones. If the original space is *anisotropic*, i.e., has no totally isotropic subbundle, then there is no hyperbolic summand and every indecomposable summand must be anisotropic. We conclude the following.

COROLLARY 1.7. *Let $(\mathcal{M}, \beta)$ be an anisotropic space over X. Then $(\mathcal{M}, \beta)$ is its own Witt kernel and is an orthogonal sum of anisotropic indecomposable spaces.*

From the corollary it follows, in particular, that any anisotropic orthogonally indecomposable space is (linearly) indecomposable.

Of course, the Witt kernel of a space need not be anisotropic. But, as every element in $W(X)$ is represented by an anisotropic space, we have the following

COROLLARY 1.8. *The Witt ring $W(X)$ is additively generated by the classes of anisotropic indecomposable spaces over X. In particular, $W(X)$ is additively generated by the classes of indecomposable spaces over X.*

Let $(\mathcal{M}, \beta)$ be a space over X. We shall say that $(\mathcal{M}, \beta)$ is of the *first kind* if every indecomposable summand of the Witt kernel of $(\mathcal{M}, \beta)$ is of the first kind. We shall say that $(\mathcal{M}, \beta)$ is of *orthogonal type* if every indecomposable summand of the Witt kernel of $(\mathcal{M}, \beta)$ is of orthogonal type.

PROPOSITION 1.9. *Let $(\mathcal{M}, \beta)$ be a space over X. If every summand in the Krull-Schmidt decomposition of $\mathcal{M}$ is self-dual of the first kind (respectively, of orthogonal type) then $(\mathcal{M}, \beta)$ is of the first kind (respectively, of orthogonal type).*

PROOF. This follows at once from the "Krull-Schmidt theorem" described above.

It follows from the proposition that an orthogonal sum of spaces of the first kind (respectively, of orthogonal type) over X is of the first kind (respectively, of orthogonal type).

We now turn to the behavior of spaces over X under base extensions. If L is an algebraic field extension of K and $Y = L \times_K X$ then, of course, a symmetric bilinear space $(\mathcal{M}, \beta)$ on X gives rise to a symmetric bilinear space on Y, the extension $L \otimes_K (\mathcal{M}, \beta)$ of $(\mathcal{M}, \beta)$ to Y.

Now let L be a finite extension of K. Let $\mathcal{N}$ be a vector bundle on Y and let $\mathcal{M} = \mathrm{tr}_{L/K}(\mathcal{N})$. Then every regular symmetric (or skew-symmetric) bilinear form γ on $\mathcal{N}$ induces a regular symmetric (or skew-symmetric) bilinear form β on $\mathcal{M}$. (Cf. [**AEJ3**] or [**K3**].) The form β is called the *trace* of γ and we write $\beta = \mathrm{tr}_{L/K}(\gamma)$. We also write $\mathrm{tr}_{L/K}(\mathcal{N}, \gamma)$ for $(\mathrm{tr}_{L/K}(\mathcal{N}), \mathrm{tr}_{L/K}(\gamma))$. With this terminology we have the following (best possible) counterpart to Proposition 1.1 for symmetric bilinear spaces. (Cf. [**AEJ3**, Proposition 5.10].)

PROPOSITION 1.10. *Let $(\mathcal{M},\beta)$ be an indecomposable symmetric bilinear space over X. Let L be a maximal field contained in $D(\mathcal{M})^+$. Here $D(\mathcal{M})^+$ is the subspace of symmetric elements in $D(\mathcal{M})$ with respect to the involution on $D(\mathcal{M})$ induced by β. Then there is an indecomposable symmetric bilinear space $(\mathcal{N},\gamma)$ over $Y = L\times_K X$ such that $(\mathcal{M},\beta) = tr_{L/K}(\mathcal{N},\gamma)$. Furthermore, $\mathcal{N}$ is of the same type as $\mathcal{M}$ and:*

(i) *If $\mathcal{M}$ is of orthogonal type then $\mathcal{N}$ is absolutely indecomposable, i.e., $D(\mathcal{N}) = L$.*
(ii) *If $\mathcal{M}$ is of symplectic type then $D(\mathcal{N})$ is a quaternion algebra with center L.*
(iii) *If $\mathcal{M}$ is of unitary type then $D(\mathcal{N})$ is a quadratic field extension of L.*

If K has cohomological 2-dimension 0, i.e., if every finite field extension of K is quadratically closed, then it follows that every indecomposable symmetric bilinear space over X is of orthogonal type. If K has cohomological 2-dimension at most 1, e.g., a finite field, then every quaternion algebra over any finite field extension of K splits and it follows that an indecomposable symmetric bilinear space over X cannot be of symplectic type.

We now come to the behavior of the kind and type of a space under base field extensions.

PROPOSITION 1.11. *Let $(\mathcal{M},\beta)$ be a space over X. Let K' be an algebraic extension of K and let $X' = K'\times_K X$. Let $(\mathcal{M}',\beta') = K'\otimes_K(\mathcal{M},\beta)$ be the extension of $(\mathcal{M},\beta)$ to X'.*

(i) *If $(\mathcal{M},\beta)$ is of the first kind then $(\mathcal{M}',\beta')$ is of the first kind.*
(ii) *If $(\mathcal{M},\beta)$ is of orthogonal type then $(\mathcal{M}',\beta')$ is of orthogonal type.*

PROOF. We may assume that $(\mathcal{M},\beta)$ is indecomposable. But then the proposition follows at once from Proposition 1.3 (respectively, Proposition 1.5) together with Proposition 1.9. □

REMARK 1.12. Let $(\mathcal{M},\beta)$ be an indecomposable space over X of the second kind such that $D(\mathcal{M})$ is a quadratic field extension L of K. Write $\mathcal{M} = \mathrm{tr}_{L/K}(\mathcal{N})$. Then $\mathcal{N}$ is absolutely indecomposable but not self-dual. It follows easily that $L\otimes_K(\mathcal{M},\beta)$ is hyperbolic, in particular of the first kind.

Let $(\mathcal{M},\beta)$ be an indecomposable space over X of symplectic type such that $D(\mathcal{M})$ is a quaternion algebra over K. Let L be a quadratic extension of K contained in $D(\mathcal{M})$. Write $\mathcal{M} = \mathrm{tr}_{L/K}(\mathcal{N})$. Then $\mathcal{N}$ is absolutely indecomposable. Also $\mathcal{N}$ is self-dual but carries no regular symmetric bilinear form. It follows easily that $L\otimes_K(\mathcal{M},\beta)$ is hyperbolic, in particular of orthogonal type.

PROPOSITION 1.13. *Let K' be a finite extension of K, let $X' = K'\times_K X$, let $(\mathcal{M}',\beta')$ be a symmetric bilinear space over X', and let $(\mathcal{M},\beta) = tr_{K'/K}(\mathcal{M}',\beta')$.*

(i) *If $(\mathcal{M}',\beta')$ is of the first kind then $(\mathcal{M},\beta)$ is of the first kind.*

(ii) *If* $(\mathcal{M}', \beta')$ *is of orthogonal type then* $(\mathcal{M}, \beta)$ *is of orthogonal type.*

PROOF. We may assume that $(\mathcal{M}', \beta')$ is indecomposable. Write $\mathcal{M}' = \mathrm{tr}_{L'/K'}(\mathcal{N}')$ with an absolutely indecomposable vector bundle $\mathcal{N}'$. Then $\mathcal{M} = \mathrm{tr}_{K'/K}(\mathcal{M}') = \mathrm{tr}_{L'/K}(\mathcal{N}')$, hence $\overline{\mathcal{M}}$ is the direct sum of the L'/K-conjugates of $\overline{\mathcal{N}'}$. Let $\mathcal{M} \cong \mathcal{M}_1 \oplus \cdots \oplus \mathcal{M}_k$ be the Krull-Schmidt decomposition of $\mathcal{M}$. Then it follows that each $\overline{\mathcal{M}_i}$ is a direct sum of some conjugates of $\overline{\mathcal{N}'}$.

As $\mathcal{M}'$ is self-dual of the first kind (in both statements), $\mathcal{N}'$ is self-dual by Proposition 1.2, hence $\overline{\mathcal{N}'}$ is self-dual and so are all its conjugates. By Proposition 1.3, it follows that each $\mathcal{M}_i$ is self-dual of the first kind. The first statement now follows by Proposition 1.9.

If $\mathcal{M}'$ is of orthogonal type then $\mathcal{N}'$ carries a regular symmetric bilinear form by Proposition 1.4, hence $\overline{\mathcal{N}'}$ carries a regular symmetric bilinear form and so do all its conjugates. By Proposition 1.5, it follows that each $\mathcal{M}_i$ is self-dual of orthogonal type. The second statement now follows by Proposition 1.9. □

Let $W_0(X)$ be the subgroup of $W(X)$ generated by the classes of indecomposable spaces of the first kind and let $W_{00}(X)$ be the subgroup of $W(X)$ generated by the classes of indecomposable spaces of orthogonal type. One can ask whether $W_0(X)$, or even $W_{00}(X)$, is all of $W(X)$. This is true if X is a complete regular curve of genus 0 because then every indecomposable space over X is of orthogonal type. (Cf. [**AEJ3**, Example 5.16].) We shall see in the next section that this is also true if X is an elliptic curve over K and in the third section that this also holds if X is a complete regular curve over a finite field. Of course, one can also ask whether $W(K)$ is *diagonalizable*, i.e., is generated by the classes of spaces of rank one. That would imply that $W(X) = W(X)_{00}$. This need not be true even when K is a local field.

COROLLARY 1.14. *Let* K' *be a finite extension of* K *and let* $X' = K' \times_K X$. *Then the natural morphism* $W(X) \to W(X')$ *maps* $W_0(X)$ *to* $W_0(X')$ *and* $W_{00}(X)$ *to* $W_{00}(X')$ *and the trace* $W(X') \to W(X)$ *maps* $W_0(X')$ *to* $W_0(X)$ *and* $W_{00}(X')$ *to* $W_{00}(X)$.

REMARK 1.15. For applications of the results of this section to the Witt ring of X the hypothesis that K is perfect is really not restrictive. Indeed, it can be shown that the Witt ring of an algebraic scheme X over a field of characteristic not equal 2 does not change under a purely inseparable base extension. As we shall see in the next section this is easy to see if X is a regular curve. The general case is somewhat harder. The proof uses generalized traces (Scharlau transfers) for schemes.

2. Elliptic curves

In this section, we shall apply the previous one to the case that X is an elliptic curve. In particular, we shall show that the Witt ring of an elliptic curve is generated by traces of spaces of rank 1.

We begin, however, with some results for a general complete regular curve X over K. Let F be the function field of X.

It is well-known that a space over X is anisotropic if and only if it is anisotropic at the generic point. Furthermore, there is an exact sequence

$$0 \to W(X) \to W(F) \to \bigoplus_{\mathfrak{p}} W(K(\mathfrak{p})) \tag{2.1}$$

where $\mathfrak{p}$ runs through the closed points of X. (Cf. [**K1**, 13.1, Satz 13.3.6].) In particular, Witt classes in $W(X)$ are those classes in $W(F)$ whose (second) residues all vanish. From this fact it follows at once that $W(X)$ does not change under purely inseparable base extensions.

We already mentioned that $W(X)$ is not only generated by the classes of indecomposable spaces over X but even by the classes of anisotropic indecomposable spaces. Often it is easier to work only with the anisotropic ones. One reason is the following proposition. It says that if the space is anisotropic and indecomposable then the underlying vector bundle is stable in the sense of Mumford [M]. Recall that the *degree* of a vector bundle on X is the degree of its determinant bundle. Also note that the underlying vector bundle of a space is self-dual and hence must have degree 0.

PROPOSITION 2.2. *Let $(\mathcal{M}, \beta)$ be an anisotropic indecomposable space over X. Then $\mathcal{M}$ has no non-trivial proper subbundle of degree ≥ 0.*

PROOF. Let $\mathcal{U}$ be a non-trivial subbundle of $\mathcal{M}$. Let ρ be the form on $\mathcal{U}$ induced by β. As β is anisotropic, it is anisotropic at the generic point. It follows that ρ is anisotropic at the generic point. In particular, ρ is regular at the generic point. It follows that $\det(\rho) : \det(\mathcal{U}) \to \det(\mathcal{U})^\vee$ is non-trivial at the generic point, hence is non-trivial. If $\det(\mathcal{U})$ had degree > 0 then there would be no non-trivial morphisms $\det(\mathcal{U}) \to \det(\mathcal{U})^\vee$. Hence $\det(\mathcal{U})$ must have degree ≤ 0. (Note that this part of the proof only uses that $(\mathcal{M}, \beta)$ is anisotropic.)

Now assume that $\det(\mathcal{U})$ has degree 0. Then also $\det(\mathcal{U})^\vee$ has degree 0. As $\det(\rho) : \det(\mathcal{U}) \to \det(\mathcal{U})^\vee$ is non-trivial, it must be an isomorphism. This means that ρ is regular on $\mathcal{U}$. But then $\mathcal{U}$ is an orthogonal summand of $(\mathcal{M}, \beta)$. As $(\mathcal{M}, \beta)$ is indecomposable, it follows that $\mathcal{U} = \mathcal{M}$. □

COROLLARY 2.3. *Let $(\mathcal{M}, \beta)$ be an anisotropic indecomposable space over X. Write $\mathcal{M} = tr_{L/K}(\mathcal{N})$ with an absolutely indecomposable vector bundle $\mathcal{N}$. Then $\mathcal{N}$ has degree 0 and has no non-trivial proper subbundle of degree ≥ 0.*

PROOF. As $\mathcal{M}$ is self-dual, $\deg(\mathcal{M}) = 0$. From [**AEJ3**], Lemma 1.4(a) we see that $\deg(\mathcal{M}) = [L : K] \deg(\mathcal{N})$ and $\operatorname{rk}(\mathcal{M}) = [L : K] \operatorname{rk}(\mathcal{N})$. In particular,

$\deg(\mathcal{N}) = 0$. If $\mathcal{N}$ contained a subbundle of degree ≥ 0 then the trace of that subbundle would be a subbundle of $\mathcal{M}$ of degree ≥ 0. The result therefore follows by the proposition. □

Now let X be an elliptic curve over K. We assume that X is described by the Weierstraß equation

$$y^2 = x^3 + b_2x^2 + b_1x + b_0\,.$$

We shall write $q(x)$ for the polynomial $x^3 + b_2x^2 + b_1x + b_0$ in x over K. We denote by $F = K(x, y)$ the function field of X over K.

Recall that X is an abelian variety over K with the infinite point O as base point (zero point). Furthermore, the group $\mathrm{Pic}^0(X)$ of isomorphism classes of line bundles of degree 0 on X is naturally isomorphic to the group $X(K)$ of K-rational points of X.

We first determine the spaces of rank 1 over X. So let $(\mathcal{M}, \beta)$ be a space with a line bundle $\mathcal{M}$ on X.

If $\mathcal{M} = \mathcal{O}_X$ then, of course, identifying ${\mathcal{O}_X}^\vee$ with $\mathcal{O}_X$ in the usual way, β is simply the multiplication by a non-zero element c in K. We denote this space by $\langle c\rangle_X$ or simply $\langle c\rangle$.

Now assume that $\mathcal{M}$ is not isomorphic to $\mathcal{O}_X$. Then, as $\mathcal{M}$ is self-dual, $\mathcal{M}$ corresponds to a K-rational point of order 2 on X, i.e., to a point Q on X of the form $(a, 0)$ for some $a \in K$. This means that we may assume that $\mathcal{M}$ is described as follows: The stalk of $\mathcal{M}$ at the generic point equals F and the stalk $\mathcal{M}_P$ of $\mathcal{M}$ at a closed point P on X equals $\mathcal{O}_{X,P}$ unless $P = O$ or $P = Q$. The stalk $\mathcal{M}_O$ is the set of functions $f \in F$ such that $\nu_O(f) \geq 1$ and the stalk $\mathcal{M}_Q$ is the set of functions $f \in F$ such that $\nu_Q(f) \geq -1$.

As $\nu_O(x-a) = -2$ and $\nu_Q(x-a) = 2$ and $\nu_P(x-a) = 0$ for $P \neq O$ or Q, we see that the maps $\mathcal{M}_P \times \mathcal{M}_P \to \mathcal{O}_{X,P}$, with $(f,g) \mapsto (x-a)fg$, define a regular symmetic bilinear form on $\mathcal{M}$. We denote this space by $\langle x-a\rangle_X$ or simply $\langle x-a\rangle$.

As $\mathrm{End}(\mathcal{M}) = K$, it now follows that $(\mathcal{M}, \beta) = \langle c(x-a)\rangle := \langle c\rangle\langle x-a\rangle$ for some non-zero element c in K.

We next consider the absolutely indecomposable spaces over X, i.e., spaces with an absolutely indecomposable underlying vector bundle.

Tillmann showed in [T] that Atiyah's classification of indecomposable vector bundles on X in the case K is algebraically closed holds in general for absolutely indecomposable vector bundles. (Cf. [**AEJ3**, §4].) For vector bundles of degree 0 we get the following proposition (cf. [**At**, Theorem 5]).

PROPOSITION 2.4.

(i) *There exists an absolutely indecomposable vector bundle $\mathcal{F}_r$ of rank r and degree 0 on X, unique up to isomorphism, such that $\mathcal{F}_r$ has non-trivial global sections. Moreover, $\mathcal{F}_1 = \mathcal{O}_X$ and for $r > 1$ there is an exact sequence*

$$0 \to \mathcal{O}_X \to \mathcal{F}_r \to \mathcal{F}_{r-1} \to 0\,.$$

(ii) *If $\mathcal{M}$ is an absolutely indecomposable vector bundle of rank r and degree 0 on X then there is a line bundle $\mathcal{L}$ of degree 0 on X, unique up to isomorphism, such that $\mathcal{M} \cong \mathcal{L} \otimes_{\mathcal{O}_X} \mathcal{F}_r$.*

As in [**At**] it follows that each $\mathcal{F}_r$ is self-dual and that for $s < r$ there is an exact sequence

$$0 \to \mathcal{F}_s \to \mathcal{F}_r \to \mathcal{F}_{r-s} \to 0\,.$$

Furthermore, this embedding of $\mathcal{F}_s$ in $\mathcal{F}_r$ as a subbundle is uniquely determined up to isomorphisms of $\mathcal{F}_s$.

Now assume that β is a regular symmetric bilinear form on $\mathcal{F}_r$. Assume also that $s \le r - s$. Then there is an embedding of $\mathcal{F}_s$ in $\mathcal{F}_{r-s}$. Using the dual of the short exact sequence above, we see that the orthogonal subbundle ${\mathcal{F}_s}^{\perp}$ of $\mathcal{F}_s$ in $(\mathcal{F}_r, \beta)$ is the inverse image under β of the image of the embedding ${\mathcal{F}_{r-s}}^{\vee} \to {\mathcal{F}_r}^{\vee}$. As $\mathcal{F}_{r-s}$ is self-dual, it follows that ${\mathcal{F}_s}^{\perp} \cong \mathcal{F}_{r-s}$. By the uniqueness of the embedding of $\mathcal{F}_s$ into $\mathcal{F}_r$, we conclude that $\mathcal{F}_s$ is a subbundle of ${\mathcal{F}_s}^{\perp}$, i.e., that $\mathcal{F}_s$ is a totally isotropic subbundle of $(\mathcal{F}_r, \beta)$, and that ${\mathcal{F}_s}^{\perp} = \mathcal{F}_{r-s}$. It follows that β induces a regular symmetric bilinear form γ on $\mathcal{F}_{r-s}/\mathcal{F}_s \cong \mathcal{F}_{r-2s}$. We know (cf. [**K2**, Theorem 3]) that the spaces $(\mathcal{F}_r, \beta)$ and $(\mathcal{F}_{r-2s}, \gamma)$ are equivalent, i.e., have the same class in $W(X)$.

If r is odd we write $r = 2s+1$ and conclude that $(\mathcal{F}_r, \beta)$ is equivalent to some space $\langle c \rangle$.

If r is even we can write $r = 2s$ and conclude that $(\mathcal{F}_r, \beta)$ is metabolic. But, in fact, r cannot be even because then we would get a regular symmetric bilinear form on $\mathcal{F}_2$. But that is impossible. Indeed, the determinant bundle of $\mathcal{F}_2$ is isomorphic to $\mathcal{O}_X$, hence the exterior product gives rise to a regular skew-symmetric bilinear form on $\mathcal{F}_2$. But, as $\mathcal{F}_2$ is absolutely indecomposable, it cannot carry both a regular symmetric bilinear form and a regular skew-symmetric bilinear form (cf. [**AEJ3**, Remark 5.9(ii)]).

The following proposition generalizes these observations.

PROPOSITION 2.5. *Let $(\mathcal{M}, \beta)$ be an absolutely indecomposable space over X. Then $(\mathcal{M}, \beta)$ is either equivalent to a space $\langle c \rangle$ with a non-zero element c in K or to a space $\langle c(x-a) \rangle$ with a non-zero element c in K and a root $a \in K$ of $q(x)$. The K-rational point corresponding to $\mathcal{M}$ under Atiyah's classification is O in the first case but $(a, 0)$ in the second case.*

PROOF. As $\mathcal{M}$ is self-dual, it has degree 0. By Proposition 2.4(ii), we may assume that $\mathcal{M} = \mathcal{L} \otimes_{\mathcal{O}_X} \mathcal{F}_r$ with a line bundle $\mathcal{L}$ on X. As $\mathcal{M}$ is self-dual, it follows from the uniqueness part of Proposition 2.4(ii) that $\mathcal{L}$ is self-dual. As $\mathcal{L}$ is a line bundle, this means that $\mathcal{L}$ carries a regular symmetric bilinear form γ. It follows that $(\mathcal{M}, \beta) = (\mathcal{L}, \gamma) \otimes_{\mathcal{O}_X} (\mathcal{F}_r, \beta_0)$ for some regular symmetric bilinear form β_0 on $\mathcal{F}_r$. From the discussion above it now follows that $(\mathcal{M}, \beta)$ is equivalent to $(\mathcal{L}, c_0\gamma)$ for some non-zero element c_0 in K. The proposition now follows from our description of spaces of rank 1 on X. □

To handle general indecomposable spaces over X we need the following result from [**T**]. (Cf. [**AEJ3**, Corollaries 4.3 and 4.4].)

PROPOSITION 2.6. *Let $\mathcal{M}$ be an indecomposable vector bundle on X. Then $D(\mathcal{M})$ is a field. Furthermore, if $\mathcal{M} = tr_{L/K}(\mathcal{N})$ with an absolutely indecomposable vector bundle $\mathcal{N}$ on $Y = L \times_K X$ then $L = D(\mathcal{M})$ is the field of definition of the point of X corresponding to $\mathcal{N}$ under Atiyah's classification.*

The following corollary follows immediately from this proposition and Proposition 1.10.

COROLLARY 2.7. *There are no indecomposable spaces of symplectic type over X. In particular, $W_{00}(X) = W_0(X)$.*

We now turn to indecomposable spaces of orthogonal type.

PROPOSITION 2.8. *Let $(\mathcal{M}, \beta)$ be an indecomposable space of orthogonal type over X. Then $(\mathcal{M}, \beta)$ is equivalent to $\langle c \rangle$ for a non-zero element c in K or to $tr_{L/K}(\langle c(x-a)\rangle)$ for a non-zero element c in L, where $L = K(a)$ with a root a of $q(x)$.*

PROOF. By Proposition 1.10 and Proposition 2.6, there is an absolutely indecomposable space $(\mathcal{N}, \gamma)$ over $Y = L \times_K X$, where $L = D(\mathcal{M})$, such that $(\mathcal{M}, \beta) = \mathrm{tr}_{L/K}(\mathcal{N}, \gamma)$. By Proposition 2.5, $(\mathcal{N}, \gamma)$ is equivalent to $\langle c \rangle$ for some non-zero element c in L or equivalent to $\langle c(x-a)\rangle$ for some non-zero element c in L and a root $a \in L$ of $q(x)$. By the second part of Proposition 2.6 and the last statement of Proposition 2.5, we have $L = K$ in the first case and $L = K(a)$ in the second case. □

REMARK 2.9. In fact, our methods give a more precise result. The space $(\mathcal{M}, \beta)$ in the proposition is isomorphic to the tensor product of one of the spaces constructed in the proof by a space $(\mathcal{F}_r, \beta_0)$.

There remains to consider the indecomposable spaces of unitary type. We begin by constructing some special ones.

Let $r \in K$ such that $t = q(r)$ is not a square in K. Let $L = K(\sqrt{t})$ and let ω be the non-trivial automorphism of L over K. Let $Y = L \times_K X$ and let $\mathcal{N}$ be the line bundle of degree 0 on Y corresponding to the L-rational point $Q = (r, \sqrt{t})$ on X. Let $\mathcal{M} = \mathrm{tr}_{L/K}(\mathcal{N})$.

We know that $L \otimes_K \mathcal{M}$ is isomorphic to $\mathcal{N} \oplus {}^{\omega}\mathcal{N}$. (Cf. [**AEJ3**, Lemma 1.4].) From ${}^{\omega}(r, \sqrt{t}) = (r, -\sqrt{t}) \neq (r, \sqrt{t})$ we see that ${}^{\omega}\mathcal{N}$ is isomorphic to $\mathcal{N}^{\vee}$ but not to $\mathcal{N}$. It follows, in particular, that $\mathrm{End}(L \otimes_K \mathcal{M}) = L \otimes_K \mathrm{End}(\mathcal{M})$ is 2-dimensional over L, hence that $\mathrm{End}(\mathcal{M})$ is 2-dimensional over K. The representation $\mathcal{M} = \mathrm{tr}_{L/K}(\mathcal{N})$ induces an action of L on $\mathcal{M}$, hence an embedding of L in $\mathrm{End}(\mathcal{M})$. We conclude that $\mathrm{End}(\mathcal{M}) = L$.

The line bundle $\mathcal{N}$ can be described as follows: The stalk of $\mathcal{N}$ at the generic point equals the function field $E = F(\sqrt{t})$ of Y and the stalk $\mathcal{N}_P$ of $\mathcal{N}$ at a closed point P on X equals $\mathcal{O}_{Y,P}$ unless $P = O$ or $P = Q$. The stalk $\mathcal{N}_O$ is the set of functions $f \in E$ such that $\nu_O(f) \geq 1$ and the stalk $\mathcal{N}_Q$ is the set of functions $f \in E$ such that $\nu_Q(f) \geq -1$. (Note that $({}^\omega\mathcal{N})_P = \mathcal{N}_{^\omega P}$.)

As $\nu_O(x-r) = -2$ and $\nu_Q(x-r) = \nu_{^\omega Q}(x-r) = 1$ and $\nu_P(x-r) = 0$ for $P \neq O$, Q or ${}^\omega Q$, we see that the maps $\mathcal{N}_{^\omega P} \times \mathcal{N}_P \to \mathcal{O}_{Y,P}$, with $(f, g) \mapsto \frac{1}{2}(x-r)\,{}^\omega f\, g$, define a regular ω-hermitian form on $\mathcal{N}$. (Cf. [**AEJ3**, Remark 5.15].) It follows that the trace of this form is a regular symmetric bilinear form β_0 on $\mathcal{M}$. (Cf. [**AEJ3**, Remark 5.15 again].) Trivial computations show that at the generic point the space $(\mathcal{M}, \beta_0)$ is isomorphic to the space $\langle x - r\rangle\langle 1, -t\rangle$ over F. By construction, the involution on $\mathrm{End}(\mathcal{M}) = L$ induced by β_0 equals ω.

Denote by $\mathcal{H}(r)_X$ or simply $\mathcal{H}(r)$ the space $(\mathcal{M}, \beta_0)$ constructed above. Then we have the following lemma.

LEMMA 2.10. *Let $(\mathcal{M}, \beta)$ be an anisotropic indecomposable space of unitary type over X such that $D(\mathcal{M})$ is a quadratic extension of K. Then $(\mathcal{M}, \beta)$ is isomorphic to $\langle c\rangle\mathcal{H}(r)$ for a non-zero element c in K and an element r in K such that $q(r)$ is not a square in K.*

PROOF. Write $L = D(\mathcal{M})$ and let ω be the non-trivial automorphism of L over K. Write $\mathcal{M} = \mathrm{tr}_{L/K}(\mathcal{N})$ with an absolutely indecomposable vector bundle $\mathcal{N}$ on $Y = L \times_K X$. From Corollary 1.7 and Proposition 2.4 it then follows that $\mathcal{N}$ is a line bundle of degree 0 on Y. By Proposition 1.2, $\mathcal{N}$ is not self-dual. On the other hand, $L \otimes_K \mathcal{M}$ is isomorphic to $\mathcal{N} \oplus {}^\omega\mathcal{N}$ by [AEJ3], Lemma 1.4. As $\mathcal{M}$ is self-dual, hence $L \otimes_K \mathcal{M}$ also, it follows that $\mathcal{N}^\vee \cong {}^\omega\mathcal{N}$.

Let $Q = (r, s)$ be the L-rational point of X corresponding to $\mathcal{N}$. Then we have $(r, -s) = ({}^\omega r, {}^\omega s)$. So r lies in K and $L = K(s)$ with $s^2 = t$ in K. As (r, s) lies on X, we have $t = q(r)$. It follows that we are in the situation considered above. So $\mathrm{End}(\mathcal{M}) = L$ and the symmetric elements in $\mathrm{End}(\mathcal{M})$ with respect to the involution on $\mathrm{End}(\mathcal{M})$ induced by β_0, where β_0 is as in the discussion above, are precisely those lying in K. It follows that $\beta = c\beta_0$ for some non-zero element c in K. □

We want to prove that $W(X) = W_{00}(X)$. To do that it suffices to prove that the class of every anisotropic indecomposable space of unitary type lies in $W_{00}(X)$. Using Proposition 1.10 and Corollary 1.14 (and writing K instead of K'), we reduce to the case of an anisotropic indecomposable space $(\mathcal{M}, \beta)$ of unitary type such that $D(\mathcal{M})$ is a quadratic extension of K.

By Lemma 2.10 (and the discussion preceding it), it therefore suffices to show that if $r \in K$ such that $t = q(r)$ is not a square in K then the class of $\mathcal{H}(r)$ lies in $W_{00}(X)$. As the natural map $W(X) \to W(F)$ is injective, this means that it suffices to show that the class of $\langle x - r\rangle\langle 1, -t\rangle$ in $W(F)$ lies in the image of $W_{00}(X) \to W(F)$.

After a translation of the x-axis we may assume that $r = 0$. Then our hypotheses imply that $t = q(0)$ is not a square in K. We want to prove that the class of $\langle x\rangle\langle 1, -t\rangle$ in $W(F)$ lies in the image of $W_{00}(X) \to W(F)$.

We first look at the case in which K contains all the roots a_1, a_2, a_3 of $q(x)$. Then $t = -a_1a_2a_3$. The two equations

$$\begin{aligned}&\langle a_3\rangle + \langle x - a_3\rangle = \langle x\rangle + \langle a_3x(x-a_3)\rangle\\ &\langle a_1a_3(a_2-a_1)\rangle\langle x-a_1\rangle + \langle a_2a_3(a_1-a_2)\rangle\langle x-a_2\rangle = \langle a_1a_2a_3x\rangle\\ &\qquad + \langle -a_3x(x-a_1)(x-a_2)\rangle\end{aligned}$$

in $W(F)$ imply that

$$\langle x\rangle\langle 1, a_1a_2a_3\rangle = \langle a_3\rangle + \langle x-a_3\rangle + \langle a_1a_3(a_2-a_1)\rangle\langle x-a_1\rangle + \langle a_2a_3(a_1-a_2)\rangle\langle x-a_2\rangle.$$

It follows that $\langle x\rangle\langle 1, -t\rangle$ lies in the image of $W_{00}(X) \to W(F)$.

We next look at the case that K contains only one root a of $q(x)$. We write $q(x) = (x-a)((x-b)^2 - d)$. Then d is not a square in K. We write $K' = K(\sqrt{d})$ and $F' = K'F$. We have $t = -a(b^2-d)$ and direct computations show that

$$\langle x\rangle\langle 1, a(b^2-d)\rangle = \mathrm{tr}_{F'/F}\Big(\langle\frac{-b+\sqrt{d}}{2\sqrt{d}}\rangle\langle x-(b+\sqrt{d})\rangle\Big) + \langle b^2-d\rangle\langle 1, a(x-a)\rangle\,.$$

It follows that $\langle x\rangle\langle 1, -t\rangle$ lies in the image of $W_{00}(X) \to W(F)$.

We now look at the remaining case that K contains no root of $q(x)$. By going up to K', where $K' = K(a)$ with a root a of $q(x)$, we reduce, using Lemma 1.5, to the cases already considered.

We now have considered every case. The conclusion is that $W_{00}(X)$ is all of $W(X)$. We state this result as follows:

THEOREM 2.11. *$W(X)$ is additively generated by the image of $W(K)$ in $W(X)$ and the classes of indecomposable spaces of the form $\mathrm{tr}_{L/K}(\langle c(x-a)\rangle)$, where $L = K(a)$ with a root a of $q(x)$ and c is a non-zero element in L.*

In short, the theorem says that $W(X)$ is additively generated by traces of spaces of rank 1.

In the case that all the points of order 2 on X are K-rational, i.e., in the case that K contains all the roots of $q(x)$, we get the following corollary.

COROLLARY 2.12. *Assume that all points of order 2 on X are K-rational. Then $W(X)$ is generated by the classes of spaces of rank 1. More precisely,*

$$W(X) = W(K) + W(K)\langle x-a_1\rangle + W(K)\langle x-a_2\rangle + W(K)\langle x-a_3\rangle$$

where a_1, a_2, a_3 are the roots of $q(x)$.

The corollary says that $W(X)$ is diagonalizable if all the points of order 2 on X are K-rational. This result was also proved by Parimala and Sujatha by a different method (cf. [**PS**, Corollary 3.5]). The hypothesis is essential. If it is not satisfied then $W(X)$ may not be diagonalizable. But in any case the spaces generating $W(X)$ by the theorem can be described explicitly.

As we are using §1, our proofs assume that K is perfect. Using that $W(X)$ does not change under a purely inseparable base extension and that the polynomial $q(x)$ is separable, we see, however, that our results hold in general.

3. Curves and cohomology

In this section X is a complete regular curve over K. Throughout, we shall assume that K is the exact constant field of X. We let F be the function field of X.

As in the case of fields, there is a connection between the Witt ring $W(X)$ of X and the étale cohomology groups $H^n(X, \mathbf{Z}/2\mathbf{Z})$.

Following Parimala, we write $I_n(X) = W(X) \cap I^n(F)$. The exact sequence (2.1) then induces exact sequences

$$0 \to I_n(X) \to I^n(F) \to \bigoplus_{\mathfrak{p}} I^{n-1}(K(\mathfrak{p})) \tag{3.1}$$

(cf. [**A**, Satz 3.1]). There is a similar sequence in cohomology, the long exact localization sequence

$$\begin{aligned} \cdots &\to H^{n-1}(X, \mathbf{Z}/2\mathbf{Z}) \to H^{n-1}(F, \mathbf{Z}/2\mathbf{Z}) \to \bigoplus_{\mathfrak{p}} H^{n-2}(K(\mathfrak{p}), \mathbf{Z}/2\mathbf{Z}) \\ &\to H^n(X, \mathbf{Z}/2\mathbf{Z}) \to H^n(F, \mathbf{Z}/2\mathbf{Z}) \to \bigoplus_{\mathfrak{p}} H^{n-1}(K(\mathfrak{p}), \mathbf{Z}/2\mathbf{Z}) \to \cdots \end{aligned} \tag{3.2}$$

If the invariants $e_F^n : I^n(F) \to H^n(F, \mathbf{Z}/2\mathbf{Z})$ and

$$e_{K(\mathfrak{p})}^{n-1} : I^{n-1}(K(\mathfrak{p})) \to H^{n-1}(K(\mathfrak{p}), \mathbf{Z}/2\mathbf{Z})$$

(cf. [**AEJ1**]) are well-defined then it follows from the commutativity of these invariants with the (second) residue maps (cf. [**A**, Satz 4.11]) that e_F^n induces a morphism e_X^n from $I_n(X)$ to the image $H^n(X, \mathbf{Z}/2\mathbf{Z})_F$ of $H^n(X, \mathbf{Z}/2\mathbf{Z})$ in $H^n(F, \mathbf{Z}/2\mathbf{Z})$. As e^n is well-defined for all fields when $n \leq 4$ (cf. [**A**], [**JR**]), this morphism e_X^n is always well defined for $n \leq 4$. For $n = 0, 1, 2$, we get the *rank index*, *discriminant*, and *Clifford invariant*, respectively. (Cf. also [**P2**].)

For this reason, we are interested in the groups $H^n(X, \mathbf{Z}/2\mathbf{Z})_F$. We want to describe them by first describing $H^n(X, \mathbf{Z}/2\mathbf{Z})$ and then describing the kernel $N^n(X, \mathbf{Z}/2\mathbf{Z})$ of the natural morphism $H^n(X, \mathbf{Z}/2\mathbf{Z}) \to H^n(F, \mathbf{Z}/2\mathbf{Z})$.

Because of the lack of a suitable reference we start by proving a result describing the étale cohomology groups $H^n(X, \mathbf{Z}/2\mathbf{Z})$. We let X_s be the curve we get by extending the base from K to K_s, where K_s is the separable closure of K. We denote by ${}_2\operatorname{Pic}(X)$ the subgroup of elements of order dividing 2 in the Picard group $\operatorname{Pic}(X)$ of X.

PROPOSITION 3.3. *Let X be a complete regular curve over K. Then there is a natural morphism $H^n(X, \mathbf{Z}/2\mathbf{Z}) \to H^{n-2}(K, \mathbf{Z}/2\mathbf{Z})$. If X has a rational point then:*

(a) *This morphism is surjective.*
(b) *The kernel $\underline{H}^n(X, \mathbf{Z}/2\mathbf{Z})$ of this morphism contains $H^n(K, \mathbf{Z}/2\mathbf{Z})$.*
(c) *The quotient $\underline{H}^n(X, \mathbf{Z}/2\mathbf{Z})/H^n(K, \mathbf{Z}/2\mathbf{Z})$ is naturally isomorphic to*

$$H^{n-1}(K, {}_2\,Pic(X_s))\,.$$

Moreover, the choice of a rational point O on X induces a decomposition

$$H^n(X, \mathbf{Z}/2\mathbf{Z}) = H^n(K, \mathbf{Z}/2\mathbf{Z}) \oplus \underline{H}^n(X, \mathbf{Z}/2\mathbf{Z})_0 \oplus H^{n-2}(K, \mathbf{Z}/2\mathbf{Z}) \cup \gamma_O$$

where $\gamma_O \in H^2(X, \mathbf{Z}/2\mathbf{Z})$ is the cycle class of O and $\underline{H}^n(X, \mathbf{Z}/2\mathbf{Z})_0$ is naturally isomorphic to $H^{n-1}(K, {}_2\,Pic(X_s))$. In particular, if ${}_2\,Pic(X)$ is all of ${}_2\,Pic(X_s)$ then $\underline{H}^n(X, \mathbf{Z}/2\mathbf{Z})_0$ is naturally isomorphic to $H^{n-1}(K, \mathbf{Z}/2\mathbf{Z}) \otimes {}_2\,Pic(X)$.

PROOF. We use the Artin (or Hochschild-Serre) spectral sequence (cf. [**Ta**, Kapitel II, §7]) relating the cohomology of X to the one of X_s. Recall that this is a spectral sequence

$$E_2^{p,q} = H^p(K, H^q(X_s, \mathbf{Z}/2\mathbf{Z})) \Rightarrow H^{p+q}(X, \mathbf{Z}/2\mathbf{Z})$$

with edge morphisms $H^n(X, \mathbf{Z}/2\mathbf{Z}) \to H^n(X_s, \mathbf{Z}/2\mathbf{Z})$ and $H^n(K, \mathbf{Z}/2\mathbf{Z}) \to H^n(X, \mathbf{Z}/2\mathbf{Z})$ induced by the natural morphism $X_s \to X$ and the structure morphism $X \to Spec(K)$, respectively.

We know that $H^q(X_s, \mathbf{Z}/2\mathbf{Z}) = 0$, hence $E_r^{p,q} = 0$, for $q > 2$. It follows that the differential d_r is trivial for $r > 3$. It also follows that $d_3 : E_3^{p,q} \to E_3^{p+3,q-2}$ is trivial for $q \neq 2$ and that $d_2 : E_2^{p,q} \to E_2^{p+2,q-1}$ is trivial for $q \neq 1, 2$.

As $E_\infty^{p,q} = 0$ for $q > 2$, the spectral sequence gives us a projection

$$H^n(X, \mathbf{Z}/2\mathbf{Z}) \to E_\infty^{n-2,2}.$$

Combining this projection with the inclusion $E_\infty^{n-2,2} \subseteq E_2^{n-2,2}$ we get a natural morphism $H^n(X, \mathbf{Z}/2\mathbf{Z}) \to H^{n-2}(K, H^2(X_s, \mathbf{Z}/2\mathbf{Z}))$. As $H^2(X_s, \mathbf{Z}/2\mathbf{Z}) \cong \mathbf{Z}/2\mathbf{Z}$, this can be interpreted as a natural morphism

$$H^n(X, \mathbf{Z}/2\mathbf{Z}) \to H^{n-2}(K, \mathbf{Z}/2\mathbf{Z}).$$

Now assume that X has a rational point. Then the canonical morphism $H^n(K, \mathbf{Z}/2\mathbf{Z}) \to H^n(X, \mathbf{Z}/2\mathbf{Z})$ is injective, allowing us to consider $H^n(K, \mathbf{Z}/2\mathbf{Z})$ as a subgroup of $H^n(X, \mathbf{Z}/2\mathbf{Z})$. This injectivity implies that the differentials $d_3 : E_3^{n-3,2} \to E_3^{n,0}$ and $d_2 : E_2^{n-2,1} \to E_2^{n,0}$ are trivial. It follows that the differential d_r is trivial for $r > 2$ and that $d_2 : E_2^{p,q} \to E_2^{p+2,q-1}$ is trivial for $q \neq 2$.

The element corresponding to 1 in $H^2(X_s, \mathbf{Z}/2\mathbf{Z}) \cong \mathbf{Z}/2\mathbf{Z}$ is the cycle class of a closed point on X_s. As X has a rational point, it follows that the canonical morphism $H^2(X, \mathbf{Z}/2\mathbf{Z}) \to H^2(X_s, \mathbf{Z}/2\mathbf{Z})$ is surjective. Using cup products with elements in $H^p(K, \mathbf{Z}/2\mathbf{Z})$, we conclude that the natural morphism

$$H^{p+2}(X, \mathbf{Z}/2\mathbf{Z}) \to H^p(K, \mathbf{Z}/2\mathbf{Z})$$

is surjective. This is (a). This surjectivity also implies that $d_2 : E_2^{p,2} \to E_2^{p+2,1}$ is trivial. It follows that the Artin spectral sequence is degenerate. We, therefore, have:

$E_\infty^{p,q} = 0$ for $q > 2$,
$E_\infty^{p,2} = E_2^{p,2} = H^p(K, H^2(X_s, \mathbf{Z}/2\mathbf{Z})) \cong H^p(K, \mathbf{Z}/2\mathbf{Z})$,
$E_\infty^{p,1} = E_2^{p,1} = H^p(K, H^1(X_s, \mathbf{Z}/2\mathbf{Z})) = H^p(K, {}_2\,\mathrm{Pic}(X_s))$,

$$E_\infty^{p,0} = E_2^{p,0} = H^p(K, H^0(X_s, \mathbf{Z}/2\mathbf{Z})) = H^p(K, \mathbf{Z}/2\mathbf{Z}).$$

From this, statements (b) and (c) follow.

The rational point O of X induces a left inverse to the canonical morphism $H^n(K, \mathbf{Z}/2\mathbf{Z}) \to H^n(X, \mathbf{Z}/2\mathbf{Z})$, the reduction morphism

$$H^n(X, \mathbf{Z}/2\mathbf{Z}) \to H^n(K, \mathbf{Z}/2\mathbf{Z})$$

at O. It follows that we have a decomposition $H^n(X, \mathbf{Z}/2\mathbf{Z}) = H^n(K, \mathbf{Z}/2\mathbf{Z}) \oplus H^n(X, \mathbf{Z}/2\mathbf{Z})_0$, where $H^n(X, \mathbf{Z}/2\mathbf{Z})_0$ is the kernel of the reduction morphism at O.

Let $\gamma_O \in H^2(X, \mathbf{Z}/2\mathbf{Z})$ be the cycle class of O. We know that the natural morphism $H^2(X, \mathbf{Z}/2\mathbf{Z}) \to H^0(K, \mathbf{Z}/2\mathbf{Z}) = \mathbf{Z}/2\mathbf{Z}$ maps γ_O to 1. It follows that cup product by γ_O is a right inverse to the natural morphism $H^n(X, \mathbf{Z}/2\mathbf{Z}) \to H^{n-2}(K, \mathbf{Z}/2\mathbf{Z})$. This gives us a decomposition $H^n(X, \mathbf{Z}/2\mathbf{Z}) = \underline{H}^n(X, \mathbf{Z}/2\mathbf{Z}) \oplus H^{n-2}(K, \mathbf{Z}/2\mathbf{Z}) \cup \gamma_O$.

Combining both decompositions we get

$$H^n(X, \mathbf{Z}/2\mathbf{Z}) = H^n(K, \mathbf{Z}/2\mathbf{Z}) \oplus \underline{H}^n(X, \mathbf{Z}/2\mathbf{Z})_0 \oplus H^{n-2}(K, \mathbf{Z}/2\mathbf{Z}) \cup \gamma_O .$$

Here $\underline{H}^n(X, \mathbf{Z}/2\mathbf{Z})_0$ is the intersection of $H^n(X, \mathbf{Z}/2\mathbf{Z})_0$ and $\underline{H}^n(X, \mathbf{Z}/2\mathbf{Z})$. By (c), $\underline{H}^n(X, \mathbf{Z}/2\mathbf{Z})_0$ is naturally isomorphic to $H^{n-1}(K, {}_2\operatorname{Pic}(X_s))$. The last statement is clear. □

There are, of course, corresponding results with the same proofs for étale cohomology mod l instead of mod 2, if l is relatively prime to the characteristic exponent of K. But then one has to use Tate twistings. For example, there is a natural morphism $H^n(X, \mu_l) \to H^{n-2}(K, \mathbf{Z}/l\mathbf{Z})$.

From the localization sequence (3.2) we get one description of $N^n(X, \mathbf{Z}/2\mathbf{Z})$. Of course, $N^0(X, \mathbf{Z}/2\mathbf{Z})$ and $N^1(X, \mathbf{Z}/2\mathbf{Z})$ are trivial. For $n = 2$ we get an isomorphism $\operatorname{Pic}(X)_2 := \operatorname{Pic}(X)/2\operatorname{Pic}(X) \to N^2(X, \mathbf{Z}/2\mathbf{Z})$, induced by the cycle class map $\operatorname{Div}(X) \to H^2(X, \mathbf{Z}/2\mathbf{Z})$.

Assume that X has a rational point O. By the above, the cycle class γ_O lies in $N^2(X, \mathbf{Z}/2\mathbf{Z})$. It follows that $N^n(X, \mathbf{Z}/2\mathbf{Z})$ contains $H^{n-2}(K, \mathbf{Z}/2\mathbf{Z}) \cup \gamma_O$. Writing $\underline{N}^n(X, \mathbf{Z}/2\mathbf{Z}) = N^n(X, \mathbf{Z}/2\mathbf{Z}) \cap \underline{H}^n(X, \mathbf{Z}/2\mathbf{Z})$, we conclude that

$$N^n(X, \mathbf{Z}/2\mathbf{Z}) = \underline{N}^n(X, \mathbf{Z}/2\mathbf{Z}) \oplus H^{n-2}(K, \mathbf{Z}/2\mathbf{Z}) \cup \gamma_O .$$

To describe $N^n(X, \mathbf{Z}/2\mathbf{Z})$ it therefore suffices to describe $\underline{N}^n(X, \mathbf{Z}/2\mathbf{Z})$.

The specialization morphism $H^n(X, \mathbf{Z}/2\mathbf{Z}) \to H^n(K, \mathbf{Z}/2\mathbf{Z})$ at O can be extended to a morphism $H^n(F, \mathbf{Z}/2\mathbf{Z}) \to H^n(K, \mathbf{Z}/2\mathbf{Z})$ (cf. [**A**, Satz 4.12]). It follows that $N^n(X, \mathbf{Z}/2\mathbf{Z})$ is contained in $H^n(X, \mathbf{Z}/2\mathbf{Z})_0 = \underline{H}^n(X, \mathbf{Z}/2\mathbf{Z})_0 \oplus H^{n-2}(K, \mathbf{Z}/2\mathbf{Z}) \cup \gamma_O$. Hence

$$\underline{N}^n(X, \mathbf{Z}/2\mathbf{Z}) \subseteq \underline{H}^n(X, \mathbf{Z}/2\mathbf{Z})_0$$

Thus describing $\underline{N}^n(X, \mathbf{Z}/2\mathbf{Z})$ is equivalent to describing its image in

$$H^{n-1}(K, {}_2\operatorname{Pic}(X_s)) .$$

REMARK 3.4. Assume that X has a rational point. We can write

$$\bigoplus_{\mathfrak{p}} H^{n-2}(K(\mathfrak{p}), \mathbf{Z}/2\mathbf{Z}) \text{ as } H^{n-2}(K, \mathrm{Div}(X_s)_2),$$

where $\mathrm{Div}(X_s)_2 = \mathrm{Div}(X_s)/2\,\mathrm{Div}(X_s)$. Using only $\mathrm{Div}^0(X_s)_2$, the localization sequence (3.2) takes the form

$$\cdots \to H^{n-2}(K, \mathrm{Div}^0(X_s)_2) \to \underline{H}^n(X, \mathbf{Z}/2\mathbf{Z}) \to H^n(F, \mathbf{Z}/2\mathbf{Z}) \to \cdots$$

For X_s this becomes just the short exact sequence

$$0 \to {}_2\mathrm{Pic}(X_s) \to H^1(K_sF, \mathbf{Z}/2\mathbf{Z}) \to \mathrm{Div}^0(X_s)_2 \to 0.$$

It seems that the Artin spectral sequence induces a morphism from the long exact sequence above to the long exact sequence of K-cohomology groups arising from this short exact sequence. In particular, the image of $\underline{N}^n(X, \mathbf{Z}/2\mathbf{Z})$ in $H^{n-1}(K, {}_2\mathrm{Pic}(X_s))$ can then be computed by using the corresponding connecting morphism.

For $n = 2$ it then also follows that the composition of the isomorphism $\mathrm{Pic}^0(X)_2 \cong \underline{N}^2(X, \mathbf{Z}/2\mathbf{Z})$ with $\underline{H}^2(X, \mathbf{Z}/2\mathbf{Z}) \to H^1(K, {}_2\mathrm{Pic}(X_s))$ is induced by the connecting morphism arising from the short exact sequence

$$0 \to {}_2\mathrm{Pic}(X_s) \to \mathrm{Pic}^0(X_s) \to \mathrm{Pic}^0(X_s) \to 0$$

Note that $\mathrm{Pic}^0(X) = H^0(K, \mathrm{Pic}^0(X_s))$, since X has a rational point (cf. [**AEJ3**, Remark 3.9]).

It is well-known that $W(X)$ is diagonalizable and is finitely generated as a ring if K is algebraically closed or real closed. Using the above, we shall show that the same is true when K is finite. We begin with a cohomological result.

THEOREM 3.5. *Let K be a finite field and let X be a complete regular curve over K. Then $H^2(X, \mathbf{Z}/2\mathbf{Z})_F = 0$.*

PROOF. This follows from the localization sequence (3.2) together with the Brauer-Hasse-Noether Theorem, which says that a central simple algebra over the global field F splits if and only if it splits locally everywhere. □

Since regular bilinear spaces over F are classified by rank, discriminant and Clifford invariant, it follows from the theorem that $I_2(X) = 0$. We, therefore, have the following

THEOREM 3.6. *Let K be a finite field and let X be a complete regular curve over K. Then elements in $W(X)$ are determined by rank index and discriminant. In particular, $W(X)$ is finite and diagonalizable.*

COROLLARY 3.7. *Let K be a finite field and let X be a complete regular curve over K. Then every anisotropic indecomposable space over X has rank 1 or 2.*

We next turn to the case that K is a local field. Then $W(X)$ may no longer be diagonalizable. For example, it can be shown that the Witt ring of the elliptic curve with Weierstraß equation $y^2 = x(x^2 - p)$ over $\mathbf{Q}_p$, p odd, is not diagonalizable (cf. [**AEJ4**]). The cohomological dimension of K is 2. But the cohomological result in the finite case does not generalize as $H^3(X, \mathbf{Z}/2\mathbf{Z})_F \neq 0$ is possible. However, $W(X)$ is still finite as shown by Parimala in [**P1**]. For completeness, we shall prove this.

PROPOSITION 3.8. *Let K be a local field and let X be a complete regular curve over X. Then $W(X)$ is finite.*

PROOF. Since K has cohomological dimension 2, we have $H^n(F, \mathbf{Z}/2\mathbf{Z}) = 0$ for $n \geq 4$. Hence the invariants e_F^n are all well defined and have kernel $I^{n+1}(F)$ (cf. [**AEJ2**, Theorem 2]) and it suffices to show that the groups $H^n(X, \mathbf{Z}/2\mathbf{Z})_F$ are finite. In fact, even the groups $H^n(X, \mathbf{Z}/2\mathbf{Z})$ are finite. This follows by using the Artin spectral sequence as in the proof of Proposition 3.3. It suffices to show that all $E_\infty^{p,q}$ are finite. Since $E_\infty^{p,q}$ is a subquotient of the finite group $E_2^{p,q} = H^p(K, H^q(X_s, \mathbf{Z}/2\mathbf{Z}))$ (cf. [**Se**, Proposition II.14]), this is immediate. □

As mentioned, when K is a local field $H^3(X, \mathbf{Z}/2\mathbf{Z})_F$ may not be trivial. In fact, $I_3(X)$ may not be trivial. The elliptic curve X with Weierstraß equation $y^2 = x(x-1)(x-4)$ over any finite extension of $\mathbf{Q}_3$ gives such an example. The Witt ring of this curve has 512 elements while the cokernel of the map $W(F) \to \bigoplus_{\mathfrak{p}} W(K(\mathfrak{p}))$ has 128 elements. So this also gives negative answers to Questions 1 and 2 in [**PS**]. We shall look at Witt rings of elliptic curves thoroughly in [**AEJ4**].

We next bound the size of these groups, at least in the case that X has a rational point.

THEOREM 3.9. *Let K be a local field and let X be a complete regular curve of genus g over K. Assume that X has a rational point. Then $H^1(K, \mathbf{Z}/2\mathbf{Z}) \cup N^2(X, \mathbf{Z}/2\mathbf{Z})$ has index at most 2^g in $H^3(X, \mathbf{Z}/2\mathbf{Z})$.*

PROOF. Let O be a rational point of X. We then have

$$N^2(X, \mathbf{Z}/2\mathbf{Z}) = \underline{N}^2(X, \mathbf{Z}/2\mathbf{Z}) \oplus \mathbf{Z}/2\mathbf{Z} \cup \gamma_O,$$

hence

$$\begin{aligned} &H^1(K, \mathbf{Z}/2\mathbf{Z}) \cup N^2(X, \mathbf{Z}/2\mathbf{Z}) \\ &\quad = H^1(K, \mathbf{Z}/2\mathbf{Z}) \cup \underline{N}^2(X, \mathbf{Z}/2\mathbf{Z}) \oplus H^1(K, \mathbf{Z}/2\mathbf{Z}) \cup \gamma_O \end{aligned}$$

and, since the cohomological dimension of K equals 2,

$$H^3(X, \mathbf{Z}/2\mathbf{Z}) = \underline{H}^3(X, \mathbf{Z}/2\mathbf{Z})_0 \oplus H^1(K, \mathbf{Z}/2\mathbf{Z}) \cup \gamma_O$$

It, therefore, suffices to show that $H^1(K, \mathbf{Z}/2\mathbf{Z}) \cup \underline{N}^2(X, \mathbf{Z}/2\mathbf{Z})$ has index at most 2^g in $\underline{H}^3(X, \mathbf{Z}/2\mathbf{Z})_0$. Using that $\underline{H}^n(X, \mathbf{Z}/2\mathbf{Z})_0 \cong H^{n-1}(K, {}_2\operatorname{Pic}(X_s))$, it then suffices to show that $H^1(K, \mathbf{Z}/2\mathbf{Z}) \cup V$ has index at most 2^g in $H^2(X, {}_2\operatorname{Pic}(X_s))$, where V is the image of $\underline{N}^2(X, \mathbf{Z}/2\mathbf{Z})$ in $H^1(X, {}_2\operatorname{Pic}(X_s))$.

By Poincaré duality, ${}_2\operatorname{Pic}(X_s)$ is self-dual, so Tate duality (cf. [**Se**, Theorem II.2]) gives a perfect pairing

$$H^n(K, {}_2\operatorname{Pic}(X_s)) \times H^{2-n}(K, {}_2\operatorname{Pic}(X_s)) \to H^2(K, \mathbf{Z}/2\mathbf{Z}) \cong \mathbf{Z}/2\mathbf{Z}.$$

Let $U \subseteq H^0(K, {}_2\operatorname{Pic}(X_s))$ be the annihilator of $H^1(K, \mathbf{Z}/2\mathbf{Z}) \cup V$ under this pairing (for $n = 0$). Then it suffices to prove that $\dim(U) \leq g$, where dim is dimension over $\mathbf{Z}/2\mathbf{Z}$. We note that, by definition, $U \cup H^1(K, \mathbf{Z}/2\mathbf{Z}) \subseteq V^\perp$, where $V^\perp$ is the annihilator of V under the Tate pairing.

We have $H^0(K, {}_2\operatorname{Pic}(X_s)) = {}_2\operatorname{Pic}(X)$ because X has a rational point. The exact sequence

$$0 \to {}_2\operatorname{Pic}(X) \to {}_2\operatorname{Pic}(X_s) \to A \to 0$$

of finite dimensional $\mathbf{Z}/2\mathbf{Z}$-vector spaces gives rise to an exact sequence

$$0 \to H^0(K, A) \to H^1(K, {}_2\operatorname{Pic}(X)) \to H^1(K, {}_2\operatorname{Pic}(X_s)) \to \cdots$$

of cohomology groups. Since $H^1(K, {}_2\operatorname{Pic}(X)) \cong H^1(K, \mathbf{Z}/2\mathbf{Z}) \otimes {}_2\operatorname{Pic}(X)$, we conclude that the kernel W of the natural map $H^1(K, \mathbf{Z}/2\mathbf{Z}) \otimes {}_2\operatorname{Pic}(X) \to H^1(K, {}_2\operatorname{Pic}(X_s))$ has dimension at most

$$\dim(A) = \dim({}_2\operatorname{Pic}(X_s)) - \dim({}_2\operatorname{Pic}(X)) = 2g - \dim({}_2\operatorname{Pic}(X)).$$

As $\dim(U \cup H^1(K, \mathbf{Z}/2\mathbf{Z})) \geq \dim(U) \cdot \dim(H^1(K, \mathbf{Z}/2\mathbf{Z})) - \dim(W)$ and $V^\perp$ contains $U \cup H^1(K, \mathbf{Z}/2\mathbf{Z})$, we get that

$$\dim(V^\perp) \geq \dim(U) \cdot \dim(H^1(K, \mathbf{Z}/2\mathbf{Z})) - 2g + \dim({}_2\operatorname{Pic}(X)).$$

We know that $V \cong \operatorname{Pic}^0(X)_2$. It is also known that

$$|\operatorname{Pic}^0(X)_2| = |{}_2\operatorname{Pic}^0(X)| \cdot |\mathfrak{o} : 2\mathfrak{o}|^g$$

where $\mathfrak{o}$ is the ring of integers in K, hence that

$$\dim(V) = \dim(\operatorname{Pic}^0(X)_2) = \dim({}_2\operatorname{Pic}^0(X)) + g \cdot (\dim(H^1(K, \mathbf{Z}/2\mathbf{Z})) - 2)$$

(Cf. [**Mi**, p. 56]). If the characteristic of K is 0 then this is an easy consequence of a theorem of Mattuck [**Ma**], which says that there is an embedding of $\mathfrak{o}^g$ into $\operatorname{Pic}^0(X)$ of finite index.

Taking the Euler characteristic of ${}_2\operatorname{Pic}(X_s)$ (cf. [**Se**, p. II–26]), we get that

$$|H^1(K, {}_2\operatorname{Pic}(X_s))| = |H^0(K, {}_2\operatorname{Pic}(X_s))| \cdot |H^2(K, {}_2\operatorname{Pic}(X_s))| \cdot |\mathfrak{o} : 2^{2g}\mathfrak{o}|.$$

As $|H^0(K, {}_2\operatorname{Pic}(X_s))| = |H^2(K, {}_2\operatorname{Pic}(X_s))|$ by duality, it follows that

$$\dim(H^1(K, {}_2\operatorname{Pic}(X_s)) = 2 \cdot \dim(H^0(K, {}_2\operatorname{Pic}(X_s))) + 2g \cdot (\dim(H^1(K, \mathbf{Z}/2\mathbf{Z})) - 2)).$$

In particular, $\dim(V) = \dim(\operatorname{Pic}^0(X)_2) = \frac{1}{2}\dim(H^1(K, {}_2\operatorname{Pic}(X_s))$. It follows that $\dim(V^\perp) = \dim(V)$. Combining this with the expression for $\dim(V)$ above and the inequality for $\dim(V^\perp)$ above we conclude that $\dim(U) \leq g$ as needed. □

The hypothesis that X has a rational point cannot be left out (although a point of odd degree suffices). Indeed, if X is a conic without a rational point then the natural map $H^3(X, \mathbf{Z}/2\mathbf{Z}) \to H^1(K, \mathbf{Z}/2\mathbf{Z})$ is onto but $N^2(X, \mathbf{Z}/2\mathbf{Z}) = \underline{N}^2(X, \mathbf{Z}/2\mathbf{Z})$.

COROLLARY 3.10. *Let K be a local field and let X be a complete regular curve of genus g over K. Assume that X has a rational point. Then $N^3(X, \mathbf{Z}/2\mathbf{Z})$ has index at most 2^g in $H^3(X, \mathbf{Z}/2\mathbf{Z})$.*

This corollary (but not the theorem) also follows from Saito's work [**Sa**]. In fact, the existence of a rational point is not needed there.

For the Witt ring of X, the corollary shows

COROLLARY 3.11. *Let K be a local field and let X be a complete regular curve of genus g over K. Assume that X has a rational point. Then $I_3(X)$ has at most 2^g elements.*

The bound in the corollary is the best possible, as can be seen by using the discussion of Mumford curves in Example 7.2 in [**Sa**].

The rank index $W(X) \to H^0(X, \mathbf{Z}/2\mathbf{Z})_F$ and discriminant

$$I(X) \to H^1(X, \mathbf{Z}/2\mathbf{Z})_F$$

are clearly surjective maps. Parimala and Sridharan have announced that the Clifford invariant $I_2(X) \to H^2(X, \mathbf{Z}/2\mathbf{Z})_F$ is not always surjective (cf. [**PSr**]). The Clifford invariant is, however, clearly surjective if

$$H^2(X, \mathbf{Z}/2\mathbf{Z}) = H^1(X, \mathbf{Z}/2\mathbf{Z}) \cup H^1(X, \mathbf{Z}/2\mathbf{Z}).$$

More generally, we have the following

PROPOSITION 3.12. *Let X be a complete regular curve over K. Assume that*

$$H^2(X, \mathbf{Z}/2\mathbf{Z})_F = \sum_{K'} cor_{F'/F}\big(H^1(X', \mathbf{Z}/2\mathbf{Z})_{F'} \cup H^1(X', \mathbf{Z}/2\mathbf{Z})_{F'})\big)$$

where K' runs through the finite extensions of K and F' is the function field of $X' = K' \times_K X$. Then the Clifford invariant $I_2 X \to H^2(X, \mathbf{Z}/2\mathbf{Z})_F$ is surjective.

PROOF. This follows from the fact that the Clifford invariant of the trace of a space is the corestriction of the Clifford invariant. (Cf. [**A**, Satz 4.18].) □

Assume that X has a rational point O. Then we can use the decomposition of Proposition 3.3. By Merkurjev's Theorem, $H^2(K, \mathbf{Z}/2\mathbf{Z}) = H^1(K, \mathbf{Z}/2\mathbf{Z}) \cup H^1(K, \mathbf{Z}/2\mathbf{Z})$, so the image of $H^2(K, \mathbf{Z}/2\mathbf{Z})$ in $H^2(X, \mathbf{Z}/2\mathbf{Z})_F$ lies in the image of the Clifford invariant. Furthermore, γ_O has trivial image in $H^2(X, \mathbf{Z}/2\mathbf{Z})_F$. There remains only to consider the image of $\underline{H}^2(X, \mathbf{Z}/2\mathbf{Z})_0$ in $H^2(X, \mathbf{Z}/2\mathbf{Z})_F$.

We have $\underline{H}^n(X, \mathbf{Z}/2\mathbf{Z})_0 \cong H^{n-1}(K, {}_2\operatorname{Pic}(X_s))$. If ${}_2\operatorname{Pic}(X)$ is all of ${}_2\operatorname{Pic}(X_s)$ then it follows that $\underline{H}^2(X, \mathbf{Z}/2\mathbf{Z})_0 = H^1(K, \mathbf{Z}/2\mathbf{Z}) \cup \underline{H}^1(X, \mathbf{Z}/2\mathbf{Z})_0$. We therefore get

COROLLARY 3.13. *Let X be a complete regular curve over K. If X has a rational point and ${}_2\operatorname{Pic}(X) = {}_2\operatorname{Pic}(X_s)$ then the Clifford invariant $I_2(X) \to H^2(X, \mathbf{Z}/2\mathbf{Z})$ is surjective.*

More generally, Proposition 3.12 implies that if $H^1(K, {}_2\operatorname{Pic}(X_s))$ is generated by the images of $H^1(K', \mathbf{Z}/2\mathbf{Z}) \cup H^0(K', {}_2\operatorname{Pic}(X_s)))$ under $cor_{K'/K}$ then the

Clifford invariant is surjective. We shall, however, use the following slightly different result.

PROPOSITION 3.14. *Let X be a complete regular curve over K with a rational point. Let L be a finite extension of K such that the Clifford invariant for $L \times_K X$ is surjective and the corestriction $H^1(L, {}_2\,Pic(X_s)) \to H^1(K, {}_2\,Pic(X_s))$ is surjective. Then the Clifford invariant for X is surjective.*

PROOF. This follows as Proposition 3.12 and the fact that the decomposition of Proposition 3.3 behaves well under corestriction (trace).

COROLLARY 3.15. *Let X be an elliptic curve over K. Then the Clifford invariant $I_2(X) \to H^2(X, \mathbf{Z}/2\mathbf{Z})$ is surjective.*

PROOF. If X has three rational points of order 2 then ${}_2\operatorname{Pic}(X)$ is all of ${}_2\operatorname{Pic}(X_s)$ and the result follows from Corollary 3.13.

Now assume that X has only one rational point of order 2. Then X has a Weierstraß equation of the form

$$y^2 = (x-a)((x-b)^2 - d)$$

where d is not a square in K. We let $L = K(\sqrt{d})$. Then ${}_2\operatorname{Pic}(X_s)$ is isomorphic to the induced module $M_{L/K}(\mathbf{Z}/2\mathbf{Z})$. It follows that the corestriction $H^1(L, {}_2\operatorname{Pic}(X_s)) \to H^1(K, {}_2\operatorname{Pic}(X_s))$ is surjective. The result now follows by the proposition, using the previous case.

Finally assume that X has no rational point of order 2. There is a separable extension L of K of degree three such that $Y = L \times_K X$ has a rational point of order 2. As $cor_{L/K} \circ res_{L/K}$ is multiplication by 3, the result follows by the proposition, using the previous cases. □

REMARK 3.16. Let X be an elliptic curve over K with only one rational point of order 2. Then ${}_2\operatorname{Pic}(X_s)$ is isomorphic to the induced module $M_{L/K}(\mathbf{Z}/2\mathbf{Z})$, using the notation of the proof of the corollary above. It follows that

$$H^{n-1}(K, {}_2\operatorname{Pic}(X_s)),$$

hence also $\underline{H}^n(X, \mathbf{Z}/2\mathbf{Z})_0$, is isomorphic to $H^{n-1}(L, \mathbf{Z}/2\mathbf{Z})$.

Scharlau raised the question whether $W(X)$ is finitely generated as a ring whenever $W(K)$ is. By the above, this is true whenever K is algebraically closed, real closed, finite, or local. But the question is still open in general. It is, of course, equivalent to the question whether $W(X)$ is finitely generated as a $W(K)$-algebra whenever $W(K)$ is finitely generated as a ring. We shall show that the hypothesis on $W(K)$ cannot be left out.

Assume now that $W(X)$ is finitely generated as a $W(K)$-algebra. Then the image of $I(K)I(X)$ in $I_2(X)/I_3(X)$ has finite index. Indeed if $\{1, \phi_i\}$ is a finite generating set with each ϕ_i lying in $I(X)$ then the images of $\phi_i\phi_j$ in $I_2(X)/I_3(X)$ generate the cosets of $I(K)I(X)/I_3(X)$.

Now assume, in addition, that the Clifford invariant is surjective. It then follows that the image of $H^1(K,\mathbf{Z}/2\mathbf{Z}) \cup H^1(X,\mathbf{Z}/2\mathbf{Z})$ in $H^2(X,\mathbf{Z}/2\mathbf{Z})_F$ is of finite index.

Now assume further that X has a rational point. We use the decomposition from Proposition 3.3. We have $H^1(X,\mathbf{Z}/2\mathbf{Z}) = H^1(K,\mathbf{Z}/2\mathbf{Z}) \oplus \underline{H}^1(X,\mathbf{Z}/2\mathbf{Z})_0$, so $H^1(K,\mathbf{Z}/2\mathbf{Z}) \cup H^1(X,\mathbf{Z}/2\mathbf{Z}) = H^2(K,\mathbf{Z}/2\mathbf{Z}) \oplus H^1(K,\mathbf{Z}/2\mathbf{Z}) \cup H^1(X,\mathbf{Z}/2\mathbf{Z})_0$. Since $\underline{N}^2(X,\mathbf{Z}/2\mathbf{Z}) \subseteq \underline{H}^2(X,\mathbf{Z}/2\mathbf{Z})_0$, we have proven the following

LEMMA 3.17. *Let X be a complete regular curve over K having a rational point. Assume that the Clifford invariant is surjective. If $W(X)$ is finitely generated as a $W(K)$-algebra then $H^1(K,\mathbf{Z}/2\mathbf{Z}) \cup H^1(X,\mathbf{Z}/2\mathbf{Z})_0 + \underline{N}^2(X,\mathbf{Z}/2\mathbf{Z})$ has finite index in $\underline{H}^2(X,\mathbf{Z}/2\mathbf{Z})_0$.*

REMARK 3.18. Note that in Lemma 3.17

$\underline{H}^1(X,\mathbf{Z}/2\mathbf{Z})_0 \cong H^0(X, {}_2\operatorname{Pic}(X_s)) = {}_2\operatorname{Pic}(X)$
$\underline{H}^2(X,\mathbf{Z}/2\mathbf{Z})_0 \cong H^1(X, {}_2\operatorname{Pic}(X_s))$
$\underline{N}^2(X,\mathbf{Z}/2\mathbf{Z}) \cong \operatorname{Pic}^0(X)_2\,.$

Thus the conclusion of this lemma implies that

> If K^*/K^{*2} is finite then the image of $\operatorname{Pic}^0(X)_2$ in $H^1(X, {}_2\operatorname{Pic}(X_s))$ is of finite index.

and that

> If $\operatorname{Pic}^0(X)_2$ is finite then $H^1(K,\mathbf{Z}/2\mathbf{Z}) \cup {}_2\operatorname{Pic}(X)$ is of finite index in $H^1(X, {}_2\operatorname{Pic}(X_s))$.

Now assume that X is an elliptic curve over K. We have seen that if X has three rational points of order 2 then $W(X)$ is a finitely generated $W(K)$-algebra. Now assume that X has exactly one rational point of order 2. So X has a Weierstraß equation of the form

$$y^2 = (x-a)((x-b)^2 - d)$$

where d is not a square in K. We let $L = K(\sqrt{d})$. Then $\underline{H}^2(X,\mathbf{Z}/2\mathbf{Z})_0$ is isomorphic to $H^1(L,\mathbf{Z}/2\mathbf{Z})$ by Remark 3.16. In this case, the conclusion of the lemma above therefore implies that if $\operatorname{Pic}^0(X)_2$ is finite then $H^1(K,\mathbf{Z}/2\mathbf{Z})$ is of finite index in $H^1(L,\mathbf{Z}/2\mathbf{Z})$.

PROPOSITION 3.19. *Let K be a global field. Let X be an elliptic curve over K with exactly one rational point of order 2. Then $W(X)$ is not finitely generated as a $W(K)$-algebra.*

PROOF. By the Weak Mordell-Weil Theorem (cf. [**Si**, Theorem VIII.1.1]), $\operatorname{Pic}^0(X)_2$ is finite. Furthermore, if L is a quadratic extension of K then the image of K^*/K^{*2} in L^*/L^{*2} is of infinite index. The proposition now follows from the discussion above. □

REFERENCES

[A] J. Arason, *Cohomologische Invarianten quadratischer Formen*, J. Algebra **36** (1975), 448–491.

[AEJ1] J. Arason, R. Elman, and B. Jacob, *The graded Witt ring and galois cohomology,* I, Can. Math. Soc. Conf. Proc. **4** (1984), 17–52.

[AEJ2] ———, *Fields of cohomogical 2-dimension three,*, Math. Ann. **274** (1986), 649–657.

[AEJ3] ———, *On indecomposable vector bundles*, Comm. Alg. **20** (1992), 1323–1351.

[AEJ4] ———, *On the Witt ring of an elliptic curve* (to appear).

[At] M. Atiyah, *Vector bundles over an elliptic curve,*, Proc. Lond. Math. Soc. **VII** (1957), 414–452.

[JR] B. Jacob and M. Rost, *Degree four cohomological invariants for quadratic forms*, Invent. Math. **96** (1989), 551–570.

[K1] M. Knebusch, *Grothendieck- und Wittringe von nichtausgearteten symmetrischen Bilinearformen*, Sitz. ber. Heidelberg, Akad. Wiss. Math.-naturw.Kl. **3** (1969/1970), 95–157.

[K2] ———, *Symmetric bilinear forms over algebraic varieties, Conference on quadratic forms 1976*, Queens's papers in pure and applied math. **46** (1977), 103–283.

[K3] ———, *Real closures of algebraic varieties, Conference on quadratic forms 1976*, Queens's papers in pure and applied math. **46** (1977), 548–568.

[Ma] A. Mattuck, *Abelian varieties over P-adic ground fields*, Ann. of Math. **62** (1955), 92–119.

[Mi] J.S. Milne, *Arithmetic Duality Theorems*, Academic Press, Boston, 1986.

[M] D. Mumford, *Geometric Invariant Theory*, Ergebnisse, Springer-Verlag, Heidelberg, 1965.

[P1] R. Parimala, *Witt groups over local fields*, Comm. Alg. **17** (1989), 2857–2863.

[P2] ———, *Witt group of affine three folds*, Duke Math. **57** (1988), 947–954.

[PSr] R. Parimala and R. Sridharan, *Graded Witt ring and unramified cohomology* (to appear).

[PS] R. Parimala and R. Sujatha, *Witt group of hyperelliptic curves*, Comment. Math Helv. **65** (1990), 559–580.

[Sa] S. Saito,, *Class field theory for curves over local fields*, J. No.Th. **21** (1985), 44–80.

[Sh] S. Shatz,, *Profinite Groups, Arithmetic, and Geometry*, Princeton University Press, 1972.

[S] W. Scharlau, *Quadratic and Hermitian Forms*, Springer-Verlag, Berlin, 1985.

[Si] J. H. Silverman, *The Arithmetic of Elliptic Curves*, Springer-Verlag, New York, 1986.

[Se] J.-P. Serre, *Cohomologie Galoisienne*, Lecture Notes in Math., vol. 5, Springer-Verlap, Berlin, 1965.

[Ta] G. Tamme, *Einführung in die étale Kohomolgie*, Der Regensburger Trichter **17** (1980).

[T] A. Tillmann, *Unzerlegbare Vektorbündel über algebraischen Kurven*, Ph.D. thesis, Fernuniversität in Hagen, 1983.

RAUNVÍSINDASTOFNUN HÁSKÓLANS, UNIVERSITY OF ICELAND, RIYKJAVIK, ICELAND

DEPARTMENT OF MATHEMATICS, UNIVERSITY OF CALIFORNIA AT LOS ANGELES, LOS ANGELES, CALIFORNIA 90024

DEPARTMENT OF MATHEMATICS, UNIVERSITY OF CALIFORNIA AT SANTA BARBARA, SANTA BARBARA, CALIFORNIA 93106

Contemporary Mathematics
Volume 155, 1994

On the Trace Formula for Quadratic Forms

EBERHARD BECKER AND THORSTEN WÖERMANN

1. Introduction

This paper deals with a variant of the trace formula for quadratic forms which allows applications to some algorithmic problems of real algebraic geometry. The formula will be applied to the counting of real zeros on 0–dimensional varieties under side constraints, as well as to the 0–dimensional case of the Bröcker–Scheiderer result about the description of basic open semi–algebraic sets. Furthermore, it can be used to give a "visible" argument for Tarski's theorem on Quantifier Elimination in the theory of real closed fields.

It is only fair to admit that our method is nothing but a modern version of old ideas of Hermite–Sylvester who had already shown how to count real zeros by calculating signatures of appropriate quadratic forms. What has been added to their approach is a certain algebraic machinery that enables us to treat multivariate problems more uniformly. In an analogous approach P. Pedersen has independently developed a similiar method to count real zeros, also starting with Hermite ideas.

2. The trace formula

Trace formulae of various kinds frequently occur in the literature on quadratic forms. In particular, M. Knebusch made an extensive study of such formulae, cf. [**K1**], [**K2**], [**K3**]. So, our topic is basically well-known. However, in our approach we will formulate a trace formula under much weaker assumptions which allows new applications to real algebraic geometry. The main difference is that degenerate forms are allowed. In addition, the ring extensions involved

1991 *Mathematics Subject Classification.* Primary: 12Y05, 14P10, 14Q99 Secondary: 11E10.

This paper is submitted in final form and no version of it will be submitted for publication elsewhere.

need not be Frobenius and, moreover, as in Mahé's work [**M**, (4.11)], the real spectrum of rings instead of the space of signatures is used.

Our framework is the following. Let A be any commutative ring, $\tau : A \to B$ any ring homomorphism rendering B into a finitely generated projective A–module. Moreover, let M be a finitely generated projective B–module, equipped with a B–bilinear map $\varphi : M \times M \to B$. Note that the non-degeneracy of φ is not assumed. If $s : B \to A$ is any A–linear map then we can define the transfer $s_*(\varphi)$ which is an A–bilinear map on M, considered as an A–module, and which is defined as follows:

$$s_*(\varphi) : M \times M \to A, (m,n) \mapsto s(\varphi(m,n))$$

We will primarily deal with the trace map

$$s = tr_{B/A} : B \to A,$$

cf. [**DeM**, I; III,§2]. Nevertheless, other linear forms occur as well and deserve a study in their own right as will be indicated at the end of this section.

The ring homomorphism $\tau : A \to B$ induces a "restriction" map $\tau^* : Sper\, B \to Sper\, A$ between the real spectra of A and B. We will freely use notions and results from the theory of the real spectrum as described in [**C-R**], [**BCR**], [**Be1**]. Note that $Sper\, A = Spec_r A = R - Spec\, A$ etc. if other notations are used. If $\alpha \in Sper\, A$ is given, any $\beta \in Sper\, B$ with $\tau^*(\beta) = \alpha$ is called an extension of α, written: $\beta \mid \alpha$. By [**C-R**, 4.3, p. 40] or [**Be1**] one knows that the canonical map $B \to B \otimes_A k(\alpha)$ yields

$$\tau_*^{-1}(\{\alpha\}) \simeq Sper(B \otimes_A k(\alpha)),$$

where $k(\alpha)$ denotes the real closed field attached to α and "$\simeq$" means "homeomorphic". In our present situation, $B \otimes_A k(\alpha)$ turns out to be a finite–dimensional $k(\alpha)$–algebra, hence $Sper(B \otimes_A k(\alpha))$ and the fiber over α are finite sets.

The final notion we need is that of the signature of a bilinear form φ on a finitely generated A–module M at a point $\alpha \in Sper\, A : sgn_\alpha(\varphi)$. By definition $sgn_\alpha \varphi := sgn(\varphi \otimes_A k(\alpha))$ where, of course, the scalar extension $\varphi \otimes_A k(\alpha)$ is defined on the finite-dimensional $k(\alpha)$–vectorspace $M \otimes_A k(\alpha)$ and the unadorned sgn is the usual signature of $k(\alpha)$, cf. [**BCR**, Ch. 15].

After all these preparations we can state

THEOREM 2.1. *(The trace formula) Under the hypothesis above the following formula holds:*

$$sgn_\alpha(tr_{B|A})_*(\varphi) = \sum_{\substack{\beta|\alpha \\ \beta \in Sper\, B}} sgn_\beta(\varphi)$$

Proof. We start from the situation $A \to B$, M a B–module equipped with the B–form φ and the A–form $tr_*(\varphi) = (tr_{B|A})_*(\varphi)$. To calculate $sgn_\alpha(tr_{B|A})_*(\varphi)$ we have to carry out a scalar extension relative to the morphism $A \to k(\alpha)$ and we then face the following situation: $k(\alpha) \to B \otimes_A k(\alpha), M \otimes_A k(\alpha)$ considered as a $B \otimes_A k(\alpha)$–module equipped with two bilinear forms:

1) the $B \otimes_A k(\alpha)$–valued form $\hat{\varphi}$ defined by

$$\hat{\varphi}(m \otimes x, n \otimes y) := \varphi(m, n) \otimes xy$$

and

2) the scalar extension $tr_*(\varphi) \otimes_A 1$ which is $k(\alpha)$–valued.

To facilitate the notations we set

$$\hat{M} := M \otimes_A k(\alpha), \hat{B} := B \otimes_A k(\alpha), \widehat{tr_*(\varphi)} := tr_*(\varphi) \otimes_A 1.$$

Being a finite–dimensional $k(\alpha)$–algebra, $\hat{B}$ decomposes uniquely into a direct sum of ideals B_i, which are local $k(\alpha)$–algebras ($\oplus$: $k(\alpha)$–direct sums):

$$\hat{B} = \bigoplus_{i=1}^{r} B_i, B_i = e_i B, e_i \text{ (indecomposable) idempotents.}$$

Since $\hat{M}$ is a $\hat{B}$–module it splits correspondingly:

$$\hat{M} = \bigoplus_{i=1}^{r} M_i, M_i = e_i \hat{M}.$$

Since $M_i = e_i \hat{M}$, distinct M_i's are orthogonal relative to $\hat{\varphi}$. Set $\varphi_i = \hat{\varphi}_{|M_i}$. Then φ_i is B_i–valued, and we claim $\widehat{tr_*(\varphi)}_{|M_i} = (tr_{B_i|k(\alpha)})_*(\varphi_i)$. To prove this first note $\widehat{tr_*(\varphi)} = (tr_{\hat{B}|k(\alpha)})_*(\hat{\varphi})$, which is due to the compatibility of trace map with scalar extensions. Consequently, $\widehat{tr_*(\varphi)}_{|M_i} = (tr_{\hat{B}|k(\alpha)})_*(\varphi_i) = (tr_{B_i|k(\alpha)})_*(\varphi_i)$, since B_i is a direct ideal summand of $\hat{B}$.

From $\widehat{tr_*(\varphi)} = (tr_{\hat{B}|k(\alpha)})_*(\hat{\varphi})$ we further get that the M_i's also provide an orthogonal decomposition relative to $\widehat{tr_*(\varphi)}$. Putting the information together we have derived so far:

$$(*) \quad sgn_\alpha(tr_*(\varphi)) = sgn\ \widehat{tr_*(\varphi)} = \sum_{i=1}^{r} sgn\ (tr_{B_i|k(\alpha)})_*(\varphi_i)$$

Setting $C = B_i, N = M_i, \psi = \varphi_i, k = k(\alpha)$ we are dealing with a finite–dimensional local k–algebra C and a C–form ψ living on a finitely generated C–module N. We have to calculate $sgn\ (tr_{C|k})_*(\varphi)$.

We start by observing that in this case N is in fact a finitely generated projective C–module since we started out from a projective B–module and all the constructions above preserve projectivity. Since C is local, N is a free C–module. Let $\mathcal{M}$ denote the maximal ideal of C. Then the reduction $\overline{\varphi} = \varphi \otimes_C 1$ is a bilinear form on $\overline{M} = M \otimes_C C/_{\mathcal{M}} = M/_{\mathcal{M}M}$ which is a finite–dimensional vector space over the field $C/_{\mathcal{M}} = \overline{C}$. We want to prove $sgn\ (tr_{C|k})_*(\varphi) = sgn\ (tr_{\overline{C}|k})_*(\overline{\varphi})$.

In our case, *char* $C/_{\mathcal{M}} \neq 2$, so $\overline{\varphi}$ has a decomposition $\overline{\varphi} = \tilde{\varphi}_1 \perp \tilde{\varphi}_2, \tilde{\varphi}_1$ non–degenerate and $\tilde{\varphi}_2$ a Null–form: $\tilde{\varphi}_2 =< 0, \dots, 0 >$. Using [**Ba**, (3.4), p. 11] we obtain a C–decomposition of $(N, \varphi) : \varphi = \varphi_1 \perp \varphi_2$ with φ_1 free and non singular and φ_2 having all its values in $\mathcal{M}$. Now, by [**Ba**, (3.5), p. 13], φ_1 admits an orthogonal basis, and therefore we are facing the following situation:

$$\varphi =< x_1 > \perp \dots \perp < x_n > \perp \varphi_2, \varphi_2(N,N) \subset \mathcal{M}, x_i \in C^*.$$

Clearly $\overline{\varphi} =< \overline{x}_1 > \perp \dots \perp < \overline{x}_n > \perp$ Null–form, and since the trace map $tr_{C|k}$ vanishes on $\mathcal{M} = rad\ C$ our last claim will follow in general once it is proved in the special case $\varphi =< x >, x \in C^*$.

We use the fact that C has a "Wedderburn–decomposition" $C = C_0 \oplus \mathcal{M}, C_0$ a subalgebra (in fact a subfield), clearly satisfying $C_0 \simeq \overline{C}$. By using $tr_{C|k}(\mathcal{M}) = 0$, and $(tr_{C|k}) \mid_{C_0} = [C : C_0] \cdot tr_{C_0|k}$ one easily checks $sgn\ tr_{C|k} < x >= sgn\ tr_{\overline{C}|k} < \overline{x} >$ which then, as said above, implies our general claim.

Going back to the equation (*) we now have to compute $sgn\ (tr_{\overline{B}_i|k(\alpha)})_*(\overline{\varphi}_i)$. Since $k(\alpha)$ is real closed there are just two cases: $\overline{B}_i = k(\alpha)$ and $\overline{B}_i = k(\alpha)(\sqrt{-1})$. In the second case, $\overline{\varphi}_i$ is a torsion form, if considered in $W(\overline{B}_i)$, and so is $(tr_{\overline{B}_i|(k(\alpha))})_*(\varphi)$ in $W(k(\alpha)) \simeq \mathbb{Z}$. Hence $sgn\ (tr_{\overline{B}_i|k(\alpha)})_*(\overline{\varphi}_i) = 0$. In the first case, $(tr_{\overline{B}_i|k(\alpha)})_*(\overline{\varphi}_i) = \overline{\varphi}_i$, and using this we get the following refinement of (*):

$$(**) \quad sgn_\alpha(\varphi) = \sum_{\overline{B}_i = k(\alpha)} sgn(\overline{\varphi}_i)$$

We next interpret the right hand side of (**) by using *Sper B*. As proved in [**C-R**, prop. 4.3, p. 40], see also [Be 1], the natural morphism $B \to B \otimes_A k(\alpha)$ induces a homeomorphism (= bijection in our case) between the fiber of *Sper* $B \to$ *Sper* A over α and $Sper(B \otimes_A k(\alpha))$. From the above decomposition $\hat{B} = \oplus B_i$ one derives that the points β_i extending α corresponds in a 1–to–1 manner to those of the B_i satisfying $\overline{B}_i/_{\mathcal{M}_i} = k(\alpha)$. More precisely, if such a B_i is given the associated β_i is defined as the homomorphism (pr = projection):

$$(***) \quad B \to B \otimes_A k(\alpha) \xrightarrow{pr} B_i \to B_i/_{\mathcal{M}_i} = k(\alpha) = k(\beta_i)$$

Here and in the sequel we consider $\alpha, \beta, \dots$ as maps. We next have to compute $(M \otimes_B k(\beta), \varphi \otimes_B k(\beta))$ if $\beta = \beta_i$ for some extension β_i. Since $k(\beta) = k(\alpha)$ we are studying the following situation:

$$\begin{array}{ccc} M & & \\ | & & \\ B & \xrightarrow{\beta} & k(\alpha) \\ \tau\uparrow & ||| & \nearrow\alpha \\ A & & \end{array}$$

Using the $k(\alpha)$–isomorphism

$$(M \otimes_A k(\alpha)) \otimes_{B \otimes_A k(\alpha)} k(\alpha) \simeq M \otimes_B k(\alpha) = M \otimes_B k(\beta)$$

$$(m \otimes x) \otimes y \mapsto m \otimes xy$$

which is a special instance of [**C-E**, prop. 2.1, p. 165] and taking into account the way β is composed $(***)$ we finally obtain:

$$(M \otimes_B k(\beta_i), \varphi \otimes_B k(\beta_i)) = (\overline{M}_i, \overline{\varphi}_i).$$

This shows that $sgn_{\beta_i}(\varphi) = sgn(\varphi \otimes_B k(\beta_i)) = sgn\ \overline{\varphi}_i$, and the proof is complete. □

In the literature, other linear forms $s : B \to A$ are considered as well. In deriving a similiar transfer formula for such more general linear forms one has to cope with the following problem. Dealing with $sgn_\alpha s_*(\varphi)$ one decomposes $M \otimes_A k(\alpha)$, $B \otimes_A k(\alpha)$, $\varphi \otimes_A 1$, $s \otimes_A 1 : \hat{B} \to k(\alpha)$ into pieces $M_i, B_i, \varphi_i, s_i : B_i \to k(\alpha)$ and studies $s_{i*}(\varphi_i)$. But, in the case of $sgn_{\beta_i}(\varphi), \beta_i$ belonging to B_i, one is concerned with $\overline{M}_i$ and $\overline{\varphi}_i$. Hence, $s_{i*}(\varphi_i)$ has to be compared to $\overline{\varphi}_i$. As shown above, $\varphi_i = < x_1 > \perp \ldots \perp < x_n > \perp \psi, x_i \in C^*, \psi(N,N) \subset \mathcal{M}_i$. From $B_i / \mathcal{M}_i = k(\alpha), \frac{1}{2} \in B_i$ and the fact that B_i is a finite dimensional local $k(\alpha)$–algebra we get $x_i = a_i \cdot \epsilon_i^2, a_i \in k(\alpha)^*, \epsilon_i \in C^*$. Thus $\varphi_i = < a_1 > \perp \ldots \perp < a_n > \perp \psi$. Invoking Frobenius–reciprocity as in [**K2**, (1.1), p. 169], which is valid for arbitrary linear forms, we obtain

$$s_{i*}(\varphi_i) = (s_{i*} < 1 >) \cdot < a_1, \ldots, a_n > \perp s_{i*}(\psi)$$

and $\overline{\varphi}_i = < a_1, \ldots, a_n > \perp$ Null–form. Thus

$$sgn\ s_{i*}(\varphi_i) = (sgn\ s_{i*} < 1 >) \cdot sgn\ \overline{\varphi}_i + sgn\ s_{i*}(\psi).$$

Setting $m(\beta_i, \alpha) = sgn\ s_{i*} < 1 > \in \mathbb{Z}$ we finally get

$$sgn_\alpha s_*(\varphi) = \sum_{\beta|\alpha} m(\beta, \alpha) sgn_\beta(\varphi) + \sum_{\beta|\alpha} sgn\ s_{\beta_*}(\psi_\beta).$$

In order to derive a transfer formula in the sense of Knebusch, cf. [**K2, K3**], one needs assumptions implying the vanishing of the second sum on the right hand side. If we either assume that $B \mid A$ is étale, forcing $B \otimes_A k(\alpha)$ to be separable, or that φ is non–degenerate then ψ does not even occur. If we assume that $s \otimes_A 1$ vanishes on the radical of $B \otimes_A k(\alpha)$ for all $\alpha \in Sper\ A$ then $s_{i*}(\psi)$ is a Null–form; e.g. this happens if s is of the type $s(x) = tr_{B|A}(xa)$ for some fixed $a \in B$. Summarizing we get

Proposition 2.2. *In each of the following cases*

- $B \mid A$ *étale,*
- *only non–degenerate forms are considered,*
- $s \otimes_A 1 = 0$ *on* $Nil(B \otimes_k k(\alpha))$ *for each* $\alpha \in Sper\ A$

there are integers $m(\beta,\alpha)$, for every $\alpha \in Sper\ A, \beta \in Sper\ B$ such that $\beta \mid \alpha$, allowing the following transfer formula:

$$sgn_\alpha s_*(\varphi) = \sum_{\beta|\alpha} m(\beta,\alpha) sgn_\beta(\varphi)$$

for any one of forms φ being considered.

The case of an étale extension is dealt with in [**M**, (4.11)]. Note that $m(\beta,\alpha) = 1$ if $s = tr_{B|A}$.

If we are neither in the first or third case of the proposition the restriction to non-degenerate forms is essential. To see this consider Knebusch's example of a Frobenius algebra in [**K3**, p. 177]:

$$A = \mathbb{R}, B = \mathbb{R}[t], t^2 = 0, \{1, t\}\ \mathbb{R}\text{-basis of } B.$$

The unique $\alpha \in Sper\ A$ has a unique extension β to B. Consider the A-linear form s given by $s(1) = 0, s(t) = 1$. Then $s_* < t > \simeq < 1, 0 >$, hence a formula $sgn_\alpha s_*(\varphi) = m \cdot sgn_\beta(\varphi)$ cannot exist.

As an example where the last condition of prop. 2.2 is deliberately violated is the famous Eisenbud–Levine formula, cf. [**E-L**]. In that case one studies linear forms $s : B \to k$, k a field with $s \neq 0$ on the Nilradical of B, and such forms are needed for the application in mind.

If $(B \mid A)$ is a Frobenius extension then Knebusch's transfer formula for signatures, cf. [**K3**],

$$\sigma(s_*\varphi) = \sum_{\tau|\sigma} n(\tau)\tau(\varphi)$$

can be derived from our formula above as he has kindly pointed out to us. To this end, let $Sign\ A$ denote the set of signatures $\sigma : W(A) \to \mathbb{Z}$ as in [**K5**, §5]. It is known that the natural map $Sper\ A \to Sign\ A$, $\alpha \mapsto \sigma_\alpha$ is surjective ($\alpha \in Sper\ A$ induces $A \to k(\alpha)$, hence $\sigma_\alpha : W(A) \to W(k(\alpha)) = \mathbb{Z}$), cf. [**K5**, §5] or [**K4**, §§1,3]. In [**K3**, (1.1), p. 169] it was shown that there is a unique choice of the multiplicities $n(\tau)$. They are all strictly positive in the case of $s = tr_{B|A}$, cf. [**K2**, (3.4), p. 72].

To derive the trace formula for signatures let $\sigma = \sigma_\alpha$ and fix α. Then, for any non-degenerate form: $\sigma(\varphi) = sgn(\varphi \otimes_A k(\alpha)) = sgn_\alpha(\varphi)$. If $\beta \mid \alpha$ then $\sigma_\beta \mid \sigma_\alpha$. If $\tau \mid \sigma$ then $n(\tau) = \sum_\beta m(\beta,\alpha)$, summing up all β's with $\sigma_\beta = \tau$. With this definition we obviously get from our trace formula

$$\sigma(s_*(\varphi)) = \sum_{\tau|\sigma} n(\tau)\tau(\varphi).$$

If $s = tr_{B|A}$ then $m(\beta,\alpha) = 1$ hence $n(\tau) \geq 0$. Moreover, $n(\tau) > 0$ in the <u>unique</u> trace formula, as cited above. Therefore, if $\sigma = \sigma_\alpha$ then any extension $\tau \mid \sigma$ arises from an extension $\beta \mid \alpha$, i.e. $\tau = \sigma_\beta$.

3. A remark on Quantifier Elimination

The results of §2 in the formulation of the following prop. 4.2 allow us to give a very short and condensed quantifier–free expression for:

(i) $\exists x : g(x) = 0 \wedge f_1(x) > 0 \wedge \ldots \wedge f_m(x) > 0$ where $g \neq 0$;

(ii) $\exists x : f_1(x) > 0 \wedge \ldots \wedge f_m(x) > 0$.

By some well known arguments of model theory this can be extended to a proof of Quantifier Elimination in the theory of real closed fields. Let R be a real closed field and $Z \subset R$ a subring. If $g = a_nX^n + a_{n-1}X^{n-1} + \ldots + a_0, f_1, \ldots, f_m \in Z[X], a_n \neq 0$, we can use Prop. 4.2 to reformulate (i) as follows.

Let $A := R[X] \big/ (g(X))$ and x the canonical image of X in A. Then (i) becomes:

$$(*) \quad \bigvee_{k=1}^{n} sgn(tr_{A|R})_* < \Pi f_i(x) > \otimes \ll f_1(x), \ldots, f_n(x) \gg = k \cdot 2^n.$$

This transfer decomposes orthogonally into transfers of the type

$$(tr_{A|R})_* < h(x) >$$

where h is expressible as $h = f_1^{\epsilon_1} \cdot \ldots \cdot f_m^{\epsilon_m}$, for $\epsilon_1, \ldots, \epsilon_m \in \{1,2\}$. We can restrict ourselves to this case.

We are now going to sketch that $(*)$ is equivalent to a disjunction of quantifier–free expressions, polynomial over $\mathbb{Z}[1/a_n]$, in the coefficients of $g, f_1, \ldots, f_m$. We note that $1, x, \ldots, x^{n-1}$ is a linear basis of the R–vectorspace A. With respect to this basis we get a matrix–presentation B for $(tr_{A|R})_* < h(x) >$ with entries

$$b_{ij} = (tr_{A|R})_*(h(x)x^ix^j) \quad i, j = 0, \ldots, n-1.$$

DEFINITION 3.1. *The polynomials $M_j \in \mathbb{Z}[Z_1, \ldots, Z_j]$ defined by the recursion formula*

$$M_j + \sum_{i=1}^{j-1} Z_iM_{j-i} + jZ_j = 0, j \in \mathbb{N}_+,$$

are called Waring–polynomials.

We need to use the following classical result.

LEMMA 3.2. *For monic g we have*

$$tr_{A|R}(x^j) = M_j(a_{n-1}, \ldots, a_0, 0, \ldots) \; j \in \mathbb{N}_+$$

$$tr_{A|R}(x^0) = n$$

Hence, if $h = c_kX^k + \ldots + c_0$ we can express the (i,j)–th entry of the matrix B, setting $M_0 := n$ as:

$$tr_{A|R}(h(x)x^ix^j) = \sum_{l=0}^{k} c_l \cdot tr_{A|R}(x^{l+i+j})$$

$$= \sum_{l=0}^{k} c_l \; M_{l+i+j}(\frac{a_{n-1}}{a_n}, \dots, \frac{a_0}{a_n}, 0, \dots).$$

There exist ways of expressing the signature of the symmetric matrix B by applying Descartes rule of signs to the characteristic polynomial χ_B of B. But these are not the most effective methods.

If $\chi_B = d_n X^n + d_{n-1}X^{n-1} + \dots + d_0$ let N_+ and N_- be the number of sign changes in the sequences

$$d_0, \dots, d_n$$

$$d_0, -d_1, \dots, (-1)^n d_n$$

respectively. Then Descartes rule implies that the signature of B is equal to the integer

$$N_+ - N_-.$$

Thus, the case i) is settled.

To deal with (ii) we use an idea of [**KK**]. Let $F := \prod_{i=1}^{m} f_i$; then we see that the polynomial

$$g = (1+F^2)^2(\frac{F}{1+F^2})' = F'(1-F^2)$$

has a root between any two adjacent roots of F, and also has a root in each of the intervals $(-\infty, \alpha_1)$, (α_2, ∞), where α_1, α_2 are the smallest and the largest root, respectively.

Using those roots of g as testing points, we can replace ii) equivalently by:

$$\exists x : g(x) = 0 \wedge f_1(x) > 0 \wedge \dots \wedge f_m(x) > 0$$

Thus, we have arrived at the first case.

4. Counting real points

In this section we are going to apply the trace formula to the counting of real points on 0–dimensional affine varieties over real closed fields. We will use the book [**BCR**] as the basic reference for real algebraic geometry.

Let R denote a real closed field. It is one of the basic computational tasks in real algebraic geometry to decide whether a given semi algebraic $S \subset R^n$ is empty or not. Here, wlog, S may be thought to be given as the set of solutions in R^n of a system of the following type:

$$F_1(X_1, \dots, X_n) = 0, \dots F_r(X_1, \dots, X_n) = 0,$$

$$G_1(X_1, \dots, X_n) > 0, \dots G_s(X_1, \dots, X_n) > 0$$

with polynomials $F_1, \dots, G_s \in R[X_1, \dots, X_n]$. The decision about "$S = \emptyset$" or "$S \neq \emptyset$" should be made by using the coefficients of the polynomials involved in the above presentation of S.

Our problem can be rephrased as follows. The equations $F_1 = 0, \dots, F_r = 0$ define an affine variety V with coordinate ring $R[X_1, \dots, X_n] \Big/ (F_1, \dots, F_r)$, and the polynomials $G_1, \dots, G_s$ gives rise to regular functions $g_1, \dots, g_s$ on V (g_i being the coset of G_i in $R[V]$). In case $K \supset R$ is a field extension, $V(K)$ denote the set of points of V with coordinates in K.

In this setting S can be described as follows:
$S = \{x \in V(R) \mid g_1(x) > 0, \dots, g_s(x) > 0\}$.

If V is a 0–dimensional variety, i.e. Krull–dim $R[V] = 0$ or, equivalently, $\#V(R(\sqrt{-1})) < \infty$, we will attach to this representation of S a quadratic form φ over R such that

$$\#S = \frac{1}{2^s} sgn\ \varphi.$$

To count real points by calculating the signature of appropriate quadratic forms is a topic really begun in the last century by Borchardt, Jacobi, Sylvester and, first of all, Hermite. They studied the case $V = \mathbb{R}$ and $V = \mathbb{R}^2$, and their method is often referred to as the Hermite–Sylvester method. A very comprehensive account of this approach, complete up to about 1939, can be found in the reprint of a survey by Krein and Naimark, cf. [**Kr-N**]. Also the books of Knebusch–Scheiderer [**K-S**] and Benedetti–Risler [**B-R**, p. 17 ff] are recommended for further information.

Our general treatment of arbitrary 0–dimensional varieties stems from a geometric interpretation of the trace formula of §2. Quite astonishingly, P. Pedersen in a nearly simultaneous and completely independent study also introduced the same quadratic forms to achieve the joint goal: counting real zeros on zero–dimensional varieties, cf. [**P1**]. A brief account of our method has already been published in [**Be2**].

Now, let V be a 0–dimensional affine variety over the real closed field R with coordinate ring $R[V]$. Each real point $x \in V(R)$ gives rise to the evaluation map $e_x : R[V] \to R$, $f \mapsto f(x)$, which in turn yields the point $\alpha_x \in Sper\ R[V]$ defined as $\alpha_x = (\mathcal{M}_x, R_+)$, $\mathcal{M}_x = ker\ e_x$. From $dim\ R[V] = 0$ one concludes $Sper\ R[V] = \{\alpha_x \mid x \in V(R)\}$. If $\varphi =< f_1, \dots, f_s >$ is any diagonizable quadratic form over $R[V]$ (as $\frac{1}{2} \in R[V]$ no difference is made between bilinear and quadratic forms) then $\varphi_x :=< f_1(x), \dots, f_s(x) >$ is a form over R and we have

$$sgn_{\alpha_x} \varphi = sgn\ < f_1(x), \dots, f_s(x) > .$$

In our geometric context, the trace formula of §1 now reads

PROPOSITION 4.1. *For any* $f_1, \dots, f_s \in R[V]$

$$sgn\left((tr_{R[V]|R})_* < f_1, \dots, f_s >\right) = \sum_{x \in V(R)} sgn\ < f_1(x), \dots, f_s(x) > .$$

We are next going to specialize this geometric trace formula. If $s = 1$, $f_1 = 1$, then $sgn\ tr_* < 1 >= \#V(R)$. For given $f_1, \dots, f_s \in R[V]$ we form the "scaled Pfisterform" $\varphi =< \prod_1^s f_i > \cdot \ll f_1, \dots, f_s \gg$. An immediate calculation shows

$$sgn\ \varphi_x = \begin{cases} 2^s, & \text{if } f_1(x) > 0, \dots, f_s(x) > 0 \\ 0, & \text{otherwise} \end{cases}$$

Summarizing these two cases we get

PROPOSITION 4.2. (i) $sgn((tr_{R[V]|R})_* < 1 >) = \#V(R)$,
(ii) $sgn((tr_{R[V]|R})_* < \prod_1^s f_i >\ll f_1, \dots, f_s \gg) = 2^s \cdot \#\{x \in V(R) \mid f_1(x) > 0, \dots, f_s(x) > 0\}$

Clearly, the right-hand sides in the last proposition give the number of points one is interested in. However, if it comes to actual computation one necessarily has to deal with the left-hand sides of the above equations in an explicit and efficient way. In the following we will outline one possible way to cope with the trace forms themselves. There are clearly other methods and best efficiency is not claimed. In a forthcoming paper by Pedersen, Roy and Spzirglas [**P3**] the complexity of all the algorithms involved will be discussed in great detail.

Let V be given by a set of polynomial equations

$$F_1(X_1, \dots, X_n) = 0, \dots, F_r(X_1, \dots, X_n) = 0,$$

i.e. by setting $\mathfrak{a} = (F_1, \dots, F_r) \lhd R[X_1, \dots, X_n]$ we have

$$R[V] = R[X_1, \dots, X_n] \big/ \mathfrak{a}$$

Note that $\mathfrak{a}$ is not assumed to be a radical ideal.

The elements f_i are represented by polynomials G_i, $i = 1, \dots, s$. Let k be the field obtained by adjoining to $\mathbb{Q}$ all the coefficients of $F_1, \dots, F_r$ and, in the second case of prop. 4.2, of $G_1, \dots, G_s$. We set $\mathfrak{a}_0 := (F_1, \dots, F_r) \lhd k[X_1, \dots, X_n]$. A careful reading of the following reasoning will show that all the necessary arithmetic operations can be carried out in k. Note $\mathfrak{a} = \mathfrak{a}_0 R[X_1, \dots, X_n]$.

In prop. 4.2 only forms over $R[V]$ admitting an orthogonal basis occur. So, it is enough to deal with the case $\psi = (tr_{R[V]|R})_* < h >$, where $h = G + \mathfrak{a}$, $G \in k[X_1, \dots, X_n]$. To derive a matrix for ψ one needs an R-basis of $R[V]$. This can be achieved by obtaining a Gröbner–basis for $\mathfrak{a}_0$ using the Buchberger algorithm which is performed inside k. From such a Gröbner–basis of $\mathfrak{a}_0$ one gets a k-basis of $k[X_1, \dots, X_n] \big/ \mathfrak{a}_0$ which remains a R-basis of $R[X_1, \dots, X_n] \big/ \mathfrak{a}$ under scalar extension. To see the details of this argumentation one may consult e.g. [**Bu**]. We also use this reference as a source for all what is needed about Gröbner bases and the Buchberger algorithm.

We will proceed differently to get finally an R-basis on an appropriate R-algebra which has more structure and allows an easier way of determining the signature. However, costs are due for deriving this other algebra.

Set $A = R[V]$. Being a finite–dimensional R–algebra, A admits a Wedderburn decomposition

$$A = J \oplus \overline{A}, J = \quad \text{Nilradical}, \quad \overline{A} \quad \text{subalgebra}$$

where the natural projection $\pi : A \to A/_J = A_{red}$ induces an R–isomorphism $\overline{A} \simeq A_{red}$. We now resume arguments presented in the proof of the trace formula. Setting $h = u \oplus \overline{h}$, $u \in J$, $\overline{h} \in \overline{A}$ we get

$$sgn\ (tr_{A|R})_* < h >= sgn\ (tr_{\overline{A}|R})_* < \overline{h} > .$$

Being only interested in signatures we should therefore continue with the determination of $sgn\ (tr_{\overline{A}|R})_* < \overline{h} >$. To this end we have to calculate $\sqrt{\mathfrak{a}}$ out of $\mathfrak{a}$ since $\overline{A} \simeq R[X_1, \dots, X_n] \big/ \sqrt{\mathfrak{a}}$. To determine $\sqrt{\mathfrak{a}}$ we use the following observation of Seidenberg, cf. **[Se]**. First note that for any $i = 1, \dots, n$

$$\mathfrak{a}_0 \cap k[X_i] \neq (0),$$

namely, if not so, then, because of $\mathfrak{a} \cap k[X_1, \dots, X_n] = \mathfrak{a}_0$, we had an injection (over k) $k[X_i] \to R[X] \big/ \mathfrak{a}$ which is impossible since $R[X] \big/ \mathfrak{a}$ is an algebraic k–algebra. Now let $g_i \neq 0$ be arbitrarily chosen in $\mathfrak{a}_0 \cap k[X_i], i = 1, \dots, n$. Using *gcd*–calculation inside $k[X_i]$ we make g_i squarefree: $\tilde{g}_i = \frac{g_i}{gcd(g_i, g_i')}$. Then clearly $\tilde{g}_i \in \sqrt{\mathfrak{a}}, i = 1, \dots, n$, and, as Seidenberg noticed,

$$\sqrt{\mathfrak{a}} = (\mathfrak{a}, \tilde{g}_1, \dots, \tilde{g}_n).$$

To prove this statement first note that each R–algebra $R[X_i] \big/ (\tilde{g}_i)$ is separable as $char\ R = 0$. Consequently, also $B = \bigotimes_{i=1}^n R[X_i] \big/ (\tilde{g}_i)$ is separable, i.e. a finite product of field extensions of R. Obviously there is a natural epimorphism

$$B \twoheadrightarrow R[X_1, \dots, X_n] \big/ (\tilde{g}_1, \dots, \tilde{g}_n) \twoheadrightarrow R[X_1, \dots, X_n] \big/ (\mathfrak{a}, \tilde{g}_1, \dots, \tilde{g}_n)$$

showing the latter algebra to be separable, hence $(\mathfrak{a}, \tilde{g}_1, \dots, \tilde{g}_n)$ is a radical ideal. This proves the claim. The same argument applied to $\mathfrak{a}_0$ shows $\sqrt{\mathfrak{a}_0} = (\mathfrak{a}_0, \tilde{g}_1, \dots, \tilde{g}_n)$.

In order to calculate $tr_{\overline{A}|R} < \overline{h} >$ one needs an R–basis of $\overline{A}$. So far, there seems to be no advantage to use $\overline{A}$ instead of A. However, it is the so-called (folklore) Shape Lemma, cf. **[GiTrZ]**, that guarantees a distinguished set of generators of $\sqrt{\mathfrak{a}_0}$, hence of $\sqrt{\mathfrak{a}} = \sqrt{\mathfrak{a}_0} \cdot R[X_1, \dots, X_n]$. Thus we get a nice basis for $\overline{A}$.

Shape Lemma. Let k be a infinite perfect field, $\mathfrak{b}$ any 0–dimensional radical ideal in $k[X_1, \dots, X_n]$. Then, possibly only after a linear change of coordinates, i. e. $Y_i = X_i, i = 1, \dots, n-1, Y_n = X_n + \sum_1^{n-1} X_i \cdot t^i$, for some $t \in k$, there are polynomials $g_1, \dots, g_{n-1}, g \in k[T], g \neq 0$, square free, degree $g_i <$ degree g such that

$$\mathfrak{b} = (Y_1 - g_1(Y_n), \dots, Y_{n-1} - g_{n-1}(Y_n), g(Y_n))$$

Proof: Set $V = \{x \in \overline{k}^n \mid F(x) = 0 \text{ for all } F \in \mathfrak{b}\}$ where $\overline{k}$ = algebraic closure. Since $dim\ \mathfrak{b} = 0$, V is finite. We consider the projection $pr : V \to \overline{k}, (x_1, \dots, x_n) \mapsto x_n$. Assume it to be injective, i.e., by definition, V to be in "general position". The Galois group $G = Gal(\overline{k} \mid k)$ operates naturally on V, and $pr(V)$ is the union of full conjugacy classes. This implies the existence of a square free polynomial (note: k perfect) $g \in k[T]$ such that $pr(V) = \{x \in k \mid g(x) = 0\}$. For each $x_n \in pr(V)$ let $(x_n^1, \dots, x_n^{n-1}, x_n)$ be the unique point in V over x_n. By using the Lagrange interpolation formula and making use of the G-action on V one actually gets polynomials $g_1, \dots, g_{n-1} \in k[X]$, degree $g_i <$ degree g such that, for all $x_n \in pr(V), (g_1(x_n), \dots, g_{n-1}(x_n), x_n)$ is the point over x_n. This shows that $\mathfrak{b}$ and $\mathfrak{b}' = (X_1 - g_1(X_n), \dots, X_{n-1} - g_{n-1}(X_n), g(X_n))$ have the same points in $\overline{k}$. Since $k[X_1, \dots, X_n] \big/ \mathfrak{b}' \simeq k[X_n] \big/ (g(X_n))$ is separable we see that also $\mathfrak{b}'$ is a radical ideal. Hence, $\mathfrak{b} = \mathfrak{b}'$ by Hilbert's Nullstellensatz.

In the case that $\mathfrak{a}$ is not in general position we have to adjust it. Using the above mentioned coordinate transformation one sees that all but finitely many t's will put $\mathfrak{a}$ into the desired general position. □

In our situation, if we had $\mathfrak{a}_0$ in general position, then

$$\overline{A} = R[X_1, \dots, X_n \big/ (X_1 - g_1(X_n), \dots, X_{n-1} - g_{n-1}(X_n), g(X_n)) \simeq R[T] \big/ (g(T))$$

with $g \in k[T]$. Also, and this is crucial, $\overline{A}$ would admit the simple basis $1, T, \dots, T^{N-1}$, N = degree g. Relative to this basis the matrix for $(tr_{\overline{A}|R})_* < \overline{h} >$ shows additional features which allows a more accessible determination of its signature. Before turning to this point we must find the generators of $\mathfrak{a}_0$ as given in the Shape Lemma. It is the key point for our calculation that this set of generators allows a conceptual characterization as was first pointed out by Gianni and Mora, cf. [**Gi-Mo**]. Using the definition of a reduced (= minimal) Gröbner basis one readily verifies that, under the above conditions on $g_1, \dots, g_{n-1}, g$,

$$X_1 - g_1(X_n), \dots, X_{n-1} - g_{n-1}(X_n), g(X_n)$$

form the reduced Gröbner basis relative to the lexicographical order satisfying $X_1 > X_2 > \dots > X_n$.

By the remarks above we derive the actual computation of $\sqrt{\mathfrak{a}_0}$ hence of $\overline{A} = R[X_1, \dots, X_n] \big/ \sqrt{\mathfrak{a}}$. One first determines $g_i \in \mathfrak{a}_0 \cap k[X_i], g_i \neq 0, i = 1, \dots, n$, e.g. by Gröbner bases techniques, cf. [**Bu**]. One then determines the reduced Gröbner basis of $(\mathfrak{a}_0, \tilde{g}_1, \dots, \tilde{g}_n) = \sqrt{\mathfrak{a}_0}$ relative to the term ordering above. If this basis is not as expected then a random choice of the parameter

t in the coordinate transformation will help to get this position and a second Gröbner basis computation will do the job.

Having done all this we are constructively given polynomials $g_1, \dots, g_{n-1}, g \in k[X_n]$ and an R–isomorphism

$$\overline{A} \to R[T] \big/ (g(T)) = \tilde{A}, X_i \mapsto g_i(T), X_n \to T (i = 1, \dots, n-1).$$

Next we use the standard basis $1, t, \dots, t^{N-1}$ where $t = T + (g(T))$, $N =$ degree of g. Given $h(T) \in k[T]$ we have to determine the matrix of $(tr_{\tilde{A}|R})_* < h(t) >$ relative to this basis. At the place (i, j) of this matrix we find

$$tr_{\tilde{A}|R}(h(t)t^{i+j-2}) = \sum_{\substack{\alpha \in R(\sqrt{-1}) \\ g(\alpha)=0}} h(\alpha)\alpha^{i+j-2}.$$

Invoking the symmetric function theorem we get that the right hand side is a $\mathbb{Z}$–polynomial in the coefficients of $h(T)$ and $g(T)$. The chapter 4 of [**P2**] is devoted to a study of algorithms for evaluating symmetric functions. One may use those methods, but there is another way to determine $tr_{\tilde{A}|R}(h(t)t^l)$. One expands the rational function $T\frac{h(T)g'(T)}{g(T)}$ in the formal power series field $k((T^{-1}))$ and passes to $\overline{k}((T^{-1})) \supset k((T^{-1}))$. Writing $g(T) = \prod(T - a)$ we get $h(T)\frac{g'(T)}{g(T)} = \sum_a \frac{h(T)}{T-a} = \sum_a \frac{h(\alpha)}{T-a} + H(T)$ for some polynomial $H \in \overline{k}[T]$. After multiplying by T we finally get: $T\frac{h(T)g'(T)}{g(T)} = TH(T) + \sum_a \frac{h(\alpha)}{1-aT^{-1}} = TH(T) + \sum_{l=0}^{\infty}(\sum_a h(a)a^l)T^{-l}$. Hence $tr_{\tilde{A}|R}(h(t)t^l)$ is the coefficient of T^{-l} in this expansion.

The resulting matrix is a so called Hankel matrix $H = (a_{i+j-2})_{i,j=1,\dots,N}$ built up from a sequence $a_0, \dots, a_{2N-2} \in k$. There are efficient methods for determing the signature of a Hankel matrix, e.g. by a theorem of Frobenius. To see the details one may consult [**G**] or [**I**].

So far we have outlined a method to compute $sgn\ (tr_{A|R})_* < h >$. This clearly implies the determination of $\#V(R) = sgn\ (tr_{A|R})_* < 1 >$. However in the case of the second statement of prop. 4.2 we would be forced to cope with as many as 2^s calculations of the type $sgn\ (tr_{A|R})_* < h_i >$. Even for fairly small values of s this would be beyond any feasible limit. In the next section we show that in fact the simultaneous inequalities $f_1 > 0, \dots, f_s > 0$ can be replaced by just a single one $h > 0$, and a single one can be handled as above. The reduction of the s inequalities to just one is not without expenses. So one should look for other methods. The fundamental B–K–R algorithm of [**BKR**] applies to several inequalities by an ingenious procedure using only one inequality in each step. For further reading one may turn to Pedersen's paper or the forthcoming one by Pedersen, Roy and Spzirglas. Also, after the reduction to the univariate case $\tilde{A} = R[T] \big/ (g(T))$ other methods are available as well, cf. [**GLRR**].

5. The 0–dimensional case of the Bröcker–Scheiderer theorem

It was proved by C. Scheiderer [**S**] and L. Bröcker (unpublished) that in any n–dimensional affine variety V over a real closed field R a basic open set $S = \{x \in V(R) \mid f_1(x) > 0, \dots, f_r(x) > 0\}$ can in fact be described by at most $\overline{n}$ inequalities ($\overline{n} = max(1, n)$):

$$S = \{x \in V(R) \mid h_1(x) > 0, \dots, h_{\overline{n}}(x) > 0\}, h_1, \dots, h_{\overline{n}} \in R[V].$$

This amazing theorem is a real challenge to computation since all known proofs do not offer constructive methods to find $h_1, \dots, h_{\overline{n}}$.

In the sequel we consider the case of dim $V = 0$ and will propose an algorithm to find h_1 starting with a description of V and polynomials representing $f_1, \dots, f_r$.

So let V be described by polynomials as

$$V : F_1 = 0, \dots, F_s = 0$$

where $F_1(X_1, \dots, X_n), \dots, F_s(X_1, \dots, X_n) \in k[X_1, \dots, X_n]$ for some $k \subset R$. We also assume $f_1, \dots, f_r \in k[X_1, \dots, X_n]$. Finally we set $\mathfrak{a}_0 = (F_1, \dots, F_s) \lhd k[X_1, \dots, X_n], \mathfrak{a} = \mathfrak{a}_0 R[X_1, \dots, X_n]$. As explained in the last section, e.g. by Gröbner–bases techniques, we find the following generators of $\sqrt{\mathfrak{a}_0}$ (possibly after a coordinate transformation)

$$X_1 - g_1(X_n), \dots, X_{n-1} - g_{n-1}(X_n), g(X_n)$$

with $g_1, \dots, g \in k[X_n]$, g square–free, $deg\ g_i < deg\ g, (i = 1, \dots, n-1)$. Setting $N(g) = \{\alpha \in R \mid g(\alpha) = 0\}$ we find an isomorphism

$$\Phi : V(R) \to N(g), (x_1, \dots, x_n) \mapsto x_n$$

sending our regular functions $f_1(X_1, \dots, X_n), \dots, f_r(X_1, \dots, X_n)$ into the univariate polynomials $f_i(g_1(T), \dots, g_{n-1}(T), T) = \overline{f}_i \in k[T], i = 1, \dots, r$. After this transformation we are concerned with the set

$$(5.1) \quad \overline{S} = \{\alpha \in R \mid g(\alpha) = 0, \overline{f}_1(\alpha) > 0, \dots, \overline{f}_r(\alpha) > 0\}.$$

Suppose we have found $h(T)$ such that $\overline{S} = \{\alpha \in R \mid g(\alpha) = 0, h(\alpha) > 0\}$. Then $\{x \in V(R) \mid f_1(x) > 0, \dots, f_r(x) > 0\} = \{x \in V(R) \mid h(x_n) > 0\}$. Hence, we have to settle the case (5.1) under the assumption that g is squarefee. We first want to point out that, in our case, the Bröcker–Scheiderer result is an immediate consequence of the geometric trace formula and some simple quadratic form theory. Setting $\varphi = <\Pi \overline{f_i}> \cdot \ll \overline{f_1}, \dots, \overline{f_r} \gg$ we will find $h \in R[T]$ with $\varphi = 2^{r-1} \cdot <h> \ll h \gg$ over $A = R[T] / (g(T))$. As g is squarefree we have $A = \prod R \times \prod R(\sqrt{-1})$ where each factor R corresponds to a point $x \in N(g)$. If some $\overline{f_i}(x) = 0$ then $\varphi_x =$ Null–form of dimension 2^r and if all $\overline{f_i}(x) \neq 0$ either φ_x hyperbolic or $\varphi_x \simeq 2^r \times <1>$. Since quadratic form theory over R and $R(\sqrt{-1})$ is completely known one easily finds $h \in R[T]$ as desired.

The algorithms we are going to propose will work in the following two cases:

(i) $R = \mathbb{R}$,

(ii) $R = \mathbb{R}\{\epsilon_1\}\dots\{\epsilon_n\}$, an iterated Puiseux–series field with ϵ_i infinitesimal small relative to $\mathbb{R}\{\epsilon_1\}\dots\{\epsilon_{i-1}\}$.

For definitions cf. [**BCR**, (1.2.3), p. 10].

1st case: $R = \mathbb{R}$

Once the binary case $r = 2$ in $(*)$ is settled a recursive procedure will cover the general case. So, consider the situation

$$g(\alpha) = 0, f_1(\alpha) > 0, f_2(\alpha) > 0, g, f_1, f_2 \in k[T].$$

and notice that we do not know the zeros of g in k. The following reasoning will make use of the $\alpha \in N(g)$ only on a conceptual level; the algorithm itself entirely deals with g, f_1, f_2, i.e. their coefficients. The set $N(g)$ being finite implies that there are polynomials, say,

$$sgn\ f_1, sgn\ f_2 \in R[T]$$

such that $(sgn\ f_i)(\alpha) = sgn(f_i(\alpha)), i = 1, 2$ holds. Now, set $H = sgn\ f_1 \cdot sgn\ f_2$ $(1 + sgn\ f_1 + sgn\ f_2)$ then one readily verifies for all $\alpha \in N(g)$:

$$\text{(5.2)} \quad H(\alpha) > 0 \Longleftrightarrow f_1(\alpha) > 0, f_2(\alpha) > 0.$$

Thus, it remains to compute $sgn\ f_i$ as defined above. As a matter of fact, we don't know of any way to do that. If one could compute the square root of positive functions in $A = R[T]/(g(T))$ then an application to $\sqrt{f^2}$ would help to find $sgn(f) \in A$. But, no method is known to us. However, looking at (5.2) and the definition of H one observes that any pair of sufficiently close approximations of $sgn(f_i), i = 1, 2$ will serve as well. In fact, if for $i = 1, 2, s_\epsilon(f_i)$ denote a polynomial with

$$\mid s_\epsilon(f_i) - sgn\ f_i \mid < \epsilon \quad \text{on} \quad N(g) \quad \text{and} \quad s_\epsilon(f_i)(\alpha) = 0 \quad \text{whenever} \quad f_i(\alpha) = 0$$

and if $0 < \epsilon < \frac{1}{3}$ then setting $H_\epsilon = s_\epsilon(f_1) \cdot s_\epsilon(f_2) \cdot (1 + s_\epsilon(f_1) + s_\epsilon(f_2))$ we also get on $N(g)$:

$$\text{(5.3)} \quad H_\epsilon > 0 \Longleftrightarrow f_1 > 0, f_2 > 0$$

Consequently, given $f \in \mathbb{R}[T]$, we have to find an approximation of $sgn(f)$ on $N(g)$ by a polynomial $s_\epsilon(f) \in \mathbb{R}[T]$. This will be basically achieved by a global Newton method applied to the ring $A_0 = \mathbb{R}[T]/(g(T))$, the latter being considered as a ring of functions on $N(g)$. To display the basic idea we assume that all of the following operations can be carried out in A_0. We are going to write down the Newton sequence for the equation $X^2 - 1 = 0$ in A_0 starting with the initial value $f_0 = f$, i.e. we get (f_k) where

$$f_{k+1} = \frac{1}{2}(f_k + \frac{1}{f_k}), \quad f_k \in A_0.$$

Looking at the parabola defined by $y = x^2 - 1$ and taking the geometric interpretation of the Newton method into account one readily checks that (f_k) converges to $sgn\ f$ on $N(g)$.

However, to carry out this idea we have to cope with the situation that f or some f_k are not units in A_0, i.e. that some f_k are not relatively prime. That this can happen accounts for a more careful approach. The basic idea will be kept nevertheless.

PROPOSITION 5.1. *Given $f, g \in \mathbb{R}[T]$ one can construct a sequence of polynomials (T_k) in $\mathbb{R}[T]$, without knowing the zero-set $N(g)$, such that*

(i) $\lim_{k\to\infty} F_k(\alpha) = sgn\ f(\alpha)$ *for every* $\alpha \in N(g)$,

(ii) $F_k(\alpha) = 0$ *for every* $\alpha \in N(g)$ *satisfying* $f(\alpha) = 0$.

Proof: **1st case:** f, g relatively prime. We are going to construct two sequences $(g_k), (F_k), k \geq 0$, in $\mathbb{R}[T]$ subject to

a) $g_0 = g, g_k \mid g_{k-1}, N(g_k) = N(g)$,

b) $F_0 = f, F_k$ and g_k relatively prime,

$$F_{k+1}(\alpha) = \frac{1}{2}\left(F_k(\alpha) + \frac{1}{F_k(\alpha)}\right) \quad \text{for all} \quad \alpha \in N(g).$$

Since, by assumption, there is no $\alpha \in N(g)$ with $f(\alpha) = 0$, condition (ii) is empty, and so the sequence (F_k) has the desired properties. Assume g_k, F_k are constructed. Then $f_k := F_k + (g_k) \in \mathbb{R}[T]/_{(g_k)} = A_k$ is a unit in A_k. Hence we can form $\tilde{f}_{k+1} = \frac{1}{2}(f_k + \frac{1}{f_k})$ in A_k. Note that f_k^{-1} can be computed by the Euclidean algorithm applied to F_k and g_k in $\mathbb{R}[T]$. Choose any $F_{k+1} \in \mathbb{R}[T]$ representing $\tilde{f}_{k+1}$ in A_k. Set $g_{k+1} = g_k/gcd(F_{k+1}, g_k)$. Since g_k, as a divisor of g, is squarefree we get that F_{k+1} and g_{k+1} are relatively prime. Clearly, $N(g_{k+1}) \subseteq N(g_k)$, but any $x \in N(g_k)\backslash N(g_{k+1})$ would satisfy $g_k(\alpha) = 0 = F_{k+1}(\alpha)$ implying $f_k^2(\alpha) + 1 = 0$ which is impossible in view of $\alpha \in \mathbb{R}$, $f_k \in \mathbb{R}[T]$. Hence, the 1st case is settled.

2nd case: Since g is squarefree the polynomials $\tilde{g} = g/gcd(g, f)$ and f are relatively prime. So we find a sequence $(\tilde{F}_k)$ doing the job on $N(\tilde{g})$ which is $N(g)\backslash N(f)$. Using again that $gcd(\tilde{g}, f) = 1$ we find by the Euclidean algorithm a polynomial, say, $(f, g)^{-1} \in k[T]$ satisfying $(f, g)^{-1} \cdot f \equiv 1 \mod \tilde{g}$. Now set

$$\chi = (f, g)^{-1} \cdot f$$

then on $N(g)$:

$$\chi(\alpha) = \begin{cases} 1 & f(\alpha) \neq 0, \quad i.e. \quad \alpha \in N(\tilde{g}) \\ 0 & f(\alpha) = 0, \quad i.e. \quad \alpha \notin N(\tilde{g}), \end{cases}$$

so χ is the characteristic function of $N(\tilde{g})$. Hence, setting $F_k := \chi \cdot \tilde{F}_k$, we get the desired sequence. □

It remains to decide when to stop the sequence (F_k) in order to get the approximation

$$| F_k(\alpha) - sgn\ f(\alpha) |< \frac{1}{3} \quad \text{for} \quad \alpha \in N(g).$$

On $N(g) \cap N(f)$ we have $F_k(\alpha) = 0$, so no problem arises. On $N(g)\backslash N(f)$ we have $F_{k+1}(\alpha) = \frac{1}{2}(F_k(\alpha) + \frac{1}{F_k(\alpha)})$. In particular, $| F_k(\alpha) |\geq 1$ for $k \geq 1$, $\alpha \in N(g)\backslash N(f)$. For those α's we get by induction, if $k \geq 1$

$$| F_{k+l}(\alpha) |\leq \frac{1}{2^l} | F_k(\alpha) | + \frac{2^l - 1}{2^l},$$

hence, choosing any bound M for $\sup\limits_{\alpha \in N(g)} | F_k(\alpha) |$, then for $k \geq 1$ one gets

$$| F_{k+l}(\alpha) - sgn\ f(\alpha) |< \frac{1}{3}$$

provided $\frac{1}{2^l}M + 1 < \frac{4}{3}$, i.e. $\log_2(3M) < l$.

The final task remains to find a bound M. There are at least two ways. Set $F = F_k$, $k \geq 1$.

I) Consider $H(X) = Res_Y(g(Y), X - f(Y))$ then for $\beta \in \mathbb{R}$:

$$H(\beta) = 0 \Longleftrightarrow \beta = f(\alpha) \quad \text{for some} \quad \alpha \in N(g).$$

As is well–known the real roots of a polynomial h can be bounded in absolute value by $1 + \|h\| := 1 + \max\{|\text{coefficients}|\}$.

II) A less sophisticated bound can be obtained as follows:

$$| F(\alpha) |=| \sum_0^N a_i\alpha^i |\leq \sum | a_i | \cdot | \alpha |^i \leq \|F\| \cdot (1 + \|g\|)^N$$

Clearly, F_k can be chosen with $deg\ F_k \leq n - 1$ where $n = deg\ g$. Hence, $M \leq \|F_k\| \cdot (1 + \|g\|)^{n-1}$.

We are going to summarize.

PROPOSITION 5.2. *If $k \geq 1$, $\sup\limits_{\alpha \in N(g)} | F_k(\alpha) |\leq M$ and the F_l are chosen with $deg\ F_l \leq n - 1$ where $n = deg\ g$ then we have*

$$| F_{k+l}(\alpha) - sgn\ f(\alpha) |< \frac{1}{3} \quad \textit{for all} \quad \alpha \in N(g)$$

provided $l > log_2(3M)$. The bound M can be obtained as described above.

<u>2nd case: $R = \mathbb{R}\{\epsilon_1\} \dots \{\epsilon_n\}$</u>

Basically, we follow the same approach as in the case $R = \mathbb{R}$. If $f \in R[T]$ there is a further polynomial denoted by $sgn\ f$ which satisfies

$$(sgn\ f)(\alpha) = sgn\ f(\alpha) \quad \text{for all roots} \quad \alpha \quad \text{of} \quad g \quad \text{in} \quad R, \quad \text{i.e.} \quad \alpha \in N(g).$$

As above, we try to find a polynomial h subject to

$$| h(\alpha) - (sgn\ f)(\alpha) | < \frac{1}{3} \quad \text{for all} \quad \alpha \in N(g).$$

To this end we want to use the Newton-method, i.e. we start by studying the sequence $f_{k+1} = \frac{1}{2}(f_k + \frac{1}{f_k}), k \geq 0, f_0 = f + (g(T))$, in $A_0 = {R[T]} \big/ {(g(T))}$. However, in our present case, even if all f_k are units, the sequence (f_k) does not necessarily converge to $sgn\ f$ on $N(g)$. This failure is due to the fact that the order of R is non–Archimedean. We will remove this problem by modifying f into $\overline{f} \in R[T]$ where $sgn\ f(\alpha) = sgn\ \overline{f}(\alpha)$ for $\alpha \in N(g)$ and $\overline{f}$ allows a convergent Newton-sequence.

To prepare the construction of $\overline{f}$ we first study the behaviour of the mapping $a, b \to a + \frac{1}{b}$ where $a, b \in R^*, ab > 0$. We will make use of the (Henselian) valuation v of R which arise from the recursive construction of R, is trivial on $\mathbb{R}$, has value group $\Gamma = \mathbb{Q} \times \ldots \times \mathbb{Q}$, n–times, lexicographically ordered in ascending order of the factors, and residue field $\mathbb{R}$. In fact, every $a \in R^*$ has a unique presentation $a = \epsilon_1^{r_1} \ldots \epsilon_n^{r_n} \cdot u, r_1, \ldots, r_n \in \mathbb{Q}$, u a unit. We get $v(a) = (r_1, \ldots, r_n) \in \Gamma$, and $a > 0$ iff the residue class of $u > 0$ in $\mathbb{R}$.

As usual, an element a is called infinitely small (resp. large) if $| a | < r$ (resp. $| a | > r$) for all $r \in \mathbb{R}, r > 0$. Equivalently, a is infinitely small (resp. large) if and only if $v(a) > 0$ (resp. $v(a) < 0$).

In particular, if a is any positive element with $v(a) \geq 0$ then $v(1 + a) = 0$. Using this one readily checks the following statements.

(∗) Assume $ab > 0$. Then
 a) $0 \geq v(a) \geq v(b) \Rightarrow v(a + \frac{1}{b}) = v(a)$.
 b) $v(a) \geq v(b) \geq 0 \Rightarrow v(a + \frac{1}{b}) = v(\frac{1}{b})$.

Consequently, for any $a \in R^*$ and setting $b = \frac{1}{2}(a + \frac{1}{a})$:

(∗∗)
 a) $v(a) < 0 \Rightarrow v(b) = v(a)$.
 b) $v(a) > 0 \Rightarrow v(b) = -v(a) < 0$.
 c) $v(a) = 0 \Rightarrow v(b) = 0$.

Now we consider the following Newton-sequence $x_{k+1} = \frac{1}{2}(x_k + \frac{1}{x_k}), x_0 = a$. Assume $v(x_0) = v(a) = 0$. Then $v(x_k) = 0$ for every $k \in \mathbb{N}$. We can write $x_k = \epsilon_k + m_k$ where $\epsilon_k \in \mathbb{R}^*, v(m_k) > 0$. Then $\epsilon_{k+1} = \frac{1}{2}(\epsilon_k + \frac{1}{\epsilon_k})$, hence (ϵ_k) converges in $\mathbb{R}$ to $sgn\ \epsilon_0 = sgn\ a$. This means we find for given $r \in \mathbb{R}^*_+$ $k(r) \in \mathbb{N}$ such that $| x_k - sgn(a) | < r$ for all $k \geq k(r)$.

Therefore, if an arbitrary element $a \in R^*$ is given and $sgn\,(a)$ should be computed via a Newton–sequence we first have to replace a by $\overline{a}$ subject to $v(\overline{a}) = 0, sgn\,(a) = sgn\,(\overline{a})$. Clearly, there is no problem if a is given explicitly. However, in our application we deal with the elements $f(\alpha), \alpha \in N(g)$ without knowing the roots $\alpha \in N(g)$, having only f, g at our disposal. As in the case $R = \mathbb{R}$ we consider $H(X) = Res_Y(g(Y), X - f(Y))$. We know that the roots of

H in R are exactly the values $f(\alpha), \alpha \in N(g)$.

Therefore we are facing the following problem: given a polynomial $h \in R[T]$ design an algorithm constructing, on the input $a \in R^*$, an element $\overline{a}$ such that for all $\alpha \in N(h), \alpha \neq 0$ we get $v(\overline{\alpha}) = 0, sgn\ \alpha = sgn\ \overline{\alpha}$. The following algorithm is based on the observations listed in $(*)$ above.

1st step. Construct a list of elements $(x_i)_{i=1,\dots N}$ of R such that

1) $x_i > 0, i = 1, \dots, N$,
2) $v(x_1) > v(x_2) > \dots > v(x_N)$,
3) for every $\alpha \in N(h), \alpha \neq 0$ there is some x_i with $v(\alpha x_i) = 0$.

2nd step. For each $a \in R^*$ compute $\overline{a}$ as the continued fraction

$$\overline{a} = [x_1 a, \dots, x_N a] = x_1 a + \cfrac{1}{x_2 a + \cfrac{1}{x_3 a + \cfrac{1}{\ddots \cfrac{1}{x_{N-1} + \cfrac{1}{x_N a}}}}}$$

We first show that $\overline{\alpha}$ has the desired properties if $h(\alpha) = 0, \alpha \in R$. Obviously, $sgn\ \overline{a} = sgn\ a$ for every $a \in R$. Let $v(\alpha x_i) = 0$. Then $v(\alpha x_1) \geq \dots \geq v(\alpha x_{i-1}) \geq v(\alpha x_i) = 0 \geq \dots \geq v(\alpha x_N)$. Let $\overline{\alpha}_i = x_i \alpha + \cfrac{1}{x_{i+1}\alpha + \cfrac{1}{\ddots \cfrac{1}{x_N \alpha}}}$ denote the "lower" part of $\overline{\alpha}$.

From $(*)$ we deduce $v(\overline{\alpha}_i) = 0$, $sgn\ \overline{\alpha}_i = sgn\ \alpha$. Again by using $(*)$ we derive $v(\overline{\alpha}) = 0$.

Thus, it remains to construct the list $(x_i)_{i=1,\dots,N}$. Let $h = \sum_{i=0}^{n} a_i X^i$, $\alpha \neq 0$ and $h(\alpha) = 0$. Since $h(\alpha) = 0$ there exist $i < j$ such that $a_i a_j \neq 0$ and $v(a_i \alpha^i) = v(a_j \alpha^j)$, i.e. $v(\alpha) = \frac{1}{j-i}(v(\frac{a_i}{a_j}))$. Therefore, if we can produce a positive element $x_{ij} \in R^*$ with $v(x_{ij}) = -\frac{1}{j-i} v(a_i/a_j)$ for each pair (i,j) such that $0 \leq i < j \leq n$ and $a_i, a_j \neq 0$ then the list (x_i) is obtained from ordering the x_{ij}'s accordingly. We have $\frac{a_j}{a_i} = \epsilon_1^{r_1} \dots \epsilon_n^{r_n} \cdot u$, u a unit. Then set $x_{ij} = \epsilon_1^{s_1} \dots \epsilon_n^{s_n}$ where $s_k = \frac{1}{j-i} r_k$, $k = 1, \dots, n$. These elements have the desired properties.

Next, we transfer this algorithm to the global setting of polynomial functions on $N(g)$. Let $f \in R[T]$ and assume first that f and g are relatively prime. We will produce a polynomial $\overline{f}$ such that $\overline{f}(\alpha) = \overline{(f(\alpha))}$ for every $\alpha \in N(g)$. As already remarked above the values $f(\alpha)$ are the zeros of $h = Res_Y(g(Y), X - f(Y))$ in R. Then construct the list (x_i) attached to h in the algorithm above. In $R(T)$ we calculate the continued fraction $[x_1 f, \dots, x_N f] =: f^*$. Using the recursion formulae for denominators of continued fractions we see that the denominators arising during the computation are of the type $f^r \cdot H$, $r = 0, 1$, $H \in R[T]$ a polynomial without zeros in R. This implies that $f^*(\alpha) = [x_1 f(\alpha), \ldots, x_N f(\alpha)] = \overline{(f(\alpha))}$ for all $\alpha \in N(g)$. Set $f^* = \frac{A}{B}$, $A, B \in R[T]$ and $\overline{g} = g/gcd(g, B)$ where we assume $B = f^r H$ as above. Then $N(g) = N(\overline{g})$ and $gcd(B, \overline{g}) = 1$ since g is squarefree. From a presentation $1 = BC + D\overline{g}$, $C, D \in R[T]$ we derive that $B(\alpha)C(\alpha) = 1$ for every $\alpha \in N(\overline{g}) = N(g)$. Hence, $f^*(\alpha) = A(\alpha)C(\alpha)$ and $\overline{f} := A \cdot C$ is one of the wanted polynomials. Clearly, this polynomial $\overline{f}$ can be further reduced modulo $\overline{g}$ without loosing the property we are interested in.

We now drop the assumption that f and g are relatively prime. Then pass to $f_0 = f$ and $\tilde{g} = g/gcd(f, g)$. Compute $\overline{f}_0$ relative to f and $\tilde{g}$ as above, i.e. $\overline{f}_0(\alpha) = \overline{(f(\alpha))}$ for every $\alpha \in N(g)$ satisfying $f(\alpha) \neq 0$. As in the case $R = \mathbb{R}$ we find a polyomial χ satisfying $\chi(\alpha) = 1$ if $g(\alpha) = 0$, $f(\alpha) \neq 0$ and $\chi(\alpha) = 0$ if $g(\alpha) = 0 = f(\alpha)$. Then set $\overline{f} = \chi \cdot \overline{f}_0$.

Thus, in both cases we have constructed $\overline{f} \in R[T]$ satisfying for every $\alpha \in N(g)$: $\overline{f}(\alpha) = 0$ if $f(\alpha) = 0$, $\overline{f}(\alpha) = \overline{(f(\alpha))}$ if $f(\alpha) \neq 0$. Now the Newton-method can be applied to $\overline{f}$ to construct a sequence (F_k) in $R[T]$ approximating $sgn\ \overline{f}$. Now, since $\overline{f}(\alpha) = 0$ if $f(\alpha) = 0$ and $sgn\ \overline{f}(\alpha) = sgn\ f(\alpha)$ otherwise, from (F_k) we obtain a polynomial $h \in R[T]$ satisfying $\mid h(\alpha) - sgn\ f(\alpha) \mid < \frac{1}{3}$ as desired. □

REFERENCES

[Ba] R. Baeza, *Quadratic forms over semilocal rings*, Lect. Notes Math., vol. 655, Springer, 1978.

[Be1] E. Becker, *On the real spectrum of a ring and its application to semi-algebraic geometry*, Bull. Amer. Math. Soc. **15** (1986), 19–60.

[Be2] ———, *Sums of squares and quadratic forms in real algebraic geometry*, Cahiers du Seminaire d'Histoire des Mathematique, 2e Serie, vol. 1, Univ. P. et M. Curie, 1991.

[BCR] J. Bochnak, M. Coste, and M.-F. Roy, *Géométrie Algébrique Réelle, Ergebnisse der Mathematik und ihrer Grenzgebiete*, (3. Folge) 12, Springer, Berlin, Heidelberg and New York, 1987.

[BKR] M. Ben-Or, D. Kozen, and J. Reif, *The complexity of elementary algebra and geometry*, J. Comp. System Sciences **32** (1986), 251–264.

[B-R] R. Benedetti, and J. J. Risler, *Real algebraic and semi-algebraic sets*, Hermann, Éditeurs des sciences et de arts, Paris, 1990.

[Bu] B. Buchberger, *Gröbner bases: An algorithmic method in polynomial ideal theory*, Multidimensional System Theory (N. K. Bose, ed.), D. Reidel Publishing Company, Dordrecht, Boston and Lancaster, 1985, pp. 184–232.

[C-E] H. Cartan, and S. Eilenberg, *Homological Algebra*, Princeton University Press, Princeton, N.J., 1956.

[C-R] M. Coste, and M.-F. Roy, *La topologie du spectre rëel*, Contemp. Math. **8** (1982), 27–59.

[DeM] F. De Meyer, and E. Ingraham, *Separable Algebras over Commutative Rings*, Springer, Berlin, Heidelberg and New York, 1971.

[E-L] D. Eisenbud, and H. I. Levine, *An algebraic formula for the degree of a C^∞ map germ*, Annals of Math. **106** (1977), 19–44.

[G] F. R. Gantmacher, *Matrizentheorie*, Springer, Berlin, Heidelberg and New York, 1986.

[Gi-Mo] P. Gianni, and T. Mora, *Algebraic solution of systems of polynomial equations using Gröbner bases*, Lect. Notes Comp. Science **356** (1987), 247–257.

[GLRR] L. Gonzalez, H. Lombardi, T. Recio, and M.-F. Roy, *Sturm–Habicht sequence*, Proc. of ISSAC–89 (1989), pp. 136–146.

[GiTrZ] P. Gianni, B. Trager, and G. Zacharias, *Gröbner bases and primary decomposition of polynomial ideals*, J. Symb. Comp. **6** (1988), 149–167.

[I] I. S. Iohvidov, *Hankel and Toeplitz Matrices and Forms*, Birkhäuser, Boston, Basel and Stuttgart, 1982.

[K1] M. Knebusch, *On the uniqueness of real closures and the existence of real places*, Comment. Math. Helv. **47** (1972), 657–673.

[K2] ———, *Real closures of commutative rings, I*, J. reine angew. Math. **274/275** (1975), 61–80.

[K3] M. Knebusch, A. Rosenberg, and R. Ware, *Signatures on Frobenius Extensions, Number Theory and Algebra*, Academic Press, New York, San Francisco and London, 1977, 167–186.

[K4] M. Knebusch, *Signaturen, reelle Stellen und reduzierte quadratische Formen*, Jahresbericht Deutsche Math. **82** (1980), 109–127.

[K5] ———, *An invitation to real spectra*, CMS Conf. Proc. 4 (Quadratic and Hermitian Forms), 1984, pp. 51–105.

[K-S] M. Knebusch and C. Scheiderer, *Einführung in die reelle Algebra*, Vieweg, Braunschweig und Wiesbaden, 1989.

[KK] E. I. Korkina and A. G. Kushnirenko, *Another proof of the Tarski–Seidenberg theorem*, Sibirskii Mathematicheskii Zhurnal **26** (1985), no. 5, 94–98.

[Kr-N] M. G. Krein and M. A. Naimark, *The method of symmetric and hermitian forms in the theory of the separation of the roots of algebraic equations*, Kharkov, 1936 (Russian)L English transl. in: Lin. Multilin. Algebra **10** (1981), 265–308.

[M] L. Mahé, *Signatures et composantes connexes*, Math. Ann. **260** (1982), 191–210.

[P1] P. Pedersen, *Generalizing Sturm's theorem to N dimensions*, Tech. Report, New York University, Courant Institute of Mathematical Science, April 1990, Dept. of Computer Science.

[P2] ———, *Counting real zeros*, Tech. Report, New York University, Courant Institute of Mathematical Science, Nov. 1990, Dept. of Computer Science.

[P3] P. Pedersen, M.-F. Roy, and A. Szpirglas, *Counting real zeros in the multivariate case*, Proc. of MEGA '92 (to appear).

[S] C. Scheiderer, *Stability index of real varieties*, Invent. Math. **97** (1989), 467–483.

[SchSt] G. Scheja and U. Storch, *Lehrbuch der Algebra*, B. G. Teubner, Stuttgart, 1988, Band 2.

[Se] A. Seidenberg, *Constructions in algebra*, Trans. AMS **197** (1974), 273–313.

MATHEMATISCHES INSTITUT DER UNIVERSITÄT DORTMUND, 4600 DORTMUND, FEDERAL REPUBLIC OF GERMANY

Contemporary Mathematics
Volume **155**, 1994

Quadratic Forms with Values in Line Bundles

W. BICHSEL AND M.-A. KNUS

1. Introduction

Let X be a scheme such that $\frac{1}{2} \in \mathcal{O}_X$ and let $\mathcal{I}$ be a line bundle over X. A quadratic space over X with values in $\mathcal{I}$ is a triple $(\mathcal{F}, h, \mathcal{I})$, where $\mathcal{F}$ is a bundle over X and h is a selfdual isomorphism

$$h : \mathcal{F} \to \mathcal{F}^* \otimes_{\mathcal{O}_X} \mathcal{I},$$

$\mathcal{F}^* = \mathcal{H}om_{\mathcal{O}_X}(\mathcal{F}, \mathcal{O}_X)$ being the dual of $\mathcal{F}$. Such quadratic spaces appear naturally in connection with forms over smooth projective curves (see [**GHKS**]), in the theory of Azumaya algebras with involutions (see [**KPSr**], [**PSr**], [**PS**]), and also in the theory of composition of quadratic forms (see [**B**], [**K**] and [**KOS**]). They were already considered by T. Kanzaki [**Ka**] and, recently, by M. Rost [**R**]. Some classical constructions for quadratic forms carry over to quadratic forms with values in line bundles. For example, it is possible to define the even Clifford algebra C_0. The odd part C_1 can be defined as a bimodule over C_0, but $C = C_0 \oplus C_1$ does not have a natural algebra structure. However there exists a map $C_1 \otimes C_1 \to C_0 \otimes \mathcal{I}$ such that $\widetilde{C}(\mathcal{F}, h, \mathcal{I}) = (C_0 \oplus C_1) \otimes L[\mathcal{I}]$, where $L[\mathcal{I}] = \oplus_{n \in \mathbb{Z}} \mathcal{I}^n$ and $\mathcal{I}^n = \mathcal{I} \otimes \cdots \otimes \mathcal{I}$ (n times), has a natural structure of a $\mathbb{Z}$-graded algebra. In fact, as observed by F. van Oystaeyen [**vO**], the $\mathcal{O}_X$-algebra $L[\mathcal{I}]$ is a natural splitting of $\mathcal{I}$ and $\widetilde{C}(\mathcal{F}, h, \mathcal{I})$ is the "classical" Clifford algebra of the quadratic space $(\mathcal{F} \otimes_{\mathcal{O}_X} L[\mathcal{I}], h \otimes 1)$.

For the sake of simplicity, we shall restrict in this paper to quadratic spaces over affine schemes $X = \operatorname{Spec}(R)$. But most of the constructions are functorial and can be easily globalized for bundles over schemes. On the other hand we shall not assume that $\frac{1}{2} \in R$. Unadorned tensor products are taken over R.

In §2, we recall some definitions and give as examples hyperbolic spaces, binary norms, quaternary norms and pfaffians. The Clifford algebra is introduced in

---

1991 *Mathematics Subject Classification.* Primary: 11E88. Secondary: 11E20 11E57 17A75.

This paper is submitted in final form and no version of it will be submitted for publication elsewhere.

§3 and some examples are computed in §4. Then we apply Clifford algebras to describe quadratic spaces of rank 4 and 6 with trivial Arf invariant. In §6, groups of similitudes for spaces of low rank are computed. With the Clifford algebra as a tool, most results of §5 and §6 are straightforward generalizations of results of [**KP**], [$\mathbf{KPS_1}$] and [$\mathbf{KPS_2}$] and we shall in some cases only sketch the proofs. We conclude in the last section with some remarks on composition of quaternary spaces. In all this paper we assume that the quadratic form is nonsingular, i.e. that the polar $h : V \to V^* \otimes I$ is an isomorphism. Nondegenerate quadratic forms, i.e. such that h is only assumed to be injective, will be treated elsewhere.

Parts of this paper are inspired by the Ph.D. thesis, [**B**], of the first author. In [**B**] the even algebra C_0 and the bimodule C_1 are constructed by descent, i.e. by splitting locally the line bundle $\mathcal{I}$ and patching. The main aim of [**B**] was the classification of quaternary spaces in relation with composition (see Section 7). Another application of the even Clifford algebra, to Azumaya algebras of rank 16 with involution, is in [**PS**], in the same volume.

We thank F. van Oystaeyen for his nice observation mentioned above. This observation lead us, in particular, to a much simpler proof of the structure theorem (3.7). Warm thanks are also due to R. Parimala, M. Ojanguren and M. Rost for remarks and discussions.

2. Quadratic modules

We begin with some definitions. Let V be a finitely generated projective R–module and let I be an invertible R-module, i.e. a line bundle over $X = \mathrm{Spec}(R)$. A map $q : V \to I$ such that

(1) $q(\lambda x) = \lambda^2 q(x),\ \lambda \in R,\ x \in V$
(2) $b_q(x, y) = q(x + y) - q(x) - q(y)$ is R–bilinear

is a *quadratic map* from V to I. We call the triple (V, q, I) a *quadratic module (with values in I)*. Let $V^* = \mathrm{Hom}_R(V, R)$ be the R-dual of V. Let

$$h : V \to \mathrm{Hom}_R(V, I) \simeq V^* \otimes I$$

be the *polar of* b_q, i.e. $h(x)(y) = b_q(x, y)$. We say that q is *nonsingular* and that (V, q, I) is a *quadratic space (with values in I)* if h is an isomorphism. We observe that the notion of a quadratic form with values in an R-module I of arbitrary rank makes sense, but that, to define nonsingularity, we need I to be invertible. For any submodule U of V, we put $U^\perp = \{x \in V \mid b_q(x, y) = 0, \forall y \in U\}$. By the rank of (V, q, I) we mean the rank of V as an R-module. Let (V, q, I) be a quadratic module with values in a free invertible R-module. Any choice of a basis element for I yields a quadratic module with values in R and different choices give forms which are proportional. Thus similitudes turn out to be the morphisms of quadratic modules with values in an invertible module. Two quadratic modules (V, q, I) and (V', q', I') are *similar* if there exist isomorphisms

$$\theta : V \xrightarrow{\sim} V' \ ,\ \eta : I \xrightarrow{\sim} I'$$

such that $q'(\theta(x)) = \eta(q(x))$ for all $x \in V$ and the pair (θ, η) is a *similitude.* If $I = I', \eta$ is given by the multiplication by a unit of R, called the *multiplier* of the similitude.

We now give examples of quadratic spaces with values in an invertible module.

2.1. Hyperbolic spaces. Let P be a finitely generated projective R-module, let I be invertible and let $V = P \oplus P^* \otimes I$. We define

$$q : V = P \oplus P^* \otimes I \to I$$

by $q(x + f \otimes \xi) = f(x)\xi$ for $x \in P$, $f \in P^*$ and $\xi \in I$. The quadratic module (V, q, I) is obviously nonsingular and P is a direct summand of V such that $P^\perp = P$. Conversely, as in the "classical case", if, for a quadratic space (V, q, I), V has a direct summand P such that $P^\perp = P$, then $(V, q, I) \simeq P \oplus P^* \otimes I$. We call such a space *hyperbolic.*

2.2. Norm forms. An R-algebra S is *quadratic* if it is projective of rank 2 as an R-module. A quadratic R-algebra has a unique R-linear involution $x \mapsto \bar{x}$ (i.e. an antiautomorphism of order ≤ 2) such that $x + \bar{x} \in R$ and $x\bar{x} \in R$ for all $x \in S$. Let next A be an R-algebra with A projective of rank 4 over R. We say that A is a *quaternion algebra* if A has an R-linear involution $x \mapsto \bar{x}$ such that $x + \bar{x} \in R$ and $x\bar{x} \in R$ for all $x \in A$. Such an involution then is unique. The quadratic map $n : x \mapsto x\bar{x}$ is the *norm* , resp. the *reduced norm* of the algebra. The algebra S, resp. A is *split* if $S \simeq R \times R$, resp. $A \simeq \mathrm{End}_R(P)$ for P projective of rank 2. We then have $n((x, y)) = xy$ for $(x, y) \in R \times R$, resp. $n(f) = \det(f)$ for $f \in \mathrm{End}_R(P)$. We call S, resp. A, *separable* if the norm n is nonsingular. Equivalently S, resp. A, is separable if and only if it is locally split for the etale topology. A separable quadratic algebra is a Galois algebra with group $\mathbb{Z}/2\mathbb{Z}$ and a separable quaternion algebra is an Azumaya algebra of rank 4. In the following, we shall assume that A is either quadratic or a quaternion algebra. Let M be a right A-module and N an R-module. A map $q : M \to N$ is a *norm form* on M if

(1) $q(xa) = q(x)n(a)$, $x \in M, a \in A$,
(2) $b_q : M \times M \to N$ defined by $b_q(x, y) = q(x+y) - q(x) - q(y)$ is R-bilinear.

One can construct a universal norm form on M:

PROPOSITION 2.3. *Let M be a right A-module. There exists an R-module $J(M)$ and a norm form $j : M \to J(M)$ such that, for any norm form $q : M \to N$, there is a unique homomorphism of R-modules $\phi : J(M) \to N$ such that $\phi \circ j = q$.*

PROOF. We construct $J(M)$ by generators and relations. Let Q be the submodule of the module $R^M \oplus M \otimes M$ generated by all elements

$$(e_{xa} - n(a)e_x, 0) \ \text{ and } \ (e_{x+y} - e_x - e_y, -x \otimes y), \ x, y \in M, \ a \in A.$$

We define

$$J(M) = (R^M \oplus M \otimes M)/Q$$

and the map j is induced by the embedding $M \to R^M, x \mapsto e_x$, composed with the canonical projection. (compare with §1.2 of [**MR**]).

We call the pair $(J(M), j)$ the *universal norm form* of M. By construction $(J(M), j)$ with the properties (1), (2) of (2.2) is unique (up to unique isomorphisms) and J commutes with scalar extensions. If $M = A$, then $J(A) \simeq R$ and j is the norm n, by uniqueness. Let P be a projective A-module of rank one, i.e. $P_p \simeq A_p$ for all $p \in \operatorname{Spec}(R)$. Since J commutes with localization, $J(P)$ is an invertible R-module and $j : P \to J(P)$ is a quadratic map. Thus $(P, j, J(P))$ is a quadratic module with values in $J(P)$. Further j is nonsingular if n is so. We describe J in the split cases: if $A \simeq R \times R$, then $P \simeq P_1 \times P_2$ for P_1, P_2 invertible R-modules and $J(P) \simeq P_1 \otimes P_2$. If $A \simeq \operatorname{End}_R(P)$, then, by Morita theory, $P \simeq P_0 \otimes P^*$ with P_0 projective of rank 2 over R and $J(P) \simeq \wedge^2 P_0 \otimes \wedge^2 P^*$.

There is a cohomological description of the functor J. Let $\mathbb{G}_m$ be the functor "units" and $\mathbb{G}_m(A)$ the functor "units of A" i.e. $\mathbb{G}_m(A)(S) = \mathbb{G}_m(A \otimes S)$ for any commutative R-algebra S. The functors $\mathbb{G}_m$ and $\mathbb{G}_m(A)$ define sheaves for the Zariski topology and the reduced norm n induces a morphism of sheaves $\mathbb{G}_m(A) \to \mathbb{G}_m$. Thus there is a map in cohomology

$$N : H^1_{\text{Zar}}(X, \mathbb{G}_m(A)) \to H^1_{\text{Zar}}(X, \mathbb{G}_m)$$

for $X = \operatorname{Spec}(R)$. The pointed set $H^1_{\text{Zar}}(X, \mathbb{G}_m(A))$ classifies projective A-modules of rank one, $H^1_{\text{Zar}}(X, \mathbb{G}_m)$ classifies invertible R-modules and the map N corresponds to the functor J. In the quaternary case, $(P, j, J(P))$ is called the *reduced norm* of P.

2.4. Pfaffians. We refer to [**KPS**$_1$] or [**Kn**] for details. Let A be an Azumaya R-algebra of rank $4m^2$ whose class in the Brauer group is of order 2 and let $\varphi : A \otimes A \to \operatorname{End}_R(M)$, M projective of rank $4m^2$, be a fixed isomorphism. The switch $A \otimes A \to A \otimes A$ induces an automorphism τ of M of order 2 and the corresponding module of alternating elements $\operatorname{Alt}(M) = \{x - \tau(x), x \in M\}$ is projective of rank $m(2m-1)$. Let n be the reduced norm of A. There exists a universal homogeneous map of degree m

$$\operatorname{pf} : \operatorname{Alt}\,(M) \to \operatorname{Pf}(M)$$

such that $\operatorname{pf}(\varphi(a \otimes a)(x)) = n(a)pf(x)$ for $x \in \operatorname{Alt}(M)$, $a \in A$, and $\operatorname{Pf}(M)$ is invertible. In the split case $A \simeq \operatorname{End}_R(V)$ and $M \simeq V \otimes V$, we have $\operatorname{Alt}(M) \simeq \wedge^2 V$ and pf is the pfaffian $\wedge^2 V \to \wedge^{2m} V$. If $m = 2$ the map pf is quadratic and $(\operatorname{Alt}(M), \operatorname{pf}, \operatorname{Pf}(M))$ is a nonsingular quadratic space.

2.5. Involutions. Let (V, q, I) be a quadratic space over R. The adjoint $h : V \to V^* \otimes I$ induces an R-linear involution σ of the algebra $\operatorname{End}_R(V)$, $\sigma(f) = h^{-1} f^* h$, for $f \in \operatorname{End}_R(V)$. Conversely, any R-linear involution of $\operatorname{End}_R(V)$, V a finitely generated projective R-module, is induced by an isomorphism $h : V \to V \otimes I$, I an invertible R-module, and $h = \varepsilon h^*$, $\varepsilon \in \mu_2(R) = \{x \in R \,|\, x^2 = 1\}$. This follows by Morita theory, observing that transpose identifies $\operatorname{End}(V)^{\text{op}}$ with

End(V^*). If $\varepsilon = 1$ and 2 is invertible in R, h is the adjoint of a nonsingular quadratic form with values in I.

Remark. Another descriptions of the reduced norm and of the pfaffian for modules can be found in [**KOS**] (for the reduced norm), [**KPS**$_1$] (for the pfaffian), [**Kn**] and in [**PS**] (this volume).

3. Clifford algebras

Let I be an invertible R-module and let $I^n = I \otimes \cdots \otimes I$, ($n$-times), $n = 1, 2, \ldots$, $I^{-1} = I^* = \mathrm{Hom}_R(I, R)$, $I^0 = R$ and $I^{-n} = (I^*)^n$. The tensor product and the canonical isomorphism

$$I \otimes I^* \xrightarrow{\sim} R, \; x \otimes f \mapsto f(x), \; x \in I, \; f \in I^*$$

define an R-algebra structure on $L[I] = \oplus_{n \in \mathbb{Z}} I^n$. We call $L[I]$ the *Laurent algebra* (in the literature $L[I]$ also appears as the *Rees algebra*) of I. Let $q : V \to I$ be a quadratic map. Let TV be the tensor algebra of V and $J(q)$ be the ideal of $TV \otimes L[I]$ generated by all elements $v \otimes v \otimes 1 - 1 \otimes q(v)$, $v \in V$. Defining a $\mathbb{Z}$-grading on $TV \otimes L[I]$ by $\partial(v \otimes 1) = 1$ for $v \in V$ and $\partial(1 \otimes x) = 2$ for $x \in I$, we get that $J(q)$ is $\mathbb{Z}$-graded. Thus the algebra

$$\widetilde{C} = \widetilde{C}(V, q, I) = TV \otimes L[I]/J(q)$$

is $\mathbb{Z}$-graded. We call $\widetilde{C}$ the *Clifford algebra* of (V, q, I). Let C_n be the submodule of $\widetilde{C}$ of elements of degree n.

LEMMA 3.1. *C_0 is a subalgebra of $\widetilde{C}$, C_1 is a C_0-bimodule and the multiplication in $\widetilde{C}$ induces an isomorphism*

$$(C_0 \oplus C_1) \otimes L[I] \xrightarrow{\sim} \widetilde{C}(V, q, I).$$

PROOF. The first two claims are obvious. Since $I \otimes I^{-1} \xrightarrow{\sim} R$, we have isomorphisms $C_n \simeq C_0 \otimes I^m$ for $n = 2m$ and $C_n \simeq C_1 \otimes I^m$ for $n = 2m + 1$.
Remark. In fact, for any $\mathbb{Z}$-graded $L[I]$-module $\widetilde{M}$, we have $M_n = M_0 \otimes I^m$ for $n = 2m$ and $M_n = M_1 \otimes I^m$ for $n = 2m + 1$. This will be used later.

In the following, we shall identify C_n with $C_0 \otimes I^m$, resp. $C_1 \otimes I^m$. We call $C_0(q) = C_0$ the *even Clifford algebra of* q and $C_1(q) = C_1$ the *Clifford module of* q. As observed by F. van Oystaeyen, the algebra $\widetilde{C} = \widetilde{C}(V, q, I)$ has the following nice interpretation:

LEMMA 3.2. *The multiplication of the algebra $L[I]$ induces an isomorphism*

$$I \otimes L[I] \xrightarrow{\sim} L[I],$$

so that $(V, q, I) \otimes L[I]$ is a quadratic space with values in $L[I]$ and

$$\widetilde{C}(V, q, I) \simeq C(V \otimes L[I], q \otimes 1).$$

The next two results are immediate consequence of (3.2).

LEMMA 3.3. *Any morphism $(\theta, \eta) : (V, q, I) \to (V', q', I')$ induces a morphism of graded algebras*

$$\widetilde{C}(\theta, \eta) : \widetilde{C}(V, q, I) \to \widetilde{C}(V', q', I')$$

such that $\widetilde{C}(\theta, \eta)(i(v)) = i(\theta(v))$ for all $v \in V$.

LEMMA 3.4. *For any commutative R-algebra S and any quadratic R-module (V, q, I), there exists a canonical isomorphism $\widetilde{C}(V, q, I) \otimes S \xrightarrow{\sim} \widetilde{C}((V, q, I) \otimes S)$.*

Let $i : V \to \widetilde{C}(V, q, I)$ be the map induced by the canonical map $V \to TV$. We have

$$i(x)^2 = 1 \otimes q(x) \in C_0 \otimes I = C_2 \quad \text{for all } \ x \in V$$

and $\widetilde{C}$ is universal with respect to the following property. Let D_0 be an R-algebra, D_1 a D_0-bimodule such that $rd = dr$ for all $r \in R$ and $d \in D_1$, and let I be an invertible R-module. Assume that there exists an R-linear map

$$\mu : D_1 \otimes D_1 \to D_0 \otimes I$$

such that

$$\widetilde{D} = (D_0 \oplus D_1) \otimes L[I]$$

is, in a natural way, a $\mathbb{Z}$-graded R-algebra, the gradation being defined as for $\widetilde{C}$. Then, for any quadratic module (V, q, I) and any R-linear map $\psi : V \to D_1$ such that

$$\mu(\psi(v) \otimes \psi(v)) = 1 \otimes q(v) \quad \text{for all } \ v \in V,$$

there exists a unique homomorphism of graded R-algebras

$$\widetilde{\psi} : \widetilde{C}(V, q, I) \to \widetilde{D}$$

such that $\widetilde{\psi}|_V = \psi$ and $\widetilde{\psi}(i(v)^2) = 1 \otimes q(v)$. By the universal property of the Clifford algebra, the automorphism $x \mapsto -x$ of V extends to an involution σ of $\widetilde{C}(V, q, I)$. We call σ the *standard involution* of $\widetilde{C}(V, q, I)$. In particular, σ restricts to an involution of $C_0(q)$.

PROPOSITION 3.5. *If V is finitely generated free with basis $\{e_1, \dots e_n\}$ and I is free with basis element t, then C_0 is a free R-module with basis*

$$\{1, e_{i_1} e_{i_2} \dots e_{i_{2m}} \otimes t^{-m}, 1 \le i_1 < i_2 < \dots < i_{2m} \le n\}$$

and C_1 is free with basis

$$\{e_{i_1} e_{i_2} \dots e_{i_{2m+1}} \otimes t^{-m}, 1 \le i_1 < i_2 < \dots < i_{2m+1} \le n\}.$$

PROOF. A variation of the Poincaré-Birkhoff-Witt theorem!

COROLLARY 3.6. *For any quadratic module (V, q, I), the canonical map $i : V \to \widetilde{C}(V, q, I)$ and the map $\lambda : L[I] \to \widetilde{C}(V, q, I)$ induced by $L[I] \to TV \otimes L[I]$ are injective.*

PROOF. By (3.4) and (3.5), since a finitely generated projective module is locally free.

Remark. Observe that for a quadratic map $q : V \to I$, with V not necessarily projective, the maps $i : V \to C_1 \subset \widetilde{C}$ and $\lambda : L[I] \to \widetilde{C}$ need not to be injective.

The centre $\widetilde{Z}$ of $\widetilde{C}$ is $\mathbb{Z}$-graded and as in (3.1) we get

$$\widetilde{Z} = \oplus_{n\in\mathbb{Z}} Z_n = (Z_0 \oplus Z_1) \otimes L[I] = \widetilde{Z}_0 \oplus \widetilde{Z}_1$$

with $\widetilde{Z}_0 = Z_0 \otimes L[I]$ and $\widetilde{Z}_1 = Z_1 \otimes L[I]$. We now describe the structure of $\widetilde{C}$ and of C_0 in the nonsingular case. Observe that, in the odd rank case, nonsingularity implies that 2 is invertible. We use the notations $\widetilde{C}_0 = C_0 \otimes L[I] = C_0(V \otimes L[I], q \otimes 1)$ and $\widetilde{C}_1 = C_1 \otimes L[I] = C_1(V \otimes L[I], q \otimes 1)$.

THEOREM 3.7. *Assume that (V, q, I) is nonsingular. Then*

(1) *If the rank of V is odd, then $Z_0 \simeq R$, Z_1 is an invertible module such that $Z_1^2 \xrightarrow{\sim} I$, C_0 is an Azumaya algebra over R and $\widetilde{C} \simeq C_0 \otimes L[Z_1]$.*
(2) *If the rank of V is even, then $Z_0 \simeq R$, $Z_1 = 0$ and $\widetilde{C}$ is an Azumaya algebra over $L[I]$. Furthermore the centre $Z(C_0)$ of C_0 is a separable quadratic algebra over R and C_0 is an Azumaya algebra over $Z(C_0)$.*

PROOF. 1) By the classical case applied to $(V \otimes L[I], q \otimes 1)$ we get that (see for example [**Kn**]) $\widetilde{C} \simeq \widetilde{C}_0 \otimes \widetilde{Z}$, $\widetilde{C}_0$ is an Azumaya algebra over $L[I]$, $\widetilde{Z}_0 = L[I]$ and the multiplication in $C_0(q \otimes 1)$ induces an isomorphism $\widetilde{Z}_1 \otimes \widetilde{Z}_1 \simeq L[I]$. The claim then follows from $C_0(q \otimes 1) = C_0 \otimes L[I]$ and $\widetilde{Z}_1 = Z_1 \otimes L[I]$.

2) follows also from the classical case applied to $C(q \otimes 1)$. In particular $\widetilde{C}$ is an Azumaya algebra over $L[I]$ and the centre $Z(\widetilde{C}_0)$ of $\widetilde{C}_0$ is a separable quadratic algebra over $L[I]$. Let $Z(\widetilde{C}_0) = Z \otimes L[I]$. Since $L[I]$ is faithfully flat over R, Z is a separable quadratic R-algebra and $Z \subset Z(C_0)$. For the same reason, and since $\widetilde{C}_0 = C_0 \otimes L[I]$, C_0 is a separable R-algebra, i.e. is a projective $C_0 \otimes C_0^{\text{op}}$-module. It follows that

$$\text{End}_{C_0 \otimes C_0^{\text{op}}}(C_0, C_0) \otimes L[I] \simeq \text{End}_{\widetilde{C}_0 \otimes_{L[I]} \widetilde{C}_0^{\text{op}}}(\widetilde{C}_0, \widetilde{C}_0).$$

Thus

$$Z(C_0) \otimes L[I] = \text{End}_{C_0 \otimes C_0^{\text{op}}}(C_0, C_0) \otimes L[I] \simeq Z(\widetilde{C}_0)$$

and, by faithfully flat descent, $Z = Z(C_0)$, so that, as claimed, $Z(C_0)$ is a separable quadratic R-algebra.

Remark. As already observed in the introduction, the even Clifford algebra C_0 and the Clifford module C_1 can also be constructed by trivializing locally the line bundle I and by patching the corresponding C_0 and C_1. This was done in [**B**] and (for C_0) in [**PS**].

4. Examples

In this section we compute the Clifford algebras of the spaces described in Section 2.

4.1. Hyperbolic spaces. Let $V = P \oplus P^* \otimes I$ be hyperbolic and let $\wedge P$ be the exterior algebra of P. Let $\lambda_x : \wedge P \otimes L[I] \to \wedge P \otimes L[I]$ be the left exterior multiplication with $x, x \in P$ and let $d_f : \wedge P \otimes L[I] \to \wedge P \otimes L[I], f \in P^* \otimes I = \operatorname{Hom}_R(P, I)$ be the derivation which extends the linear form $f : P \to I$. The map $(x, f) \mapsto \lambda_x + d_f \in \operatorname{End}_R(\wedge P) \otimes L[I]$ induces, as in the classical case (see [**Kn**] or [**Ba**]), an isomorphism

$$\widetilde{C}(V, q, I) \xrightarrow{\sim} \operatorname{End}_R(\wedge P) \otimes L[I].$$

4.2. Binary norms. Let S be a separable quadratic R-algebra with norm n and let M be a projective right S-module of rank one. Let $j : M \to J(M)$ be the universal norm on M. We have $C_0(j) = S$, $C_1(j) = M$ and $\mu : C_1 \otimes C_1 \to C_0 \otimes J(M)$ is locally given by $\mu(ex \otimes ey) = j(e)\bar{x}y$, where $x \mapsto \bar{x}$ is the involution of S and e is (locally) a basis of M as an S-module. Conversely, any quadratic space (V, q, I) of rank 2 is similar to the norm form $(M, j, J(M))$ for some rank one module over a separable quadratic R-algebra S. We take $M = C_1(q) = V$ and $S = C_0(q)$.

4.3. Quaternary norms. Let A be a separable quaternion algebra and let P be a projective right A-module of rank one. To compute the Clifford algebra of the reduced norm of P, we shall use another description of the space $(P, j, J(P))$. Let $P^{(*)} = \operatorname{Hom}_A(P, A)$ be the A-dual of P. The algebra A has an involution $\sigma_A : a \mapsto \bar{a}$ given by $\bar{a} = \operatorname{tr}(a) - a$, where tr is the reduced trace of A. We view $P^{(*)}$, which has a natural structure of left A-module, as a right A-module through σ_A. Similarly $B = \operatorname{End}_A(P)$, which is also an Azumaya R-algebra of rank 4, has an involution σ_B and we view $P^{(*)}$ as a left B-module through σ_B. Thus P and $P^{(*)}$ are A-B-bimodules. The R-module

$$I(P) = \operatorname{Hom}_{A\text{-}B}(P^{(*)}, P)$$

is invertible and the evaluation induces an isomorphism of A-B-bimodules

$$\theta : P^{(*)} \otimes I(P) \xrightarrow{\sim} P.$$

Let $\pi = \theta^{-1}$. We define a quadratic map $q : P \to A \otimes I(P)$ by $q(x) = \pi(x)(x)$. The map q is such that $q(xa) = q(x)n(a)$, $x \in P$, $a \in A$. We claim that q has values in $I(P) = R \otimes I(P)$. Localizing, we may assume that P is free over A with generator e, $P = eA$. Then $P^* = Af$, with $f(e) = 1$, and $I(P)$ is free with generator $u, u(f) = e$. The map θ is given by $\theta(f \otimes u) = e$ and $q(ea) = \bar{a}a \otimes u = 1 \otimes n(a)u$. Similarly, we have a quadratic map

$$q' : P \to B \otimes I(P)$$

given by

$$q'(x) = x\pi(x).$$

The map q' also has values in I and the above computation for P free shows that $q'(x) = q(x), x \in P$. By the universal property of $J(P)$, there exists a linear map $\gamma : J(P) \to I(P)$ such that $q = \gamma \circ j$. Localizing, we see that $(1, \gamma) : (P, j, J(P)) \to (P, q, I(P))$ is an isomorphism. Similarly, there is an isomorphism

$$(1, \gamma') : (P, j, J(P)) \xrightarrow{\sim} (P, q', I(P)),$$

where the quadratic space on the right is obtained by considering P as a B-module. We shall use this description of the quaternary norm to compute its even Clifford algebra and its Clifford module. Let D_0 be the algebra $A \times B$, with $B = \operatorname{End}_A(P)$ as above, let D_1 be the D_0-bimodule $P \oplus P^{(*)} \otimes I(P)$ with the operation

$$(a, b)(x, f \otimes \xi)(a', b') = (bxa', afb' \otimes \xi) \tag{4.4}$$

for $a, a' \in A$, $b, b' \in B$, $x \in P, f \in P^{(*)}$ and $\xi \in I(P)$. It is convenient to use matrix notations:

$$D_0 = \begin{pmatrix} A & 0 \\ 0 & B \end{pmatrix} \quad \text{and} \quad D_1 = \begin{pmatrix} 0 & P^{(*)} \otimes I(P) \\ P & 0 \end{pmatrix}.$$

The map

$$\mu : D_1 \otimes D_1 \to D_0 \otimes I(P)$$

given by

$$((x, f \otimes \xi), (u, g \otimes \eta)) \mapsto (f(u), 0) \otimes \xi + (0, xg) \otimes \eta$$

for $u, x \in P$, $f, g \in P^*$ and $\xi, \eta \in I(P)$, induces naturally a graded algebra structure on $\widetilde{D} = (D_0 \oplus D_1) \otimes L[I(P)]$. The map $\iota : P \to D_1$, $x \mapsto (x, \pi(x))$ is such that

$$\mu(\iota(x) \otimes \iota(x)) = (\pi(x)(x), x\pi(x)) = (1, 1)q(x).$$

By the universal property of the Clifford algebra of $\widetilde{C}(P, q, I(P))$, there exists a homomorphism

$$\tilde{\iota} : \widetilde{C}(P, q, I(P)) \to \widetilde{D}$$

such that $\tilde{\iota}|_P = \iota$. It follows from (3.7) that $\tilde{\iota}$ is an isomorphism. Thus, identifying $(P, j, J(P))$ with $(P, q, I(P))$ as above, we get

PROPOSITION 4.5. *Let A be an Azumaya R-algebra of rank 4 and let P be a projective A-module of rank one. The even Clifford algebra and the Clifford module of the reduced norm $(P, j, J(P))$ are given by*

$$C_0 \simeq A \times \operatorname{End}_A(P) \quad \textit{and} \quad C_1 \simeq P \oplus P^{(*)} \otimes I(P)$$

with the bimodule operation as in (4.4). In particular, the centre of C_0 is a split quadratic algebra.

We shall identify $\widetilde{C}$ with $\widetilde{D}$ and use the matrix notation for $D_0 \oplus D_1$. The standard involution of $\widetilde{C}$ then is induced by

$$\sigma : \begin{pmatrix} a & f \otimes \xi \\ p & b \end{pmatrix} \mapsto \begin{pmatrix} \sigma_A(a) & -\pi(p) \\ --\pi^{-1}(f \otimes \xi) & \sigma_B(b) \end{pmatrix} \tag{4.6}$$

4.7. Pfaffians. Let A be an Azumaya algebra of rank 16 with a fixed isomorphism $\varphi : A \otimes A \xrightarrow{\sim} \mathrm{End}_R(M)$. We now compute the Clifford algebra of the quadratic space $(\mathrm{Alt}(M), \mathrm{pf}, \mathrm{Pf}(M))$ as defined in (2.4). We regard M as a left A-module (or as an right A^{op}-module) through the action $ax = \varphi(a \otimes 1)x$, $x \in M$, $a \in A$ and denote by $M^{(*)}$ the A-dual of M. Let

$$\varphi^* : A^{\mathrm{op}} \otimes A^{\mathrm{op}} \xrightarrow{\sim} \mathrm{End}_R(M^{(*)})$$

be given by $\varphi^*(a \otimes a')(f)(x) = f(\varphi(1 \otimes a')x)a$ and let $\mathrm{Alt}(M^{(*)})$ be the corresponding set of alternating elements. In view of [**KPS**$_1$**, Proposition 4.2**] there exists a unique R-linear isomorphism

$$\pi : \ \mathrm{Alt}(M) \xrightarrow{\sim} \mathrm{Alt}(M^{(*)}) \otimes \mathrm{Pf}(M)$$

such that $\pi(x)(x) = \mathrm{pf}(x)$ for all $x \in \mathrm{Alt}(M)$. Furthermore we have $x\pi(x) = 1 \otimes \mathrm{pf}(x)$ in $M \otimes_{A^{\mathrm{op}}} \otimes M^{(*)} \otimes \mathrm{Pf}(M) = \mathrm{End}_{A^{\mathrm{op}}}(M) \otimes \mathrm{Pf}(M)$. The construction of the Clifford algebra now is very similar to the construction for quaternary forms given in (4.3). We put $B = \ \mathrm{End}_{A^{\mathrm{op}}}(M)$ and define

$$D_0 = \begin{pmatrix} A^{\mathrm{op}} & 0 \\ 0 & B \end{pmatrix} \quad \text{and} \quad D_1 = \begin{pmatrix} 0 & M^{(*)} \otimes \mathrm{Pf}(M) \\ M & 0 \end{pmatrix}.$$

As above, $D_0 \oplus D_1$ extends to a graded algebra $\widetilde{D}$. The map $\iota : \mathrm{Alt}(M) \to D_1$, $x \mapsto (x, \pi(x))$ is such that $\mu(\iota(x) \otimes \iota(x)) = (\pi(x)(x), x\pi(x)) = (1,1) \cdot \mathrm{pf}(x)$. By the universal property of the Clifford algebra of $\widetilde{C}(\mathrm{Alt}(M), \mathrm{pf}, \mathrm{Pf}(M))$, there exists a homomorphism

$$\tilde{\iota} : \widetilde{C}(\mathrm{Alt}(M), \mathrm{pf}, \mathrm{Pf}(M)) \to \widetilde{D}$$

such that $\tilde{\iota}|_{\mathrm{Alt}(M)} = \iota$. It follows from (3.7) that $\tilde{\iota}$ is an isomorphism. Thus

PROPOSITION 4.8. *Let A be a Azumaya R-algebra of rank* 16 *with a fixed isomorphism $\varphi : A \otimes A \xrightarrow{\sim} \mathrm{End}_R(M)$. The even Clifford algebra and the Clifford module of the pfaffian* $(\mathrm{Alt}(M), \mathrm{pf}, \mathrm{Pf}(M))$ *are given by*

$$C_0 = A^{\mathrm{op}} \times \mathrm{End}_{A^{\mathrm{op}}}(M) \quad \textit{and} \quad C_1 = M \oplus M^{(*)} \otimes_R \mathrm{Pf}(M)$$

In particular, the centre of C_0 is a split quadratic algebra.

Let $\psi : A^{\mathrm{op}} \xrightarrow{\sim} \mathrm{End}_{A^{\mathrm{op}}}(M)$ be given by $a' \mapsto \varphi(1 \otimes a')$. Identifying $\widetilde{C}$ with $\widetilde{D}$, the standard involution of $\widetilde{C}$ is given on $D_0 \oplus D_1$ by

$$\begin{pmatrix} a & n \otimes \xi \\ m & b \end{pmatrix} \mapsto \begin{pmatrix} \psi^{-1}(b) & \tau_0(n) \otimes \xi \\ \tau(m) & \psi(a) \end{pmatrix}, \tag{4.9}$$

where τ is the automorphism of order 2 of M defined in (2.4) and τ_0 is the corresponding automorphism of $M^{(*)}$ (with respect to φ^*).

4.10. Involutions. Let A be an Azumaya R-algebra with an R-linear involution σ of orthogonal type, i.e A is locally isomorphic (for the etale topology) to an algebra $\mathrm{End}_R(V)$ and σ is induced by a quadratic form (V, q, I) (we assume that 2 is invertible in R). Then $C_0(V, q, I)$ glues to an R-algebra $C_0(A, \sigma)$, the *even Clifford algebra* of (A, σ). This algebra was introduced over fields by N. Jacobson, [**J**] (see also [**T**]). Its generalization to Azumaya algebras over algebraic schemes is described in [**PS**]. The Clifford module $C_1(V, q, I)$ cannot be globalized, but, as observed by M. Rost ([**R**]), $V \otimes C_1(V, q, I)$ can be globalized.

5. Spaces with trivial Arf invariant

In this section, we prove converses to (4.5) and (4.8). We refer to [**KP**] and [**KPS$_1$**] for details (in the "classical" case).

THEOREM 5.1. *Let (V, q, I) be a quadratic space of rank 4 with trivial Arf invariant. Then there exists an Azumaya R-algebra A of rank 4 and a projective A-module P of rank one such that (V, q, I) is similar to the reduced norm $(P, j, J(P))$.*

PROOF. Let e be an idempotent generating the centre $Z \simeq R \times R$ of C_0. We put $A = C_0 e$ (as Re-algebra) and $P = C_1 e$. Then (5.1) follows as in the proof of (6.4) of [**KP**].
Remark. The algebra A and the module P are clearly not uniquely determined in (5.1) since, for example, A can be replaced by $B = \mathrm{End}_A(P)$ and (independently) P can be replaced by $P^* \otimes_R J(P)$. More generally, if (A', P') is another pair satisfying (5.1), one can show that there is a decomposition $R = R_1 \times R_2$ of R such that, for the induced decompositions

$$A = A_1 \times A_2, A' = A_1' \times A_2', P = P_1 \times P_2 \text{ and } P' = P_1' \times P_2'$$

there exist algebra isomorphisms $A_1' \simeq A_1, A_2' \simeq \mathrm{End}_{A_2}(P_2)$ and corresponding semilinear isomorphisms $P_1' \simeq P_1, P_2' \simeq P_2^* \otimes J(P_2)$.

THEOREM 5.2. *Let (V, q, I) be a quadratic space of rank 6 with trivial Arf invariant. There exist an Azumaya R-algebra A of rank 16, a projective R-module M of rank 16 and an isomorphism $\varphi : A \otimes A \xrightarrow{\sim} \mathrm{End}_R(M)$ such that (V, q, I) is similar to the pfaffian* $(\mathrm{Alt}(M), \mathrm{pf}, \mathrm{Pf}(M))$.

PROOF. Let e be an idempotent generating the centre $Z \simeq R \times R$ of C_0 and let $f = 1 - e$. We put $A = C_0 e$ (as Re-algebra), $M = C_1 f$ and define $\varphi : C_0 e \otimes C_0 e \to \mathrm{End}_R(C_1 f)$ by $\varphi(ae \otimes be)(xf) = aexf\sigma(be) = ax\sigma(b)f$, where σ is the standard involution of C_0. The involutary map τ of M is the restriction of the standard involution to $C_1 f$. We have $M^{(*)} = C_1 e \otimes I^{-1}$ and

$$\pi : \mathrm{Alt}(M) = \mathrm{Alt}(C_1 f) \xrightarrow{\sim} \mathrm{Alt}(M^{(*)}) \otimes \mathrm{Pf}(M) = \mathrm{Alt}(C_1 e)$$

is given by $xf \mapsto xe$, $x \in \mathrm{Alt}(C_1 f)$.
Remark. As in (5.1), the algebra A and the module M are not uniquely determined. A corresponding discussion (for the "classical" case) is in [**KPS$_1$**].

6. Groups of similitudes

Let (V, q, I) be a quadratic space of even rank and $GO(q)$ be the group of similitudes of (V, q, I). Any similitude $u \in GO(q)$ induces an automorphism of $C_0(q)$, hence an automorphism $\gamma(u)$ of its centre $Z = Z(C_0(q))$. We call u a *direct similitude* if $\gamma(u) = 1$. Thus we have an exact sequence

$$1 \longrightarrow GO_+(q) \longrightarrow GO(q) \longrightarrow \operatorname{Aut}(Z)$$

denoting by $GO_+(q)$ the group of direct similitudes of (V, q, I). Observe that $\operatorname{Aut}(Z) = \mathbb{Z}/2\mathbb{Z}$ if R is connected. We now describe $GO_+(q)$ for V of rank 2, 4 and 6.

6.1. Binary forms. We may assume that a rank 2 quadratic space (V, q, I) is a norm form $(M, j, J(M))$ for a rank one projective module M over a separable quadratic R-algebra S (see (4.2)). Let $u \in GO_+(q)$. Since u is the identity on $Z(C_0) = C_0 = S$, the map u is an S-linear automorphism of M. Thus

$$GO_+(q) = \mathbb{G}_m(S).$$

6.2. Quaternary forms. If (V, q, I) is a quadratic space of rank 4 we get as in [**KPS$_2$**] an exact sequence

$$1 \longrightarrow \{(n_0(z), z^{-1}) \mid z \in \mathbb{G}_m Z\} \longrightarrow \mathbb{G}_m(R) \times \mathbb{G}_m(C_0) \xrightarrow{\alpha} GO_+(q) \xrightarrow{\beta} \operatorname{Pic}(Z),$$

where n_0 is the norm on Z. The map α is defined as $\alpha(\lambda, c)(x) = \lambda c x \sigma(c)$, where σ is the standard involution of C_0 and β is given by

$$\beta(u) = \{y \in C_0 \mid u(y)c = cy \quad \text{for all } c \in C_0\}.$$

If the Arf invariant of (V, q, I) is trivial, we can assume by (5.1) that the quadratic space is a norm form on a rank one module P over a separable quaternion algebra A. Then $C_0 \simeq A \times B$ with $B = \operatorname{End}_A(P)$ (see (4.5)), the exact sequence reduces to

$$1 \longrightarrow \{(\lambda, \lambda^{-1}) \mid \lambda \in \mathbb{G}_m(R)\} \longrightarrow \mathbb{G}_m(B) \times \mathbb{G}_m(A) \xrightarrow{\alpha} GO_+(q) \xrightarrow{\beta} \operatorname{Pic}(R \times R)$$

and the map α is given by $\alpha(b, a)(x) = bx\sigma_A(a)$.

6.3. Pfaffians. For a quadratic space (V, q, I) of rank 6 we define

$$GU(C_0) = \{c \in C_0 \mid \sigma(c)c \in \mathbb{G}_m(Z)\}.$$

Let now

$$H = \{(z, c) \in \mathbb{G}_m(Z) \times GU(C_0) \mid \sigma_0(z)z^{-1} = n_{C_0}(c)(\sigma(c)c)^{-2}\},$$

where σ_0 is the involution of Z and n_{C_0} is the reduced norm of C_0. We get, as in [**KPS$_2$**], an exact sequence

$$1 \longrightarrow \{(z^2, z^{-1}) \mid z \in \mathbb{G}_m(Z)\} \longrightarrow H \xrightarrow{\alpha} GO_+(q) \xrightarrow{\beta} \operatorname{Pic}(Z),$$

where $\alpha(z,c)(x) = zcx\sigma(c)$. If the Arf invariant of the quadratic space is trivial, we can assume, by (5.2) that q is the pfaffian $(\mathrm{Alt}(M), \mathrm{pf}, \mathrm{Pf}(M))$ for some Azumaya algebra A of rank 16 over R and, with the notations of (4.7), the sequence reduces to

$$1 \longrightarrow \{(\lambda^2, \lambda^{-1}) \mid \lambda \in \mathbb{G}_m(R)\} \longrightarrow \mathbb{G}_m(R) \times \mathbb{G}_m(A) \xrightarrow{\alpha} GO_+(q) \xrightarrow{\beta} \mathrm{Pic}(R \times R)$$

where $\alpha(\rho, a)(x) = \rho\varphi(a \otimes a)(x)$.

7. Composition of quaternary forms

Let A be an Azumaya R-algebra of rank 4 and let P be a projective right A-module of rank one. Let $B = \mathrm{End}_A(P)$, so that P is a B-A-bimodule. Further let Q be a projective left A-module of rank one, hence a projective right C-module of rank one with $C = \mathrm{End}_A(Q)^{\mathrm{op}} \simeq \mathrm{End}_A(Q)$. The tensor product $M = P \otimes_A Q$ is a projective right B-, left C-module of rank one. Let $q_1 : P \to I(P)$, $q_2 : Q \to I(Q)$ be fixed reduced norms. Then

$$q_3 : P \otimes_A Q \to I(P) \otimes_R I(Q),$$

given by

$$q_3(x \otimes y) = q_1(x) \otimes q_2(y),$$

is the reduced norm of $P \otimes_A Q$ and $I(P \otimes Q) \simeq I(P) \otimes I(Q)$.

Let (P_i, q_i, I_i), $i = 1, 2, 3$, be quadratic spaces of rank 4 with values in invertible ideals. A *composition* $P_1 \times P_2 \to P_3$ is an R-bilinear map $\mu : P_1 \times P_2 \to P_3$ together with an isomorphism

$$\mu' : I_1 \otimes_R I_2 \xrightarrow{\sim} I_3$$

such that

$$q_3(\mu(x_1, x_2)) = \mu'(q_1(x_1) \otimes q_2(x_2)), \; x_1 \in P_1, \; x_2 \in P_2.$$

Thus the canonical map $P \times Q \to P \otimes_A Q$ is a composition of reduced norms. We say that (P_1, q_1, I_1) is *composable* if there exist quadratic spaces (P_2, q_2, I_2), (P_3, q_3, I_3) and a composition $\mu : P_1 \times P_2 \to P_3$.

THEOREM 7.1. *Let (P_1, q_1, I_1) be a quadratic space of rank 4. The following conditions are equivalent :*

(1) *(P_1, q_1, I_1) has trivial Arf invariant.*
(2) *There exists an Azumaya R-algebra A of rank 4 such that P_1 is a projective A-module of rank one and $q_1 : P_1 \to I_1$ is (similar to) the reduced norm of P_1 with respect to A.*
(3) *(P_1, q_1, I_1) is composable.*

PROOF. (1)$\Leftrightarrow$ (2) is (4.5) and (5.1). To show (2) $\Rightarrow$ (3), we may take, for example, $(P_2, q_2, I_2) = (A, n, R)$. To check (3) $\Rightarrow$ (2), we use the idea of the proof of Theorem (2.10) of [K+]. Let (P_2, q_2, I_2) be such that there exists a composition $\mu : P_1 \times P_2 \to P_3$. Let A be the set of pairs (s_1, s_2) of morphisms s_1 of (P_1, q_1, I_1), resp. s_2 of (P_2, q_2, I_2) satisfying $\mu(s_1 x_1, x_2) = \mu(x_1, s_2 x_2)$. Then A is an R-algebra with the multiplication $(s_1, s_2)(s'_1, s'_2) = (s'_1 \circ s_1, s_2 \circ s'_2)$. One can check as in **[K+]** that A is an Azumaya algebra of rank 4, that P_1 is a projective right A-module of rank one and that q_1 is the reduced norm of P_1.

References

[B] W. Bichsel, *Quadratische Räume mit Werten in invertierbaren Moduln*, ETH, Zürich, 1985.

[Ba] H. Bass, *Lectures on topics on algebraic K-Theory*, Tata Institute of Fundamental Research, Bombay, 1967.

[GHKS] W.-D. Geyer, G. Harder, M. Knebusch, and W. Scharlau, *Ein Residuensatz für symmetrischen Bilinearformen*, Inventiones math. **11** (1970), 319 – 328.

[J] N. Jacobson, *Clifford algebras for algebras with involution of type D*, J. Algebra **1** (1964), 288 –300.

[Ka] T. Kanzaki, *On bilinear module and Witt ring over a commutative ring*, Osaka J. Math. **8** (1971), 485 – 496.

[K] M. Kneser, *Composition of binary quadratic forms*, J. of Number Theory **15** (1982), 406 – 413.

[K+] M.Kneser, M.-A. Knus, M. Ojanguren, R. Parimala, and R. Sridharan, *Composition of quaternary quadratic forms*, Compositio Mathematica **60** (1986), 133 – 150.

[Kn] M.-A. Knus, *Quadratic and hermitian forms over rings*, Springer-Verlag, Heidelberg, New York, 1991.

[KOS] M.-A. Knus, M. Ojanguren, and R. Sridharan, *Quadratic forms and Azumaya algebras*, J. reine und angewandte Mathematik **303/304** (1978), 231 – 248.

[KP] M.-A. Knus and A. Paques, *Quadratic spaces with trivial Arf invariant*, J. of Algebra **93** (1985), 267 – 291.

[KPS$_1$] M.-A. Knus, R. Parimala, and R. Sridharan, *A classification of rank 6 quadratic spaces via pfaffians*, J. reine und angewandte Mathematik **398** (1989), 187 – 218.

[KPS$_2$] ———, *Pfaffians, central simple algebras and similitudes*, Math. Z. **206** (1991), 589 – 604.

[KPSr] M.-A. Knus, R. Parimala, and V. Srinivas, *Azumaya algebras with involutions*, J. Algebra **130** (1990), 65 – 82.

[MR] A. Micali and Ph. Revoy, *Modules quadratiques*, Bull. Soc. Math. France, Mémoire **63** (1979).

[PS] R. Parimala and R. Sridharan, *Reduced norms and pfaffians via Brauer Severi schemes*, Contemporary Mathematics (This volume), 1992.

[PSr] R. Parimala and V. Srinivas, *Analogues of the Brauer groups for algebras with involution*, Duke Math. J. (to appear).

[R] M. Rost, *Oral communication, November 1991.*

[T] J. Tits, *Formes quadratiques, groupes orthogonaux et algèbres de Clifford*, Inventiones math. **5** (1968), 19 – 41.

[vO] F. van Oystaeyen, *Oral communication, October 1991.*

Mathematik, ETH, 8092 Zürich, Switzerland

Contemporary Mathematics
Volume **155**, 1994

On Annhilators in Graded Witt Rings And in Milnor's K–Theory

MARTIN KRÜSKEMPER

ABSTRACT. In this paper the main problem under investigation is the determination of the annihilators of Pfister forms in the graded Witt ring of a field.

Let F be a field with $char\, F \neq 2$. With F we associate the Witt ring $W(F)$ and its ideals $I^n(F)$ where $I(F)$ is the ideal of all evendimensional form classes, the Milnor K–theory ring $\bigoplus_{n=0}^{\infty} k_n(F)$, the graded Witt ring $\bigoplus_{n=0}^{\infty} \bar{I}^n(F)$ and the Galois cohomology ring $\bigoplus_{n=0}^{\infty} H^n(G, \mathbb{Z}/2)$ as defined in [Mi]. Several problems concerning those rings have been examined in recent years: Milnor conjectured that the latter three rings are always isomorphic. Important is also the behaviour of the functors $k_n, H^n, \bar{I}^n, I^n$ by passing to quadratic extensions or, more generally, by passing to function fields of quadrics. See [**AEJ**]. Conjectures on real local–global principles for the functors have been examined in [**Ma**] and [**K**]. A well known theorem of Pfister and Witt describes the annihilators of Pfister forms in the Witt ring. In this paper we want to study annihilators of generators in $k_n(F)$, $I^n(F)$ and $\bar{I}^n(F)$. We will introduce conjectures describing those annihilators. For instance, let $x = l(a_1)\cdots l(a_m) \in k_m(F)$ and let $ann_n(x)$ denote the subgroup of $k_n(F)$ of all annihilators of x. Then we conjecture $ann_n(x) = ann_1(x)k_{n-1}(F)$. We can show that the conjecture holds in many special cases. We further study connections between the problems above. In particular, there is a close connection between the conjectures on real local-global principles and the ones on annihilators.

Parts of the paper were written during a stay at the U.C. Berkeley and the author gratefully acknowledges the hospitality there.

1. Introduction and basic results

Let F be a field with $char\ F \neq 2$. Then $W(F)$ denotes the Witt ring of F and $I(F)$ the ideal of even dimensional form classes. Let $G = G_F$ denote the absolute

---

1991 *Mathematics Subject Classification.* Primary 11E81 19G12; Secondary 12G05 19D45.
The author was supported by DFG (Deutsche Forschungsgemeinschaft).
This paper is submitted in final form and no version of it will be submitted for publication elsewhere.

Galois group of F. The main objects of study in this paper are the ideals $I^n(F)$ of $W(F)$ and the graded rings $\bigoplus_{n=0}^{\infty} k_n(F)$, $\bigoplus_{n=0}^{\infty}(\bar{I}^n(F) := I^n(F)/I^{n+1}(F))$ and $\bigoplus_{n=0}^{\infty} H^n(G, \mathbb{Z}/2)$ as defined by Milnor in [**Mi**]. Most of the results we get hold only for the functors $k_n, I^n, \bar{I}^n$, but they would also hold for H^n if it was known that $H^n(G, \mathbb{Z}/2)$ is generated by cupproducts (see [**AEJ**]).

We will first introduce some terminology and recall some well kown open problems. We will use the standard notations of quadratic form theory as can be found in [**S**] for instance. If ψ is a quadratic form over F, which is always tacitly assumed to be nondegenerate, then $D(\psi)$ is the set of all elements in F^* represented by ψ. If φ is another form over F then $\psi \cong \varphi$ means ψ is isometric to φ. If ψ is Witt equivalent to φ we write $\psi = \varphi$. For $a_1, \dots, a_n \in F^*$ we use the symbols $\langle a_1, \dots, a_n \rangle$, $\langle\langle a_1, \dots, a_n \rangle\rangle$, $l(a_1) \cdots l(a_n)$, $(a_1) \cup \cdots \cup (a_n)$ as defined in [**Mi**]. Let $P_n(F) := \{ \langle\langle a_1, \dots, a_n \rangle\rangle \mid a_i \in F^* \}$ and let $\bar{P}_n(F)$ be the image of $P_n(F)$ in $\bar{I}^n(F)$. The elements $\langle\langle a_1, \dots, a_n \rangle\rangle$ of $P_n(F)$ are called n-fold Pfister forms. Let $G_n(F) := \{ l(a_1) \cdots l(a_n) \in k_n(F) \mid a_i \in F^* \}$ and $C_n(F) := \{ (a_1) \cup \cdots \cup (a_n) \in H^n(G, \mathbb{Z}/2) \mid a_i \in F^* \}$.

We furthermore have the maps $s_n : k_n(F) \to \bar{I}^n(F)$ and $h_n : k_n(F) \to H^n(G, \mathbb{Z}/2)$ which map $l(a_1) \cdots l(a_n)$ to $\langle\langle -a_1, \dots, -a_n \rangle\rangle$ and to $(a_1) \cup \cdots \cup (a_n)$. For details see [**Mi**]. Both maps are conjectured to be injective. Note that s_n is surjective and h_n is surjective if and only if $H^n(G, \mathbb{Z}/2)$ is generated by $C_n(F)$. It is known that s_n, h_n are isomorphisms for $n \leq 3$ (see [**Mi**], [**MS**], [**R**]). By [**EL**, 3.2], the map s_n is injective on $G_n(F)$. Furthermore by [**K**, corollary 5], s_n, h_n are injective for F and n if $n \geq n(F)$. Here, $n(F)$ denotes the nilpotence degree of $I(F)$, that is the least $n \leq \infty$ such that $I^n(F(\sqrt{-1}) = 0$. By [**EP**, theorem 3.3] and [**K**, theorem 1] we have:

PROPOSITION (1). *Let F be a field. Then $n(F) \leq n$ if and only if $I^n(F)$ is torsion free and $I^n(F) = 2I^{n-1}(F)$.*

Proposition 1 answers a question asked in [**AEJ**, see 5.20]. If X_F denotes the space of all orderings of F (see [**L**] or [, 3.5] for the definition and some basic facts), then proposition 1 shows $k_n(F) \cong \bar{I}^n(F) \cong Cont(X_F, \mathbb{Z}/2)$ for $n \geq n(F)$ where $Cont(X_F, \mathbb{Z}/2)$ denotes the set of all continuous functions from X_F to $\mathbb{Z}/2$.

Important is also the behaviour of the Milnor functors under quadratic extensions. Let $L = F(\sqrt{a})$ be a quadratic extension of F. Let $s : L \to F$ be the F-linear map defined by $s(1) = 0$ and $s(\sqrt{a}) = 1$. Then by [A], we have two zero sequences

$$I^n(F) \to I^n(L) \to^{s} I^n(F) \to^{\langle\langle --a \rangle\rangle} I^{n+1}(F) \to I^{n+1}(L)$$

and

$$\bar{I}^n(F) \to \bar{I}^n(L) \to^{s} \bar{I}^n(F) \to^{\langle\langle -a \rangle\rangle} \bar{I}^{n+1}(F) \to \bar{I}^{n+1}(L)\,.$$

It is conjectured that both sequences are exact. By [**AEJ**, 1.17, 1.24, 1.26], the sequences are exact for small n and they are exact for all n on Pfister forms. The corresponding sequence for H^n is exact (see [**A**, 4.6]).

Next, we introduce the important notion of function field of a quadratic form. Let F be a field and let $\varphi = \langle a_1, \dots, a_n \rangle$ be a form over F such that $n \geq 2$ and φ is not isometric to $\langle 1, -1 \rangle$. Then the field $F(\varphi)$ which is defined to be the quotient field of $F[X_1, \dots, X_{n-1}]/(a_1 + a_2 X_1^2 + \cdots a_n X_{n-1}^2)$ is called the function field of φ. The transcendence degree of $F(\varphi)$ over F is $n-2$. Note that here we are working with the small function field which behaves in many ways like the (big) function field as defined in [**S**].

The determination of the kernel of the map $\Gamma_n(F) \to \Gamma_n(F(\varphi))$ for the functors $\Gamma_n = k_n, I^n, \bar{I}^n, H^n$ seems to be difficult. For $a_i \in F^*$ let $\Gamma_m(a_1, \dots, a_m)$ denote the element $l(a_1) \cdots l(a_m), \langle\langle a_1, \dots, a_m \rangle\rangle, \langle\langle a_1, \dots, a_m \rangle\rangle, (a_1) \cup \cdots \cup (a_m)$ resp. if $\Gamma_m = k_m, I^m, \bar{I}^m, H^m$. In this paper we are especially interested in the case where $\varphi = \langle\langle a_1, \dots, a_m \rangle\rangle$ for some $a_1, \dots, a_m \in F^*$. One hopes that in this case the kernel equals $\Gamma_{n-m}(F)\langle\langle a_1, \dots, a_m \rangle\rangle$ if $\Gamma_n = I^n, \bar{I}^n$ and equals $\Gamma_{n-m}(F)\Gamma_m(-a_1, \dots, -a_m)$ if $\Gamma_n = k_n, H^n$. See [**A**, section 5]. Arason showed that the conjecture holds for $n \leq 3$ and $\Gamma_n = H^n$ and hence also for $\Gamma_n = k_n, \bar{I}^n$. See furthermore [**JR**, theorem 4.1].

The corresponding statement for the Witt ring functor holds: If φ is a Pfister form over F and $\rho \in W(F)$ is 0 in $W(F(\varphi))$ then ρ is divisible by φ, that is there exists some $\psi \in W(F)$ such that $\rho = \psi\varphi$. Thus we get:

Remark (1). Let $\varphi \in P_m(F)$. If $m < n$ then $ker(I^n(F) \to I^n(F(\varphi))) = \varphi I^{n-m}(F)$ if and only if for any $\rho \in I^n(F)$ divisible by φ there exists some $\psi \in I^{n-m}(F)$ such that $\rho = \psi\varphi$. If $n \leq m$ then $ker(I^n(F) \to I^n(F(\varphi))) = \varphi W(F)$. Furthermore, if $n = m + s, s = 1, 2$ then $ker(I^n(F) \to I^n(F(\varphi))) = \varphi I^s(F)$.

The last statement in remark 1 immediately follows from the main theorem of [**AP**]. Thus we see that the conjecture holds for $n \leq 4$ and $\Gamma_n = I^n$ as well since the case $m = 1, n = 4$ is covered by [**AEJ**, 1.24]. Next we show that the conjecture holds on generators.

THEOREM (1). *Let $\Gamma_n = k_n, \bar{I}^n, I^n$, $m \leq n$. Let $\varphi = \langle\langle a_1, \dots, a_m \rangle\rangle$. Assume $x = 0$ in $\Gamma_n(F(\varphi))$ for $x = \Gamma_n(c_1, \dots, c_n)$. Then there exist $w_i \in F^*$ such that $x = \langle\langle a_1, \dots, a_m \rangle\rangle \Gamma_{n-m}(w_1, \dots, w_{n-m})$ if $\Gamma_n = I^n, \bar{I}^n$ and $x = l(-a_1) \cdots l(-a_m)$ $\Gamma_{n-m}(w_1, \dots, w_{n-m})$ if $\Gamma_n = k_n$.*

PROOF. Let $\Gamma_n = I^n$ and $x = \langle\langle c_1, \dots, c_n \rangle\rangle$. By [**S**, 4.5.4], there exists a 2^{n-m} dimensional form τ over F with $x \cong \varphi\tau$. Then apply [**A**, lemma 1.4]. Next, assume $x \in I^{n+m+1}(F(\varphi))$. Then [**AP**, the main theorem] implies $x = 0$ in $W(F(\varphi))$ and we can again apply [**A**, 1.4]. Assume now $x = 0$ in $k_n(F(\varphi))$ for $x = l(c_1) \cdots l(c_n)$. Then $s_n(x) = 0$ in $I^n(F(\varphi))$ and hence $\langle\langle -c_1, \dots, -c_n \rangle\rangle = \langle\langle a_1, \dots, a_m, w_1, \dots, w_{n-m} \rangle\rangle$ modulo $I^{n+1}(F)$ for $w_i \in F^*$. By [**EL**, 3.2], $x = l(-a_1) \cdots l(-a_m) l(-w_1) \cdots l(-w_{n-m})$. □

Now assume F is formally real. For every ordering $P \in X_F$ let F_P denote the real closure of F with respect to P. If T is a preordering of F then X/T denotes the space of all orderings P of F satisfying $T \subset P$. Recall that $W_T(F)$ is the ideal of $W(F)$ generated by all forms $\langle 1, -t \rangle$, $t \in T$ and $k_{nT}(F)$ is the subgroup of $k_n(F)$ generated by all $l(t)l(c_1)\cdots l(c_{n-1})$, $t \in T$, $c_i \in F^*$. Keeping the notations of [**K**] we recall the following conjectures: (1): $ker(k_n(F) \to \prod_{P \in X/T} k_n(F_P)) = k_{nT}(F)$, (2): $I^n(F) \cap W_T(F) = I^{n-1}(F)W_T(F)$ and (3): $ker(\bar{I}^n(F) \to \prod_{P \in X/T} \bar{I}^n(F_P)) = I^n(F) \cap W_T(F)/I^{n+1}(F) \cap W_T(F)$. For results concerning these conjectures we refer to [**Ma**], [**K**].

We now introduce a conjecture concerning the Milnor functors that will be our main topic of study. A theorem of Pfister and Witt (see [**S**, 2.10.13]) describes the annihilators of Pfister forms in the Witt ring of a field. In this paper we want to study annihilators of generators for the Milnor functors. In the following let $\Delta_n(F) = G_n(F), P_n(F), \bar{P}_n(F), C_n(F)$ respectively, if $\Gamma_n = k_n, I^n, \bar{I}^n, H^n$ respectively. Let $m, n \in I\!N$. For $x \in \Gamma_m(F)$ let $ann_n(x) := \{\, y \in \Gamma_n(F) \mid yx = 0 \; in \Gamma_{n+m}(F) \,\}$. Then we conjecture:
$ann_n(x)$ is generated by $ann_n(x) \cap \Delta_n(F)$ for all $x \in \Delta_m(F)$.

We have to remark here that if $\Gamma_n = I^n$ then "generated" means generated as $W(F)$ ideal, otherwise it means generated as group. The conjecture holds on generators by definition, hence it holds for $n = 1$. In order to give a complete description of the annihilators we have to look at the following stronger conjecture:
$ann_n(x) = ann_1(x)\Gamma_{n-1}(F)$ for all $x \in \Delta_m(F)$.

If the latter conjecture was verified we would have a complete description of the annihilators. Let $\varphi = \langle\langle a_1, \ldots, a_m \rangle\rangle$ and let $\eta = l(-a_1)\cdots l(-a_m)$. Then $ann_1(\varphi)$ is the ideal generated by $\langle\langle -a \rangle\rangle$ resp. the subgroup of $\bar{I}(F)$ generated by $\langle\langle -a \rangle\rangle + I^2(F)$ for $a \in D(\varphi)$ if $\Gamma_n = I^n$ resp. $\Gamma_n = \bar{I}^n$. The latter statement follows from the main theorem of [**AP**]. Furthermore by [**EL**, 3.2], $ann_1(\eta)$ is generated by $l(a)$, $a \in D(\varphi)$ if $\Gamma_n = k_n$. Obviously, for $m = 1$ both conjectures coincide. As a matter of fact, both conjectures coincide in general for $\Gamma_n = k_n, I^n, \bar{I}^n$ as theorem 2 will show.

EXAMPLE. Let $F = Q(X, Y)$. Let $U := -1 - Y^2 - 3X$. Then by [**AP2**, Beispiel 1], the form $\rho = \langle\langle X, U \rangle\rangle$ is a torsion form over F which is not strongly balanced (see [**AP2**]), which means that $D(\rho)$ does not contain nonzero totally negative elements. Let $n \in I\!N$ such that $2^n\langle 1 \rangle \rho = 0$ (actually, we have $4\rho = 0$). Then $\rho \in ann_2(2^n\langle 1 \rangle) \cap P_2(F)$, but there exists no Pfister form $\tau \in ann_1(2^n\langle 1 \rangle)I(F)$ such that $\rho = \tau$. Theorem 2 or remark 3 will however show, that $\rho \in ann_1(2^n\langle 1 \rangle)I(F)$.

The latter conjecture can be reformulated as follows. As usual, (see [**S**, 4.1.1]) if $\varphi \in P_m(F)$, then φ' denotes the pure subform of φ, that is the uniquely determined form that satisfies $\varphi \cong \langle 1 \rangle + \varphi'$.

Remark (2). Let $\varphi = \langle\langle a_1, \ldots, a_m \rangle\rangle$. Then $ann_n(\varphi) = ann_1(\varphi)I^{n-1}(F)$ if and only if the zero sequence

$$\bigoplus_{a\in D(\varphi')} I^n(F(\sqrt{a})) \to^s I^n(F) \to^\varphi I^{n+m}(F)$$

is exact. A corresponding statement holds also for k_n and $\bar{I}^n$.

PROOF. Apply [**S**, 2.14.8]. □

Now we show that both conjectures actually coincide. We will prove a little bit more than that:

THEOREM (2). *Let* $a_1,\dots,a_m,b_1,\dots,b_n,c_1,\dots,c_n \in F^*$. *Let* $\varphi = \langle\langle a_1,\dots,a_m\rangle\rangle$ *and let* $\eta = l(-a_1)\cdots l(-a_m)$. *Then the following statements are equivalent:*

(i) $\langle\langle b_1,\dots,b_n\rangle\rangle = \langle\langle c_1,\dots,c_n\rangle\rangle$ *modulo* $ann_1(\varphi)W(F)$.
(ii) $\langle\langle b_1,\dots,b_n\rangle\rangle = \langle\langle c_1,\dots,c_n\rangle\rangle$ *modulo* $ann_1(\varphi)I^{n-1}(F)$.
(iii) $l(-b_1)\cdots l(-b_n) = l(-c_1)\cdots l(-c_n)$ *modulo* $ann_1(\eta)k_{n-1}(F)$.

In particular, if $\Gamma_n = k_n, I^n, \bar{I}^n$ *and* $x \in \Delta_m(F)$, *then* $ann_n(x) = ann_1(x)\Gamma_{n-1}(F)$ *if and only if* $ann_n(x)$ *is generated by* $ann_n(x) \cap \Delta_n(F)$.

Compare theorem 2 with [**Ma**, theorem 8] and [**EL**, 3.2]. To prove the theorem we generalize the arguments in the proof of [**Ma**, theorem 8] which are in turn a modification of the arguments used to prove [**EL**, 3.2]. We need to apply the theory of abstract Witt rings. We will see that the proof gives a nice example of an application of the theory of abstract Witt rings to the theory of Witt rings of fields. We use the term abstract Witt ring as defined in [**Ma2**]. For details we refer to [**Ma2**].

Fix some $\varphi = \langle\langle a_1,\dots,a_m\rangle\rangle \in P_m(F)$. Let $\eta = l(-a_1)\cdots l(-a_m)$. Furthermore, let $\mathcal{A} = \{\rho \in W(F) \mid \rho\varphi = 0\}$. Then by [**S**, 2.10.13], we have $\mathcal{A} = ann_1(\varphi)W(F)$. It is well known that φ is multiplicative, that is $D(\varphi)$ is a subgroup of the square class group F^*/F^{*2} (see [**S**, 2.10.4]. Then [**Ma2**, 4.24] states that $(W(F)/\mathcal{A}, F^*/D(\varphi))$ is an abstract Witt ring in the sense of [**Ma2**]. Thus we can consider forms over $F^*/D(\varphi)$. Note that two forms $\langle b_1,\dots,b_m\rangle$ and $\langle c_1,\dots,c_n\rangle$ are isometric over $F^*/D(\varphi)$ if and only if $\langle b_1,\dots,b_m\rangle = \langle c_1,\dots,c_n\rangle$ in $W(F)/\mathcal{A}$ and $m=n$. A quadratic form $\langle b_1,\dots,b_n\rangle$ over F is said to represent $u \in F^*$ modulo $\mathcal{A}$ if there exist $w_1,\dots,w_n \in D(\varphi)\cup\{0\}$ such that $u = \sum_{i=1}^n b_i w_i$. The next lemma follows from [**Ma2**, 4.24] (compare [**Ma**, lemma 1]):

LEMMA (1). *A form* $\langle b_1,\dots,b_n\rangle$ *over* F *represents* $u \in F^*$ *modulo* $\mathcal{A}$ *if and only if there exist* $w_2,\dots,w_n \in F^*$ *such that* $\langle b_1,\dots,b_n\rangle \cong \langle u,w_2,\dots,w_n\rangle$ *over* $F^*/D(\varphi)$.

Using lemma 1 one can use the same arguments as in the proof of [**Ma**, lemma 2], which are in turn the same arguments as can be found in [**S**, 4.1.4], to show:

LEMMA (2). *Let* $\rho = \langle\langle b_1, \ldots, b_n \rangle\rangle$ *and* $\xi = l(-b_1)\cdots l(-b_n)$. *Suppose* ρ' *represents* $u \in F^*$ *modulo* $\mathcal{A}$. *Then there exist* $w_2, \ldots, w_n \in F^*$ *such that* $\rho = \langle\langle u, w_2, \ldots, w_n \rangle\rangle$ *modulo* $ann_1(\varphi)I^{n-1}(F)$ *and* $\xi = l(-u)l(-w_2)\cdots l(-w_n)$ *modulo* $ann_1(\eta)k_{n-1}(F)$.

We now prove theorem 2 by generalizing the arguments used in the proof of [**Ma**, theorem 8]:

PROOF OF THEOREM 2. $(iii) \to (ii) \to (i)$ are obvious. $(i) \to (iii)$: We proceed by induction on n. The case $n = 1$ is trivial. By lemma 2, there exist $w_2, \ldots w_n \in F^*$ such that $l(-b_1)\cdots l(-b_n) = l(-c_1)l(-w_2)\cdots l(-w_n)$ modulo $ann_1(\eta)k_{n-1}(F)$. By induction hypothesis, $l(-w_2)\cdots l(-w_n) = l(-c_2)\cdots l(-c_n)$ modulo $ann_1(l(-c_1)\eta)k_{n-2}(F)$. Let $l(x) \in ann_1(l(-c_1)\eta)$. We will show $l(x)l(-c_1)\tau \in ann_1(\eta)k_{n-1}(F)$ for any $\tau = l(u_1)\cdots l(u_{n-2})$ and thus $l(-b_1)\cdots l(-b_n) = l(-c_1)\cdots l(-c_n)$ modulo $ann_1(\eta)k_{n-1}(F)$. Let $\tau = l(u_1)\cdots l(u_{n-2})$. We have $\langle\langle -x \rangle\rangle \in ann_1(\langle\langle c_1 \rangle\rangle\varphi)$ and hence $x \in D(\langle\langle c_1 \rangle\rangle\varphi)$. Choose $y, z \in D(\varphi) \cup \{0\}$ such that $x = y + c_1 z$. If $z = 0$ there is nothing to prove. Let $y, z \in D(\varphi)$. Then by lemma 2 again, $l(x)l(-c_1)\tau = l(y + c_1 z)l(-c_1 z)\tau = l(y)l(-yxc_1)\tau = 0$ modulo $ann_1(\eta)k_{n-1}(F)$. If $y = 0, z \in D(\varphi)$ then $l(c_1 z)l(-c_1)\tau = l(z)l(-c_1)\tau \in ann_1(\eta)k_{n-1}(F)$.

The second statement is now obvious for $\Gamma_n = I^n$ and it is valid for $\Gamma_n = \bar{I}^n$ by the main theorem of [**AP**]. To prove the statement for $\Gamma_n = k_n$ apply again [**EL**, theorem 3.2] (use the same arguments as in the last part of the proof of theorem 1). □

Keeping the notations above set $R = W(F)/\mathcal{A}$. Then $I^n(R)$ denotes the ideal of R generated by all n-fold Pfister form classes. We have projection maps $p : I^n(F) \to I^n(R)$ and $\bar{p} : \bar{I}^n(F) \to \bar{I}^n(R)$. Then $ann_n(\varphi) = ann_1(\varphi)I^{n-1}(F)$ if and only if the kernel of p equals $\mathcal{A}I^{n-1}(F)$. A corresponding statement also holds for $\bar{p}, \bar{I}^n$. If ψ is a form over $F^*/D(\varphi)$, then we can also define $D_R(\psi)$ in the obvious way. Any form ψ over $F^*/D(\varphi)$ is decomposable into $\psi_{an} + \psi_h$ where ψ_{an} is anisotropic over $F^*/D(\varphi)$ and $\psi_h \cong m\langle 1, -1 \rangle$ over $F^*/D(\varphi)$. The uniquely determined m is called the Witt index of ψ over $F^*/D(\varphi)$.

Let $\Gamma_n = k_n, I^n, \bar{I}^n$ and $x \in \Delta_m(F)$. Theorem 2 shows that if $y \in ann_n(x)$ can be written as a sum of two generators then $y \in ann_1(x)\Gamma_{n-1}(F)$. Using results of Elman and Lam on the linkage of Pfister forms we can show the same for sums of three generators:

THEOREM (3). *Let* $\Gamma_n = k_n, I^n, \bar{I}^n$. *Let* $x \in \Delta_m(F)$ *and* $y \in ann_n(x)$. *Assume* $y = y_1 + y_2 + y_3$ *for* $y_i \in \Delta_n(F)$ *if* $\Gamma_n = k_n, \bar{I}^n$ *and* $y_i = \langle a_i \rangle z_i$, $a_i \in F^*$, $z_i \in P_n(F)$ *if* $\Gamma_n = I^n$. *Then* $y_i \in ann_1(x)\Gamma_{n-1}(F)$.

For the proof of theorem 3 we need the following lemma. For a proof use the same arguments as in the proof of [**EL**, 4.4]. The arguments of Elman and Lam remain valid for the abstract Witt ring. We keep the notations above.

LEMMA (4). *Let ψ, ψ^* be n-fold Pfister forms over $F^*/D(\varphi)$. Then the following statements are equivalent:*

(i) *The Witt index of $\psi + \langle -1\rangle\psi^*$ over $F^*/D(\varphi)$ is $\geq 2^r$.*

(ii) *There exist an r-fold Pfister form ρ over $F^*/D(\varphi)$ and two $n-r$-fold Pfister forms τ, τ^* over $F^*/D(\varphi)$ such that $\psi \cong \rho\tau$ and $\psi^* \cong \rho\tau^*$ over $F^*/D(\varphi)$.*

PROOF OF THEOREM 3. By [**EL**, 6.1], if the statement of theorem 3 holds for $\Gamma_n = \bar{I}^n$, then it also holds for $\Gamma_n = k_n$. Assume $x \in P_m(F)$ and $yx \in I^{n+m+1}(F)$. Then by [**EL**, theorem 4.8] we have $yx = 0$ in $W(F)$. Thus it suffices to show that if $x \in P_m(F)$ and $y = y_1 + \langle a\rangle y_2 + \langle b\rangle y_3$ with $yx = 0$, then $y \in ann_1(x)I^{n-1}(F)$. Setting $\varphi := x$ and keeping the notations above we then have $y_1 + \langle a\rangle y_2 = \langle -b\rangle y_3$ in R. In particular, there exist $c \in D_R(y_1)$ and $d \in D_R(y_2)$ such that $c+ad = 0$. Since the Pfister forms y_1 and y_2 over $F^*/D(\varphi)$ are multiplicative by [**Ma2**, 3.1], we have $y_1 \cong \langle -ad\rangle y_1$ and $\langle a\rangle y_2 \cong \langle ad\rangle y_2$ over $F^*/D(\varphi)$. This shows that the form $y_1 + \langle -1\rangle y_2$ has Witt index $\geq 2^{n-1}$ in R. Hence by lemma 4, there exist $z \in P_{n-1}(F)$ and $u, v \in F^*$ such that $y_1 = z\langle\langle u\rangle\rangle$ and $y_2 = z\langle\langle v\rangle\rangle$ over $F^*/D(\varphi)$. This implies $\langle b\rangle y_3 \cong \langle adu, -adv\rangle z$ over $F^*/D(\varphi)$. By theorem 2 the formulas above also hold in $W(F)$ modulo $ann_1(\varphi)I^{n-1}(F)$. Hence, $y = z\langle ad\rangle(\langle -1\rangle\langle\langle u\rangle\rangle + \langle\langle v\rangle\rangle + \langle u, -v\rangle) = 0$ modulo $ann_1(\varphi)I^{n-1}(F)$. □

Unfortunately, it is not known whether the conjecture holds for small n like $n = 2, 3$. We can only show applying the theorem of Pfister, Witt:

Remark (3). Let F be a field and $\Gamma_n = I^n$. Let $x \in P_m(F)$. Then $ann_2(x) = ann_1(x)I(F)$.

PROOF. Let $\varphi \in P_m(F)$ and $\rho \in I^2(F)$ such that $\rho\varphi = 0$. By [**S**, 2.10.13], there exist $a_i, b_i \in F^*$ such that $\rho = \sum\langle a_i, a_ib_i\rangle$ with $\langle a_i, a_ib_i\rangle\varphi = 0$. Then $\langle a_i, a_ib_i\rangle = \langle\langle a_i, b_i\rangle\rangle - \langle 1, b_i\rangle$ and $\sum\langle 1, b_i\rangle = \langle\langle b_1, b_2\rangle\rangle + \langle\langle -b_1b_2, b_3\rangle\rangle + \cdots$. Hence, ρ equals a sum of forms $\rho_i \in ann_1(\varphi)I(F)$. □

2. Connections between the conjectures

In the following we want to point out some connections between the conjectures introduced in §1. Then we give examples of field classes for which the conjectures on annihilators hold. The following remarks will be useful throughout this section:

Remark (4).

(a) Assume that $ann_n(\varphi)$ is generated by $ann_n(\varphi) \cap P_n(F)$ and the sequence $\varphi I^{n+1}(F) \to I^{n+m+1}(F) \to I^{n+m+1}(F(\varphi))$ is exact for $\varphi \in P_m(F)$. Then $ann_n(\varphi + I^{n+1}(F))$ is generated by $ann_n(\varphi + I^{n+1}(F)) \cap \bar{P}_n(F)$. In particular, if for a quadratic extension L/F the sequences $I^n(L) \to I^n(F) \to I^{n+1}(F)$ and $I^n(F) \to I^{n+1}(F) \to I^{n+1}(L)$ are exact then $\bar{I}^n(L) \to \bar{I}^n(F) \to \bar{I}^{n+1}(F)$ is exact as well.

(b) Assume $ann_n(\varphi) = ann_1(\varphi)\bar{I}^{n-1}(F)$ for m, all $\varphi \in \bar{P}_m(F)$, $n \leq s$. Then the sequence $\varphi I^n(F) \to I^{n+m}(F) \to I^{n+m}(F(\varphi))$ is exact for all $n \leq s$, $\varphi \in P_m(F)$.

(c) Assume $ann_n(\varphi) = ann_1(\varphi)\bar{I}^{n-1}(F)$ for all $\varphi \in \bar{P}_m(F)$, $n \geq r$ and there exists some $s \geq r$ such that $ann_s(\varphi) = ann_1(\varphi)I^{s-1}(F)$ for all $\varphi \in P_m(F)$. Then for any n, $r \leq n \leq s$ and $\varphi \in P_m(F)$ we have $ann_n(\varphi) = ann_1(\varphi)I^{n-1}(F)$.

PROOF.

(a) Assume $\rho\varphi \in I^{n+m+1}(F)$ for $\rho \in I^n(F)$. Then by assumption and remark 1, there exists some $\psi \in I^{n+1}(F)$ such that $\rho\varphi = \psi\varphi$ and hence $\rho - \psi = \sum \rho_i, \rho_i \in ann_n(\varphi) \cap P_n(F)$. For the second statement apply remarks 1,2.

(b) Assume $\rho \in I^{n+m}(F)$ is divisible by $\varphi \in P_m(F)$. Let $t \leq n$ be maximal such that there exists some $\psi \in I^t(F)$ such that $\rho = \psi\varphi$. Assume $t < n$. Then by assumption, there exist $\tau \in I^{t+1}(F)$ and $\psi' \in ann_1(\varphi)I^{t-1}(F)$ such that $\psi = \psi' + \tau$. Thus $\tau\varphi = \rho$ which is a contradiction.

(c) Is easy and left to the reader. □

Similarly, one can show:

Remark (5). Let L/F be a quadratic extension.

(a) Assume the sequences $I^n(F) \to I^n(L) \to I^n(F)$ and $I^{n+1}(L) \to I^{n+1}(F) \to I^{n+2}(F)$ are exact. Then the sequence $\bar{I}^n(F) \to \bar{I}^n(L) \to \bar{I}^n(F)$ is exact as well.

(b) Assume the sequences $I^{n-1}(F) \to I^n(F) \to I^n(L)$ and $I^{n+1}(F) \to I^{n+1}(L) \to I^{n+1}(F)$ are exact. Then the sequence $\bar{I}^{n-1}(F) \to \bar{I}^n(F) \to \bar{I}^n(L)$ is exact as well.

(c) Assume the sequence $\bar{I}^{m-1}(F) \to \bar{I}^m(F) \to \bar{I}^m(L)$ is exact for all $m \leq n-1$. Then the sequence $I^n(F) \to I^n(L) \to I^n(F)$ is exact.

PROOF. The proofs for (a) and (b) are easy and left to the reader. (c): Apply [**A**, 2.4(3)]. □

In the following, $S_F = S$ will always denote the set of all sums of squares in F. The theorem of Pfister and Witt which describes the annihilators of Pfister forms in the Witt ring implies a local-global principle as is pointed out in [**S**, 2.10.14] for instance. Similarly, if the conjecture on annihilators for the Milnor functors holds, then the local-global principles for the Milnor functors (1), (2), (3) hold as well. More precisely:

PROPOSITION (2). *Assume that for F, $n \in \mathbb{N}$ and for all $x \in \Delta_m(F)$ we have $ann_n(x) = ann_1(x)\Gamma_{n-1}(F)$ for $\Gamma_n = k_n, \bar{I}^n, I^n$ respectively. Then for all preorderings T of F the conjecture* (1) *holds,* (2) *and* (3) *hold,* (2) *holds for n respectively.*

PROOF. Assume first that there exist $a_1, \ldots, a_m \in F^*$ such that $T = S[a_1, \ldots, a_m]$ and $\Gamma_n = k_n$. Let $x \in k_n(F)$ such that $x = 0$ in $k_n(F_P)$ for all $P \in X_F/T$. Then by [**K**, corollary 4], there exists some s such that $l(-1)^s l(-a_1) \cdots l(-a_m)x = 0$ in $k_{n+m+s}(F)$. By assumption, there exist $t_i \in D(2^s\langle\langle a_1, \ldots, a_m\rangle\rangle) \subset T$ and $y_i \in G_{n-1}(F)$ such that $x = \sum l(t_i)y_i$. Thus $x \in k_{nT}(F)$. Now let T be an arbitrary preordering and let $x \in k_n(F)$ such that $x = 0$ in $k_n(F_P)$ for all $P \in X_F/T$. It is easy to see that there exist $a_1, \ldots, a_m \in T$ such that $x = 0$ in $k_n(F_P)$ for all $P \in X/T'$, $T' = S[a_1, \ldots, a_m]$. Thus by the first part of the proof, $x \in k_{nT'}(F) \subset k_{nT}(F)$. The proofs for $\Gamma_n = \bar{I}^n, I^n$ work in the same way. □

A converse of proposition 2 holds in the following special case.

THEOREM (4). *Assume $I^r(F)$ is torsion free.*

(a) *Then the following statements are equivalent:*
   (i) (2) *and* (3) *hold (resp.* (1) *holds, resp.* (2) *holds) for all T, $n \geq r-1$.*
   (ii) *For all $m \in I\!N$, $x \in \Delta_m(F)$ and $n \geq r-1$ we have $ann_n(x) = ann_1(x)\Gamma_{n-1}(F)$ for $\Gamma_n = \bar{I}^n$ (resp. $\Gamma_n = k_n$, resp. $\Gamma_n = I^n$).*

(b) *Furthermore,* (2) *and* (3) *hold for $T = S, n \geq r-1$ if and only if for $L = F(\sqrt{-1})$ and for all $n \geq r-1$ the sequence $\bar{I}^n(L) \to \bar{I}^n(F) \to \bar{I}^{n+1}(F)$ is exact. Moreover,* (2) *holds for $S, n = r-1$ if and only if $I^{r-1}(L) \to I^{r-1}(F) \to I^r(F)$ is exact.*

PROOF.

(a) $(ii) \to (i)$ follows from Proposition 2. $(i) \to (ii)$: Let $\varphi = \langle\langle a_1, \ldots, a_m\rangle\rangle$ and let $\rho \in I^n(F)$ such that $\rho\varphi \in I^{n+m+1}(F)$. Set $T = S[a_1, \ldots, a_m]$. Then $sgn_P(\rho)$ is divisible by 2^{n+1} for all $P \in X/T$. By assumption there exist $\rho_i \in P_n(F) \cap I^{n-1}(F)W_T(F)$ such that $\rho = \sum \rho_i$ modulo $I^{n+1}(F)$. Then we have $sgn_P(\varphi\rho_i) = 0$ for all $P \in X_F$ and hence $\varphi\rho_i = 0$ for all i. The proofs for the second and third statement work in the same way.

(b) Assume for all $n \geq r-1$ the sequence $\bar{I}^n(L) \to \bar{I}^n(F) \to \bar{I}^{n+1}(F)$ is exact. Let $\rho \in I^n(F)$ such that $sgn_P(\rho)$ is divisible by 2^{n+1} for all $P \in X_F$. We have to show $\rho \in I^{n-1}(F)W_S(F)$ modulo $I^{n+1}(F)$. Replacing ρ by $2^s\rho$ if necessary, by [**K**, corollary 4] we may assume $2\rho \in I^n(F)W_S(F) = 0$. Thus by assumption, $\rho \in s(I^n(L) \subset I^{n-1}(F)W_S(F)$ modulo $I^{n+1}(F)$ by [**S**, 2.14.8]. Conversely, if $2\rho \in I^{n+2}(F)$ then the signature values of ρ are divisible by 2^{n+1}. Hence $\rho \in I^{n-1}(F)W_S(F)$ modulo $I^{n+1}(F)$ by assumption. By [**AEJ**, 1.17], every torsionform in $P_n(F)$ is Witt equivalent to some $s(\psi), \psi \in P_n(L)$ and hence $s(I^n(L)) = I^{n-1}(F)W_S(F)$. The second statement follows from $s(I^{r-1}(L)) = W_S(F)I^{r-2}(F)$ also. □

In the following we need the notion of the stability index $st(X_F)$ of a space of orderings X_F or X_F/T for a preordering T of F. For the definition and details see [**L**]. From theorem 4 and [**AEJ**, 1.24] we conclude:

COROLLARY (1). *Assume $I^4(F)$ is torsion free. Then (2) holds for $n = 3$, $T = S$.*

COROLLARY (2). (a) *Assume $n \geq n(F) - 1$ and $m \in \mathbb{N}$. Then for $\Gamma_n = k_n, I^n, \bar{I}^n$, and for all $x \in \Delta_m(F)$ we have $ann_n(x) = ann_1(x)\Gamma_{n-1}(F)$.*

(b) *Assume $n \geq st(X_F)$ and $I^{n+m}(F)$ is torsion free. Then for $\Gamma_n = k_n, I^n, \bar{I}^n$ and for all $x \in \Delta_m(F)$ we have $ann_n(x) = ann_1(x)\Gamma_{n-1}(F)$.*

PROOF. For all preorderings T of F we have $st(X_F/T) \leq st(X_F)$. By [**K**, theorem 1], (1), (2) and (3) hold for all $T, n \geq st(F)$. Then (a) follows from theorem 4. To show (b) use the same arguments as in the proof of theorem 4(a). □

We can prove a corresponding statement for the conjecture on function fields of quadratic forms. As usual, for $b_1, \dots, b_n \in F^*$ let $H(b_1, \dots, b_n) = \{P \in X_F \mid b_1, \dots, b_n \in P\}$.

PROPOSITION (3). *Assume $n \geq st(X_F)$ and $I^{n+m}(F)$ is torsion free. Then the sequence $\Gamma_n(F) \to \Gamma_{n+m}(F) \to \Gamma_{n+m}(F(\varphi))$ is exact for all*

$$\varphi = \langle\langle a_1, \dots, a_m \rangle\rangle \Gamma_n = I^n, \bar{I}^n, k_n\,.$$

PROOF. Let $\rho \in I^{n+m}(F)$ such that $\rho = 0$ in $W(F(\varphi))$. By [**K**, lemma 1], we may assume $\rho = \sum l_i \rho_i$, $l_i \in \mathbb{Z}$ such that for $\rho_i = \langle\langle c_1^i, \dots, c_{n+m}^i \rangle\rangle$ we have $H_i := H(c_1^i, \dots c_{n+m}^i)$ are pairwise disjoint in X_F. Recall that for an ordering $P \in X_F$ there exists an extension to $F(\varphi)$ if and only if $P \in X_F - H(a_1, \dots, a_m)$. This follows from [**S**, 3.1.10], since obviously there exist extensions of any $P \in X_F$ to purely transcendental extensions of F. Hence $\rho = 0$ in $W(F(\varphi))$ implies $H_i \subset H(a_1, \dots, a_m)$ for any i. Since $st(X_F) \leq n$, there exist $d_j^i \in F^*$ such that $H_i = H(d_1^i, \dots, d_n^i)$. Thus the signature values of ρ_i and $\psi_i := \langle\langle d_1^i, \dots, d_n^i \rangle\rangle \varphi$ coincide and thus $\rho = \sum l_i \psi_i$. Now assume $\rho \in I^{n+m+1}(F(\varphi))$. Then $sgn_P(\rho)$ is divisible by 2^{n+m+1} for all $P \in X_F - H(a_1, \dots, a_m)$. Thus there exists some $\psi \in I^{n+m+1}(F)$ such that $sgn_P(\rho + \psi) = 0$ for all $P \in X_F - H(a_1, \dots, a_m)$. Then we can use the same arguments as above to show that $\rho + \psi = \varphi\tau$ for some $\tau \in I^n(F)$. Since the hypothesis implies $n(F) \leq m + n$ we have $k_{n+m}(F) \cong \bar{I}^{n+m}(F)$. Let $x \in k_{n+m}(F)$ such that $x = 0$ in $k_{n+m}(F(\varphi))$. Then $s_{n+m}(x) = \varphi \sum_i \langle\langle c_1^i, \dots, c_n^i \rangle\rangle$ for some $c_j^i \in F^*$. Thus $x = l(-a_1) \cdots l(-a_m) \sum_i l(-c_1^i) \cdots l(-c_n^i)$. □

We now summarize the results of theorem 2, proposition 3 and corollary 2:

COROLLARY (3). *Let $m, n \in \mathbb{N}$. Suppose $n \geq st(X_F)$ and $I^{n+m}(F)$ is torsion free. Then for $\Gamma_n = k_n, I^n, \bar{I}^n$ and for any $\varphi = \langle\langle a_1, \dots, a_m \rangle\rangle$ the sequence*

$$\bigoplus_{a \in D(\varphi')} \Gamma_n(F(\sqrt{a})) \to \Gamma_n(F) \to \Gamma_{n+m}(F) \to \Gamma_{n+m}(F(\varphi))$$

is exact.

The next corollary improves the result [**AEJ**, 2.11] slightly:

COROLLARY (4). *Let $n = n(F)$ and $\Gamma_r = k_r, \bar{I}^r, I^r$. Let L/F be a quadratic extension. Then the long sequence*

$$\Gamma_{n-1}(L) \to \Gamma_{n-1}(F) \to \Gamma_n(F) \to \Gamma_n(L) \to \Gamma_n(F) \to \Gamma_{n+1}(F) \to \cdots$$

is exact.

PROOF. By [**EL2**, 6.3], we have $n(L) = n(F)$. Then by proposition 1 and [**AEJ**, theorem 2.11], the above sequence is exact for $\Gamma_r = I^r$ except possibly at the first two spots. The sequence is exact at the first two spots for the functors $k_r, \bar{I}^r, I^r$ by corollary 3. Applying the remarks above we see that the above sequence is exact for $\Gamma_r = \bar{I}^r$ and hence also for $\Gamma_r = k_r$, since the Milnor map is an isomorphism for $n \geq n(F)$. □

Up to now only little is known concerning conjecture (3). The only known result is (the more or less trivial fact) that if $st(X_F) \leq k$ then conjecture (3) holds for $F, n \geq k-1$. A proof follows immediately from the fact that every clopen set $H \subset X_F$ can be written as a disjoint union of finitely many basic clopen sets $H(a_1, \ldots, a_k)$ if $st(X_F) \leq k$ (see [**K**, corollary 1]). Thus the following corollary seems to be important:

COROLLARY (5). *Assume $I^3(F)$ is torsion free and $st(X_F) \leq 6$. Then conjecture* (3) *holds for F, $T = S$ and all n.*

PROOF. By theorem 4, we need to show that for $L = F(\sqrt{-1})$ the sequence $\bar{I}^n(L) \to \bar{I}^n(F) \to \bar{I}^{n+1}(F)$ is exact for all n. By corollary 4, this holds for $n \geq 5$. By [**AEJ**, 1.24, 1.26], the results of [**MS**], [**R**], [**JR**] and remark 4(a), the sequence is exact for $n \leq 4$ as well. □

It is well known that weak versions of the conjectures (1), (2) and (3) hold. See [**Ma**, theorem 2] or [**K**, corollary 4]. It is easy to generalize those statements:

PROPOSITION (4). *Let $\Gamma_n = k_n, I^n, \bar{I}^n$. Let $x \in \Delta_m(F)$. Suppose $y \in ann_n(x)$. Then there exists some $r \in \mathbb{N}$ such that $2^r y \in ann_1(x)\Gamma_{r+n-1}(F)$ if $\Gamma_n = I^n, \bar{I}^n$ and $l(-1)^r y \in ann_1(x)\Gamma_{r+n-1}(F)$ if $\Gamma_n = k_n$.*

PROOF. Let $\Gamma_n = I^n$. Choose $x = \langle\langle a_1, \ldots, a_m\rangle\rangle, y$ as above. Let $T = S[a_1, \ldots, a_m]$. By [**Ma**, theorem 2] or [**K**, corollary 4(ii)], there exists some $s \in \mathbb{N}$ such that $2^s y \in W_T(F)I^{s+n-1}(F)$. Hence, there exist $y_i = \langle\langle t_i, c_i^1, \ldots, c_i^{s+n-1}\rangle\rangle$, $t_i \in -T, c_i^j \in F^*$, such that $2^s y = \sum_i y_i$ and $x\langle\langle t_i\rangle\rangle$ are torsion forms. Choose some $s' \in \mathbb{N}$ such that $2^{s'} x\langle\langle t_i\rangle\rangle = 0$ for all i. Set $r = s + s'$. Then $2^r y = 2^{s'} \sum_i y_i \in ann_1(x)I^{r+n-1}(F)$. The proofs for $\Gamma_n = \bar{I}^n, k_n$ work in the same way. □

Another example of fields for which we are able to prove something are the so called linked fields. Let $n \geq 2$. Then $I^n(F)$ is called linked if for each pair $\varphi, \psi \in P_n(F)$ there exist $c, d \in F.$ and $\tau \in P_{n-1}(F)$ such that $\varphi \cong \langle\langle c\rangle\rangle\tau$ and

$\psi \cong \langle\langle d\rangle\rangle\tau$. See [**AEJ**, 4.1] for some basic results on n-linked fields. If $I^n(F)$ is linked we have $\bar{I}^n(F) = \bar{P}_n(F)$ and thus by [**EL**, 3.2], $k_n(F) \cong \bar{I}^n(F)$.

PROPOSITION (5). *Suppose $I^s(F)$ is linked. Let $\Gamma_n = k_n, I^n, \bar{I}^n$ and let $m \in I\!N, n \geq s$. Let $\varphi = \langle\langle a_1, \ldots, a_m\rangle\rangle$. Then $ker(\Gamma_n(F)) \to \Gamma_n(F(\varphi))) = \Gamma_m(a_1, \ldots, a_m)\Gamma_{n-m}(F)$ if $\Gamma_n = \bar{I}^n, I^n$ and $ker(\Gamma_n(F)) \to \Gamma_n(F(\varphi))) = \Gamma_m (-a_1, \ldots, -a_m)\Gamma_{n-m}(F)$ if $\Gamma_n = k_n$. Furthermore, for all $x \in \Delta_m(F)$ we have $ann_n(x) = ann_1(x)\Gamma_{n-1}(F)$.*

PROOF. If $I^s(F)$ is linked, then so is $I^n(F)$ by [**A**, 4.2]. By assumption and by theorem 1, the statements are valid for $\Gamma_n = \bar{I}^n$ and hence also for $\Gamma_n = k_n$. Let $\Gamma_n = I^n$. Let $\rho \in I^n(F)$ such that $\rho = 0$ in $W(F(\varphi))$. By [**AEJ**, 4.3], there exist $\rho_i \in P_{n+i}(F)$ such that $\rho = \sum\langle\pm 1\rangle\rho_i$, $i = 0, \ldots, r$. By [**AP**, main theorem] and induction, we have $\rho_i = 0$ in $W(F(\varphi))$. Then apply theorem 1. To show the second statement use the same arguments as above or apply remark 4(c), [**AEJ**, 4.8] and corollary 2. □

One of the main results of [**Ma**][Ma] (see [**Ma**, theorem 7]) states that if one of the conjectures (1), (2), (3) holds for F and the Witt ring of K is a group extension (in the category of abstract Witt rings) of $W(F)$, then the conjecture holds for K as well. Essentially, this means that if K is a field with a complete, discrete valuation v such that (1), (2), (3) resp. holds for the residue class field F, then (1), (2), (3) resp. holds also for K. Next, we want to show the corresponding statement for the conjectures on annihilators.

THEOREM (5). *Let K be a field with a complete, discrete valuation v such that char $F \neq 2$ for the residue class field F with respect to v. Let $m, n \in I\!N$ and $\Gamma_n = k_n, \bar{I}^n, I^n$. Suppose $ann_s(\Gamma_t(a_1, \ldots, a_t))$ is generated by $ann_s(\Gamma_t(a_1, \ldots, a_t)) \cap \Delta_s(F)$ for $s = n-1, n$, for $t = m-1, m$ and for any $\Gamma_t(a_1, \ldots, a_t) \in \Delta_t(F)$. Then $ann_n(\Gamma_m(a_1, \ldots, a_m))$ is generated by $ann_n(\Gamma_m(a_1, \ldots, a_m)) \cap \Delta_n(K)$ for any $\Gamma_m(a_1, \ldots, a_m) \in \Delta_m(K)$, $a_i \in K^*$.*

PROOF. Let $\pi \in K^*$ be a prime element. Let $\Gamma_n = I^n$ and let $\varphi \in P_m(K)$. Suppose $\rho\varphi = 0$ in $W(K)$ for $\rho \in I^n(K)$. By a well known result of Springer (see [**S**, 6.2.6]), there exists a ring isomorhism $W(K) \cong W(F)[\mathbb{Z}/2]$. Choose $b^i_l, c_j, c^j_k u_i, v_j \in F^*$ such that $\rho = \sum_i \psi_i + \sum_j \tau_j\langle\langle c_j\pi\rangle\rangle$ where $\psi_i := \langle u_i\rangle \langle\langle b^i_1, \ldots, b^i_n\rangle\rangle$ and $\tau_j := \langle v_j\rangle\langle\langle c^j_1, \ldots, c^j_{n-1}\rangle\rangle$. We now need to consider two different cases:

Case 1: There exist $a_i \in F^*$ such that $\varphi = \langle\langle a_1, \ldots, a_m\rangle\rangle$. Then comparing residue class forms shows (a) $\varphi(\sum_i \psi_i + \sum_j \tau_j) = 0$ and (b) $\sum_j \varphi\langle c_j\rangle\tau_j = 0$. Thus by assumption, there exist $\rho_k \in ann_n(\varphi) \cap P_n(F)$ and $w_k \in F^*$ such that $\sum_i \psi_i + \sum_j\langle\langle c_j\rangle\rangle\tau_j = \sum_k\langle w_k\rangle\rho_k$. Hence, we have $\rho = \sum_k\langle w_k\rangle\rho_k + \sum_j\langle c_j\pi, -- c_j\rangle\tau_j$ in $W(K)$. Furthermore, by (b) and assumption, there exist $w'_l \in F^*$ and $\rho^*_l \in ann_{n-1}(\varphi) \cap P_{n-1}(F)$ such that $\sum_j\langle c_j\rangle\tau_j = \sum_l\langle w'_l\rangle\rho^*_l$.

Case 2: There exist $a_1, \ldots, a_m \in F^*$ such that $\varphi = \langle\langle a_1, \ldots, a_{m-1}, a_m\pi \rangle\rangle$. Then let $\lambda = \langle\langle a_1, \ldots, a_{m-1}\rangle\rangle$. Let Ω be the $W(K)$ ideal generated by $ann_n(\varphi) \cap P_n(K)$. We have $\rho\varphi = 0$ and thus by computing the first residue class form of $\rho\varphi$ we get $\lambda(\sum_i \psi_i + \sum_j \tau_j \langle\langle a_m c_j \rangle\rangle) = 0$. Hence by assumption, $\sum_i \psi_i + \sum_j \tau_j \langle\langle a_m c_j\rangle\rangle \in \Omega$. Thus, $\rho = \sum_j \tau_j \langle -a_m c_j, \pi c_j \rangle$ modulo Ω. Further, $\varphi\tau_j\langle -a_m c_j, \pi c_j\rangle = \lambda\langle\langle \pi a_m\rangle\rangle\langle\langle -\pi a_m \rangle\rangle\langle -c_j a_m\rangle \tau_j = 0$ for all j which proves the claim. The proofs for the cases $\Gamma_n = k_n, \bar{I}^n$ work in the same way. □

Remark (6). If the assumptions for F in theorem 4 are satisfied for all $m, n \in \mathbb{N}$, then by remark 4(b), the sequence $\varphi I^n(K) \to I^{n+m}(K) \to I^{n+m}(K(\varphi))$ is exact for all $m, n \in \mathbb{N}$ and for all $\varphi \in P_m(K)$.

Finally, we want to find a large category of field classes for which the conjecture on annihilators holds:

THEOREM (6). *Let $\mathcal{C}$ be the smallest subcategory of the category of abstract Witt rings that contains all abstract Witt rings R for which $I^4(R) = 2I^3(R)$ is torsion free and that is closed under formation of finite direct sums and finite group extensions. Let F be a field such that $W(F) \in \mathcal{C}$. Let $\Gamma_n = k_n, I^n, \bar{I}^n$. Then for all $m, n \in \mathbb{N}$ and $x \in \Delta_m(F)$ we have $ann_n(x) = ann_1(x)\Gamma_{n-1}(F)$. Furthermore, the sequences $\varphi I^n(F) \to I^{n+m}(F) \to I^{n+m}(F(\varphi))$ are exact.*

PROOF. By [**AEJ2**, theorem 10], we have $k_n(F) \cong \bar{I}^n(F)$. By theorem 5 (or more precisely, by an obvious generalization of it), we may assume $I^4(F) = 2I^3(F)$, that is $n(F) = 4$. If $n \geq 3$ apply corollary 2. For $n = 2$ and $\Gamma_n = I^n$ apply remark 3. For $n = 2$ and $\Gamma_n = \bar{I}^n$ apply remark 4(a) and proposition 3. For the last statement apply remark 4(b). □

Summarizing our results we might say that there is as much evidence for the validity of the conjecture on annihilators as there is for the validity of the other conjectures mentioned in §1. Unfortunately however, it is unknown whether the conjecture on annihilators holds for $n = 2$, $\Gamma_n = k_n, \bar{I}^n$ and $m \geq 2$ in general.

REFERENCES

[A] J. Kr. Arason, *Cohomologische Invarianten quadratischer Formen*, J. Algebra **36** (1975), 448–491.

[AEJ] J. Kr. Arason, R. Elman, and B. Jacob, *The graded Witt ring and Galois cohomology* I, Can. Math. Soc. Conf. Proc., vol. 4, American Mathematical Society, Providence, 1986, pp. 17–44.

[AEJ2] ———, *On quadratic forms and Galois cohomology*. Rocky Mountain J. Math. **19**, vol. 3 (1989), 575–588.

[AP] J. Kr. Arason and A. Pfister, *Beweis des Krullschen Durchschnittsatzes für den Wittring*, Invent. Math. **12** (1971), 173–176.

[AP2] ———, *Zur Theorie der quadratischen Formen über formal reellen Körpern*, Math. Z. **153** (1977), 289–296.

[EL] R. Elman and T. Y. Lam, *Pfister forms and K-theory of fields*, J. Algebra **23** (1972), 181–213.

[EL2] ———, *Quadratic forms under algebraic extensions*, Math. Ann. **219** (1976), 21–42.

[EP] R. Elman and A. Prestel, *Reduced stability of the Witt ring of a field and its Pythagorean closure*, Am. J. Math. **106** (1984), 1237–1260.

[JR] B. Jacob and M. Rost, *Degree four cohomological invariants for quadratic forms*, Invent. Math. **96** (1989), 551–570.

[K] M. Krüskemper, *On real local-global principles*, Math. Z. **204** (1990), 145–151.

[L] T. Y. Lam, *Orderings, valuations and quadratic forms*, Am. Math. Soc. Series Math. **52** (1983).

[Ma] M. Marshall, *Some local-global principles for formally real fields*, Can. J. Math. **29** (1977), 606–614.

[Ma2] ———, *Abstract Witt rings*, Queen's Papers Pure Appl. Math. **57** (1980).

[MS] A. S. Mercurjev and A. A. Suslin, *On the K_3 of a field*, LOMI preprint, K–2–87 (1987).

[Mi] J. Milnor, *Algebraic K–theory and quadratic forms*, Inv. Math. **9** (1970), 318–344.

[R] M. Rost, *Hilbert 90 for K_3 for degree two extensions* (to appear).

[S] W. Scharlau, *Quadratic and hermitian forms*, Springer, Berlin, Heidelberg, and New York, 1984.

Mathematisches Institut der Universität, Einsteinstrasse 62, D-4400 Münster

Contemporary Mathematics
Volume **155**, 1994

An Application of the Theory of Order Completions

KA HIN LEUNG

In the paper [**Sc**], Scot commented that no application of the theory of order completion was apparent. In this paper, we shall provide apparently the first application of such theory. Namely, we shall use the idea of completing ordered fields and groups to give an algebraic proof of Tamhankar's result [**T**]. Our approach is different from Tamhankar's proof as it involves a lot of combinatorial ideas. Moreover, our proof gives us a better understanding of the underlying reasons why Tamhankar's result is true.

This work represents a portion of the author's Ph.D. dissertation done at the University of California, Berkeley. The author would like to thank his advisor Professor T. Y. Lam for his guidance and encouragment.

Let R be a ring and $P \subset R$. We say P is an ordering if $P + P \subset P, P \cdot P \subset P, P \cup (-P) = R$ and $P \cap (-P) = \{0\}$. If P is an ordering on R, we then say (R, P) is an ordered ring. Obviously, the elements in R can be totally ordered by setting $a > b$ iff $a - b \in P$ for all $a, b \in R$.

It was proved by A. A. Albert that in an ordered division ring, any element algebraic over the center is central. V. Tamhankar [**T**] then generalized it to ordered rings.

THEOREM 1. *Let (A, P) be an ordered ring and α be an element of A. If α is a root of a polynomial $f(t) \in (Z(A) \cup \mathbb{Z})[t]$ and $f'(\alpha)$ is regular, then α is central. Here $f'(t)$ is the formal derivative of $f(t)$.*

Our strategy is to rephrase the problem differently. We shall first fix some terminologies. We say that $A < B$ if A, B are subsets of an ordered set such that for any $a \in A$, there exists $b \in B$ with $a < b$. Note that we assume the ordering

1991 *Mathematics Subject Classification.* 12J10, 12J15, 16W80.
This paper is submitted in final form and no version of it will be submitted for publication elsewhere.

concerned is fixed. Next, we introduce some new terminologies.

DEFINITION 2. *Let* (R, P) *be an ordered ring and* (M, Q) *be an additive ordered abelian group. We call* (M, Q) *an* (R, P)*-bimodule if* M *is a* (R, R)*-bimodule and* P, Q *are compatible, i.e.* $P \cdot Q \subset Q$ *and* $Q \cdot P \subset Q$.

For our purpose, we shall assume R is commutative. We define $C_R(M) = \{a \in R : a \cdot u = u \cdot a \ \forall u \in M\}$ which is clearly a subring of R. We define

$$J = \{a \in R : a \text{ is a zero divisor}\} \cup \{a \in R : a \cdot u = 0 \text{ or } u \cdot a = 0 \text{ for some nonzero } u \in M\}.$$

It is easy to see that $RJ \subset J$ and J is convex. Therefore, J is closed under addition. Hence, we conclude that J is a convex ideal. Our goal is to prove the following:

THEOREM 3. *If* $\alpha \in R$ *is a root of a polynomial* $f(t) = \lambda_n t^n + \lambda_{n-1} t^{n-1} + \cdots + \lambda_0 \in (C_R(M) \cup \mathbb{Z})[t]$ *and* $f'(\alpha) \notin J$, *then* $\alpha \in C_R(M)$. *Again,* $f'(t)$ *is the formal derivative of* $f(t)$.

REMARK:

(i) By putting $M = R = A$ and $C_R(M)$ to be the center of A, we see that Theorem 3 clearly implies Theorem 1.

(ii) If $f(t) = t^n$, then $\alpha \in C_R(M)$. This is because for all $u \in M$, we have then

$$0 = (\alpha^r \cdot u - u \cdot \alpha^r) = \sum_{j=0}^{r-1} \alpha^{r-1-j} \cdot (\alpha \cdot u - u \cdot \alpha) \cdot \alpha^j.$$

Since all summands on the right hand side are of the same sign, $\alpha \cdot u - u \cdot \alpha$ must be 0.

For convenience, we write R' for $C_R(M)$ and assume α and $f(t)$ be as defined in the above theorem. Note that $S := R' \backslash J$ is nonempty, as by assumption, $f'(\alpha) \notin J$ implies not all λ_i's are in J. We claim that every element in S can be assumed invertible.

Clearly, S is multiplicatively closed and every element in S is regular. By [**F**, p. 109, Theorem 3], P can be extended to an ordering $PS^{-1} := \{as^{-1} : a \in R, s \in S \text{ and } sa \in P\}$ on RS^{-1}. Similarly, it can be proved that $S^{-1}Q := \{s^{-1}u : u \in M, s \in S \text{ and } su \in Q\}$ is also an ordering on the (RS^{-1}, RS^{-1})-bimodule $S^{-1}M$. Obviously, PS^{-1} and $S^{-1}Q$ are compatible and $C_{RS^{-1}}(S^{-1}M) = R'S^{-1}$. Replacing R, M by $S^{-1}R, S^{-1}M$ if necessary, the desired claim follows.

In order to prove Theorem 3, we can clearly assume $R = R'[\alpha]$. As J is a convex ideal, $\bar{P} := \{a + J : a \in P\}$ is an ordering on R/J. As R/J is a finite extension of the field R'/J, it follows from [**L**, Proposition 2.8] that $R/J < R'/J$. In particular, we have $R < R'$. Thus $R < R' < R' \backslash J = S$. For any nonzero $z \in M$, we define $M(z)$ to be the (R, R)-bimodule generated by z

and $I(z) := \{u \in M(z) : |u| < |az| \; \forall a \in S\}$. As $\mathbb{Q} \subset S$ and every element in S is invertible, we see that $I(z)$ is a convex subgroup of $M(z)$ and $SI(z) = I(z)$. Since $RI(z) + I(z)R < SI(z)$, we have $RI(z) + I(z)R \subset I(z)$. In other words, $I(z)$ is an (R, R)-bimodule. On the other hand, it is obvious that $Jz + zJ \subset I(z)$ and $M(z) < Sz$. Hence $JM(z) + M(z)J < JSz \subset I(z)$. Thus, $JM(z) + M(z)J$ is in $I(z)$. So, we can view $\bar{M}(z) := M(z)/I(z)$ as an $(R/J, R/J)$-bimodule. Finally, as $\bar{P}$ is compatible with the ordering $Q(z) := (M(z) \cap Q)/I(z)$, we conclude that $(\bar{M}(z), Q(z))$ is an $(R/J, \bar{P})$-bimodule.

LEMMA 4. *For any* $u \in M$, $\alpha \in C_R(M(u))$ *if* $\alpha + J \in C_{R/J}\left(\bar{M}(\alpha \cdot u - u \cdot \alpha)\right)$.

PROOF. Let $f(t) = \lambda_n t^n + \lambda_{n-1} t^{n-1} + \ldots + \lambda_0$ be as defined in Theorem 3. For $i \geq 2$, $(\alpha^i \cdot u - u \cdot \alpha^i) = \sum_{j=0}^{i-1} \alpha^{i-1-j} \cdot (\alpha \cdot u - u \cdot \alpha) \cdot \alpha^j$. Suppose $\alpha + J \in C_{R/J}\left(\bar{M}(\alpha \cdot u - u \cdot \alpha)\right)$. Then $\alpha \cdot (\alpha \cdot u - u \cdot \alpha) - (\alpha \cdot u - u \cdot \alpha) \cdot \alpha \in I(\alpha \cdot u - u \cdot \alpha)$. Hence

$$
\begin{aligned}
0 = f(\alpha) \cdot u - u \cdot f(\alpha) &\equiv \textstyle\sum_{i=1}^{n} i \lambda_i \alpha^{i-1} (\alpha \cdot u - u \cdot \alpha) \\
&\equiv f'(\alpha) \cdot (\alpha \cdot u - u \cdot \alpha) \bmod I(\alpha \cdot u - u \cdot \alpha).
\end{aligned}
$$

By assumption $f'(\alpha) + J \neq 0$, so it follows that $\alpha \cdot u - u \cdot \alpha \in I(\alpha \cdot u - u \cdot \alpha)$. This is impossible by the definition of $I(\alpha \cdot u - u \cdot \alpha)$ unless $\alpha \cdot u - u \cdot \alpha = 0$. □

In view of the previous lemma, it is clear that Theorem 3 is an easy consequence of the following:

THEOREM 5. *Let* (M, Q) *be an* (K, P)*-bimodule. If* F *is a subfield of* $C_K(M)$, *then any element algebraic over* F *is in* $C_K(M)$.

To prove Theorem 5, we may assume $K = F[\alpha]$ where $\alpha \in K$ is algebraic over F. Consequently, we may also assume F is finitely generated over $\mathbb{Q}$. Thus rank $(K, P) = n$ is finite by [**KN**, Proposition 1.2]. Recall that in [**KN**], the rank of an ordered field (K, P) is defined to be the rank of $A(K, P) := \{a \in K : |a| < n \text{ for some } n \in \mathbb{N}\}$. We shall prove Theorem 5 by induction on the rank of the ordered field (K, P).

Since the rank of (K, P) is finite, there is a rank one valuation v compatible with P. Let $(\tilde{K}, \tilde{v})$ be the completion of (K, v). As $(\tilde{K}, \tilde{v})$ is an immediate extension of (K, v), by [**P**, p. 92, Satz 11], P extends uniquely to an ordering $\tilde{P}$ compatible with $\tilde{v}$ on $\tilde{K}$. It is well known that $\tilde{K}$ is topologically complete with respect to the valued topology $T_{\tilde{v}}$ which is the same as the interval topology $T_{\tilde{P}}$ by [**L**, Proposition 5.8]. So $(\tilde{K}, \tilde{P})$ is in fact an order completion of (K, P) defined in [**Sc**].

PROPOSITION 6. *Let* $(\tilde{K}, \tilde{P})$ *and* F *be as defined before. Suppose* $\tilde{F}$ *is the closure of* F *in* $\tilde{K}$. *Then there exists* $x \in \tilde{K}$ *such that* x *is algebraic over a subfield* k *of rank* $\leq n - 1$ *in* $\tilde{F}$ *and that* $\tilde{K}$ *is a radical extension of* $\tilde{F}[x]$.

PROOF. Let $\tilde{v}'$ and v' be the restriction of $\tilde{v}$ on $\tilde{F}$ and F respectively. By [E: p.12, Corollary 3], $(\tilde{F}, \tilde{v}')$ is a completion of (F, v'). As v' is of rank one, $(\tilde{K}, \tilde{v})$ and $(\tilde{F}, \tilde{v}')$ are Henselian [**E**, Theorem 17.18]. Note that the residue fields of $\tilde{K}$ and $\tilde{F}$ can be identified with $\bar{K}$ and $\bar{F}$ respectively. Let $R_{\tilde{v}}$ be the valuation ring of $\tilde{v}$ and $M_{\tilde{v}}$ the maximal ideal of $R_{\tilde{v}}$. We define η to be the natural projection of $R_{\tilde{v}}$ onto $R_{\tilde{v}}/M_{\tilde{v}} = \bar{K}$. By [**S**, p. 218, Corollary 1], there exist subfields $k' \subset \tilde{F} \cap R_{\tilde{v}}, k \subset \tilde{K} \cap R_{\tilde{v}}$ such that $k' \subset k$; $\eta|_k : k \to \bar{K}$ is an isomorphism and $\eta(k') = \bar{F}$. As $\bar{K}$ is a finite extension of $\bar{F}$, it follows that k is also finite extension of k'. Therefore, $k = k'[x]$ for some $x \in k'$.

Since $\tilde{v}$ is compatible with $\tilde{P}$, $\tilde{P}$ induces an ordering $\bar{\tilde{P}}$ on $\bar{K}$. By [**KN**, Proposition 1.5], rank $(\bar{K}, \bar{P}) = n - 1$. On the other hand, it is clear that $\eta|_k$ is an order isomorphism. Therefore the rank of $(k', k' \cap \tilde{P})$ is $n-1$. Lastly, observe that $(\tilde{K}, \tilde{v})$ is totally ramified over $\tilde{F}[x]$. Hence by [**S**, p. 64, Theorem 3], $\tilde{K}$ is a radical extension of $\tilde{F}[x]$. □

PROOF OF THEOREM 5. As before, we define $M(z), I(z)$ and $\bar{M}(z)$ by replacing R, R', J with K, F and $\{0\}$ respectively. By Lemma 4, it suffices to show that $\alpha \in C_K(\bar{M}(z))$ for all $z \in M$. We denote T_P and $T_{Q(z)}$ to be the interval topologies on K and $\bar{M}(z)$ respectively. As we have proved before, $M(z) < Fz$. So, we conclude that for any $u \in \bar{M}(z)$, $u \neq 0$ iff $az + I(z) < |u| < bz + I(z)$ for some $a, b \in F \cap P$. This shows that the left action and right action mappings are continuous with respect to the product topology $T_P \times T_{Q(z)}$. Let $(\tilde{K}, \tilde{P})$ be as defined before and $(\tilde{M}(z), \tilde{Q}(z))$ the order completion of $(\bar{M}(z), Q(z))$ as defined in [**CG**]. As $\tilde{K}$ and $\tilde{M}(z)$ are the topological completions of K and $\bar{M}(z)$ respectively, by [**B**, Chapter 3, §6.6], the left action and the right action mappings extend continuously to a left action and a right action mappings from $\tilde{K}$ to $\tilde{M}(z)$. It follows that $\tilde{P}$ and $\tilde{Q}(z)$ are compatible and $\tilde{F} \subset C_{\tilde{K}}\left(\tilde{M}(z)\right)$. Therefore $(\tilde{M}(z), \tilde{Q}(z))$ is a $(\tilde{K}, \tilde{P})$-bimodule.

It remains to prove that $\tilde{K} = C_{\tilde{K}}\left(\tilde{M}(z)\right)$ When rank$(K, P) = 0$, (K, P) is an archimedean field. Clearly, $\tilde{K} = \tilde{F}$ and they are order isomorphic to $\mathbb{R}$. In particular, $\tilde{K} = C_{\tilde{K}}\left(\tilde{M}(z)\right)$. Suppose now rank$(K, P) = n$. Let $x \in \tilde{K}$ be as defined in Proposition 6. By induction, $x \in C_{\tilde{K}}(\tilde{M}(z))$. Since $\tilde{K}$ is now a radical extension of $\tilde{F}[x]$, it follows from the remark after Theorem 3 that $\tilde{K} = C_{\tilde{K}}\left(\tilde{M}(z)\right)$. □

REFERENCES

[B] N. Bourbaki, *General topology*, Addison-Wesley, 1966.

[CG] L. W. Cohen and C. Goffman, *The topology of ordered abelian groups*, Trans. Amer. Math. Soc. **67** (1949), 310–319

[E] O. Endler, *Valuation theory*, Springer Verlag, Berlin and New York, 1972.

[F] L. Fuchs, *Partially ordered algebraic systems*, Pergamon Press, Oxford, London, New York, and Paris, 1963.

[KN] D. Kijima and M. Nishi, *Maximal ordered fields of rank n*, Hiroshima Math. J. **17** (1987), 157–167.

[L] T. Y. Lam, *The theory of ordered fields*, Ring Theory and Algebra, III (B. MacDonald, ed.), Lecture Notes in Pure and Applied Math., vol. 55, Dekker, New York, pp. 1–152.

[P] S. Prieß-Crampe, *Angeordnete Strukturen: Gruppen, Körper, Projektive Ebenen*, Ergebnisse der Mathematik und ihrer Grenzgebiete, vol. 98, Springer-Verlag, Berlin, Heidelberg, New York, and Tokyo, 1983.

[S] O. Schilling, *The Theory of Valuations*, Amer. Math. Soc., Mathematical Survey **4**, 1950.

[Sc] D. Scott, *On completing ordered fields*, In: Applications of Model Theory to Algebra, Analysis and Probability (Internat. Sympos., Pasadena, CA, 1967), Holt, Rinehart and Winston, New York, 1969, pp. 274–278.

[T] M. V. Tamhankar, *On algebraic extensions of subrings in an ordered ring*, Algebra Universalis **14** (1982), 25–35.

DEPARTMENT OF MATHEMATICS, NATIONAL UNIVERSITY OF SINGAPORE, SINGAPORE, 0511
E-mail address: MATLKH@NUSVM.bitnet

Contemporary Mathematics
Volume **155**, 1994

Growth of the u-Invariant Under Algebraic Extensions

DAVID B. LEEP AND A. S. MERKURJEV

1. Introduction

There has been renewed interest in studying the u-invariant of fields as a result of the second author's construction of fields having u-invariant equal to any given even number [**Me**]. An important and still unanswered question concerns how much the u-invariant can grow under a finite algebraic extension. (See [**La**, p. 333, question 5].) In particular, if F is a field and $[K : F] = n$, the problem is to find a good upper bound for $u(K)$ in terms of $u(F)$ and n. The best result so far states $u(K) \leq \frac{n+1}{2}u(F)$ ([**Le**] or [**Sch**, p. 104]). This result was shown in [**Le**] to be best possible when $n = 1$ or 3. Little else has been proved. In fact, the only examples of u-invariant growth have been cases where $u(K) = 2u(F)$ and $u(F)$ is a power of 2. (For example, let F be the quadratic closure of the rational numbers $\mathbf{Q}$ and let $[K : F] \geq 3$.)

This paper reports on new constructions of u-invariant growth. In particular, corollary 2.3 shows the u-invariant estimate above is also best possible when $n = 2$.

We follow standard terminology of quadratic form theory as found in [**La**] and [**Sch**]. In particular, the u-invariant, $u(F)$, of a field is the supremum of the dimensions of anisotropic quadratic forms defined over F. We assume all fields in this paper have characteristic $\neq 2$. We let $[x]$ denote the greatest integer $\leq x$. Results on central simple algebras and the Brauer group can be found in [**D**].

We express thanks to T. Y. Lam for help in preparing this paper, to Adrian Wadsworth for discussions concerning Lemmas 3.1 and 4.2, and to Bill Jacob who sent us related unpublished work in this area.

1991 *Mathematics Subject Classification.* 11E04, 11E81.

This paper is submitted in final form and no version of it will be submitted for publication elsewhere.

2. The main theorems

In this section we state the two main theorems and derive some corollaries. Theorem 2.1 is proved in §3 and Theorem 2.2 is proved in §4.

THEOREM 2.1. *For every $r, n \in \mathbf{Z}$, $r \geq 1$, $n \geq 2$, there exists a cyclic extension of fields K/F such that $[K:F] = 2r$, $u(F) = 2n$ and $u(K) \geq 2n+2+2[\frac{n-2}{2r}]$.*

THEOREM 2.2. *Let r, n be positive integers and assume r is not a power of 2. Then there exists a cyclic extension K/F such that $[K:F] = r$, $u(F) = 2n$, and $u(K) \geq 4n$.*

Bill Jacob independently proved a weaker version of Theorem 2.1 using slightly different techniques [**J**]. Under the same assumptions Jacob showed $u(K) \geq 2n+2$. In view of Theorem 2.2, it is clear that Theorem 2.1 is useful only when $[K:F]$ is a power of 2.

COROLLARY 2.3.
(1) *For each even integer $n \geq 2$, there exists a quadratic extension K/F such that $u(F) = 2n$, $u(K) = 3n$.*
(2) *For each odd integer $n \geq 3$, there exists a quadratic extension K/F such that $u(F) = 2n$, $u(K) \in \{3n-1, 3n\}$.*

PROOF. We use Theorem 2.1 in the case $r = 1$ to construct a quadratic extension K/F with $u(F) = 2n$ and

$$u(K) \geq 2n+2+2[\frac{n-2}{2}] = \begin{cases} 3n & \text{if } n \text{ is even} \\ 3n-1 & \text{if } n \text{ is odd} \end{cases}$$

Since $u(K) \leq \frac{3}{2}u(F) = 3n$ by [**Le**] or [**EL**, Theorem 4.3], (1) and (2) follow easily. □

Mináč showed in [**Mi**] that there are examples in Corollary 2.3(2) where $u(K) = 3n-1$ for each odd integer $n \geq 3$. For $n = 3$, he also pointed out that $u(F) = 6$ implies $u(K) \neq 9$. It remains open whether $u(K) = 3n$ is possible for odd $n \geq 5$.

COROLLARY 2.4. *For every positive integer n, there exists a cyclic cubic extension K/F such that $u(F) = 2n$, and $u(K) = 4n$.*

PROOF. Apply Theorem 2.2 with $r = 3$ and note $u(K) \leq 2u(F)$ by [**Le**]. □

The proofs of Theorems 2.1, 2.2 depend on the following theorem. A proof can be found in [**T**].

THEOREM 2.5 (MERKURJEV). *Let F be a field, char $F \neq 2$. Let q be an odd-dimensional quadratic form defined over F, dim $q \geq 3$, and let $F(q)$ denote the function field of q. Let $C_0(q)$ denote the even part of the Clifford algebra of q. Let D be a finite dimensional F-central division algebra. Then $D \bigotimes_F F(q)$ fails to be a division algebra if and only if $D \cong D_0 \bigotimes C_0(q)$ for some division algebra D_0 defined over F.*

We will also make frequent use of the following result which is proved in the appendix of [**T**].

PROPOSION 2.6. *Let A be a tensor product of s quaternion algebras defined over F. If A is a division algebra, then $u(F) \geq 2s + 2$.*

3. The proof of theorem 2.1

LEMMA 3.1. *Let $r, n, m \in \mathbf{Z}$, $r \geq 1$, $m \geq n \geq 2$. There exists a cyclic extension K/F, $[K : F] = 2r$, an F-central division algebra D, and a K-central division algebra T satisfying the following three properties.*

(1) *ind$D = 2^{n-1}$, and D is a tensor product of $n-1$ quaternion algebras defined over F.*
(2) *ind$T = 2^{m-1}$, and T is a tensor product of $m-1$ quaternion algebras defined over K.*
(3) *ind$(Cor_{K/F}T) \geq 2^{n-1}$.*

PROOF. Let $K = \mathbf{Q}(\{x_{ij}\}, \{y_l\}), 1 \leq i \leq 2(m-1), 1 \leq j \leq 2r, 1 \leq l \leq 2(n-1)$, where the x's and y's are indeterminates. Let G be the cyclic group of order $2r$ generated by $\sigma \in autK$ where $\sigma(x_{ij}) = x_{i,j+1}$ for $1 \leq j \leq 2r-1$, $\sigma(x_{i,2r}) = x_{i,1}$, and $\sigma(y_l) = y_l$. Let $F = K^G$, the fixed field of K under G. Then $[K : F] = 2r$. Let

$$D = \bigotimes_{k=1}^{n-1} (y_{2k-1}, y_{2k})_F$$

and let

$$T = \bigotimes_{k=1}^{m-1} (x_{2k-1,1}, x_{2k,1})_K.$$

Then (1) and (2) are satisfied. Now,

$$(Cor_{K/F}T)_K \cong T \bigotimes T^{\sigma} \bigotimes \cdots \bigotimes T^{\sigma^{2r-1}} \cong \bigotimes_{j=1}^{2r} \bigotimes_{k=1}^{m-1} (x_{2k-1,j}, x_{2k,j})_K.$$

It follows $(Cor_{K/F}T)_K$ is a K-central division algebra having index $2^{2r(m-1)}$. Thus,

$$2^{2r(m-1)} = \operatorname{ind}(Cor_{K/F}T)_K \leq \operatorname{ind}(Cor_{K/F}T) \leq (\operatorname{ind} T)^{2r} = 2^{2r(m-1)}.$$

Therefore, ind$(Cor_{K/F}T) = 2^{2r(m-1)}$, (and in fact, $Cor_{K/F}T$ is an F-central division algebra). Since $2r(m-1) \geq n-1$, this proves (3). □

LEMMA 3.2. *Let F, K, D, T be as in Lemma 3.1. Let q be an anisotropic quadratic form defined over F of dimension $2n+1$. Let $F(q)$, $K(q)$ denote the function fields of q over F, K respectively. Then*

(1) *$D_{F(q)}$ is a division algebra,*

(2) *$ind(Cor_{K(q)/F(q)}(T_{K(q)})) \geq 2^{n-1}$,*

(3) *if $m = n+1+[\frac{n-2}{2r}]$, then $T_{K(q)}$ is a division algebra.*

PROOF. (1) follows from Theorem 2.5 since $\dim C_0(q) = 2^{2n} > 2^{2n-2} = \dim D$.

Since $F(q)$ is a pure transcendental extension of F followed by a quadratic extension, it follows

$$\operatorname{ind}(Cor_{K(q)/F(q)}(T_{K(q)})) = \operatorname{ind}(Cor_{K/F}T)_{F(q)} \geq \frac{\operatorname{ind}(Cor_{K/F}T)}{2}.$$

If $\operatorname{ind}(Cor_{K/F}T) \geq 2^n$, then (2) is clear. Now assume $\operatorname{ind}(Cor_{K/F}T) = 2^{n-1}$ and let E be the division algebra in the Brauer class of $Cor_{K/F}T$. Then $\dim E = 2^{2(n-1)}$ and $E_{F(q)}$ remains a division algebra by Theorem 2.5. Thus $\operatorname{ind} E_{F(q)} \geq 2^{n-1}$ and (2) follows.

Now let $m = n+1+[\frac{n-2}{2r}]$ and assume $T_{K(q)}$ is not a division algebra. Then Theorem 2.5 implies $T \cong C_0(q)_K \bigotimes B$ for some division algebra B defined over K. In particular, $C_0(q)_K$ is a division algebra and $\operatorname{ind}(C_0(q)_K) = 2^n$. Thus $\operatorname{ind} B = 2^{m-n-1}$. Since $C_0(q)$ is defined over F, we have $Cor_{K/F}(C_0(q)_K) = [K:F]C_0(q) = 0$ in $Br(F)$ since $[K:F]$ is even. Therefore,

$$2^{n-1} \leq \operatorname{ind}(Cor_{K/F}T) = \operatorname{ind}(Cor_{K/F}B) \leq (2^{m-n-1})^{2r},$$

$$n-1 \leq 2r(m-n-1),$$

$$(2r+1)n + (2r-1) \leq 2rm,$$

$$n+1+\frac{n-1}{2r} \leq m = n+1+[\frac{n-2}{2r}].$$

This is a contradiction so $T_{K(q)}$ is a division algebra and this proves (3). □

PROOF OF THEOREM 2.1. Let F,K,D,T be as in Lemma 3.1. We construct a field F' over F having $u(F') = 2n$ in the "usual" way: Let $F = F_0 \subset F_1 \subset F_2 \subset \cdots$ be an ascending chain of fields where F_{i+1} is a compositum of $\{F_i(q_\alpha)\}$ where q_α ranges over all the $(2n+1)$–dimensional forms over F_i. Let $F' = \cup_{i=0}^{\infty} F_i$ and let $K' = KF'$. Then K'/F' is a cyclic extension of degree $2r$ since K and $F(q)$ are linearly disjoint over F. Clearly $u(F') \leq 2n$. Since Lemma 3.2 implies D, T remain division algebras over F', K' respectively, Proposition 2.6 implies $u(F') \geq 2n$ and $u(K') \geq 2m$. Therefore $u(F') = 2n$ and $u(K') \geq 2n+2+2[\frac{n-1}{2r}]$. □

4. The proof of theorem 2.2

LEMMA 4.1. *Let G be a cyclic group of order r generated by σ. Consider the group ring $A = \mathbf{Z}/p\mathbf{Z}[G]$ where p is a prime number. Then $1+\sigma$ is nilpotent in A if and only if $p = 2$ and r is a power of 2.*

PROOF. Let $r = p^e s$, $e \geq 0$, $p \not| s$. First we note $p^{\phi(s)} \equiv 1 \pmod{s}$ implies $p^e p^{\phi(s)} \equiv p^e \pmod{p^e s}$, so $p^{e+\phi(s)} \equiv p^e \pmod{r}$. Then

$$(1+\sigma)^{p^{e+\phi(s)}} = 1 + \sigma^{p^{e+\phi(s)}} = 1 + \sigma^{p^e} = (1+\sigma)^{p^e}.$$

If $1+\sigma$ is nilpotent, then $(1+\sigma)^{p^{e+\phi(s)}} = 0$. (If $(1+\sigma)^N = 0$, then $(1+\sigma)^{N'} = 0$ for some $N' \leq p^{e+\phi(s)}$.) It follows $1 + \sigma^{p^e} = 0$. This forces $\sigma^{p^e} = 1$, $p = 2$. This implies $s = 1$ and $r = p^e = 2^e$. Conversely, if $r = 2^e$, $p = 2$, then $(1+\sigma)^{2^e} = 1 + \sigma^{2^e} = 1 + 1 = 0$. □

Let K/F be a cyclic extension, $[K : F] = r$, and let $G = Gal(K/F) = <\sigma>$ The group ring $\mathbf{Z}/2\mathbf{Z}[G]$ acts on $Br_2(K)$, the subgroup of the Brauer group generated by elements of order 2. By the multiplicative set generated by $1+\sigma$, we mean the set $\{1, 1+\sigma, (1+\sigma)^2, (1+\sigma)^3, \cdots\}$ in $\mathbf{Z}/2\mathbf{Z}[G]$.

LEMMA 4.2. *Let r, n be positive integers and assume r is not a power of 2. There exists a cyclic extension K/F, $[K : F] = r$, $G = Gal(K/F) = <\sigma>$, and a K-central division algebra D satisfying* (1), (2) *below.*

(1) *D is a tensor product of $2n-1$ quaternion algebras defined over K, $ind(D) = 2^{2n-1}$.*
(2) *$ind(f(\sigma)D) = 2^{2n-1}$ for all elements $f(\sigma) \in \mathbf{Z}/2\mathbf{Z}[G]$ that lie in the multiplicative set generated by $1+\sigma$.*

PROOF. Let $K = \mathbf{Q}(\{x_{ij}\}, \{y_i\})$, $1 \leq i \leq 2n-1$, $0 \leq j \leq r-1$, where the x's and y's are indeterminates. Let G be the cyclic group of order r generated by $\sigma \in autK$ where for all i, $\sigma(x_{ij}) = x_{i,j+1}$, $0 \leq j \leq r-2$, $\sigma(x_{i,r-1}) = x_{i,0}$, $\sigma(y_i) = y_i$. Let $F = K^G$, the fixed field of K under G. Then K/F is a cyclic extension, $[K : F] = r$.

Let $D = \bigotimes_{i=1}^{2n-1}(x_{i,0}, y_i)_K$ and let $f(\sigma) \in \mathbf{Z}/2\mathbf{Z}[G]$, $f(\sigma) \neq 0$. Then $f(\sigma) = \sum_{i=0}^{r-1} e_i\sigma^i$, $e_i \in \{0,1\}$, some $e_i \neq 0$. In $Br(K)$, we have $f(\sigma)D = \bigotimes_{i=1}^{2n-1}(z_i, y_i)_K$ where $z_i = \prod_{j=0}^{r-1} x_{i,j}^{e_j}$. Since $L := \mathbf{Q}(\{z_i, y_i\})$, $1 \leq i \leq 2n-1$, is a pure transcendental extension of $\mathbf{Q}$, it follows $\text{ind}(f(\sigma)D) = 2^{2n-1}$ when considered as an element in $Br(L)$. Since K/L is also a pure transcendental extension, it follows $\text{ind}(f(\sigma)D) = 2^{2n-1}$ considered in $Br(K)$. Since the multiplicative set generated by $1+\sigma$ doesn't contain 0, by Lemma 4.1, it follows (1), (2) hold. □

LEMMA 4.3. *Let r, n be positive integers, and assume r is not a power of 2. Let K/F be a cyclic extension, $[K : F] = r$, $G = Gal(K/F)$, and let D be a K-central division algebra satisfying* (1), (2) *of Lemma* 4.2. *Let q be a quadratic form defined over F of dimension $2n+1$. Then* (1), (2) *below hold.*

(1) *$D_{K(q)}$ is a division algebra.*

(2) *ind*$(f(\sigma)D_{K(q)}) = 2^{2n-1}$ *for all* $f(\sigma) \in \mathbf{Z}/2\mathbf{Z}[G]$ *that lie in the multiplicative set generated by* $1+\sigma$.

PROOF. Let T be a division algebra in the Brauer class of $f(\sigma)D$ in $Br(K)$ where $f(\sigma) = (1+\sigma)^j$ for some $j \geq 0$. By assumption, $\operatorname{ind} T = 2^{2n-1}$. Suppose $T_{K(q)}$ is not a division algebra. Then Theorem 2.5 implies $T \cong_K C_0(q_K) \bigotimes_K S$ for some division algebra S defined over K. Since $C_0(q_K)$ is a division algebra, it follows $\operatorname{ind}(C_0(q_K)) = 2^n$ and $\operatorname{ind} S = 2^{n-1}$. We have $C_0(q_K)$ has order 2 in $Br(K)$ and $C_0(q_K) \cong \sigma(C_0(q_K))$, since q is defined over F. Therefore,

$$\operatorname{ind}(1+\sigma)T = \operatorname{ind}(1+\sigma)S \leq (2^{n-1})^2 = 2^{2n-2}.$$

On the other hand, in $Br(K)$ we have $(1+\sigma)f(\sigma)D = (1+\sigma)^{j+1}D$ and by assumption (Lemma 4.2(2)) we have $\operatorname{ind}(1+\sigma)T = 2^{2n-1}$. This is a contradiction so $T_{K(q)}$ is a division algebra. This proves (1) and (2). □

PROOF OF THEOREM 2.2. Let F, K, D be as in Lemma 4.2. Let $F = F_0 \subset F_1 \subset F_2 \subset \cdots$ be an ascending chain of fields where F_{i+1} is a compositum of $\{F_i(q_\alpha)\}$ and where q_α ranges over all the $(2n+1)$-dimensional quadratic forms defined over F_i. Let $F' = \cup_{i=0}^{\infty} F_i$ and let $K' = KF'$. Then K'/F' is a cyclic extension, $[K' : F'] = r$, and $u(F') \leq 2n$ since K and $F(q)$ are linearly disjoint over F. Since $D_{K'}$ is a division algebra by repeated applications of Lemma 4.3(1), it follows $u(K') \geq 2(2n-1)+2 = 4n$ by Proposition 2.6. To show $u(F') = 2n$, consider the division algebra $D_1 := \bigotimes_{i=1}^{n-1}(y_{2i-1}, y_{2i})_F$ defined over F. Theorem 2.5 implies D_1 remains a division algebra over F'. Thus, $u(F') \geq 2(n-1)+2 = 2n$ by Proposition 2.6. □

REFERENCES

[D] P. K. Draxl, *Skew fields*, London Math. Soc. Lecture Note Series, vol. 81, Cambridge University Press, 1983.

[EL] R. Elman and T. Y. Lam, *Quadratic forms and the u-invariant, I*, Math. Zeit. **131** (1973), 283–304.

[J] W. Jacob, unpublished manuscript.

[La] T. Y. Lam, *The Algebraic Theory of Quadratic Forms*, second printing with revisions (1980), W. A. Benjamin, Addison-Wesley, Reading, Mass., 1973.

[Le] D. B. Leep, *Systems of quadratic forms*, J. reine angew. Math. **350** (1984), 109–116.

[Me] A. Merkurjev, *Simple algebras and quadratic forms*, Izv. Akad. Nauk. Ser. Math **55** (1991), 218–224 (In Russian).

[Mi] J. Mináč, *Remarks on the u-invariant.*

[Sch] W. Scharlau, *Quadratic and hermitian forms*, Grundlehren der Math. Wiss., vol. 270, Springer-Verlag, Berlin- Heidelberg-New York, 1985.

[T] J.-P. Tignol, *Réduction de l'indice d'une algèbre simple centrale sur le corps des fonctions d'une quadrique*, Bull. Soc. Math. Belg. **42** (1990), 2, ser. A, 735–745.

UNIVERSITY OF KENTUCKY

ST. PETERSBURG STATE UNIVERSITY

E-mail address: leep@ms.uky.edu

Contemporary Mathematics
Volume 155, 1994

Remarks on Merkurjev's Investigations of the u-Invariant

JÁN MINÁČ

Quite recently Merkurjev made a spectacular breakthrough in the study of the u-invariant. He proved that every positive even integer is the u-invariant of some field. Here u is the well known Kaplansky invariant [**Lam2**], [**Kap**].

It is known that the u-invariant of a field cannot be 3, 5 or 7. Therefore the first unknown value is 9.

In this paper we suggest a possible approach to the question of whether $u(F) = 2n+1$, $n \geq 4$, is possible. In particular we show that certain universal forms of dimensions 9, 11, 13 or 15 do not exist.

We also strengthen one of Merkurjev's results, namely Theorem 1 below, on growth of u-invariants under quadratic extensions (cf. Theorem 2).

We assume that all fields investigated in this paper are not formally real and that their characteristic is not 2.

We follow standard terminology of quadratic form theory and central simple algebras as found in [**Lam1**], [**Lam2**] and [**Sch**].

ACKNOWLEDGEMENT. I am grateful to Adrian Wadsworth for apprising me of recent related results and his kind interest in my work; Bill Jacob for helping me with the exposition; the referee for pointing out stronger versions of my previous theorems as well as for his simplifications of my proofs; and Clive Reis for torturing me with the proper use of "a," "an," and "the" as well as for allowing me to share with him our admiration and enthusiasm for quadratic form theory.

Suppose that K/F is a quadratic extension. It is known that $u(K) \leq \frac{3}{2}u(F)$, [**EL**], [**Lee**]. Therefore if $u(F) = 4$ then $u(K) \leq 6$. It was shown independently

1991 *Mathematics Subject Classification.* 11E04.
Supported in part by the Natural Sciences and Engineering Research Council of Canada.
This paper is submitted in final form and no version of it will be submitted for publication elsewhere.

by B. Jacob, D. Leep and A. S. Merkurjev that $u(K) = 6$ is actually possible [**Lee2**]. If $u(F) = 6$ then $u(K) \leq 9$. But we claim that $u(K) = 9$ is impossible. Indeed since $u(F) = 6$ we have $I^3(F) = \{0\}$, where $I(F)$ is the fundamental ideal of the Witt ring $W(F)$. (See [**Lam1**].) From Arason's theorem [**Ara**, Satz 3.6] we see that $I^3(K) = \{0\}$ as well. Moreover since $u(K) \leq 9$, it is, in particular, finite and so by Lemma 4.9, Chapter 11 in [**Lam1**] it follows that $u(K)$ is even. Thus $u(K) \neq 9$ as we claimed.

The argument above can be used in order to strengthen the following theorem due to Merkurjev. (See [**LM**, Corollary 2.3]. Observe also that Leep obtained a generalization of this theorem. See Theorem 2.1, *ibid.*)

THEOREM 1. (*A. S. Merkurjev*)

(1) *For each even integer $n \geq 2$, there exists a quadratic extension K/F such that $u(F) = 2n$ and $u(K) = 3n$.*
(2) *For each odd integer $n \geq 3$, there exists a quadratic extension K/F such that $u(F) = 2n$ and $u(K) \in \{3n-1, 3n\}$.*

We can strengthen Merkurjev's theorem as follows.

THEOREM 2. *Suppose that n is an odd integer ≥ 3. Then there exists a quadratic extension K/F such that $u(F) = 2n$ and $u(K) = 3n - 1$.*

PROOF. We shall use a slightly modified version of Merkurjev's construction. For details of his construction we refer the reader to [**LM**]. Set $m = \dfrac{3n-1}{2}$ and let K_0/F_0 be any quadratic extension, $K_0 = F_0(\sqrt{a})$, such that there exists a central division algebra D over F_0 of index $2^{(n-1)}$ which is a product of $(n-1)$ quaternion algebras and a central division algebra T over K_0 of index $2^{(m-1)}$, which is in turn a product of $m-1$ quaternion algebras such that $ind(cor(T)) \geq 2^{(n-1)}$ (Here $cor(T)$ means the corestriction of T (see [**Sch**]). The existence of such an extension K_0/F_0 and the corresponding algebra T follows from Lemma 3.1 in [**LM**].

Let q be a quadratic form over F_0 of dimension $2n+1$, and $F_0(q), K_0(q)$ its function fields over F_0 and K_0, respectively. Then Merkurjev showed that D and T remain division algebras over $F_0(q)$ and $K_0(q)$, respectively (cf. Lemma 3.2 in [**LM**]).

On the other hand from Theorem 1 in [**Mer1**] it follows (as Merkurjev showed himself) that for every form $\sigma = \langle 1, -a, -b, ab, -c\rangle, a, b, c \in F_0, D$ and T remain division algebras over $F_0(\sigma)$ and $K_0(\sigma)$, respectively.

Using this one can construct by induction a tower of fields $F_0 \subset F_1 \subset F_2 \subset \cdots$ as follows (see [**Mer1**]).

Suppose F_i is already constructed. Set F_i' to be the compositum of all fields $F_i(q)$ where q runs over all quadratic forms of dimension $(2n+1)$ over F_i. Then set F_{i+1} to be the compositum of all fields $F_i'(\sigma)$, where σ is a form of the type $\langle 1, -a, -b, ab, -c\rangle, a, b, c \in F_i'$.

Finally set $F = \cup F_i$ and $K = F(\sqrt{a}) = \cup K_i$, where $K_i = F_i(\sqrt{a})$. Then by [**Lee2**], $u(F) = 2n$ and $u(K) \geq 2m$.

Moreover from [**EL**] we have a $u(K) = 2m$ or $(2m+1)$. However, from our construction we also have $I^3(F) = \{0\}$ and from Arason's theorem quoted above [**Ara**, Satz 3.6], we have $I^3(K) = \{0\}$. Therefore $u(K) = 2m$ as we claimed. □

REMARK. In a recent joint work [**MW**] with Adrian Wadsworth we determined all possible pairs $(u(F), u(K))$; where K/F is a quadratic extension and $I^3(F) = \{0\}$.

I don't know whether there exists a quadratic extension K/F where $u(F) = 2n$; $u(K) = 3n$ for n odd, $5 \leq n$.

The existence of a field F, $char F \neq 2$ with $u(F) = 2n+1, 4 \leq n$, is an outstanding problem in quadratic form theory. If one believes that there exists no such field, then one could consider the theorem below as a first induction step in proving it. We shall assume that the reader is familiar with the Witt invariant $c(\sigma)$ of the quadratic form σ (see [**Lam1**, pp. 120–121]). Witt's invariant $c(\sigma)$ is an element of order at most 2 of the Brauer group.

It can be seen from the definition of $c(\sigma)$ that $c(\sigma)$ is the Brauer equivalent to the tensor product of quaternion algebras. In the next paragraph we shall work in the Brauer group of F, in particular all equalities below mean that the underlying central simple algebras are Brauer equivalent.

LEMMA 3. *Suppose that $u(F) = 2n+1, 4 \leq n$. Let σ be an anisotropic quadratic form defined over F, $\dim \sigma = 2n+1$, $c(\sigma) = \bigotimes_{i=1}^{t} (a_i, b_i)_F$, with t smallest possible ($(a_i, b_i)_F$ are quaternion algebras with slots a_i, b_i over F). Then $0 \leq t \leq n-1$.*

PROOF. Since σ is universal, it represents $d = d_{\pm}(\sigma)$, the signed determinant of σ. Write $\sigma \cong \langle d \rangle \perp \tau$. Then $\dim \tau = 2n$ and $d_{\pm}\tau = 1$.

Therefore from [**Lam1**, p. 121], we conclude that $c(\tau) = c(\sigma)$. On the other hand, as Merkurjev observed [**Mer1**], $c(\tau)$ is Brauer equivalent to a product of $(n-1)$ quaternion algebras over F. Therefore $t \leq (n-1)$ as we claimed. □

THEOREM 4. *Keep the assumptions above and assume in addition that $I^4(F) = \{0\}$. Then $t \neq 0$.*

REMARK. Observe that if $u(F) \in \{9, 11, 13, 15\}$, then $I^4(F) = \{0\}$ and therefore $t \neq 0$. This means that in this case there exists no anisotropic universal form σ such that $\dim \sigma = u(F)$ and $c(\sigma) = 1$.

PROOF. Set $u = u(F)$ and assume that σ is an anisotropic u-dimensional form such that $c(\sigma) = 1$.

For each $a \in \dot{F}$ and each $\psi \in I^3(F)$ we have $\psi\langle 1, -a \rangle = 0$ in $W(F)$ (because $I^4(F) = \{0\}$). Thus $\psi \cong a\psi$ and ψ is universal.

On the other hand set $d = d_{\pm}(\sigma)$, the signed determinant of σ. Then $\sigma = \langle d \rangle \perp \psi$ for some ψ since σ is universal. Then $d_{\pm}(\psi) = 1$ and $\dim \psi = u - 1$ is even. Therefore $\psi \in I^2(F)$.

On the other hand, we have (see [**Lam1**, p. 121])

$$\begin{aligned}1 &= c(\sigma)\\ &= c(\langle d\rangle \perp \psi)\\ &= c(\psi)(-d, d_{\pm}\psi)\\ &= c(\psi).\end{aligned}$$

Therefore from Merkurjev's well known theorem [**Mer2**] we conclude that $\psi \in I^3(F)$. Thus ψ is universal and hence σ is isotropic, a contradiction, proving our claim that $t \neq 0$. $\square$

REMARK. Merkurjev's theorem quoted above is rather deep and its proof quite subtle. In the case when $\dim \psi \leq 12$ (this means $\dim \sigma \leq 13$), we can refer the reader to Pfister's elementary proof [**Pfi**].

Theorem 4 solves our question about the existence of an anisotropic form σ over F, $\dim \sigma = u(F) = 2n+1, n \geq 4$ in the case when $t = t(\sigma) = 0$.

In the case $t(\sigma) = 1$ we can prove the following. (I am grateful for the formulation of Proposition 5 below as well as for its proof to the referee. Originally I had a weaker statement with a less elementary proof.)

PROPOSITION 5. *Suppose that* $u(F) = 2n+1, 4 \leq n$ *and that* σ *is an anisotropic quadratic form over* F, $\dim \sigma = 2n+1$ *and* $t(\sigma) = 1, c(\sigma) = (a,b)_F$. *Finally assume* $I^4(F) = \{0\}$.

Then $\sigma \cong \sigma_1 \perp \sigma_2$ *where*

(1) $\dim \sigma_1 = 3$,

(2) σ_1 *is a subform of* $\langle\langle -a, -b\rangle\rangle$.

In particular, σ *is isotropic over* $K = F(\langle 1, -a, -b\rangle)$, *the function field of* $\langle 1, -a, -b\rangle$ *over* F.

PROOF. Suppose that σ is as above and set

$$\tau_1 = \sigma \perp - \langle\langle -a, -b\rangle\rangle .$$

Then

$$\begin{aligned}c(\tau_1) &= c(\sigma)c(- \langle\langle -a, -b\rangle\rangle), &&\text{see [Lam1, p. 121]},\\ &= c(\sigma)(a,b)_F, &&\text{see [Lam1, pp. 116 and 121]},\\ &= 1.\end{aligned}$$

Also $\dim \tau_1 = 2n+5$ and $u(F) = 2n+1$. Hence there exists a form τ_2 such that

$$\tau_1 \cong \tau_2 \perp \mathbb{H} \perp \mathbb{H}$$

where $\mathbb{H}$ is a hyperbolic plane. Also

$$1 = c(\tau_1) = c(\tau_2), \quad \text{see [Lam1, p. 121]}.$$

Therefore from Theorem 4 we conclude that τ_2 is isotropic over F. This means that there exists a form τ_3 over F such that

$$\tau_1 \cong \tau_3 \perp 3\mathbb{H}.$$

Or equivalently,

$$\sigma \perp - \langle\langle -a, -b \rangle\rangle \cong \tau_3 \perp 3\mathbb{H}.$$

However both σ and $-\langle\langle -a, -b \rangle\rangle$ are anisotropic over F. Therefore there exists $d_1 \in D_F(\sigma) \cap D_F(\langle\langle -a, -b \rangle\rangle)$. Thus

$$\sigma \perp - \langle\langle -a, -b \rangle\rangle \cong \langle d_1, -d_1 \rangle \perp \sigma' \perp \mu$$

where σ' is a subform of σ and μ is a subform of $-\langle\langle -a, -b \rangle\rangle$. Using Witt cancellation [**Lam1**, Ch. 1, §4] we see that

$$\sigma' \perp \mu \cong \tau_3 \perp 2\mathbb{H}.$$

Using the same argument two more times, we find that there exists a form $\sigma_1 = \langle d_1, d_2, d_3 \rangle$ over F which is a subform of both σ and $\langle\langle -a, -b \rangle\rangle$. Thus (1) and (2) are proved.

Consider now $K = F(\langle 1, -a, -b \rangle)$ the function field of $\langle 1, -a, -b \rangle$ over F. Then $\langle\langle -a, -b \rangle\rangle_K$ is hyperbolic over K. (See [**Sch**, p. 155].) Since $\sigma_{1,K} := \sigma_1 \otimes_F K$ is a ternary subform of $\langle\langle -a, -b \rangle\rangle_K$ we see that $\sigma_{1,K}$ is isotropic over K as we claimed. □

Finally we finish with the following example.

Example. Suppose that σ is a 9–dimensional form $\sigma = \langle\langle -a, -b \rangle\rangle \perp \psi$, such that $c(\sigma) = (a, b)_F$. Then ψ and consequently σ are isotropic.

Proof. Let $d(\psi)$ be the determinant of ψ. Then we have

$$\begin{aligned} (a, b)_F &= c(\sigma) \\ &= c(\psi) c(-d(\psi) \langle\langle -a, -b \rangle\rangle) \\ &= c(\psi) c(\langle\langle -a, -b \rangle\rangle)(-d(\psi), 1)_F \\ &= c(\psi)(a, b)_F. \end{aligned}$$

Thus $c(\psi) = 1$. Write $\psi = \langle s \rangle \perp \Theta$, where $s \in \dot{F}$ and $\dim \Theta = 4$. Then

$$\begin{aligned} 1 &= c(\psi) \\ &= c(\Theta)(-s, d(\Theta))_F. \end{aligned}$$

Set $L = F(\sqrt{d(\Theta)})$. Then $c(\Theta_L) = 1$. Thus from Theorem 14.3(ii), p. 88, in [**Sch**], we conclude that Θ is isotropic. Hence also ψ and σ are isotropic as we claimed. □

References

[Ara] J. Arason, *Cohomologische Invarianten Quadratischer Formen*, Journal of Algebra **36** (1975), 448–491.

[EL] R. Elman and T. Y. Lam, *Quadratic forms and the u-invariant I*, Math. Zeit **131** (1973), 283–304.

[Kap] I. Kaplansky, *Quadratic forms*, J. Math. Soc. Japan **5** (1953), 200–207.

[Lam1] T. Y. Lam, *The Algebraic Theory of Quadratic Forms*, W. A. Benjamin, Inc., Reading, MA, 1973 (revised printing 1980).

[Lam2] T. Y. Lam, *Fields of u-invariant* 6 *after A. Merkurjev*, Israel Mathematical Conference Proceedings, Ring Theory (L. Rowen, ed.), 1989, pp. 12–30.

[Lee] D. Leep, *Systems of quadratic forms*, J. Reine angew. Math **350** (1984), 109–116.

[LM] D. Leep and A. S. Merkurjev, *Growth of the u-invariant under algebraic extensions* (these proceedings).

[Mer1] A. Merkurjev, *Simple algebras and quadratic forms*, Izv. Akad. Nauk. Ser. Math **55** (1991), no. 1, 218–224. (In Russian.)

[Mer2] A. Merkurjev, *On the norm residue symbol of degree* 2, Dokl. Akad. Nauk. SSSR **261** (1981), 542–547; English transl. in Sov. Math. Dokl. **24** (1981), 546-551.

[MW] J. Mináč and A. Wadsworth, *u-invariants of algebraic extensions*, Preprint.

[Pfi] A. Pfister, *Quadratische Formen in beliebigen Körpern*, Invent. Math **1** (1966), 116–132.

[Sch] W. Scharlau, *Quadratic and Hermitian Forms*, vol. 270, Springer-Verlag, 1985.

Department of Mathematics, Middlesex College, The University of Western Ontario, London, Ontario, Canada

E-mail address: 7116_440@uwovax.uwo.ca

Contemporary Mathematics
Volume **155**, 1994

On the Canonical Class of a Curve and the Extension Property for Quadratic Forms

R. PARIMALA AND W. SCHARLAU

December 1990

Let k be a field and X a smooth projective curve over k with function field $F = k(X)$. Let $\Omega_X = \Omega \in Cl(X)$ be the canonical class. We write the divisor class group $Cl(X)$ additively. A divisor class Θ with $2\Theta = \Omega$ is called a theta characteristic of X. We call Ω even, if such a Θ exists (over k) and odd otherwise.

In the theory of quadratic forms it is important to know whether Ω is even because in this case there exists a reciprocity law for quadratic forms over F with respect to the second residue class forms (see §3). This, in turn, implies the following extension property: If $p \in X$ is a *rational point* and φ a regular form on the curve $X - p$, then φ can be extended to a regular form on X.

Not very much is known about the parity of the canonical class. In many cases Ω is even but examples with Ω odd are known. However, these examples are quite trivial: Ω cannot be even if $deg(\Omega) = 4d - 2$ and all points of X are of even degree (see also [**FST**]).

In this paper we discuss the case of hyperelliptic curves and give in §2.4 and §2.9 an (almost) complete answer to the question when Ω is even. This also gives for the first time examples where X does not have the extension property (see §3.2). Finally, we discuss the case of local fields. It turns out (§4.1) that in this case Ω being even is equivalent to the extension property. For other fields this is generally not true.

1991 *Mathematics Subject Classification.* 14C20, 14H05, 11E81, 11E88.

The authors were supported in part by the NSF and the DFG (Deutsche Forschungsgemeinschaft.

This paper is submitted in final form and no version of it will be submitted for publication elsewhere.

For all basic terminology on quadratic forms and Witt groups, we refer to [**S**].

This work was done during the "special year on real algebraic geometry and quadratic forms" at the University of California, Berkeley. We are grateful to this institution, the NSF and DFG for financial support.

1. Known results

To put our results in proper perspective, we give a brief summary of what is known about the parity of the canonical class. Let g denote the genus of X.

(1) Let k be a perfect field of characteristic 2. Then Ω is even. In fact, there is a canonical Θ with $2\Theta = \Omega$ (this result is completely elementary, see [**M**]).

(2) If k is algebraically closed, then Ω is even (because deg $\Omega = 2g - 2$ is even and $Cl_0(X)$ is divisible).

(3) If k is real closed, then Ω is even if and only if g is odd or there is a rational point. In particular, Ω is odd if X is the "anisotropic conic," the non-rational curve of genus 0. (This is essentially due to Witt.)

(4) If k is finite, then Ω is even. (This follows from class field theory : see [**W**, p. 291], [**KS**], [**S**, Chap. 6, §7].)

(5) If k is quasi-finite, that is, every algebraic extension of k is cyclic, then Ω is even, if X has a rational point. (This result is due to Atiyah [**At**], Mumford [**M**], and Serre.)

We assume char $(k) \neq 2$ from now on.

2. The canonical class of hyperelliptic curves

A *hyperelliptic curve* X is a degree 2 covering of a curve Y of genus 0; that is, Y is either the projective line or an anisotropic conic. The function field $k(Y)$ of Y is thus either the rational function field $k(T)$ or a field $k(T)(\sqrt{aT^2+b})$ where (a, b) is a non-split quaternion algebra over k ([Wi]). In the first case, we determine the parity of Ω in all cases. In the second case there remains a (small) gap. Our solution will be expressed in terms of the genus and the ramification points of the covering $\pi : X \to Y$. We note that the covering $\pi : X \to Y$ is uniquely determined, up to isomorphism by X if the genus of X is at least 2 [**Ch**, Ch. IV, §9], so that our results are intrinsic.

We begin with the case $\pi : X \to P^1$ and introduce some notation. Let $E/F, F = k(T)$ be the corresponding function fields. Then $E = F(\sqrt{p(T)})$ where $p(T)$ is a square free polynomial. Without loss of generality we may assume that ∞ is unramified, so that deg $(p(T)) = 2d$ and $g = d - 1$. We factor $p(T)$ into monic irreducible factors

$$p(T) = \eta p_1(T) \dots p_n(T), \ \eta \in k^*.$$

Let $p_1, \dots, p_n$ be the points of $Y = P^1$ corresponding to these polynomials.

They are precisely the ramified points: let $\pi(P_i) = p_i$, $P_i \in X$. The point ∞ is decomposed or inert depending on whether η is a square or not; let $\infty_X = \pi^*(\infty)$.

LEMMA (2.1). *The divisor $(g-1)\infty_X$ is a canonical divisor of X.*

PROOF. (See also [**PS**].) We have $\Omega_Y = -2\infty$. By a theorem of Hurwitz,

$$\Omega_X = \pi^*\Omega_Y + [R],$$

where R is the ramification divisor, $[\,.\,]$ denoting the divisor class. Hence

$$\Omega_X = [-2\infty_X + P_1 + \cdots + P_n].$$

Since

$$(\sqrt{p(T)})_X = -d\infty_X + P_1 + \cdots + P_n,$$

we see that $(d-2)\infty_X$ is a canonical divisor. □

COROLLARY (2.2). *If g is odd, then Ω_X is even.* □

COROLLARY (2.3). *Suppose there is a ramification point of odd degree. Then Ω_X is even.*

PROOF. We may assume that g is even. If $\deg p_1 = 2k+1$, then $(p_1(T))_X = 2P_1 - (2k+1)\infty_X$ and ∞_X is even. □

In view of 2.2 and 2.3 we are left with the case where g is even and all ramification points have even degree. In this case, the precise condition for Ω_X to be even is more complicated:

THEOREM (2.4). *Let X be a hyperelliptic curve of even genus. Let $\pi : X \to P^1$ be a covering of degree two such that all ramification points have even degree. Then the following statements are equivalent:*

(1) *Ω_X is even.*
(2) *There is a nonsquare $\varepsilon \in k$ such that the quaternion algebra $(\varepsilon, p(T))$ splits over $k(T)$.*
(3) *There exists a quadratic extension $\ell = k(\sqrt{\varepsilon})$ such that ℓ is contained in the residue class field $k(p_i)$ of all ramified points, and if ∞ is inert and $k(\infty) = k(\sqrt{\eta})$, then η is a norm of ℓ/k.*

PROOF. The equivalence of (2) and (3) follows from a theorem of Harder [**S**, Ch. 6, §3]: The quaternion algebra $(\varepsilon, p(T))$ splits if and only if $<1, -\varepsilon, -p(T), \varepsilon p(T)>$ is hyperbolic. This is true if and only if all the second residue forms are 0 and the first residue form at the rational point ∞ is 0. This is a reformulation of (3).

We keep the notation introduced above and prove that (1) $\Rightarrow$ (2). Suppose that Ω_X is even. Then there exists a divisor Θ and a function $f \in E$ such that

$$\infty_X = 2\Theta + (f)_X. \qquad (*)$$

Taking norms we get the following equation in Div(Y):

$$2\infty = 2N(\Theta) + (N(f))_Y.$$

In particular, $N(f)$ has even order everywhere. Hence there exist $h \in F, \varepsilon \in k$ such that

$$N(f) = h^2\varepsilon. \qquad (**)$$

We first show that ε cannot be a square in k. Otherwise, modifying f by a scalar we could assume $N(f) = h^2$. By Hilbert 90 there exists $g \in E$ with

$$f = h(\sigma(g)g^{-1}) = h(\sigma(g)g)g^{-2} = f'g^{-2}$$

with $f' \in F$ (and σ is the non-trivial automorphism of E/F). Hence $(f)_X = (f')_X - 2(g)_X$. Substituting this in $(*)$ and replacing Θ by $\Theta - (g)$, we may assume that $f \in F$ in $(*)$. Let

$$(f)_Y = e_\infty\infty + e_1p_1 + \cdots + e_np_n + \sum_{p\neq p_i,\infty} e_pp. \qquad (***)$$

Since $(f)_X$ has even coefficients at all $p \neq \infty$, all $e_p, p \neq p_i, \infty$ are even and e_∞ is odd by $(*)$. Since all p_i have even degree, the right hand side of $(**\ *)$ has odd degree leading to a contradiction. Now $(**)$ implies $<1, -p> \simeq\ <\varepsilon, -\varepsilon p>$ which is assertion (2).

We now prove that (2) $\Rightarrow$ (1). Suppose $(\varepsilon, p(t))$ is split over $k(T)$. Then there exist $r(T), s(T) \in F$ with

$$r(T)^2 - \varepsilon s(T)^2 = p(T).$$

By the Cassels-Pfister theorem [**S**, 4.3.2] we can assume that $r(T), s(T)$ are polynomials. We consider the divisor of

$$f = r(T) + \sqrt{p(T)} \in E.$$

It is positive except at ∞_X. We have

$$N(f) = r(T)^2 - p(T) = \varepsilon s(T)^2.$$

Therefore $(f)_X$ can have zeros only at points q corresponding to an irreducible factor $q(T)$ of $s(T)$. Let us consider such a point. Then $p(T)$ is a square mod $q(T)$ and therefore q decomposes in two points $Q, \overline{Q}$ in X. We note also that $p(T)$ is a unit at q and hence $\sqrt{p(T)}$ is a unit at Q and $\overline{Q}$. If v is the valuation at Q, we have

$$v((r + \sqrt{p})(r - \sqrt{p})) = v((r + \sqrt{p})(r + \sqrt{p} - 2\sqrt{p})) = 2k.$$

If $v(r + \sqrt{p}) > 0$, it follows that

$$v(r - \sqrt{p}) = 0,\ v(r + \sqrt{p}) = 2k.$$

Therefore $(f)_X$ is even except at ∞.

If ∞ is inert, then $p(T)$ and $s(T)^2$ have the same even degree $2d$. Since the genus is even, d is odd. Therefore

$$v_\infty(r + \sqrt{p}) = v_\infty(r - \sqrt{p}) = -d.$$

We see that the canonical divisor

$$\Omega_X = (g-1)\infty_X + (f)_X$$

is even. If ∞ is decomposed and v_∞ is the valuation at one of the two infinite points of X, then $v_\infty(\sqrt{p}) = -d$. If $v_\infty(r+\sqrt{p}) > -d$, it follows that $v_\infty(r - \sqrt{p}) = -d$. Hence the multiplicity of f at one of the infinite points is odd and therefore also at the other (because the rest is even). Thus once again the canonical divisor Ω_X is even. $\square$

The choice of f in the above proof, which is rather crucial, is in fact quite canonical. Namely

$$s_* < f > \cong < 1, -\varepsilon >,$$

where $s : E \to F$ is the trace $s(1) = 0, s(\sqrt{p}) = 1$. We work in fact in the Elman-Lam exact triangle.

If there is a theta characteristic Θ, then all others are elements of $\Theta +_2 Cl(X)$ where ${}_2Cl(X)$ denotes the subgroup of elements of order ≤ 2 in the class group. Thus the number of theta characteristics defined over k is equal to the order of ${}_2Cl(X)$.

Remark (2.5).

$$|{}_2\, Cl(X)\,| \quad \begin{array}{ll} \geq 2^{n-1} & \text{if all } p_i \text{ are of even degree} \\ = 2^{n-2} & \text{if there is a } p_i \text{ of odd degree.} \end{array}$$

Proof. Let $f(T)$ be an arbitrary product of some of the $p_i(T)$ but containing an even number of odd degree p_i. Then $(f(T))_X$ is an even divisor (also at ∞) and $\frac{1}{2}(f(T))_X$ is principal only if $f(T) = p(T)$ or 1. From this we get the inequality $\geq$ in both cases.

Assume now that p_1 is of odd degree and let D be a divisor such that $2D$ is principal : $2D = (f)_X$. We repeat the argument from the proof of the last theorem. Taking norms we get the equation (**). Then ε is a square mod p_1 and therefore a square in k. It follows that $f = f'g^{-2}$ with $f' \in F, g \in E$. Since $(f)_X$ is even, f' is up to squares a product of some of the p_i. So there are no other elements in ${}_2Cl(X)$ than those coming from products of the p_i. $\square$

We come now to the case where X is a 2-fold covering of an anisotropic conic $Y : ax^2 + by^2 = z^2$. Let F be its function field, that is, $F = k(T)(\sqrt{aT^2 + b})$. All points of Y and therefore all divisors of X are of even degree. So we get immediately

Remark (2.6). Let $\pi : X \to Y$ be an arbitrary covering with genus X even. Then Ω_X is odd.

PROOF. $\deg(\Omega_X) = 2g - 2 \equiv 2(\bmod 4)$. □

Let us therefore study a 2-fold covering of odd genus. We assume again that ∞_Y is unramified. Let $E = F(\sqrt{p}), p \in F$ be the function field of X. Changing p by a suitable square we may assume

$$(p)_Y = -2d\infty_Y + p_1 + \cdots + p_n$$

where $p_1 \ldots p_n$ are the ramification points. Let $P_1 \ldots P_n$ be the corresponding points of X. As in the case $Y = P_1$ we get (with $\Omega_Y = [-\infty_Y]$)

$$\Omega_X = [-\infty_X + P_1 + \cdots + P_n]$$

$$(\sqrt{p})_X = -d\infty_X + P_1 + \cdots + P_n$$

$$\Omega_X = [(d-1)\infty_X],$$

so that $g = genus(X) = 2d - 1$.

COROLLARY (2.7). *If $g \equiv 1$ (mod 4), then Ω_X is even.* □

So we are left with the case $g \equiv 3$ (mod 4). Since d is now even, we will change our notation and replace d by $2d$. So we have

$$\deg_X(P_1 + \cdots + P_n) = 8d$$

$$\Omega_X \equiv [\infty_X] \text{ (mod even divisors).}$$

Suppose there is a ramification point say p_1 of degree $2t$ with t odd. There is a function $f_1 \in F$ with

$$(f_1)_Y = -t\infty_Y + p_1, \ (f_1)_X = -t\infty_X + 2P_1.$$

COROLLARY (2.8). *If there is a ramification point of degree $2t, t$ odd, then Ω_X is even.* □

So we are left with the following situation

$$\Omega_X = [(2d-1)\infty_X] \equiv [\infty_X] \text{ (mod even divisors),}$$

$$(\sqrt{p})_X = -2d\infty_X + P_1 + \cdots + P_n,$$

$$\text{genus } (X) = 4d - 1,$$

$$\deg(P_i) = 4d_i.$$

Unfortunately, we can prove in this case only one direction of an analogue of 2.4.

PROPOSITION (2.9). *In the situation described above, assume there is no non-square ε of k such that $(\varepsilon, p)_F$ splits. Then Ω_X is odd.*

PROOF. This is proved in exactly the same way as (1) ⇒ (2) in 2.4. □

Remark (2.10). It is perhaps not entirely obvious but in fact easy to see, that there are examples satisfying the conditions of the last proposition. For example, one can proceed as follows: Let $k = \mathbb{Q}$ and choose a non-split quaternion algebra (a, b). Let $F = k(T)(\sqrt{aT^2 + b})$ be the function field of this quaternion algebra. If we choose non-squares $c_1 \dots c_n \in \mathbb{Q}$ not belonging to the same square class and choose

$$p_i(T) = T^4 - c_i, \; p(T) = p_1(T) \dots p_n(T),$$

then $(\varepsilon, p(T))$ is non-split over $k(T)$ because ε is a non-square at one of the ramified points. But we want more, namely $(\varepsilon, p(T))_F \neq 0$. This is true, for example, when all p_i decompose in $F/\mathbb{Q}(T)$. The residue class fields at the ramified points are

$$\mathbb{Q}(\sqrt[4]{c_i}, \sqrt{a\sqrt{c_i} + b}).$$

We choose c_i now so that $a\sqrt{c_i} + b$ is a square in $\mathbb{Q}(\sqrt{c_i})$. It is easy to do this in many ways. We have the conditions

$$a\sqrt{c_i} + b = (x_i + y_i\sqrt{c})^2$$

$$a = 2x_iy_i, b = x_i^2 + y_i^2c_i.$$

Now try suitable x_i, y_i.

(To be really specific take $a = 2, b = 3; c_1 = 2, c_2 = -6.9, c_3 = -13.16, c_4 = -22.25, c_5 = -33.36, c_6 = -46.49, \dots$.)

QUESTION (2.11). Is there a curve of genus $g \equiv 1 \pmod 4$ with odd canonical class? Our results so far give only examples with $g \not\equiv 1 \pmod 4$. For $g = 1$ the canonical class is trivial and therefore even.

3. The extension property for hyperelliptic curves

We say that a curve X or its function field $k(X)$ *admits a reciprocity law* if there exist for every point p a uniformizing parameter π_p and a k-linear map $s_p : k(p) \to k$, $s_p \neq 0$ such that for every $\varphi \in W(k(X))$

$$\sum_p s_{p*}\partial_p(\varphi) = 0.$$

where $\partial_p = \partial_p^2$ is the second residue form with respect to π_p. The main result of **[GHKS]** states that X admits a reciprocity law if Ω_X is even. The converse is not true: For example, one can take a real curve of even genus without real points. Then $s_{p*}\partial_p(\varphi) = 0$ for all p, so that the reciprocity law holds trivially.

It is fairly obvious and well-known that the existence of a reciprocity law implies the extension property for quadratic spaces (as formulated in the introduction).

We now restrict ourselves to hyperelliptic curves and consider the situation discussed earlier. We construct examples of curves which do not have the extension property. This can happen only if Ω_X is odd. So we may assume that we have the situation of 2.4. Let us also assume without loss of generality that ∞ is decomposed in X, that is $p(T)$ is a monic polynomial. We have (if Ω_X is odd)

$$\ll -\varepsilon, -p \gg_{k(T)} \neq 0$$

for all non-squares $\varepsilon \in k$, $\ll a_1, \ldots, a_n \gg$ denoting the n-fold Pfister form $<1, a_1> \otimes <1, a_2> \cdots \otimes <1, a_n>$. The following proposition shows that similar conditions will be relevant with respect to the extension property.

PROPOSITION (3.1). *Assume that X does not have the extension property. Then there exists a form $\varphi_0 \in I^2(k), \varphi_0 \neq 0$ such that $\varphi_0 \otimes <1, -p> = 0$.*

PROOF. We may assume that the rational point at which the extension property fails is one of the infinite points of X, say ∞_1 with respect to a covering $\pi : X \to P^1$. Then there exists a form $\varphi \in W(k(X))$ such that $\partial_P(\varphi) \neq 0$ only for $P = \infty_1$. We consider the Elman-Lam exact triangle [**S**, 2.5.10]:

$$\begin{array}{ccc} & W(E) & \\ r^* \nearrow & & \searrow s_* \\ W(F) & \stackrel{t}{\leftarrow} & W(F) \end{array}$$

where $s : E \to F$ is defined by $s(1) = 0, s(\sqrt{p}) = 1$ and $t(\varphi) = \varphi \otimes < 1, -p >$ and r^* is induced by the inclusion of F in E. Let $\varphi_0 = s_*(\varphi)$ so that $\varphi_0 \otimes <1, -p> = 0$. Since $\partial_P(\varphi) = 0$ for all finite points, well-known formulae for $\partial_P(s_*)$ imply that $\partial_P(\varphi_0) = 0$ for all finite points of P^1 (see e.g. [**PS**, Lemma 2.1]). Therefore $\varphi_0 \in W(k)$. By the same argument (see 3.3) and the assumption,

$$\varphi_0 = \partial^1_\infty(\varphi_0) = \partial^2_{\infty_1}(\varphi) \neq 0.$$

Since image $s_* \subset I(F)$, dim (φ_0) is even. Let ε be the discriminant of φ_0. Computation of the Witt invariant shows that $0 = c(\varphi_0 \otimes < 1, -p >) = (\varepsilon, p)$. Since Ω_X is odd, ε is a square. Therefore $\varphi_0 \in I^2(k)$. $\square$

We now give a criterion for hyperelliptic curves to have no extension property. We shall later see that this criterion can be verified for suitable k so that we have examples of curves without the extension property.

PROPOSITION (3.2). *Let $X \to P^1$ be a hyperelliptic curve with Ω_X odd, ∞ decomposed and with ramified points $p_1, p_2, \ldots, p_n \in P^1$ of even degrees. Assume in addition*

(1) *there exists $0 \neq \varphi_0 \in I^2(k)$ with $\varphi_0 \otimes <1, -p> = 0$,*

(2) *the "trace" map $s : \bigoplus_{i=1}^{n} W(k(p_i)) \to I(k)$ is surjective.*

Then X does not have the extension property.

PROOF. We use a well known refinement of the Elman-Lam exact triangle, namely the following exact sequence:

$$I^2(F) \xrightarrow{r^*} I^2(E) \xrightarrow{s_*} I^2(F) \xrightarrow{t} I^3(F).$$

By assumption (1) there exists a form $\varphi \in I^2(E)$ with $s_*(\varphi) = \varphi_0$. We have to compute the second residue forms of φ which are all even dimensional.

(1) Let $q \in P^1$ be a finite inert point corresponding to the monic prime $q(T)$ and with $Q \mid q$. At q, Q we take second residue forms with respect to the parameter $q(T)$. Then we have the following commutative diagram with exact rows.

$$\begin{array}{ccccc} W(F) & \xrightarrow{r^*} & W(E) & \xrightarrow{s_*} & W(F) \\ \partial_q \downarrow & & \downarrow \partial_Q & & \downarrow \partial_q \\ W(k(q)) & \xrightarrow{r_q^*} & W(k(Q)) & \xrightarrow{s_{q*}} & W(k(q)). \end{array}$$

Since $\partial_q(s_*\varphi) = 0$, there exists a form $\chi_q \in W(k(q))$ with $r_q^*(\chi_q) = \partial_Q(\varphi)$.

(2) Let q be decomposed and finite with Q_1, Q_2 above q. Then we have the same situation with $W(k(q)) \oplus W(k(q))$ instead of $W(k(Q))$ and

$$r_q^*(\chi) = (\chi, \chi), \quad s_{q*}(\chi_1, \chi_2) = \chi_1 - \chi_2$$

in the lower row. Again there exists a form χ_q with $\chi_q = \partial_{Q_1}(\varphi) = \partial_{Q_2}(\varphi)$.

(3) If q is ramified and $Q \mid q$ then we have (for any choice of a uniformizing parameter) the same diagram with $r_q^* = 0, s_{q^*}$ an isomorphism. In particular $\partial_Q(\varphi) = 0$.

We now choose $\chi_\infty \in W(k)$ so that the reciprocity formula holds and apply [S, 6.3.5]: There exists a form $\chi \in W(F)$ having precisely the χ_q as second residue class forms. Since all χ_q are even dimensional it follows that $d(\chi) \in k$. Adding a suitable form in $W(k)$ to χ we may assume that $\chi \in I^2(F)$. We have $s_*(\varphi - r^*(\chi)) = \varphi_0$ and replacing φ by $\varphi - r^*(\chi)$, we may assume without loss of generality that φ has second residue forms $\neq 0$ only at the two infinite points ∞_1, ∞_2.

(4) We now consider the second residue forms ∂^2 at the infinite points with respect to the uniformizing parameter T^{-1}. Then we have the following formulae for any $\varphi \in W(E)$ (see 3.3)

$$\partial^1_\infty(s_*\varphi) = \partial^2_{\infty_1}(\varphi) - \partial^2_{\infty_2}(\varphi), \; \partial^2_\infty(s_*(\varphi)) = \partial^1_{\infty_1}(\varphi) - \partial^1_{\infty_2}(\varphi).$$

Thus we have $\varphi_0 = \chi_1 - \chi_2$ where χ_1, χ_2 denote the second residue forms of φ at ∞_1 and ∞_2.

Since $\chi_2 \in I(k)$, there exist by assumption (2) forms $\psi_i \in W(k(p_i))$ with

$$\sum_{i=1}^n s_{i^*}(\psi_i) + \chi_2 = 0.$$

Then there exists a form $\psi \in W(F)$ with second residue forms ψ_i at p_i and χ_2 at ∞ and no others. We replace again φ by $\varphi - r^*(\psi)$. Since $\partial_P(r^*\psi) = 0$

for P ramified, this form still has second residue forms 0 at all finite points, in particular at the ramified points. Moreover

$$\partial^2_{\infty_2}(\varphi - r^*(\psi)) = 0,\ \partial^2_{\infty_1}(\varphi - r^*(\psi)) = \varphi_0 \neq 0.$$

So the extension property does not hold for X. □

We have yet to prove a lemma concerning the computation of the residue forms at infinity.

LEMMA (3.3). *Let $F = k(T), p(T)$ a monic polynomial of degree $2d$ with d odd and $E = k(T)(\sqrt{p(T)})$. Let ∞_1, ∞_2 be the two infinite primes of E. We take second residue forms with respect to the uniformizing parameter T^{-1}. Then for $\phi \in W(E)$, we have*

$$\partial^1_\infty(s_*\varphi) = \partial^2_{\infty_1}(\varphi) - \partial^2_{\infty_2}(\varphi),\ \partial^2_\infty(s_*(\varphi)) = \partial^1_{\infty_1}(\varphi) - \partial^1_{\infty_2}(\varphi).$$

PROOF. It is sufficient to consider $\varphi = <a + b\sqrt{p}>$ where a, b are polynomials. Then $s_*(\varphi) = <b, b(pb^2 - a^2)>$ if $b \neq 0$ and $s_*(\varphi) = 0$ if $b = 0$. The proof follows from a tedious case by case calculation. We consider one case, all others being handled in the same way. Suppose b is of even degree and deg $(a) <$ deg $(b) + d$. Then the power series of $a \pm b\sqrt{p}$ in the parameter T^{-1} begins with a term $\pm\beta T^{-l}$ with l odd. We then have,

$$\partial^1_\infty(s_*\varphi) = \partial^1_\infty <b, b(pb^2 - a^2)> = <\beta, \beta>,\ \partial^2_{\infty_1}(\varphi) = <\beta>,\ \partial^2_{\infty_2}(\varphi) = <-\beta>$$

$$\partial^2_\infty(s_*\varphi) = 0,\ \partial^1_{\infty_1}(\varphi) = \partial^1_{\infty_2}(\varphi) = 0.$$ □

The conditions of the last proposition are not very practical; in particular (2) is usually difficult to verify in concrete cases. But, remarkably, we get both for free in the case of local fields as will be shown in the next section.

4. Local fields

THEOREM (4.1). *Let k be a local field of characteristic not 2 and $X \to P^1$ a hyperelliptic curve with a rational point. Then the following conditions are equivalent:*

(1) *Ω_X is even.*
(2) *X admits a reciprocity law.*
(3) *X has the extension property.*

PROOF. We know (1) ⇒ (2) ⇒ (3) . We prove (3) ⇒ (1). Assume that Ω_X is odd. We show that X does not have the extension property. We have the situation of 2.4 and we use the same notation. Let φ_0 be the unique anisotropic quaternion norm form over k. We claim $\varphi_0 \otimes <1, -p> = 0$, that is $\varphi_0 \simeq p\varphi_0$. The form φ_0 has trivial second residue forms everywhere. The form $p\varphi_0$ can have non-trivial ones at most at the ramified points p_i, where

$$\partial^2_{p_i}(p\varphi_0) = \alpha_i\varphi_0,\ \alpha_i \in k(p_i).$$

By local class field theory, φ_0 splits in the even degree extension $k(p_i)/k$. At ∞ both the forms φ_0 and $p\varphi_0$ have first residue form φ_0. By Harder's theorem $\varphi_0 \simeq p\varphi_0$. Thus condition (1) of 3.2 is satisfied. For condition (2) we use the results of [**LW**], where the transfer ideal

$$T(k(p_i)/k) = im(s_* : W(k(p_i)) \to W(k))$$

is computed for local fields as follows: Let ℓ_i be the maximal multi-quadratic intermediate field $k(p_i) \supset \ell_i \supset k$. Then $T(k(p_i)/k) = T(\ell_i/k)$. If $\ell = k(\sqrt{\alpha_1}, \ldots, \sqrt{\alpha_r})$, then

$$T(\ell/k) = \bigcap_{i=1}^{r} \operatorname{ann} \langle 1, -\alpha_i \rangle .$$

The ideal ann $\langle 1, -a \rangle$ is generated by binary forms $\langle 1, -b \rangle$ such that $\langle 1, -a \rangle \otimes \langle 1, -b \rangle = 0$; that is, $(a, b) = 0$. The Hilbert symbol

$$(\ ,\ ) : k\,/k^{\cdot 2} \times k^{\cdot}/k^{\cdot 2} \to \mathbb{Z}/2\mathbb{Z}$$

is nonsingular. Since by assumption Ω_X is odd, by 2.4 we have $\bigcap \ell_i = k$. These two facts imply easily that a form of even dimension and arbitrary determinant can be written as $\perp \varphi_i$ with φ_i in the image of the transfer ideal of $k(p_i)/k$ or ℓ_i/k. This is exactly condition (2) of 3.2. □

Remark (4.2). Let X be hyperelliptic and assume it has good reduction to the residue field $\bar{k}$. Since $\Omega_{\bar{X}}$ is always even, it is interesting to know whether Ω_X can be odd. We leave it to the reader to check that this in fact cannot happen: If X has good reduction then Ω_X is even, if char $\bar{k} \neq 2$.

Remark (4.3). One may also ask whether there is something like a *local global principle.* So assume k is a global field. If $\mathcal{P}$ is a prime and $k_{\mathcal{P}}$ the completion at $\mathcal{P}$ we denote by $X_{\mathcal{P}}, \Omega_{\mathcal{P}}$ etc. the corresponding objects over $k_{\mathcal{P}}$. If Ω is even, of course, $\Omega_{\mathcal{P}}$ is even for all $\mathcal{P}$. But the converse is far from being true. In fact, it seems that even in the case Ω odd, $\Omega_{\mathcal{P}}$ is usually even for all $\mathcal{P}$. One has to investigate how $p(T)$ splits over $k_{\mathcal{P}}$ at the finitely many $\mathcal{P}$ where X does not have good reduction. We want to give a specific example. Let $p_1, p_2, \ldots, p_n$ be distinct primes $\equiv 1 \pmod 4$ with n odd and such that

$$\left(\frac{p_i}{p_1}\right) = \left(\frac{p_1}{p_i}\right) = 1, \; i = 2, \ldots, n,$$

for example, $p_1 = 5, p_2 = 29, p_3 = 41, \ldots$. Take $k = \mathbb{Q}$ and

$$p(T) = (T^2 - p_1) \ldots (T^2 - p_n).$$

Then Ω is odd, but, over every completion $\mathbb{Q}_p$, the polynomial $p(T)$ has either a linear factor, or the quadratic factors define the unique unramified quadratic extension of $\mathbb{Q}_p$. Therefore Ω_p is even for all p.

Remark (4.4). Examples of quadratic forms on $X - p$ which do not extend to X (X a hyperelliptic curve over a local field, p a rational point) give also examples of elements in the Brauer group $Br(X)$ which do not arise as Witt invariants of quadratic forms on X, in view of [**PSr**, §4]. This implies that Merkurjev's theorem does not generalize to curves.

References

[At] M. Atiyah, *Riemann surfaces and spin structures*, Ann. Scient. Ec. Norm. Sup., 4^e serie 4 (1971), 47–62.

[Ch] C. Chevalley, *Introduction to the theory of algebraic functions of one variable*, Math. Surveys., Amer. Math. Soc., Providence, R.I. 1951.

[FST] A. Fröhlich, J.-P. Serre, and J. Tate, *A different with an odd class*, J. reine angew. Math. **209** (1962), 6–7.

[GHKS] W.-D. Geyer, G. Harder, M. Knebusch, and W. Scharlau, *Ein Residuensatz für symmetrische Bilinearformen*, Invent. Math. **11** (1970), 319–328.

[KS] M. Knebusch and W. Scharlau, *Quadratische Formen und quadratische Reziprozitätsgesetze*, Math. Z. **121** (1971), 346–368.

[LW] D. Leep and A. Wadsworth, *The transfer ideal of quadratic forms and a Hasse norm theorem mod squares*, AMS Transactions **315** (1989), 415–431.

[M] D. Mumford, *Theta characteristics of an algebraic curve*, Ann. Scient. Ec. Norm. Sup., 4^e serie 4 (1971), 181–192.

[P] R. Parimala, *Witt groups of conics, elliptic and hyperelliptic curves*, J. Number Theory **28** (1988), 69–93.

[PSr] R. Parimala and R. Sridharan, *Graded Witt ring and unramified cohomology*, *K*-Theory 6 (1982), 29–44.

[PS] R. Parimala and R. Sujatha, *Witt group of hyperelliptic curves*, Comment. Math. Helvetici **65** (1990), 559–580.

[S] W. Scharlau, *Quadratic and hermitian forms*, Grundlehren der mathematischen Wissenschaften, vol. 270, Berlin, Heidelberg, New York, 1985.

[W] A. Weil, *Basic number theory*, Grundlehren der mathematischen Wissenschaften, vol. 144, Berlin, Heidelberg, New York, 1967.

[Wi] E. Witt, *Über ein Gegenbeispiel zum Normensatz*, Math. Z. **39** (1935), 462–467.

School of Mathematics, Tata Institute of Fundamental Research, Homi Bhabha Road, Bombay 400 005, India

E-mail address: PARIMALA@tifrvax.BITNET

Mathematisches Institut der Universität, Einsteinstrasse 62, 4400 Münster, Germany

Contemporary Mathematics
Volume **155**, 1994

Reduced Norms and Pfaffians Via Brauer-Severi Schemes

R. PARIMALA AND R. SRIDHARAN

Introduction

Let A be a sheaf of Azumaya algebras of (constant) rank n^2 over a scheme S. For any integer $i, 1 \leq i \leq n$, Azumaya algebras $\lambda^i A$ Brauer equivalent to $\otimes^i A$ were defined by Suslin in [**Su**, §5] as the direct images of the "exterior products" of the pull back of A to the Brauer-Severi scheme of A over S. We define in this paper functors λ^i from the category of A-modules to the category of $\lambda^i A$-modules. The functor $\lambda^n : A - \text{Mod} \to \mathcal{O}_S - \text{Mod}$ coincides with the reduced norm functor Nrd defined in [**KOS**]. Let rank $A = 4m^2$ and let the Brauer-class of A be 2-torsion. If $\varphi : A \otimes_{\mathcal{O}_S} A \xrightarrow{\sim} \text{End}_{\mathcal{O}_S} P$ is an isomorphism of $\mathcal{O}_S$-algebras, P being a locally free sheaf over S, there exists an $\mathcal{O}_S$-linear endomorphism ψ of P such that the inner conjugation by ψ is the transport under φ of the switch map of $A \otimes A$. Let $\mathcal{S}(P)$ denote the submodule of P consisting of the "alternating elements" for ψ. Using Brauer-Severi-schemes, we give an elegant way of constructing the Pfaffian map $\mathcal{S}(P) \xrightarrow{\text{Pf}} \text{Pf}(P), \text{Pf}(P)$ denoting the invertible sheaf Pfaffian associated to the datum (A, P, φ) (cf. [**KPS1**]) for an earlier construction for affine schemes S, using faithfully flat descent). This leads to a canonical nonsingular pairing $\text{Pf}(P) \otimes \text{Pf}(P) \xrightarrow{d} \text{Nrd}(P)$ such that $d(\text{Pf}(x) \otimes \text{Pf}(x)) = \text{Nrd}(x)$ for all $x \in \mathcal{S}(P)$. The notion of the Pfaffian discriminant of an Azumaya algebra with involution over a commutative ring defined in [**KPS3**, §3] can now be interpreted as a special case of the above pairing. We also define the Clifford algebra of an algebra with an orthogonal involution over any scheme which coincides with that of Jacobson for fields (cf. [**J**]. We give an alternate proof (without the use of Pfaffians) of the fact [**KPS2**, Th. 5.2] that any orthogonal involution of trivial

1991 *Mathematics Subject Classification.* 11E88, 13A20, 14M20.

This paper is submitted in final form and no version of it will be submitted for publication elsewhere.

Pfaffian discriminant of an Azumaya algebra of rank 16 splits canonically into a tensor product of the standard involutions of two quaternion subalgebras. We conclude with an example of a rank 16 Azumaya algebra with an even symplectic involution which is indecomposable.

Throughout this paper, by a module over a scheme S, we mean a quasi-coherent sheaf on S. We use "Morita Theory" for modules over arbitrary algebraic schemes, without explicit reference. The Morita theory for rings generalizes verbatim to a theory over schemes.

Part of this work was done while the authors were visiting the Department of Mathematics, University of California, Berkeley during the special year on real algebraic geometry and quadratic forms. We thank this Institute, especially T. Y. Lam for the hospitality which we enjoyed during our stay. Our warm thanks are due to M.-A. Knus and M. Ojanguren for valuable comments and discussions.

1. λ-functors on modules over Azumaya algebras

Let S be any algebraic scheme. Let A be a sheaf of Azumaya algebras of (constant) rank n^2 over S. We recall the definition due to Suslin [**Su**, §5] of the exterior operations $\lambda^i A$. Let X/S denote the Brauer-Severi scheme corresponding to A. The structure map $f : X \to S$ is proper with every geometric fibre of f isomorphic to the projective n-space. There is a canonical locally free $\mathcal{O}_X$-module J of rank n such that $f^*A \simeq (\mathrm{End}_{\mathcal{O}_X} J)^{\mathrm{op}}$ (cf. [**Q**, §4]). If $A = \mathrm{End}_{\mathcal{O}_S}$, E being a locally free module of rank n over $\mathcal{O}_S$, then $X = \mathbb{P}_S(E^\vee)$ is the projective bundle associated to $E^\vee = \mathrm{Hom}_{\mathcal{O}_S}(E, \mathcal{O}_S)$ and $J = \mathcal{O}_X(-1) \otimes_{\mathcal{O}_S} E^\vee$. The algebras $\lambda^i A$ are defined by

$$\lambda^i A = f_*(\mathrm{End}_{\mathcal{O}_X}(\wedge^i J))^{\mathrm{op}}.$$

It is shown in [**Su**, Proposition 5.1] that $\lambda^i A$ are Azumaya algebras over S of rank $\binom{n}{i}^2$, Brauer equivalent to $\overset{i}{\otimes} A$. Further, $\lambda^1 A = A, \lambda^n A = \mathcal{O}_S$. The ith-exterior map $\wedge^i : (\mathrm{End}_{\mathcal{O}_X} J)^{\mathrm{op}} \to (\mathrm{End}_{\mathcal{O}_X} \wedge^i J)^{\mathrm{op}}$ gives rise to a polynomial map of degree i, $\lambda^i = f_*(\wedge^i) : A \to \lambda^i A$ which is multiplicative. The map $\lambda^n : A \to \mathcal{O}_S$ is simply the reduced norm.

For any sheaf of algebras B on S, let B - Mod denote the category of coherent $\mathcal{O}_S$-sheaves which are left B-modules. We construct functors

$$\lambda^i : A\text{ - Mod} \to \lambda^i A\text{ - Mod}$$

for $1 \le i \le n$. Let M be a sheaf which is a left A–module. Then f^*M is a left module over $f^*A = (\mathrm{End}_{\mathcal{O}_X} J)^{\mathrm{op}}$ which we identify with $\mathrm{End}_{\mathcal{O}_X} J^\vee, J^\vee = \mathrm{Hom}_{\mathcal{O}_X}(J, \mathcal{O}_X)$ through transposition. By Morita theory, there exists an $\mathcal{O}_X$-module N such that

$$f^*M = J^\vee \otimes_{\mathcal{O}_X} N$$

with f^*A operating through the action of $\mathrm{End}_{\mathcal{O}_X} J^\vee$ on $J^\vee$. The module $\wedge^i J^\vee \otimes_{\mathcal{O}_X} \wedge^i N$ is a left $\mathrm{End}_{\mathcal{O}_X}(\wedge^i J^\vee)$-module. We define

$$\lambda^i M = f_*(\wedge^i J^\vee \otimes_{\mathcal{O}_X} \wedge^i N).$$

This is a left module over $f_*(\mathrm{End}_{\mathcal{O}_X} \wedge^i J^\vee) = f_*(\mathrm{End}\ \wedge^i J)^{\mathrm{op}} = \lambda^i A$. Let $h : M \to M'$ be a homomorphism of A–modules. Let $f^*M = J^\vee \otimes_{\mathcal{O}_X} N, f^*M' = J^\vee \otimes_{\mathcal{O}_X} N'$. By Morita theory, there exists an $\mathcal{O}_X$-linear map $\widetilde{h}: N \to N'$ such that $f^*h = 1 \otimes \widetilde{h}$. We define $\lambda^i h = f_*(1 \otimes \wedge^i \widetilde{h})$. This defines the functor $M \to \lambda^i M$. The exterior map $J^\vee \otimes_{\mathcal{O}_X} N \xrightarrow{\sim} \mathrm{Hom}_{\mathcal{O}_X}(J, N) \xrightarrow{\wedge^i} \mathrm{Hom}_{\mathcal{O}_X}(\wedge^i J, \wedge^i N) = \wedge^i J^\vee \otimes \wedge^i N$ yields a map $\lambda^i : M \to \lambda^i M$ which is simply the composite

$$M \to f_* f^* M \xrightarrow{f_*(\wedge^i)} \lambda^i M.$$

The following properties of the functors λ^i determine them uniquely.

PROPOSITION (1.1). *The functors $\lambda^i, 1 \leq i \leq n$ have the following properties:*

1) *If $M = A$, regarded as a left A-module, $\lambda^i M = \lambda^i A$ regarded as a left module over itself and the map $\lambda^i : A \to \lambda^i A$ is the polynomial map defined earlier.*
2) *λ^i commutes with base change.*
3) *If $A = End_{\mathcal{O}_S} E$, E being a locally free module of rank n over $\mathcal{O}_S, M = E \otimes_{\mathcal{O}_S} N$, then, $\lambda^i M = \wedge^i E \otimes_{\mathcal{O}_S} \wedge^i N$ and the map $\lambda^i : M \to \lambda^i M$ is the composite*

$$E \otimes_{\mathcal{O}_S} N \xrightarrow{\sim} Hom_{\mathcal{O}_S}(E^\vee, N) \to Hom_{\mathcal{O}_S}(\wedge^i E^\vee, \wedge^i N) \xrightarrow{\sim} \wedge^i E \otimes_{\mathcal{O}_S} \wedge^i N.$$

   Further, if $M' = E \otimes_{\mathcal{O}_S} N'$ and $f : M \to M'$ is a homomorphism of A-modules, which, by Morita theory, is of the form $f = 1 \otimes f_0, f_0 : N \to N'$ being a homomorphism of $\mathcal{O}_S$-modules, $\lambda^i f = 1 \otimes \wedge^i f_0$.
4) *If P is a locally free A-module whose rank as an $\mathcal{O}_S$-module is m, then m is of the form $n \cdot k$, k an integer and $\lambda^i P$ is a locally free $\lambda^i A$–module whose rank as an $\mathcal{O}_S$–module is $\binom{n}{i}\binom{k}{i}$. In particular, if P is of rank 1 over A, then $\lambda^n P$ is a rank one $\mathcal{O}_S$-module.*

PROOF. Property 1) follows from the definition of λ^i and 2) is a consequence of the fact that direct image commutes with base change. To verify 3), we observe that if $A = \mathrm{End}_{\mathcal{O}_S} E$, then $X = \mathbb{P}_S(E^\vee), f^*A = \mathrm{End}_{\mathcal{O}_X} f^*E, J = \mathcal{O}_X(-1) \otimes_{\mathcal{O}_S} E^\vee$ and $f^*M = f^*E \otimes_{\mathcal{O}_X} f^*N = J^\vee \otimes_{\mathcal{O}_X} (\mathcal{O}_X(-1) \otimes_{\mathcal{O}_X} f^*N)$. We have,

$$\begin{aligned}
\lambda^i M &= f_*(\wedge^i J^\vee \otimes_{\mathcal{O}_X} (\wedge^i(\mathcal{O}_X(-1) \otimes_{\mathcal{O}_X} f^*N)) \\
&= f_*(\wedge^i J^\vee \otimes_{\mathcal{O}_X} \mathcal{O}_X(-i) \otimes_{\mathcal{O}_X} \wedge^i(f^*N)) \\
&= f_*(\wedge^i(f^*E) \otimes_{\mathcal{O}_X} (\wedge^i(f^*N)) \\
&= \wedge^i E \otimes_{\mathcal{O}_S} \wedge^i N,
\end{aligned}$$

using projection formula and noting that f^* commutes with $\wedge^i$. It suffices to check 4) locally for the faithfully flat topology and hence it follows from 3). □

COROLLARY (1.2). *If $S = \operatorname{Spec} R$ is affine, the functor λ^n coincides with the reduced norm functor* Nrd *constructed in* [**KOS**, Th. 2.1].

2. The Pfaffian

Let A be an Azumaya algebra of rank $4n^2$ over S such that its class in $Br(S)$ is 2–torsion. Let $\varphi : A \otimes_{\mathcal{O}_S} A \to \operatorname{End}_{\mathcal{O}_S} P$ be an isomorphism of $\mathcal{O}_S$-algebras, P being a locally free sheaf of rank $2n$ over S. We give a construction of the Pfaffian associated to the triple (A, P, φ) using the Brauer-Severi scheme of A which, even in the case $S = \operatorname{Spec} R$ puts in a neat perspective, the construction given in [**KPS1**, Th. 2.2].

Following [**KPS1**, §2], we call the triple (A, P, φ) a *2-torsion datum*. The switch map $\omega : A \otimes A \to A \otimes A$ induced by $\omega(a \otimes b) = b \otimes a$ is inner [**KO**, Prop. 4.1, p. 112] and in fact, there is a "canonical" unit $u \in H^0(S, A \otimes_{\mathcal{O}_S} A)$ such that $u^2 = 1$ and Int $u = \omega$. The image of u under the isomorphism $A^{\mathrm{op}} \otimes_{\mathcal{O}_S} A \simeq \operatorname{End}_{\mathcal{O}_S} A$ is the linear map given by the reduced trace $\operatorname{Trd} : A \to \mathcal{O}_S \hookrightarrow A$. Let $\varphi(u) = \psi$. Then $\psi : P \to P$ is an $\mathcal{O}_S$-linear map with $\psi^2 = 1$ and is called the *module involution* on P. Let $\mathcal{S}(P)$ denote the sheaf of alternating elements for ψ; i.e., $\mathcal{S}(P)$ is the image of $1 - \psi$. Then the fact that $\mathcal{S}(P)$ is a locally free module of rank $n(2n-1)$ over S may be verified locally for the faithfully flat topology (cf. [**KPS1**]). We associate to the datum (A, P, φ) a locally free sheaf $\operatorname{Pf}(P)$ of rank one over S and a polynomial map $\operatorname{Pf} : \mathcal{S}(P) \to \operatorname{Pf}(P)$ of degree n, which is functorial.

Suppose $A = \operatorname{End}_{\mathcal{O}_S} E$, E being locally free of rank $2n$ over $S, P = E \otimes_{\mathcal{O}_S} E$ and $\varphi = \operatorname{Can} : \operatorname{End}_{\mathcal{O}_S} E \otimes_{\mathcal{O}_S} \operatorname{End}_{\mathcal{O}_S} E \xrightarrow{\sim} \operatorname{End}_{\mathcal{O}_S}(E \otimes_{\mathcal{O}_S} E)$ the canonical isomorphism. The datum $(\operatorname{End}_{\mathcal{O}_S} E,\ E \otimes_{\mathcal{O}_S} E,\ \operatorname{Can})$ is called a *split datum.* In view of [**KPS1**, Lemma 2.1] any datum is isomorphic to a split datum locally for the faithfully flat topology. We first construct the Pfaffian map for the split datum. The module involution $\psi : E \otimes_{\mathcal{O}_S} E \to E \otimes_{\mathcal{O}_S} E$ is simply the switch so that $\mathcal{S}(E \otimes_{\mathcal{O}_S} E) = \wedge^2 E$. We define $\operatorname{Pf}(E \otimes_{\mathcal{O}_S} E) = \wedge^{2n} E$ and the map $\operatorname{Pf} : \mathcal{S}(E \otimes_{\mathcal{O}_S} E) = \wedge^2 E \to \operatorname{Pf}(E \otimes_{\mathcal{O}_S} E) = \wedge^{2n} E$ to be the classical Pfaffian. With respect to an ordered basis $\{e_i\}$ of E locally, we have

$$\operatorname{Pf}(\sum_{i<j} a_{ij}(e_i \wedge e_j)) = \operatorname{Pf}(A)(e_1 \wedge e_2 \cdots \wedge e_{2n}),$$

where A is the alternating matrix of size $2n \times 2n$ with entries a_{ij} for $i < j$.

Let now (A, P, φ) be any datum. Let $f : X \to S$ be the Brauer-Severi scheme corresponding to A. We pull back the datum (A, P, φ) to X through f. We have an isomorphism

$$f^*\varphi : \operatorname{End}_{\mathcal{O}_X} J^\vee \otimes_{\mathcal{O}_X} \operatorname{End}_{\mathcal{O}_X} J^\vee \to \operatorname{End}_{\mathcal{O}_X} f^*P.$$

By Morita theory, there is an invertible sheaf $\mathcal{L}$ on X and an isomorphism

$$\widetilde{\varphi}\colon J^\vee \otimes_{\mathcal{O}_X} J^\vee \otimes_{\mathcal{O}_X} \mathcal{L} \xrightarrow{\sim} f^*P$$

such that the induced map

$$\mathrm{End}_{\mathcal{O}_X}(J^\vee \otimes J^\vee) = \mathrm{End}_{\mathcal{O}_X}(J^\vee \otimes_{\mathcal{O}_X} J^\vee \otimes_{\mathcal{O}_X} \mathcal{L}) \to \mathrm{End}_{\mathcal{O}_X}(f^*P)$$

is simply $f^*\varphi$. The diagram

$$\begin{array}{ccc} J^\vee \otimes_{\mathcal{O}_X} J^\vee \otimes_{\mathcal{O}_X} \mathcal{L} & \overset{\widetilde{\varphi}}{\to} & f^*P \\ \xi \otimes 1 \downarrow & & \downarrow f^*\psi \\ J^\vee \otimes_{\mathcal{O}_X} J^\vee \otimes_{\mathcal{O}_X} \mathcal{L} & \overset{\widetilde{\varphi}}{\to} & f^*P \end{array}$$

is commutative, where ξ is the switch map on $J^\vee \otimes_{\mathcal{O}_X} J^\vee$. Hence $\widetilde{\varphi}$ induces an isomorphism of $\wedge^2 J^\vee \otimes_{\mathcal{O}_X} \mathcal{L}$ with the module of alternating elements for $f^*\psi$, which is simply $f^*(\mathcal{S}(P))$, since f is flat and $\mathcal{S}(P)$ is locally free. We denote the isomorphism $\wedge^2 J^\vee \otimes_{\mathcal{O}_X} \mathcal{L} \to f^*(\mathcal{S}(P))$ again by $\widetilde{\varphi}$. We have the "classical" Pfaffian $\wedge^2 J^\vee \to \wedge^{2n} J^\vee$ defined earlier and this yields a polynomial map $\mathrm{Pf} : \wedge^2 J^\vee \otimes_{\mathcal{O}_X} \mathcal{L} \to \wedge^{2n} J^\vee \otimes_{\mathcal{O}_X} \mathcal{L}^n$ of degree n.

PROPOSITION (2.1). *The $\mathcal{O}_S$-module $\mathrm{Pf}(P) = f_*(\wedge^{2n} J^\vee \otimes_{\mathcal{O}_X} \mathcal{L}^n)$ is locally free of rank one and the map*

$$f_*(\mathrm{Pf} \circ \widetilde{\varphi}^{-1}) : \mathcal{S}(P) = f_* f^* \mathcal{S}(P) \to f_*(\wedge^{2n} J^\vee \otimes_{\mathcal{O}_X} \mathcal{L}^n)$$

is a polynomial map of degree n. Further, if $(\mathrm{End}_{\mathcal{O}_S} E,\ E \otimes_{\mathcal{O}_S} E,\ \mathrm{Can})$ is the split datum, $\mathcal{S}(E \otimes_{\mathcal{O}_S} E) = \wedge^2 E, \wedge^{2n} J^\vee \otimes_{\mathcal{O}_X} \mathcal{L}^n = f^*(\wedge^{2n} E)$, and $f_*(\mathrm{Pf} \circ \widetilde{\varphi}^{-1}) : \wedge^2 E \to \wedge^{2n} E$ is simply the classical Pfaffian.

PROOF. Since it is sufficient to verify the proposition locally for the faithfully flat topology, we may assume, without loss of generality, that the given datum is the split datum $(\mathrm{End}_{\mathcal{O}_S} E,\ E \otimes_{\mathcal{O}_S} E,\ \mathrm{Can})$. We have,

$$X = \mathbb{P}_S(E^\vee), J = \mathcal{O}_X(-1) \otimes_{\mathcal{O}_S} E^\vee,\ f^*A = \mathrm{End}_{\mathcal{O}_X} f^*E$$

and

$$f^*(\mathrm{Can}) : \mathrm{End}_{\mathcal{O}_X}(f^*E) \otimes_{\mathcal{O}_X} \mathrm{End}_{\mathcal{O}_X}(f^*E) \to \mathrm{End}_{\mathcal{O}_X} f^*(E \otimes_{\mathcal{O}_S} E)$$

is the canonical map. Thus, $\mathcal{L} = \mathcal{O}(-2)$ and

$$\widetilde{\varphi}\colon J^\vee \otimes J^\vee \otimes_{\mathcal{O}_X} \mathcal{L} \to f^*(E \otimes_{\mathcal{O}_S} E)$$

is the natural map under the identification $J^\vee = f^*E \otimes_{\mathcal{O}_X} \mathcal{O}(1)$. We have, $\wedge^{2n} J^\vee \otimes_{\mathcal{O}_X} \mathcal{L}^n = f^*(\wedge^{2n} E)$ and the map

$$\mathrm{Pf} \circ \widetilde{\varphi}^{-1} : f^*(\wedge^2 E) \to \wedge^2 J^\vee \otimes_{\mathcal{O}_X} \mathcal{O}_X(-2) \to \wedge^{2n} J^\vee \otimes_{\mathcal{O}_X} \mathcal{O}_X(-2n) = f^*(\wedge^{2n} E)$$

is the pul! back of the Pfaffian $\mathrm{Pf} : \wedge^2 E \to \wedge^{2n} E$ under f. Thus,

$$f_* f^*(Pf \circ \widetilde{\varphi}^{-1}) : \wedge^2 E \to \wedge^{2n} E$$

is the classical Pfaffian and we have the following: $\square$

PROPOSITION (2.2).

1) *The association* $(A, P, \varphi) \to (\mathcal{S}(P), Pf(P), \mathrm{Pf})$ *commutes with base change.*
2) *For the split datum* $(End\, E,\ E \otimes_{\mathcal{O}_S} E,\ \mathrm{Can})$, $\mathrm{Pf}(E \otimes_{\mathcal{O}_S} E) = \wedge^{2n} E$, $\mathcal{S}(E \otimes_{\mathcal{O}_S} E) = \wedge^2 E$ *and* $\mathrm{Pf} : \wedge^2 E \to \wedge^{2n} E$ *is the classical Pfaffian.*

COROLLARY (2.3). *If* $S = \mathrm{Spec}\, R$, *the Pfaffian constructed above coincides with that defined in* [**KPS1**, Th. 2.2].

3. Reduced norm as a square of the Pfaffian

Let A be an Azumaya algebra of rank $4n^2$ over S whose class in $\mathrm{Br}(S)$ is 2–torsion. Let $\varphi : A \otimes_{\mathcal{O}_S} A \to \mathrm{End}_{\mathcal{O}_S} P$ be an isomorphism of algebras. We regard P as a left A-module through $\varphi(A \otimes_{\mathcal{O}_S} 1)$. Then P is a locally free A-module of rank 1 and we have a degree $2n$ map $\mathrm{Nrd} = \lambda^n : P \to \mathrm{Nrd}(P)$. On the other hand, we have the Pfaffian map $\mathrm{Pf} : \mathcal{S}(P) \to \mathrm{Pf}(P)$ defined on the set of ψ-alternating elements of P, which is of degree n. The following theorem compares these two maps, abstracting the general philosophy "the reduced norm is the Pfaffian scared."

THEOREM (3.1). *Let* (A, P, φ) *be a 2–torsion datum as above. There exists a functorial pairing* $d : \mathrm{Pf}(P) \otimes_{\mathcal{O}_S} \mathrm{Pf}(P) \to \mathrm{Nrd}(P)$ *such that for* x *a local section of* $\mathcal{S}(P)$, $d(\mathrm{Pf}(x) \otimes \mathrm{Pf}(x)) = \mathrm{Nrd}(x)$.

The rest of the section is devoted to proving this theorem.

Let $f^*\varphi : \mathrm{End}_{\mathcal{O}_X} J^\vee \otimes_{\mathcal{O}_X} \mathrm{End}_{\mathcal{O}_X} J^\vee \to \mathrm{End}_{\mathcal{O}_X}(f^*P)$ be induced by

$$\widetilde{\varphi}\colon J^\vee \otimes_{\mathcal{O}_X} J^\vee \otimes_{\mathcal{O}_X} \mathcal{L} \xrightarrow{\sim} f^*P,$$

$\mathcal{L}$ being an invertible sheaf on X. We have,

$$\begin{aligned} \mathrm{Nrd}(P) &= f_*(\wedge^{2n} J^\vee \otimes_{\mathcal{O}_X} \wedge^{2n}(J^\vee \otimes_{\mathcal{O}_X} \mathcal{L})) \\ &= f_*(\wedge^{2n} J^\vee \otimes_{\mathcal{O}_X} \wedge^{2n} J^\vee \otimes_{\mathcal{O}_X} \mathcal{L}^{2n}) \end{aligned}$$

Let $c : \wedge^{2n} J^\vee \otimes_{\mathcal{O}_X} \mathcal{L}^n \otimes_{\mathcal{O}_X} \wedge^{2n} J^\vee \otimes_{\mathcal{O}_X} \mathcal{L}^n \to \wedge^{2n} J^\vee \otimes_{\mathcal{O}_X} \wedge^{2n} J^\vee \otimes_{\mathcal{O}_X} \mathcal{L}^{2n}$ be the natural map. This induces an isomorphism

$$f_*(c) : f_*(\wedge^{2n} J^\vee \otimes_{\mathcal{O}_X} \mathcal{L}^n \otimes_{\mathcal{O}_X} \wedge^{2n} J^\vee \otimes_{\mathcal{O}_X} \mathcal{L}^n) \to f_*(\wedge^{2n} J^\vee \otimes_{\mathcal{O}_X} \wedge^{2n} J^\vee \otimes_{\mathcal{O}_X} \mathcal{L}^{2n}).$$

LEMMA (3.2). *The natural map*

$$\begin{aligned} &f_*(\wedge^{2n} J^\vee \otimes_{\mathcal{O}_X} \mathcal{L}^n) \otimes_{\mathcal{O}_S} f_*(\wedge^{2n} J^\vee \otimes_{\mathcal{O}_X} \mathcal{L}^n) \\ &\qquad \to f_*(\wedge^{2n} J^\vee \otimes_{\mathcal{O}_X} \mathcal{L}^n \otimes_{\mathcal{O}_X} \wedge^{2n} J^\vee \otimes_{\mathcal{O}_X} \mathcal{L}^n) \end{aligned}$$

is an isomorphism.

PROOF. It suffices to prove this locally for the faithfully flat topology of S. We therefore assume that $A = \mathrm{End}_{\mathcal{O}_S} E, P = E \otimes_{\mathcal{O}_S} E$, $\varphi = \mathrm{Can}$. In this case, $\mathcal{L} = \mathcal{O}(-2)$, $J = \mathcal{O}(-1) \otimes_{\mathcal{O}_S} E^\vee, \wedge^{2n} J^\vee \otimes_{\mathcal{O}_X} \mathcal{L}^n = f^*(\wedge^{2n} E)$ and the map in the lemma is simply the map

$$f_*(f^*(\wedge^{2n}E)) \otimes_{\mathcal{O}_S} f_*(f^*(\wedge^{2n}E)) \to f_*(f^*(\wedge^{2n}E \otimes_{\mathcal{O}_S} \wedge^{2n}E)),$$

which reduces to the identity map

$$\wedge^{2n}E \otimes_{\mathcal{O}_S} \wedge^{2n}E \to \wedge^{2n}E \otimes_{\mathcal{O}_S} \wedge^{2n}E. \quad \square$$

We thus have a pairing $d : \mathrm{Pf}(P) \otimes_{\mathcal{O}_S} \mathrm{Pf}(P) \to \mathrm{Nrd}(P)$ which, in the case of a split datum, $(\mathrm{End}_{\mathcal{O}_S} E, E \otimes_{\mathcal{O}_S} E, \mathrm{Can})$ coincides with the identity map $\wedge^{2n}E \otimes_{\mathcal{O}_S} \wedge^{2n}E \to \wedge^{2n}E \otimes_{\mathcal{O}_S} \wedge^{2n}E$. By definition, d commutes with base change and hence is uniquely determined.

PROPOSITION (3.3). *For x a local section of $\mathcal{S}(P)$, $d(\mathrm{Pf}(x) \otimes_{\mathcal{O}_S} \mathrm{Pf}(x)) = \mathrm{Nrd}(x)$.*

PROOF. It is enough to check this locally for the faithfully flat topology. Hence we may assume that $S = \mathrm{Spec}(R)$ and the given datum is the split datum $(\mathrm{End}_{\mathcal{O}_S} E,\ E \otimes_{\mathcal{O}_S} E,\ \mathrm{Can})$ with E free of rank $2n$. The map $\mathrm{Pf} : \wedge^2 E \to \wedge^{2n} E$ is the classical Pfaffian, the reduced norm $\mathrm{Nrd} : E \otimes_{\mathcal{O}_S} E \to \wedge^{2n}E \otimes_{\mathcal{O}_S} \wedge^{2n}E$ is the composite

$$E \otimes_{\mathcal{O}_S} E \xrightarrow{\sim} \mathrm{Hom}_{\mathcal{O}_S}(E^\vee,\ E) \to \mathrm{Hom}_{\mathcal{O}_S}(\wedge^{2n}E^\vee, \wedge^{2n}E) \xrightarrow{\sim} \wedge^{2n}E \otimes_{\mathcal{O}_S} \wedge^{2n}E$$

and the pairing $\mathrm{Pf}(E \otimes_{\mathcal{O}_S} E) \otimes_{\mathcal{O}_S} \mathrm{Pf}(E \otimes_{\mathcal{O}_S} E) \to \mathrm{Nrd}(E \otimes_{\mathcal{O}_S} E)$ is the identity. We identify $E \otimes_{\mathcal{O}_S} E$ with $E^\vee \otimes_{\mathcal{O}_S} E = \mathrm{End}_{\mathcal{O}_S} E$ through the choice of a basis $(e_1, e_2, \dots, e_{2n})$ of E and the dual basis $(e_1^*, e_2^*, \dots, e_{2n}^*)$ of $E^\vee$. With this identification, $E \otimes_{\mathcal{O}_S} E = M_{2n}(R)$, $\wedge^2 E = \mathrm{Alt}_{2n}(R)$, $\wedge^{2n}E = R$, $\mathrm{Pf} : \mathrm{Alt}_{2n} R \to R$ is the Pfaffian on the set of alternating elements $\mathrm{Alt}_{2n} R, \mathrm{Nrd} : M_{2n}(R) \to R$ is the determinant and $(\mathrm{Pf}(x))^2 = \det(x)$, for $x \in \mathrm{Alt}_{2n}(R)$. This proves the proposition and completes the proof of the theorem. $\square$

4. Quadratic modules with values in a line bundle

Let S be an algebraic scheme. Suppose 2 is invertible in $\mathcal{O}_S$. Let (E, q) be a quadratic space over $\mathcal{O}_S$ of rank $2n$ with values in a line bundle $\mathcal{L}$. (By a *quadratic space* we mean that the adjoint map $b = b_q : E \to E^\vee \otimes_{\mathcal{O}_S} \mathcal{L}$ is an isomorphism.) We have an induced isomorphism

$$\wedge^{2n} b : \wedge^{2n}E \xrightarrow{\sim} \wedge^{2n}E^\vee \otimes_{\mathcal{O}_S} \mathcal{L}^{2n}$$

which yields a discriminant module $(\wedge^{2n}E \otimes_{\mathcal{O}_S} \mathcal{L}^{-n}, (-1)^{n(2n-1)} \wedge^{2n} b \otimes 1)$. We call this the *discriminant* of (E, q) and denote it by $\mathrm{disc}(E, q)$. Locally, for the choice of a basis e for $\mathcal{L}$ and the dual basis $e^\vee$ of $\mathcal{L}^\vee$, if (E, q) is identified with a quadratic space with values in $\mathcal{O}_S$, the discriminant coincides with the (signed) discriminant of (E, q).

We associate to the pair (E, q) an *even Clifford algebra* $C_0(q)$ which is an Azumaya algebra over its center Z which is separable quadratic over S. If $S = \mathrm{Spec}(R)$, this algebra coincides with the grade zero subalgebra of the Clifford algebra $\widetilde{C}(q)$ of (E, q) defined in [**BK**, §3]. Let $\{U_i\}$ be an open covering of S which trivializes both E and $\mathcal{L}$. Let $\varphi_i : \mathcal{L}_{|U_i} \xrightarrow{\sim} \mathcal{O}_{U_i}$ be trivialisations for $\mathcal{L}$ and let $u_{ij} = \varphi_j^{-1}\varphi_i \in \mathcal{O}^*_{U_i \cap U_j}$, $*$ denoting the group of invertible sections. Let $b_i : E_{|U_i} \to (E_{|U_i})^\vee$ be the map induced by b, identifying $\mathcal{L}_{|U_i}$ with $\mathcal{O}_{U_i}$ through φ_i. The identity map $E_{|U_i \cap U_j} \to E_{|U_i \cap U_j}$ is a similarity of $(E_{|U_i}, b_i)$ with $(E_{|U_j}, b_j)$ on $U_i \cap U_j$ with factor of similarity u_{ij}. Hence the identity map induces isomorphisms [**Kn**, p. 407], $\lambda_{ij} = C_0(\mathrm{Identity}) : C_0(E_{|U_i}, b_i) \to C_0(E_{|U_j}, b_j)$ on $U_i \cap U_j$ with the property that $C_0(\mathrm{Identity})(x \cdot y) = u_{ij}^{-1} \cdot x \cdot y$ for $x, y \in H^0(U_i \cap U_j, E)$. The maps λ_{ij} patch to yield an algebra on S, denoted $C_0(E, q)$ and we call it the *even Clifford algebra* of (E, q). The Clifford algebra involutions on $C_0(E_{|U_i})$ patch to yield an involution on $C_0(E, q)$.

Let Z be the center of $C_0(E, q)$. The fact that Z is a separable quadratic algebra over $\mathcal{O}_S$ and $C_0(E, q)$ is Azumaya over Z may be verified locally. On U_i, $C_0(E_{|U_i}, b_i)$ is the even Clifford algebra of a genuine quadratic space over U_i with values in $\mathcal{O}_{U_i}$ and hence $Z_{|U_i}$ has the required properties locally. Let $\mathrm{Tr} : \mathcal{O}_Z \to \mathcal{O}_S$ be the trace; namely $\mathrm{Tr} = 1 + \tau$, τ denoting the nontrivial automorphism of $\mathcal{O}_Z$ over $\mathcal{O}_S$. Let $\mathcal{L}$ be the kernel of the trace Tr. The norm on $\mathcal{O}_Z$ restricted to $\mathcal{L}$ yields a discriminant module $(\mathcal{L}, N_{\mathcal{L}})$ over $\mathcal{O}_S$. Using arguments similar to [**PS**, Lemma 4, §2], one can show that $(\mathcal{L}, -N_{\mathcal{L}})$ is isometric to $\mathrm{disc}(E, q)$. Thus, if $\mathrm{disc}(E, q)$ is trivial, $\mathcal{O}_Z \simeq \mathcal{O}_S \times \mathcal{O}_S$.

Let A be a rank 4 Azumaya algebra over S and E an A-module locally free of rank one. Let $\mathrm{Nrd}(E)$ be the reduced norm of E as an A-module. The map $\mathrm{Nrd} : E \to \mathrm{Nrd}(E)$ is a nonsingular quadratic form q of rank 4 over S. We recall some results on the classification of rank 4 quadratic spaces with values in line bundles from [**BK**] in a global setting.

PROPOSITION (4.1). *The even Clifford algebra $C_0(E, q)$ is isomorphic to $A \times \mathrm{End}_A E$.*

PROOF. Let (U_i) be an open covering of S which trivializes both E and $\mathrm{Nrd}\, E$. Let $E_i = E_{|_{U_i}}$, $q_i = q_{|U_i} : E_i \to E_i^\vee$. Let $\tilde{f}_i$ denote the isomorphism of E_i with its A-dual induced by q_i through the trace, i.e. it is the composite of the following maps,

$$E_i \xrightarrow{q_i} E_i^\vee \xrightarrow{\mathrm{Tr}^{-1}} \mathrm{Hom}_{A_{|U_i}}(E_i, A_{|U_i}).$$

The map

$$E_i \to \mathrm{End}_{A_{|U_i}}(A_{|U_i} \oplus E_i)$$

given locally by

$$x \mapsto \begin{pmatrix} 0 & x \\ \tilde{f}_i(x) & 0 \end{pmatrix}$$

induces an isomorphism

$$\psi : C(E_i, q_i) \simeq \mathrm{End}_{A_{|U_i}}(A_{|U_i} \oplus E_i)$$

which restricts to an isomorphism

$$\psi_i : C_0(E_i, q_i) \simeq A_{|U_i} \times (\mathrm{End}_{A_{|U_i}} E_i).$$

It is easy to check that ψ_i patch to yield an isomorphism

$$C_0(E, q) \simeq A \times \mathrm{End}_A E. \quad \square$$

PROPOSITION (4.2). *Let (E, q) be a quadratic space of rank 4 with values in a line bundle $\mathcal{L}$ and of trivial discriminant. Let $A \to \mathrm{End}_{\mathcal{O}_S} E$ be a quaternion subalgebra such that*

$$b_q(ax, y) = b_q(x, \bar{a}y)$$

for local sections x, y of E and a of A respectively, bar denoting the canonical involution $Tr - 1$ of A. Then (E, q) is isometric to $(E$, Nrd).

PROOF. The map b_q induces an A-linear isomorphism

$$E \to \mathrm{Hom}_A(E, A) \otimes_{\mathcal{O}_S} \mathcal{L}\ .$$

By [**KOS**, 3.1], there is an isomorphism $\eta : \mathcal{L} \to \mathrm{Nrd}(E)$, such that η induces an isometry $(E, q) \simeq (E, \mathrm{Nrd}\ E)$. $\square$

PROPOSITION (4.3). *Let (E, q) be a quadratic space of rank 4 with values in a line bundle $\mathcal{L}$ and of trivial discriminant over S. Then there exists a quaternion algebra (Azumaya algebra of rank 4) A over S such that E is a left A-module, locally free of rank one. Further, if* $\mathrm{Nrd}(E)$ *denotes the line bundle on S given by the reduced norm functor, then (E, q) is isometric to $(E$, Nrd). In particular, if b_q denotes the adjoint of q, $b_q(ax, y) = b_q(x, \bar{a}y)$, for x, y local sections of E and a a local section of A, bar denoting the canonical involution* $\mathrm{Tr} - 1$ *on A.*

PROOF. Since (E, q) has trivial discriminant, there is an idempotent $e \in H^0(S, \mathcal{O}_Z)$ which generates $\mathcal{O}_Z$ over $\mathcal{O}_S$. We set $A = C_0(E, q)e$. Let $\{U_i\}$ be a covering of S as chosen earlier, trivialising E and $\mathcal{L}$. Let $(E_i, q_i) = (E, q)$ restricted to U_i and $e_i \in H^0(U_i, \mathcal{O}_Z)$ the restriction of e. We have an isomorphism (cf. [**KOS**, Proof of (4.5)])

$$\alpha_i : C_0(E_i, q_i)e_i \to \mathrm{End}(C_1(E_i, q_i)e_i)$$

such that

$$\alpha_i(xe_i)(ye_i) = xye_i.$$

The maps β_i which are defined as the composites

$$E_i \to C_1(E_i, q_i) \to C_1(E_i, q_i)e_i,$$

the first map being the inclusion in the Clifford algebra and the second one being the projection through the idempotent e_i, are isomorphisms and give rise to isomorphisms

$$\tilde{\beta}_i\colon \text{End } E_i \to \text{End}(C_1(E_i, q_i)e_i).$$

Let $\tau_i : C_0(E_i, q_i)e_i \to \text{End } E_i$ be the composite $\tau_i = \tilde{\beta}_i^{-1} \circ \alpha_i$. Using the fact that e is globally defined, one verifies that $\{\tau_i\}$ patch to give a homomorphism $\tau : A \to \text{End } E$. The Clifford algebra involution on $C_0(E, q)$ restricts to the canonical involution on A, which we denote by bar. Then the adjoint b_q of q verifies the condition

$$b_q(ax, y) = b_q(x, \bar{a}y)$$

for x, y local sections of E and a a local section of A. This is easily verified locally for the faithfully flat topology. Thus by (4.2), (E, q) is isometric to $(E, \text{ Nrd})$. $\square$

5. The discriminant and the Clifford algebra of an involution

Let A be an Azumaya algebra of rank $4n^2$ over a scheme S and σ an involution on A. Using the pairing of the Pfaffian with the reduced norm described in §3, we give an interpretation of the (Pfaffian) discriminant associated to the pair (A, σ) defined in [**KPS1**, Proposition 3.2] having values in the discriminant group $\text{Disc}(S)$.

To the pair (A, σ) is asssociated a canonical datum $\varphi_\sigma : A \otimes_{\mathcal{O}_S} A \to \text{End}_{\mathcal{O}_S} A$, defined locally by $\varphi_\sigma(x \otimes y) = L_x \circ R_{\sigma y}, L$ and R denoting the left and right multiplications respectively. The module involution $\psi_\sigma : A \to A$ associated to φ_σ is simply $\varepsilon \cdot \sigma$ where $\varepsilon \in \mu_2(A)$ is the type of σ (cf. [**KPS1**, §3]). The module $\mathcal{S}(A) = \text{Im } (1 - \varepsilon\sigma)$ is the sub-sheaf of ε-alternating elements of A. The left A-module structure of A through $\varphi_\sigma(A \otimes 1)$ is the standard left A-module structure on A and $\text{Nrd} : A \to \text{Nrd}(A) = \mathcal{O}_S$ is the reduced norm. The pairing $d = d_\sigma : \text{Pf}_\sigma(A) \otimes \text{Pf}_\sigma(A) \to \text{Nrd}(A) = \mathcal{O}_S$ gives a discriminant module $(\text{Pf}_\sigma(A), d_\sigma)$ in $\text{Disc}(S)$. (Here $\text{Pf}_\sigma(A)$ denotes the Pfaffian for the datum (A, A, φ_σ).) The pairing d_σ has the following explicit description. The involution $f^*\sigma$ on $f^*A = \text{End}_{\mathcal{O}_X} J^\vee$ is induced by an isomorphism $q : J^\vee \to J \otimes_{\mathcal{O}_X} M$ for some line bundle M on X such that $q^\vee : (J \otimes_{\mathcal{O}_X} M)^\vee = J^\vee \otimes_{\mathcal{O}_X} M^\vee \to (J^\vee)^\vee \xrightarrow{\sim} J \otimes_{\mathcal{O}_X} M \otimes_{\mathcal{O}_X} M^\vee$ is $q \otimes 1$. The $(2n)^{\text{th}}$ exterior map $\wedge^{2n} q : \wedge^{2n} J^\vee \to \wedge^{2n} J \otimes_{\mathcal{O}_X} M^{2n}$ yields an isomorphism $(\wedge^{2n} q \otimes 1) : \wedge^{2n} J^\vee \otimes_{\mathcal{O}_X} M^{-n} \to \wedge^{2n} J \otimes_{\mathcal{O}_X} M^{2n} \otimes_{\mathcal{O}_X} M^{-n} \simeq (\wedge^{2n} J^\vee \otimes_{\mathcal{O}_X} M^{-n})^\vee$ which is a discriminant module over X. It is an easy verification that $f_*(\wedge^{2n} J^\vee \otimes_{\mathcal{O}_X} M^{-n}, \wedge^{2n} q \otimes 1)$ is simply the discriminant module Pf_σ defined earlier. If $A = \text{End}_{\mathcal{O}_S} E$ is the split algebra, σ is given by a symmetric pairing $b : E \to E^\vee \otimes_{\mathcal{O}_S} \mathcal{L}, \mathcal{L}$ being a line bundle on S and then the discriminant of σ is simply $(\wedge^{2n} E \otimes_{\mathcal{O}_S} \mathcal{L}^{-n}, \wedge^{2n} b \otimes 1)$ where $\wedge^{2n} b \otimes 1$ is the obvious map.

We now define the Clifford algebra of an orthogonal involution σ on A which for the case of algebras over fields coincides with the definition in [**J**]. If $(A, \sigma) = (\text{End}_{\mathcal{O}_S} E, \tau)$, where τ is induced by a quadratic form $b_\tau : E \to E^\vee \otimes_{\mathcal{O}_S} \mathcal{L}$, $\mathcal{L}$

denoting an invertible sheaf on S, we define $C(A,\sigma) = C_0(E, b_\tau)$. Then $C(A,\sigma)$ is an Azumaya algebra over its center which is a separable quadratic algebra over S defined by the discriminant of the involution (cf. §4). In the general case, we pull back (A,σ) to the Brauer-Severi scheme X of A. Suppose $f^*\sigma$ on $f^*A = \mathrm{End}_{\mathcal{O}_X} J^\vee$ is induced by $q : J^\vee \to J \otimes \mathcal{M}$. We define $C(A,\sigma) = f_* C_0(J^\vee, q)$. The fact that $C(A,\sigma)$ commutes with base change and coincides in the split case with the definition given above uniquely determines the algebra. It follows that $C(A,\sigma)$ is an Azumaya algebra over its centre which is a separable quadratic algebra over S defined by the discriminant of σ.

6. Orthogonal involutions of rank 16 algebras

In this section, we prove a global version of [**KPS2**, Th. 5.2].

PROPOSITION (6.1). *Let S be a scheme with 2 invertible in $\mathcal{O}_S$. Let τ be an orthogonal involution on a rank 16 endomorphism algebra $\mathrm{End}_{\mathcal{O}_S} E$ with trivial Pfaffian discriminant. Then, there exists a rank 4 Azumaya algebra A over S and an injection $A \to \mathrm{End}_{\mathcal{O}_S} E$ such that τ restricts to the canonical involution on A. If B is an Azumaya algebra of rank 4 satisfying the same property as A, then B is isomorphic either to A or to $\mathrm{End}_A E$.*

PROOF. The involution τ on $\mathrm{End}_{\mathcal{O}_S} E$ is induced by a symmetric bilinear form $b_\tau : E \to E^\vee \otimes_{\mathcal{O}_S} \mathcal{L}$ whose discriminant (cf. §4) is trivial. By (4.3), there exists a rank 4 Azumaya algebra A such that E is a left A-module, locally free of rank one over A and the quadratic form q is given by the reduced norm on E regarded as a left A-module. More precisely, there exists an isomorphism $\mathcal{L} \xrightarrow{\eta} \mathrm{Nrd}(E)$ such that the diagram

$$\begin{array}{ccc} E & \xrightarrow{\widetilde{q}} & \mathcal{L} \\ \mathrm{Nrd} \searrow & & \swarrow \eta \\ & \mathrm{Nrd}\, E & \end{array}$$

is commutative, $\widetilde{q}$ denoting the quadratic form whose adjoint is b_τ (noting that 2 invertible in $\mathcal{O}_S$). Thus, we may assume, without loss of generality, that the involution τ on $\mathrm{End}\ {}_{\mathcal{O}_S} E$ is induced by the adjoint b_τ of the reduced norm $\mathrm{Nrd} : E \to \mathrm{Nrd}\ E$. The pairing b_τ has the property

$$b_\tau(ax, y) = b_\tau(x, \overline{a}y)$$

for x, y local sections of E and a a local section of A, by (4.3). This implies that the involution τ on $\mathrm{End}_{\mathcal{O}_S} E$ restricts on A to the standard involution $a \to \overline{a}$. If B is a quaternion subalgebra of $\mathrm{End}_{\mathcal{O}_S} E$ with τ restricting to the canonical involution on B, then $b_\tau(ax, y) = b_\tau(x, \overline{a}y)$ for x, y local sections of E and a a local section of B. By (4.3), (E, q) is isometric to the reduced norm of E as a B–module. Comparing the Clifford algebras, by (4.1), $B \simeq A$ or $B \simeq \mathrm{End}_A E$. $\square$

THEOREM (6.2). *Let A be a rank 16 Azumaya algebra over S with an orthogonal involution σ with Pf_σ trivial. Then (A,σ) splits as $(A_1,\sigma_1)\otimes_{\mathcal{O}_S}(A_2,\sigma_2)$ with A_i quaternion algebras and σ_i the standard involutions on A_i. This decomposition is unique upto switch of the factors.*

PROOF. Let $f : X \to S$ be the Brauer-Severi scheme corresponding to A. The involution $f^*\sigma$ on $f^*A = \mathrm{End}_{\mathcal{O}_X} J^\vee$ is given by a symmetric bilinear form $b : J^\vee \to J \otimes_{\mathcal{O}_X} \mathcal{L}$, $\mathcal{L}$ being an invertible sheaf on $\mathcal{O}_X$. Since Pf_σ is trivial, $\mathrm{Pf}_{f^*\sigma}(f^*A)$ is trivial. By (6.1), there exists an Azumaya algebra $\mathcal{B}$ on $\mathcal{O}_X$ of rank 4 and an injection $\eta : \mathcal{B} \to \mathrm{End}_{\mathcal{O}_X} J^\vee$ such that $f^*\sigma$ restricted to $\mathcal{B}$ is the standard involution on $\mathcal{B}$ and b is simply the adjoint of the reduced norm on $J^\vee$ as a $\mathcal{B}$-module. Let $B = f_*\mathcal{B}$ and $\eta_0 = f_*(\eta) : B \to A$. It is easily seen, by going over to a local faithfully flat splitting and using (6.1) that B is an Azumaya algebra over S and η_0 identifies the restriction of σ to B with the standard involution on B. Let B^0 the centralizer of B in A. Then σ restricts to an even symplectic involution on B^0, σ being orthogonal and its restriction to B being even symplectic. Since an even symplectic involution on a quaternion algebra is unique, it is the standard involution.

We now prove the uniqueness of the decomposition upto switch. If $(A,\sigma) = (B_1,\tau_1)\otimes_{\mathcal{O}_S}(B_2,\tau_2)$ is another decomposition with the same property, by the uniqueness statement of Proposition 6.1, it follows that $f^*(B_i,\tau_i)$ is isomorphic to $\mathcal{B}$ with the standard involution on it over X, for $i = 1$ or 2, so that

$$(B_i,\tau_i) = f_*f^*(B_i,\tau_i) \simeq f_*\mathcal{B} = B,$$

along with the standard involution on B, for $i = 1$ or 2. This completes the proof of the theorem. □

Remark (6.3). We observe that the quaternion algebras A_i, $i = 1, 2$, occuring in the factorization of (A,σ) are precisely the components of $C_0(A,\sigma) \simeq A_1 \times A_2$. In the split case $(\mathrm{End}_{\mathcal{O}_S}(E),\tau)$, let τ be induced by a bilinear form $b_\tau : E \to E^\vee \otimes_{\mathcal{O}_S} \mathcal{L}$. Since the discriminant of τ is trivial, $C_0(E,q_\tau) \simeq A_1 \times A_2$, q_τ being the quadratic form associated to b_τ. In this case, (E,q_τ) corresponds to the reduced norm of E as an A-module (cf. §4) and there is a representation $A_1 \to \mathrm{End}_{\mathcal{O}_S}(E)$ such that τ restricts to the canonical involution of A_1. The centralizer of A_1 in $\mathrm{End}_{\mathcal{O}_S}(E)$ is, by uniqueness, A_2. The general case follows from the very construction of the factors, as direct images of the corresponding ones over the Brauer-Severi scheme. The result for algebras over fields is due to [**T**].

Remark (6.4). The discriminant of an even symplectic involution is always trivial (cf. [**KPS1**, Proposition 3.4]). It is also true that over a field K, (including $\mathrm{Char}\, K = 2$) every even-symplectic involution on a rank 16 Azumaya algebra splits as a tensor product of involutions on quaternion subalgebras (cf. [**KPS3**] and [**R**]). We have however the following example of a rank 16 Azumaya algebra over the polynomial ring $\mathbb{R}[X,Y]$, with an even symplectic involution which is indecomposable.

EXAMPLE (6.5). Let K be any field which admits a rank 5 quadratic form q_0 with discriminant 1 (e.g. $< 1,1,1,1,1 >$ over $\mathbb{R}$). Let q be a rank 5 quadratic space over $K[X,Y]$ which is indecomposable, whose reduction modulo (X,Y) is q_0 (cf. [**P**, Th. 3.2]). Then the standard involution on $C(q)$ restricts to an even symplectic involution τ on $C_0(q)$ [**K**, 5.4.5] which is a rank 16 Azumaya algebra over $K[X,Y]$. The algebra $C_0(q)$ itself is indecomposable, i.e., it cannot be written as a tensor product of quaternion subalgebras (cf. [**KPS2**, §5, Remark 1]). In particular, the involution τ on $C_0(q)$ is indecomposable.

REFERENCES

[BK] W. Bichsel and M.-A. Knus, *Quadratic forms with values in line bundles* (toappear).

[J] N. Jacobson, *Clifford algebras for algebras with involution of type D*, J. Algebra **1** (1964), 288–300.

[Kn] M. Kneser, *Composition of binary quadratic forms*, J. Number Theory **15** (1982), 406–413.

[K] M.-A. Knus, *Quadratic and Hermitian Forms Over Rings*, Grundlehren, Springer-Verlag, Berlin-Heidelberg-New York, 1991.

[KO] M.-A. Knus and M. Ojanguren, *Théorie de la descente et algèbres d'Azumaya*, SLN 389, Berlin-Heidelberg-New York, 1974.

[KOS] M.-A. Knus, M. Ojanguren, and R. Sridharan, *Quadratic forms and Azumaya algebras*, J. reine angew. Math. **303/304** (1978), 231–248.

[KPS1] M.-A. Knus, R. Parimala, and R. Sridharan, *A classification of rank* 6 *quadratic spaces via Pfaffians*, J. reine angew. Math. **398** (1989), 187–218.

[KPS2] ———, *Pfaffians, central simple algebras and similitudes*, Math. Z, **206** (1991), 589–604.

[KPS3] ———, *Involutions on rank* 16 *central simple algebras*, J. Ind. Math. Soc. 57 (1991), 143–151.

[P] R. Parimala, *Indecomposable Quadratic spaces over the affine plane*, Advances in Math. **62** (1986), 1–6.

[PS] R. Parimala and V. Srinivas, *Analogues of the Brauer group for Algebras with Involution*, Duke Math J. 66 (1992), 207–237.

[Q] D. Quillen, *Higher Algebraic K-Theory* I, SLN 341, 1973, pp. 85–147.

[R] L. H. Rowen, *Central simple algebras*, Israel J. Math. **29** (1978), 285–301.

[Su] A. A. Suslin, *K-Theory and K-Cohomology of Certain Group Varieties*, Univ. Bielefeld, 1990 (preprint 90–026).

[T] D. Tao, Ph. D. Thesis, U.C. at San Diego, 1991.

SCHOOL OF MATHEMATICS, TATA INSTITUTE OF FUNDAMENTAL RESEARCH, HOMI BHABHA ROAD, BOMBAY 400 005, INDIA

E-mail address: PARIMALA@tifrvax.BITNET

Contemporary Mathematics
Volume **155**, 1994

Matching Witts With Global Fields

R. PERLIS, K. SZYMICZEK, P. E. CONNER, AND R. LITHERLAND

Dedicated to the memory of E. Witt

1. Introduction

In 1937 Witt introduced the subsequently named *Witt ring W(F)* into the study of bilinear forms over a field F (see [**W**]). Since then, the abstract theory of bilinear forms over fields has evolved, to a large extent, into the study of the Witt ring. In 1970 Harrison found a useful criterion for two fields to have isomorphic Witt rings (see his unpublished notes [**H**] and §2 below).

Today two fields with isomorphic Witt rings are called *Witt equivalent*. The fields studied in this paper are global fields, *i.e.*, number fields and function fields in one variable over finite fields. We are interested in determining when global fields are Witt equivalent. In 1985 Baeza and Moresi showed, among other things, that any two global fields of characteristic 2 are Witt equivalent (see [**B-M**]). It is not difficult to see that a global field of characteristic 2 is never Witt equivalent to a global field of characteristic different from 2 (see §2). Henceforth in this introduction, the term *global field* will exclude fields of characteristic 2.

We define a *reciprocity equivalence* between global fields K and L to be a pair of maps (t, T), where t is a group isomorphism

$$t : K^*/K^{*2} \longrightarrow L^*/L^{*2}$$

between the square-class groups of K and L, and T is a bijection

$$T : \Omega_K \longrightarrow \Omega_L$$

1991 *Mathematics Subject Classification.* Primary: 11E12, Secondary: 11E08.

We would like to thank C. U. Jensen for his remarks when we were first getting started. The first author gratefully acknowledges support from the University of Geneva in1987 and agin in 1988, support from the University of Bordeaux (I) in 1988, and support from RAGSQUAD in 1991. The second author gratefully acknowledges support from Louisiana State University while in residence as a Visiting Professor in 1985–86, and partial support from the State Committee for Scientific Research (KBN) of Poland.

This paper is submitted in final form and no version of it will be submitted for publication elsewhere

between the sets Ω_K, Ω_L of non-trivial places, with (t, T) preserving Hilbert symbols in the sense that

$$(a,b)_P = (ta, tb)_{TP}$$

for all a, b in K^*/K^{*2} and all P in Ω_K.

Section 4 contains our first main result: *Two global fields are Witt equivalent if and only if they are reciprocity equivalent.*

We prove that, as a consequence, a number field is never Witt equivalent to a global function field. Secondly, two Witt equivalent number fields have the same degree over the field Q of rational numbers. This means that the field Q is distinguished from all other global fields by the structure of the Witt ring $W(Q)$.

Section 5 separates reciprocity equivalences into two types, called *tame* and *wild.* It will be shown in a later paper that tame equivalence preserves arithmetic properties of the underlying fields (2-ranks of class groups, etc.). We do not pursue these questions in this paper, but with these later applications in mind, we develop the subsequent theory taking tame and wild into account.

In §6 we consider the problem of constructing a reciprocity equivalence (t, T). If we work directly from the definition, this is a doubly infinite task, since t and T are maps of infinite sets. We introduce the notion of a *small equivalence* from K to L. Small equivalences are finite objects, involving a finite set S of primes. When a small equivalence exists, it can often be established explicitly. Our goal in §6 is to prove our second main result: when a small equivalence exists, it extends to a reciprocity equivalence that can be chosen to be tame outside the initial set S.

Carpenter has taken the finitely-many (but complicated) conditions for a small equivalence and simplified them greatly (see [**C1**], [**C2**]). She proves that there is a small equivalence between global fields K and L if and only if three simple conditions are satisfied:

a) -1 is a global square in both K and L or in neither,
b) K and L have the same number of real embeddings,
c) there is a bijection from the dyadic places of K to those of L such that, if the dyadic place P of K corresponds to the dyadic place Q of L, then the local degrees $[K_P : Q_2] = [L_Q : Q_2]$ agree and -1 is a local square in both K_P and L_Q or in neither.

Conditions b) and c) are vacuously true for function fields, showing:

Global function fields of odd characteristic are Witt equivalent if and only if they have the same level.

With [**B-M**], this settles the classification of global function fields up to Witt equivalence.

Combining Carpenter's results with the results of this paper, one obtains without effort a *Hasse principle* for isomorphisms of Witt rings of global fields:

Two global fields have isomorphic Witt rings if and only if the completions of these fields can be paired so that corresponding completions have isomorphic Witt rings.

In contrast to the classical Hasse principle for quadratic forms, neither implication is trivial in the principle stated above. See the end of §6.

Section 7 contains some examples.

A few comments are in order.

We began our work in 1985 and obtained all of these results by the end of 1988, with more cumbersome proofs. We circulated several early versions of this paper and talked about our results at the Corvallis meeting in 1986, at Oberwolfach in July, 1989, at the Ninth Czechoslovak Colloquium on Number Theory in September, 1989, and at RAGSQUAD in January, 1991.

Several other mathematicians have worked on related questions, and we briefly mention some of their contributions. Czogała obtained a complete classification of *quadratic* fields up to tame reciprocity equivalence. There are infinitely many tame classes. However, there are exactly 7 classes of quadratic fields if one classifies up to reciprocity equivalence, as Czogała and Carpenter have shown independently (see [**C2**], [**Cz2**]). Palfrey has proved that the maps t and T in a reciprocity equivalence mutually determine each other, and he has quantified the extent to which a reciprocity equivalence can be wild (see [**Pa**]). We have already referred to the work of Carpenter. Her work depends on the results published in the present paper. While the present paper refers to her work in order to obtain the local-global principle stated above, all our other results are independent of what she has proved. Her three conditions for small equivalence, combined with our §6, give a straight-forward method for deciding when two number fields are Witt equivalent.

The problem of counting the Witt equivalence classes of number fields of given degree n is solved in [**Sz2**]. An expository article on all of the above is found in [**Sz1**].

Questions concerning Witt equivalence of quadratic extensions of given global fields are discussed in [**Cz3**], [**J-Mar**], and [**Sz3**]. Our original proof of our first main theorem, based on papers of Ware [**Wa**] and Arason-Elman-Jacob [**A-E-J**], can be found in [**Sz4**].

We address this paper to a general audience. We keep the exposition as elementary as possible, and ask the reader's indulgence if we explain certain facts that are well-known to the experts.

2. Preliminaries

Our general references for basic assertions concerning Witt rings are [**Lam**] and [**M-H**]. Let K be an arbitrary field. The fundamental ideal I_K, consisting of Witt classes of even rank, is the unique ideal of index 2 in the Witt ring $W(K)$. A ring isomorphism $\phi : W(K) \longrightarrow W(L)$ from the Witt ring of K to the Witt ring of another field L takes I_K to I_L and powers of I_K to corresponding powers of I_L. For a global field K, the square of the fundamental ideal vanishes,

$$I_K^2 = 0,$$

if K has characteristic 2 (see [**M-H**, Th. (5.10)]) and does not vanish for global fields of any other characteristic (see [**M-H**, Ch. III]). Hence, if K and L are Witt equivalent global fields and one field has characteristic 2, the other does also.

Baeza and Moresi have investigated Witt equivalence among arbitrary fields of characteristic 2. The dimension of such a field E over the subfield E^2 of squares is 2^τ for τ in $\{0, 1, \ldots, \infty\}$. This number τ is also the minimal positive integer n with $I_E^{n+1} = 0$ (see [**M-H**, Th. (5.10)]). It follows that Witt equivalence between fields of characteristic 2 preserves τ. Baeza and Moresi prove (see [**B-M**, Theorems (2.9) and (2.10)]) that all fields with $\tau = 0$ are Witt equivalent; when $\tau = 1$, two fields are Witt equivalent if and only if they have the same cardinality; and when $\tau > 1$, two fields are Witt equivalent if and only if they are isomorphic. Global function fields of characteristic 2 are countably infinite and have $\tau = 1$. Hence:

If two global fields are Witt equivalent, then both have or neither has characteristic 2, *and any two global fields of characteristic* 2 *are Witt equivalent.*

For the remainder of this paper, we exclude fields of characteristic 2 from our discussion. In particular, the term *field* will mean a field of characteristic not 2.

From above, a Witt ring isomorphism $\phi : W(K) \longrightarrow W(L)$ induces additive group isomorphisms

$$I_K/I_K^2 \xrightarrow{t_\phi} I_L/I_L^2 \quad \text{and} \quad I_K^2/I_K^3 \xrightarrow{u_\phi} I_L^2/I_L^3.$$

Combining this with the pairing

$$I_K/I_K^2 \times I_K/I_K^2 \longrightarrow I_K^2/I_K^3$$

coming from multiplication in the Witt ring yields the following commutative diagram:

$$\begin{array}{ccc} I_K/I_K^2 \times I_K/I_K^2 & \longrightarrow & I_K^2/I_K^3 \\ t_\phi\downarrow \qquad \downarrow t_\phi & & \downarrow u_\phi \\ I_L/I_L^2 \times I_L/I_L^2 & \longrightarrow & I_L^2/I_L^3 \end{array} \tag{1}$$

The horizontal bilinear maps can be degenerate; the degeneracy is measured by

$$\{a \in K^*/K^{*2} | \textit{for every } b \in K^*/K^{*2},\ \langle 1, -a, -b, ab\rangle \textit{ lies in } I_K^3\}.$$

This turns out to be nothing other than the so-called *Kaplansky radical* of K.

For a global field K, the Kaplansky radical is the trivial subgroup of K^*/K^{*2}. For our purposes in this section, it does not matter whether the radical is trivial or not.

Let us recall two classical Witt class invariants. Let $\langle a_1, a_2, \ldots, a_n\rangle$ be a diagonalized representative of a given Witt class X, chosen so $n \equiv 0$ or $1 \pmod 8$. Then the *discriminant* of X is the class of $dis\ X = a_1 \cdots a_n$ in K^*/K^{*2}. This is well-defined on Witt classes and induces a canonical isomorphism

$$dis : I_K/I_K^2 \cong K^*/K^{*2}$$

associating the square-class a with the class of $\langle 1, -a\rangle$ $mod\ I_K^2$. With this identification, the map t_ϕ sends -1 in K^*/K^{*2} to -1 in L^*/L^{*2}, since $t_\phi(-1) = dis\ \phi\langle 1,1\rangle = dis(\phi\langle 1\rangle + \phi\langle 1\rangle) = dis\langle 1,1\rangle = -1$.

There is also the *Hasse-Witt invariant* $h(X)$ of the class X. The Hasse-Witt invariant is defined by choosing a diagonalized representative of X of rank $n \equiv 0, 1 \pmod 8$ with $n > 1$, and declaring $h(X)$ to be the Brauer class of the tensor product of quaternion algebras $\otimes(a_i, a_j)_K$, the product running over all pairs of subscripts with $i < j$.

Using the identification between I_K/I_K^2 and square classes, diagram (1) above translates into a commutative diagram

$$\begin{array}{ccccc} K^*/K^{*2} & \times & K^*/K^{*2} & \longrightarrow & I_K^2/I_K^3 \\ t\downarrow & & \downarrow t & & \downarrow u \\ L^*/L^{*2} & \times & L^*/L^{*2} & \longrightarrow & I_L^2/I_L^3 \end{array} \tag{2}$$

where we have written t and u in place of t_ϕ and u_ϕ. Explicitly, two square-classes a, b in K pair to the coset of $\langle 1, -a\rangle \otimes \langle 1, -b\rangle$ $mod\ I_K^3$.

Harrison's Criterion: Let K and L be fields of characteristic different from 2. Then the following are equivalent:

a) $W(K) \cong W(L)$ as rings.
b) There is an isomorphism $t : K^*/K^{*2} \longrightarrow L^*/L^{*2}$ of square-class groups sending -1 to -1, and an isomorphism $u : I_K^2/I_K^3 \longrightarrow I_L^2/I_L^3$ such that diagram (2) commutes.
c) There is an isomorphism $t : K^*/K^{*2} \longrightarrow L^*/L^{*2}$ of square-class groups taking -1 to -1 and such that a binary form $\langle a, b\rangle$ represents 1 over K if and only if $\langle ta, tb\rangle$ represents 1 over L.

We preface the proof of Harrison's Criterion with some known, elementary facts.

LEMMA 1.

a) *The Hasse-Witt invariant vanishes on* I_K^3.
b) $h(\langle 1, -a, -b, ab\rangle) = h(\langle a, b\rangle)$ *is the Brauer class of the quaternion algebra* $(a, b)_K$.
c) *The Witt class of* $\langle 1, -a, -b, ab\rangle$ *lies in* I_K^3 *if and only if* $\langle a, b\rangle$ *represents* 1 *over* K.

PROOF OF LEMMA. I_K^3 is additively generated by 3-fold Pfister forms. Parts a) and b) follow by direct computation. For part c), suppose $\langle a, b\rangle$ represents 1 over K. Then $\langle 1, -a, -b, ab\rangle$ is isotropic, hence hyperbolic (see [**Lam**, *loc. cit.*]). So $\langle 1, -a, -b, ab\rangle$ is 0 in $W(K)$, and therefore lies in I_K^3. The converse follows immediately from the *Hauptsatz* of Arason-Pfister (see [**Lam**, p. 289]); but we shall give a low-brow argument. If $\langle 1, -a, -b, ab\rangle$ lies in I_K^3, then taking the Hasse-Witt invariant shows that the quaternion algebra $(a, b)_K$ is trivial in the Brauer group $Br(K)$, by parts a) and b) above. Then, [**Lam**, Theorem 2.7, p. 58], shows that $\langle a, b\rangle$ represents 1 over K.

PROOF OF HARRISON'S CRITERION. The discussion preceding the statement of Harrison's criterion proves the implication *a*) $\Rightarrow$ *b*). It remains to show *b*) $\Rightarrow$ *c*) $\Rightarrow$ *a*). For the first of these, let maps t and u be given as in *b*). Then a binary form $\langle a, b\rangle$ represents 1 over K if and only if the corresponding 2-fold Pfister form $\langle 1, -a\rangle \otimes \langle 1, -b\rangle$ lies in I_K^3, by part c) of Lemma 1. By the commutative diagram (2), the u-image of this 2-fold Pfister form lies in I_L^3. So the Witt class of $\langle 1, -ta, -tb, tatb\rangle$ lies in I_L^3, and this means that $\langle ta, tb\rangle$ represents 1 over L, by part c) of Lemma 1. This proves *b*) $\Rightarrow$ *c*).

Next, let t be a map as in *c*). Define a map ϕ on diagonal forms by mapping $\langle a_1, \ldots, a_n\rangle$ to $\langle ta_1, \ldots, ta_n\rangle$. This map is independent of the chosen diagonalization, as we now show.

The case of rank $n = 1$ is trivial. For $n = 2$, suppose $\langle a, b\rangle \sim \langle c, d\rangle$ are two diagonalizations of a binary form. Taking discriminants shows that $ab = cd$ as products of square-classes, so $(ta)(tb) = (tc)(td)$. Scaling the form by a yields $\langle 1, ab\rangle \sim \langle ac, ad\rangle$. Hence $\langle ac, ad\rangle$ represents 1 over K, and thus $\langle tatc, tatd\rangle$ represents 1 over L. So we have $\langle tatc, tatd\rangle \sim \langle 1, s\rangle$, with $s = tatctatd = tctd$. Scaling by ta then gives $\langle tc, td\rangle \sim \langle ta, tatctd\rangle = \langle ta, tb\rangle$, as desired. For $n > 2$, take a chain equivalence between the two diagonalizations. That is, select a chain of equivalent diagonalized forms leading from the first to the second, each form differing from the preceding in at most two adjacent places. The binary case above then shows that ϕ is well-defined on diagonal forms of any rank n.

Since t takes -1 to -1, it follows that ϕ takes hyperbolic planes to hyperbolic planes, thereby inducing a well-defined map on Witt classes of diagonal forms. Since every Witt class is represented by a diagonal form, ϕ is a map on Witt rings that is clearly additive. Then ϕ is multiplicative as well, since ϕ is multiplicative on 1-dimensionals, and these additively generate the Witt ring. The inverse of the square-class isomorphism t induces a map on Witt rings that is inverse to ϕ, showing that ϕ is bijective. This proves Harrison's Criterion.

Of the following remarks, only the first two are essential for the sequel.

REMARKS.

1. Merkur'ev has shown that I_K^2/I_K^3 is isomorphic to $Br_2(K)$, the 2-torsion subgroup of the Brauer group (see [**Me**]). So diagram (2) can be replaced by a more-convenient diagram with $Br_2(K)$ replacing I_K^2/I_K^3, in which a pair a, b of square classes map to the Brauer class of the quaternion

algebra $(a, b)_K$. In the following section, we will utilize this connection to the Brauer group. However, our results do not depend on Merkur'ev's theorem, since we will be concerned exclusively with global fields, and for these, the isomorphism $I_K^2/I_K^3 \cong Br_2(K)$ was proved in 1970 (see [**M**]).

2. Let P be a place of a global field K and let Q be a place of a global field L, neither field of characteristic 2. The reader is invited to show, by applying Harrison's Criterion, above, that

$$W(K_P) \cong W(L_Q)$$

if and only if there is a Hilbert-symbol-preserving isomorphism between the local square-class groups. In the sequel, we will occasionally refer to local Witt ring isomorphisms instead of talking about local symbol-preserving square-class maps. The reader may also wish to show that two local fields are Witt equivalent if and only if they have the same level and the same number of square classes. See also [**Ma**, pp. 95–97].

3. The condition $t(-1) = -1$ does not follow from the remaining conditions of either b) or c) of Harrison's Criterion, as can be seen by taking $K = Z/3Z$ and $L = Z/5Z$. For these fields, there are only two square-classes, and the unique square-class isomorphism fails to map -1 to -1, but satisfies the remaining assertions. The interested reader is invited to show that every square-class isomorphism preserving the value sets of binary forms *does* map -1 to -1 if and only if the Kaplansky radical of each of the underlying fields is trivial.

4. The equivalence of a) with c) of what we are calling *Harrison's Criterion* was Harrison's original statement in his unpublished notes ([**H**, p. 21]). We will refer to condition c) as *Harrison's condition,* and any square-class map t satisfying condition c) as a *Harrison map.* Cordes has modified condition c) to

   d). There is a square-class map t taking -1 to -1 and satisfying $tD_K\langle 1, a\rangle = D_L\langle 1, ta\rangle$ for all square-classes a, where $D_K\langle 1, a\rangle$ is the set of square-classes in K^*/K^{*2} represented over K by the form $\langle 1, a\rangle$.

   (see [**Co**]). Condition d) has become known as the *Harrison-Cordes* condition. We will not need d) in this paper.

5. The paper [**B-M**] shows that, with the proper interpretations, the equivalences a) through d) hold also in characteristic 2.

6. We have seen that any Witt ring isomorphism ϕ canonically induces a Harrison map $t = t_\phi$, and that any Harrison map t canonically induces a Witt ring isomorphism $\phi = \phi_t$. These are not inverse operations. In general $t \to \phi_t \to t_{\phi_t}$ returns to the original Harrison map t. But several different Witt ring isomorphisms ϕ can give rise to the same Harrison map t_ϕ. If we take the case $L = K$, there is a split exact sequence

$$1 \longrightarrow Kernel \longrightarrow Aut\ W(K) \rightleftarrows Harrison(K) \longrightarrow 1$$

from the group of automorphisms of $W(K)$ to the group of Harrison maps on K^*/K^{*2}. The kernel has been described in [**Le-Ma**, Prop. (3.4)]. For global fields, the kernel can be identified with the set of group homomorphisms from $K^*/\pm K^{*2}$ to I_K^2.

We close this section by using Harrison's Criterion to draw a consequence for global fields. The definition of reciprocity equivalence is given in the Introduction.

COROLLARY 1. *If global fields K and L are reciprocity equivalent, then they are Witt equivalent.*

PROOF. Let (t, T) be a reciprocity equivalence between K and L. If a binary form $\langle a, b\rangle$ represents 1 over K, then each Hilbert symbol $(a,b)_P$ equals 1 as P runs over the places of K. But $(a,b)_P = (ta, tb)_{TP}$, so all Hilbert symbols $(ta, tb)_{TP} = 1$. This means that the binary form $\langle ta, tb\rangle$ represents 1 everywhere locally, so by the Hasse principle, $\langle ta, tb\rangle$ represents 1 over L.

Next, we show that $t(-1) = -1$; from this, by Harrison's condition, the fields K and L are Witt equivalent. Now, for any square-class x in K^*/K^{*2}, the symbol $(x, -x)_P = 1$. Suppose that $t(-1) = c$ in L^*/L^{*2}. Then $1 = (x, -x)_P = (tx, ctx)_{TP} = (tx, -tx)_{TP}(tx, -c)_{TP} = (tx, -c)_{TP}$. So as tx varies over L^*/L^{*2} and as TP varies over the places of L the symbol $(tx, -c)_{TP} = 1$. By the non-degeneracy of Hilbert symbols, $-c = 1$ is a square locally everywhere. So, by the Global Square Theorem, $c = -1$. This proves the Corollary.

3. Localization

The previous section was concerned with general fields. Now we turn our attention to global fields of characteristic different from 2 and prove that every Witt ring isomorphism ϕ induces a canonical bijection between the sets of non-complex places of the underlying fields. We call this *localization.* It represents the essential step for our first main theorem, which will be proved in §4. The basic idea is simple. Fix a non-complex place P of K and let Θ_P be the canonical surjection from K^*/K^{*2} to the square-class group of the completion. Then $t = t_\phi$ maps $ker\Theta_P$ to a subgroup of L^*/L^{*2}. We will prove that this subgroup has the form $ker\Theta_Q$ for a unique non-complex place Q of L. Sending P to Q defines a bijection T' of non-complex places. It is not difficult to extend T' to a bijection T mapping the complex places as well. As simple as this idea is, it takes some work to carry it out. We do not *directly* show that t maps $ker\Theta_P$ to $ker\Theta_Q$ for some Q. Rather, we collect together those places P at which a given quaternion algebra $(a,b)_K$ is nonsplit, and show that t maps the union $\cup_P\ ker\Theta_P$ correctly. Then we use a combinatorial argument to show that t maps each constituent kernel to a corresponding kernel.

The details of this approach to the Localization Lemma are based on the following description of the group $Br_2(K)$ of a global field K. The reader should read Remark 1 near the end of §2 before proceeding. Let Γ_K denote the collection of finite even-order subsets of the non-complex places in the set Ω_K of all non-trivial places of K. The *symmetric difference* $A\Delta B = A \cup B \setminus A \cap B$ of two elements of Γ_K again lies in Γ_K, and (Γ_K, Δ) is an infinite abelian 2-torsion group, the identity element being the empty set. Identifying the Brauer class of a quaternion algebra with the set of places where it is nonsplit gives an isomorphism between $Br_2(K)$ and Γ_K. Namely, every class in $Br_2(K)$ is represented by a single quaternion algebra (see [**Lam**, Cor. 3.8, p. 171]), and the theorem of Brauer-Albert-Hasse-Noether shows that a quaternionic Brauer class is completely determined by the collection of places where it is nonsplit. By the *slash notation* $(a|b)_K = \{P \in \Omega_K \mid (a,b)_K \text{ is nonsplit locally at } P\}$ we denote the set in Γ_K associated with the quaternion algebra $(a,b)_K$.

Let an element Γ of Γ_K be given, and let a be a global square-class. We will say that a *represents* Γ when there is a global square-class b with $(a|b)_K = \Gamma$. For a square-class a to represent Γ, it is necessary and sufficient for a to be a local non-square at every $P \in \Gamma$. (A proof of this fact for number fields is found in [**O'M**, Cor. 71:19a]; the same proof is valid for global function fields of odd characteristic). The empty set is represented by any square-class whatsoever.

Now let $t : K^*/K^{*2} \longrightarrow L^*/L^{*2}$ and $u : \Gamma_K \longrightarrow \Gamma_L$ be any two group isomorphisms satisfying

$$(ta|tb)_L = u(a|b)_K$$

for all a, b in K^*/K^{*2}. Then there is the following commutative diagram that is reminiscent of diagram (2) in §2:

$$\begin{array}{ccc} K^*/K^{*2} \times K^*/K^{*2} & \longrightarrow & \Gamma_K \\ t\downarrow \quad \times \quad \downarrow t & & \downarrow u \\ L^*/L^{*2} \times L^*/L^{*2} & \longrightarrow & \Gamma_L \end{array} \tag{3}$$

Fix a non-complex place P of K and let

$$\Theta_P : K^*/K^{*2} \longrightarrow K_P^*/K_P^{*2}$$

be the canonical surjection. Fix Γ in Γ_K and consider the *non-representing set* N_Γ defined by

$$N_\Gamma = \{a \in K^*/K^{*2} | \ for\ all\ b \in K^*/K^{*2},\ \ (a|b)_K \neq \Gamma\}.$$

LEMMA 2. *Let K and L be global fields and let t and u be any group isomorphisms giving a commutative diagram (3). Then*

a) $t(N_\Gamma) = N_{u(\Gamma)}$.
b) *For non-complex places $P \neq Q$ of K we have $ker\Theta_P \neq ker\Theta_Q$* .
c) *If $\Gamma \in \Gamma_K$ is non-empty, then $N_\Gamma = \cup_{P\in\Gamma}\, ker\Theta_P$.*

PROOF. When Γ is empty, a) is vacuously true. When Γ is non-empty, then a) follows from the commutativity of diagram (3). With the Approximation

Theorem of valuation theory, there is a global square-class in K^*/K^{*2} that is locally a square at P but not at Q, proving b). And c) is a reformulation of the necessary and sufficient conditions stated above for a square-class to represent Γ. This proves Lemma 2.

For Γ in Γ_K, the non-representing set N_Γ is usually not itself a group, but is a union of subgroups of K^*/K^{*2} by Lemma 2 c) above. We say that a subset of N_Γ is a *maximal subgroup* if the subset is maximal in N_Γ with respect to being a subgroup of K^*/K^{*2} .

LEMMA 3. *Fix a non-empty* $\Gamma \in \Gamma_K$.

a) *For* $P \in \Gamma$, $ker\Theta_P$ *is a maximal subgroup of* N_Γ .
b) *If* $\Gamma = \{P, Q\}$ *then* $ker\Theta_P$ *and* $ker\Theta_Q$ *are the only maximal subgroups of* N_Γ .

PROOF. For a), suppose that $ker\Theta_P$ is not maximal. Then there is a square-class a in N_Γ but not in $ker\Theta_P$ such that the entire coset $a(ker\Theta_P)$ is contained in N_Γ . Choose a' in K^*/K^{*2} such that $\Theta_P(a') = \Theta_P(a) \neq 1$ and such that a' is a non-square locally at the remaining places in Γ. By part c) of Lemma 2, the square-class a' does not lie in N_Γ , contradicting the fact that a' must lie in $a(ker\Theta_P) \subseteq N_\Gamma$. Hence $ker\Theta_P$ is a maximal subgroup of N_Γ , proving a). Suppose b) is false. Then there is a subgroup G of K^*/K^{*2} contained in N_Γ but not contained in either $ker\Theta_P$ or in $ker\Theta_Q$. By part c) of Lemma 2, G can be written as a union $G = (G \cap ker\Theta_P) \cup (G \cap ker\Theta_Q)$ of *two* proper subgroups, which is impossible. This proves b) and Lemma 3.

The next lemma is the key for proving Theorem 1, found in the next section. Note that the hypotheses of the lemma are fulfilled whenever K and L are global fields with isomorphic Witt rings.

LOCALIZATION LEMMA. *Let* t *and* u *be any group isomorphisms giving a commutative diagram* (3). *Then there is a canonical 1-to-1 correspondence* T' *between the sets of non-complex places of* K *and of* L *such that*

$$u(\Gamma) = \{T'(P) \mid P \in \Gamma\}$$

for every Γ *in* Γ_K .

PROOF. Before we define T' let us observe that Lemma 3, together with part a) of Lemma 2 imply

$$|\Gamma| = 2 \Leftrightarrow |u(\Gamma)| = 2.$$

Now fix four distinct non-complex places P, Q, R, V of K. Then

$$\{P, Q\} \overset{u}{\mapsto} \{Q_1, Q_2\}$$

and

$$\{P, R\} \overset{u}{\mapsto} \{Q_3, Q_4\}$$

for places $Q_1 \neq Q_2$ and $Q_3 \neq Q_4$ of L. Taking symmetric differences, we see that u maps $\{Q, R\} = \{P, Q\}\Delta\{P, R\}$ to $\{Q_1, Q_2\}\Delta\{Q_3, Q_4\}$ which must therefore have cardinality two. It follows that $\{Q_1, Q_2\}$ and $\{Q_3, Q_4\}$ intersect in exactly one element, which we may assume to be $Q_1 = Q_4$, with Q_1, Q_2, Q_3 pairwise distinct. We have proved:

Whenever two 2-element sets in Γ_K have exactly one place in common, then their u-images have exactly one place in common.

Now consider $\{P, V\}$. This set has one element in common with each of $\{P, Q\}$ and $\{P, R\}$. So $u(\{P, V\})$ must be either $\{Q_1, Q_5\}$ or $\{Q_2, Q_3\}$. Suppose the latter. Then $u(\{P, Q\}\Delta\{P, R\}\Delta\{P, V\}) = \{Q_1, Q_2\}\Delta\{Q_1, Q_3\}\Delta\{Q_2, Q_3\} = 1$, while $\{P, Q\}\Delta\{P, R\}\Delta\{P, V\} = \{P, Q, R, V\} \neq 1$. Hence $u(\{P, V\} = \{Q_1, Q_5\}$. This shows that, as V varies, the sets $u(\{P, V\})$ all have Q_1 in common. Now define T' by $T'(P) = Q_1$, and for $V \neq P$ define $T'(V)$ to be the place different from Q_1 occurring in $u(\{P, V\})$. So T' is a map of (non-complex) places. Now, T' is a 1-to-1 correspondence, for if P_1 and P_2 are two places of K distinct from P, then $\{P_1, P_2\} = \{P, P_1\}\Delta\{P, P_2\}$ so $u(\{P_1, P_2\}) = \{Q_1, T'(P_1)\}\Delta\{Q_1, T'(P_2)\} = \{T'(P_1), T'(P_2)\}$ has cardinality two, showing that T' is injective. And T' is surjective: for Q_1 is $T'(P)$ and if Q' is distinct from Q_1, then $\{Q_1, Q'\}$ is the u-image of some two-element set Γ in Γ_K. If Γ contains P, then Q' is the T'-image of the other element. If Γ does not contain P, then write $\Gamma = \{P_2, P_3\} = \{P, P_2\}\Delta\{P, P_3\}$ so

$$\begin{aligned}\{Q_1, Q\} &= u(\{P, P_2\})\Delta u(\{P, P_3\}) \\ &= \{Q_1, T'(P_2)\}\Delta\{Q_1, T'(P_3)\} \\ &= \{T'(P_2), T'(P_3)\}\end{aligned}$$

So Q lies in the image of T'. Hence T' is a bijection between the non-complex places of K and of L. Finally, $u(\Gamma) = \{T'(P) \mid P \in \Gamma\}$ whenever Γ has cardinality two. Since such sets Γ generate the group Γ_K, and since both u and T' respect symmetric differences, it follows that $u(\Gamma) = \{T'(P) \mid P \in \Gamma\}$ for all Γ in Γ_K. This proves the Localization Lemma.

4. First main theorem and consequences

Throughout, global fields have characteristic not equal to 2.

Let t be an isomorphism between the square-class groups of two global fields K and L, and let P be a place, archimedean or not, of K. Pick a place, which we will call TP, of L and suppose that t preserves Hilbert symbols *locally* at P, *i.e.*,

$$(a, b)_P = (ta, tb)_{TP}$$

for all a, b in K^*/K^{*2}. Consider the diagram

$$
\begin{array}{ccccccc}
ker\Theta_P & \longrightarrow & K^*/K^{*2} & \xrightarrow{\Theta_P} & K_P^*/K_P^{*2} & \rightarrow & 1 \\
(4) & & \downarrow t & & \downarrow t_P & & \\
ker\Theta_{TP} & \longrightarrow & L^*/L^{*2} & \xrightarrow{\Theta_{TP}} & L_{TP}^*/L_{TP}^{*2} & \rightarrow & 1
\end{array}
$$

in which the vertical map t_P will be defined in a moment. Now, $ker\Theta_P$ consists of all square-classes a with $(a, x)_P = 1$ for all classes x in K^*/K^{*2}. Since the map t preserves Hilbert symbols, it follows that t sends $ker\Theta_P$ to $ker\Theta_{TP}$. This induces a local symbol-preserving isomorphism t_P of local square-class groups

$$K_P^*/K_P^{*2} \xrightarrow{t_P} L_{TP}^*/L_{TP}^{*2}$$

making diagram (4) commute.

Recall that the local square-class group K_P^*/K_P^{*2} has order

$$
(5) \qquad
\begin{array}{ll}
1 & \text{when } P \text{ is complex} \\
2 & \text{when } P \text{ is real} \\
4 & \text{when } P \text{ is non-archimedean and non-dyadic} \\
2^{d+2} > 4 & \text{when } P \text{ is dyadic, where } d = [K_P : Q_2]
\end{array}
$$

(see [**Lam**, Ch. 6, Th. 2.22, p. 161] when P is non-archimedean). Thus the local isomorphism t_P forces both P and TP to be real places, or both complex places, or both dyadic places. Moreover, when both P and TP are dyadic, then K and L are both number fields and the local degrees $[K_P : Q_2]$ and $[L_{TP} : Q_2]$ coincide. With this, we are ready to prove the following lemma.

LEMMA 4. *Let K and L be global fields, let t be an isomorphism between their square-class groups, and let T' be any bijection between the non-complex places such that the pair (t, T') preserves Hilbert symbols. Then*

a) *The map t induces local symbol-preserving isomorphisms*

$$K_P^*/K_P^{*2} \xrightarrow{t_P} L_{T'P}^*/L_{T'P}^{*2}.$$

b) *T' restricts to a bijection on the real places, and a bijection on the dyadic places.*
c) *K and L are both number fields or both function fields.*
d) *When K and L are number fields, then $[K : Q] = [L : Q]$.*
e) *T' sends primes P of K at which -1 is a local square to primes $T'P$ of L at which -1 is a local square.*
f) *The square-class map t uniquely determines the bijection T'.*

PROOF. Statements a) and b) follow from the comments preceding the lemma. Since global function fields of odd characteristic have no dyadic primes, and number fields always do, c) follows. For number fields, the local square-class isomorphisms at the dyadic primes imply that the local dyadic degrees coincide. Summing the local degrees at the dyadic primes shows that the global degrees are equal, proving d). For e), observe that t necessarily maps the square-class of -1 in K to -1 in L (copy the proof of Corollary 1 at the end of §2). If -1 is a local square at P then $(x,-1)_P = 1$ for all x. Hence $(tx,-1)_{T'P} = 1$ for all square-classes tx in L^*/L^{*2}, and this forces -1 to be a local square at $T'P$. For f), suppose that (t, T_1') and (t, T_2') are two symbol-preserving pairs, with the same square-class map t. Composing the first with the inverse of the second yields a symbol-preserving *self-equivalence* on K, of the form (id, T''). If T'' is not identically the identity, then choose a prime $P \neq T''P$. Pick a to be a local square at P but not at $T''P$. Then all Hilbert symbols $(a,x)_P$ are trivial while some symbols $(a,x)_{T''P}$ are not trivial. This contradicts the fact that (id, T'') preserves Hilbert symbols, proving f) and Lemma 4.

This brings us to our first main theorem.

THEOREM 1. *Two global fields K, L are Witt equivalent if and only if they are reciprocity equivalent.*

PROOF. Corollary 1 at the end of §2 shows that reciprocity equivalent number fields are Witt equivalent. For the other direction, let ϕ be an isomorphism from $W(K)$ to $W(L)$. Let $t = t_\phi$ be the canonically induced square-class map (see §2) and let T' be the canonical bijection between the non-complex places given by the Localization Lemma. By Lemma 4, both K and L are function fields or neither are. If both are, then (there being no complex primes) (t, T') is a reciprocity equivalence, as desired. If both fields are number fields, then they have the same degree over Q, and the same number, r_1, of real places, by Lemma 4. Hence they have the same number, r_2, of complex places as well, since $r_1 + 2r_2$ equals the global degree. Extend T' to a bijection T of all places by choosing any bijection between the complex places. All Hilbert symbols vanish at complex places, so (t, T) is the desired reciprocity equivalence.

For future reference, we observe that Theorem 1 together with part d) of Lemma 4 imply

COROLLARY 2. *The field Q of rational numbers is distinguished from among all other global fields, including those of characteristic two, by the structure of the ring $W(Q)$.*

REMARKS.

1. We want to explicitly point out the difference between a global Harrison map and a reciprocity equivalence. By definition, a Harrison map is a global square-class map t satisfying $t(-1) = -1$ and, given a pair a, b of square-classes in K, then *all* Hilbert symbols $(a,b)_P$ vanish precisely when *all* Hilbert symbols $(ta, tb)_Q$ vanish. This is *formally less* than a reciprocity equivalence, which requires there to be a bijection T so that corresponding Hilbert symbols agree, regardless of whether all symbols vanish or not. Our proof of Theorem 1 shows

that every Harrison map t is the *front end* of a reciprocity equivalence. That is, given a Harrison map t, then there is always a bijection T such that (t, T) is a reciprocity equivalence.

2. It is worthwhile to observe explicitly that *any* Witt equivalent fields have the same *level.* The additive order of the multiplicative identity $\langle 1 \rangle$ in $W(F)$ is twice the level of the field F (see [M-H], Th. (4.5), Ch. III), so even an isomorphism between the additive Witt groups preserves the level of the underlying fields.

We invite the reader to solve

Exercise:

Call a global field *lonely* if it is not Witt equivalent to any other global field. Prove that Q and $Q(\sqrt{-1})$ are the only lonely global fields.

5. Tame and wild

This short section is simply to introduce a refinement into the notion of reciprocity equivalence. We separate reciprocity equivalences into two types, called *tame* and *wild.* This has nothing to do with tame and wild ramification, but after some initial hesitation, we decided that these names are appropriate. It will be shown in a later paper that tame reciprocity equivalence preserves many arithmetic properties of the underlying fields. While we will not pursue these topics here, we will be careful in the next section to set up the basic theory in such a way that tame and wild are taken into consideration.

We say that a reciprocity equivalence (t, T) is *tame at the non-archimedean place P* when

$$ord_P(a) \equiv ord_{TP}(ta) \pmod 2$$

for all square-classes a; otherwise the equivalence is *wild* at P. Thus, to say (t, T) is tame at P means that when a global square-class $\overline{x}$ contains a *local* prime element x at P, then $t(\overline{x})$ contains a *local* prime element at TP.

The equivalence (t, T) is *tame* when it is tame for *all* non-archimedean P. Otherwise, (t, T) is *wild*, and the *wild set* of (t, T) consists of all P where the equivalence is wild. Carpenter has shown that when K and L are reciprocity equivalent, then there is an equivalence with a finite wild set (see [**C2**]). Under the same hypothesis, Palfrey has shown that there is always an equivalence with an infinite wild set (see [**Pa**]).

6. Small equivalence

We turn our attention in this section to the problem of producing reciprocity equivalences. We would like to also keep in mind the problem of making the equivalence *tame*, or at least tame outside a finite set of primes. Let S be a non-empty finite set of primes of K containing all archimedean and dyadic primes. We say that S is *sufficiently large* when the class number of the ring

$$O_K^S = \{x \in K \mid ord_P(x) \geq 0 \text{ for all } P \text{ outside } S\}$$

of S-integers of K is odd. To force this latter condition, it is sufficient to take any set of generators of the Sylow 2-subgroup of the S-class group and add these additional primes to S. We let U_K^S denote the unit group of O_K^S.

A small S-equivalence between K and L is a composite object consisting of the following four items:

1. A 1-1 correspondence T from a sufficiently large set S of primes of K to a sufficiently large set TS of primes of L;
2. A group isomorphism $t_S : U_K^S/(U_K^S)^2 \longrightarrow U_L^{TS}/(U_L^{TS})^2$;
3. For each prime P in S a symbol-preserving isomorphism t_P from K_P^*/K_P^{*2} onto L_{TP}^*/L_{TP}^{*2};
4. A commutative diagram

$$\begin{array}{ccc} U_K^S/(U_K^S)^2 & \xrightarrow{\text{diag}} & \prod_{P\in S} K_P^*/K_P^{*2} \\ \downarrow t_S & & \downarrow \Pi t_P \\ U_L^{TS}/(U_L^{TS})^2 & \xrightarrow{\text{diag}} & \prod_{P\in S} L_{TP}^*/L_{TP}^{*2} \end{array}$$

Observe that for any x, y in $U_K^S/(U_K^S)^2$ there is a small reciprocity law, in the sense that

$$\prod_{P\in S} (x,y)_P = 1,$$

since all symbols with x and y at primes outside of S vanish.

We say the small S-equivalence is *tame* when each of the local maps t_P is tame, *i.e.*, when each t_P preserves P-orders mod 2. As always, complex infinite primes play no role whatsoever, and can be ignored. Although the definition of small equivalence requires a lot, every set and group involved is finite, so a small equivalence, when one exists, is a finite object.

Our second main result is

THEOREM 2. *A small S-equivalence between global fields K and L can be extended to a reciprocity equivalence. The extension can always be chosen to be tame outside S.*

LEMMA 5. *Let S be a sufficiently large set of primes, as defined above.*

a) *A square-class a in K^*/K^{*2} lies in $U_K^S/(U_K^S)^2$ if and only if $ord_P(a) \equiv 0 \ (mod\ 2)$ for every P outside of S.*
b) *The map $U_K^S/(U_K^S)^2 \longrightarrow \prod_{P\in S} K_P^*/K_P^{*2}$ is injective.*

PROOF.

a). One implication is obvious. For the other, the O_K^S-ideal generated by a is, by assumption, a product

$$aO_K^S = \prod Q_i^{2m_i}$$

with prime ideals Q_i of O_K^S appearing to *even* powers. Since S is sufficiently large, the class number $h = h_S$ of O_K^S is odd. Take h-th powers. Then a and a^h lie in the same square-class, and $Q_i^h = q_i O_K^S$ is principal. So

$$a \cdot square = u \cdot \prod (q_i^{m_i})^2$$

for some S-unit u. So the square-class of a is represented by the S-unit u, proving a).

b). The S-class number is the order of Hilbert S-class field of K, that is, the order of the maximal abelian unramified extension of K in which the primes in S split completely. Take u in the kernel of the given map. Then u is a local square at every P in S, and this means that the primes in S split completely in the extension $K(\sqrt{u})/K$. Now u is a local unit at any prime outside of S. Any prime outside S is finite and non-dyadic, and therefore is unramified in $K(\sqrt{u})$. Thus $K(\sqrt{u})/K$ lies in the Hilbert S-class field of odd degree over K. Hence $K(\sqrt{u}) = K$, so the square-class u is trivial, showing the map to be injective.

Before embarking on the proof of Theorem 2, we remark that the process of extending a small equivalence is not unique. We will take efforts to insure that our extension does not introduce any new primes to the wild set. In [**Pa**], Palfrey has quantified the extent to which it is possible to extend wildly, producing exotic reciprocity equivalences with infinite wild sets (see [**Pa**]). Now we turn to the proof of the theorem.

PROOF OF THEOREM 2. Let Q be a finite prime of K outside S, and set $S' = S \cup \{Q\}$. We will carefully choose a prime Q' outside of TS and show that the given S-equivalence extends to an S'-equivalence mapping Q to Q'. Once this is established, the theorem will follow easily.

The class number $h = h_K(S)$ is odd, since S is given to be sufficiently large and Q^h is a principal O_K^S-ideal: $Q^h = qO_K^S$. Thus

$$ord_Q(q) \equiv 1 \pmod 2 \quad \text{and} \quad ord_P(q) \equiv 0 \pmod 2$$

for all finite primes P outside of $S' = S \cup \{Q\}$. Therefore

$$U_K^{S'}/(U_K^{S'})^2 = U_K^S/(U_K^S)^2 \ \cup \ q(U_K^S/(U_K^S)^2).$$

The element q lies in each completion K_P, so for P in S there is the local square-class of $t_P(q)$ in L_{TP}^*/L_{TP}^{*2}. Two elements that are locally near each other lie in the same local square-class, so by the Approximation Theorem ([**L**], p. 35) there is α in L^* whose local square-class is $t_P(q)$ for each P in S. Define a cycle m of L by

$$m = 4 \cdot \prod TP,$$

P running over S.

The group $I(m)$ of fractional O_L-ideals prime to the cycle m corresponds to the group of fractional O_L^{TS} ideals. So the ideal αO_L^{TS} gives rise to a class X in the generalized ideal class group $I(m)/P_m$. Like every generalized ideal class, X contains infinitely many prime ideals for number fields, (see [**L**, p. 166–167] for number fields, or [**We**, XIII, §11], in general). Let Q' be a prime in the class X; we view Q' as a prime in O_L^{TS}. By definition,

$$Q' = (\alpha\lambda)O_L^{TS}$$

for some λ in L^*, with $\lambda \equiv 1 \pmod{^* m}$. Then λ is a local square at each prime TP with P in S, by [**O'M**, Cor. 63:1a] (the presence of the factor 4 in m makes things work when TP is dyadic). Define $q' = \alpha\lambda$. Then q' lies in the square-class $t_P(q)$ for each P in S, and q' satisfies $ord_{Q'}(q') = 1$ and $ord_B(q') = 0$ for all finite primes B of L outside of $TS \cup \{Q'\}$. We set $TQ = Q'$ and $TS' = TS \cup \{TQ\}$. Then the TS'-units, modulo squares, are given by

$$U_L^{TS'}/(U_L^{TS'})^2 = U_L^{TS}/(U_L^{TS})^2 \;\cup\; q'(U_L^{TS}/(U_L^{TS})^2).$$

We want to show that the given S-equivalence t_S extends to an S'-equivalence $t_{S'}$ that is tame at Q.

We set $t_{S'}(q) = q'$ and $t_{S'}(x) = t_S(x)$ for x in $U_K^S/(U_K^S)^2$. It remains to define a local map t_Q, to check that this map preserves the local symbol at Q, and then to check that the appropriate diagram commutes.

Now the primes Q and $TQ = Q'$ are finite and non-dyadic, so the local square-class groups are given by

$$K_Q^*/K_Q^{*2} \;=\; \{1, v, q, vq\} \quad \text{and} \quad L_{TQ}^*/L_{TQ}^{*2} = \{1, v', q', v'q'\}$$

where v and v' are local non-square units of K_Q and L_{TQ}, respectively. We define the local square-class isomorphism t_Q by sending $1 \to 1, v \to v', q \to q'$, and $vq \to v'q'$. Then t_Q is tame, and preserves symbols: Since v and v' necessarily generate the respective quadratic unramified extensions of K and of L, and since any local unit is a norm from an unramified extension, it follows that each of the symbols $(v,v)_Q$, $(v',v')_{TQ}$ is 1. By the same reasoning, the symbols $(v,q)_Q$, $(v',q')_{TQ}$ equal -1. This forces $(v,vq)_Q = (v',v'q')_{TQ} = -1$, and the only symbols remaining to check are $(q,q)_Q = (q,-1)_Q$ and $(q',q')_{TQ} = (q',-1)_{TQ}$. Now by reciprocity

$$(q,-1)_Q \prod (q,-1)_P = 1 \quad \text{and} \quad (q',-1)_{TP} \prod (q',-1)_{TP} = 1$$

as P runs over S, since the symbols are 1 at primes other than these. But the element q' was chosen to lie in the same local square-class as $t_P(q)$ for each P in S, from which it follows that

$$(q,-1)_P = (q',-1)_{TP}.$$

To establish that we have an S'-equivalence, it remains to check commutativity of the diagram

$$\begin{array}{ccccc} U_K^S/(U_K^S)^2 & \cup & q(U_K^S/(U_K^S)^2) & \xrightarrow{\text{diag}} & K_Q^*/K_Q^{*2} \times \prod_{P\in S} K_P^*/K_P^{*2} \\ t_S\downarrow & & \downarrow t_S & & t_Q\downarrow \qquad \downarrow \Pi t_P \\ U_L^{TS}/(U_L^{TS})^2 & \cup & q'(U_L^{TS}/(U_L^{TS})^2) & \xrightarrow{\text{diag}} & L_{TQ}^*/L_{TQ}^*2 \times \prod_{P\in S} L_{TP}^*/L_{TP}^{*2} \end{array}$$

Now take a square-class x in $U_K^S/(U_K^S)^2$. We already have

$$t_{S'}(x) = t_S(x) = t_P(x)$$

for each prime P in S. At the prime Q, the local square-class of x is either 1 or v, depending on whether the symbol $(x, q)_Q$ is 1 or -1. Similarly, $t_{S'}(x) = t_S(x)$ is either 1 or v', depending on whether the symbol $(t_S(x), q')_{TQ}$ is 1 or -1. So we want these symbols to be equal. By reciprocity, we have

$$\begin{aligned} (x,q)_Q = \prod_{P\in S}(x,q)_P &= \prod_{P\in S}(t_P(x), t_P(q))_{TP} \\ &= \prod_{P\in S}(t_{S'}(x), q')_{TP} = (t_{S'}(x), q')_{TQ}. \end{aligned}$$

Next, take the square-class of q. Surely

$$t_{S'}(q) = t_Q(q) = q' \text{ in } L_{TQ}^*/L_{TQ}^{*2} \,.$$

And for a prime P in S, we have $t_{S'}(q) = q' = t_P(q)$ by construction of q'. This completes the proof that the given small S-equivalence extends to a small S'-equivalence tame outside of S.

Now, add a prime of L outside ot TS' and tamely extend the inverse equivalence $(t_{S'}^{-1}, T^{-1})$ to a small equivalence on this larger set of primes. Then add a prime of K and extend the equivalence. If we imagine the remaining primes of K and of L written out in a sequence, and if we continue to extend the equivalence by adding the first remaining prime of K, extending, and then the first remaining prime of L and extending, then we see that we have set up at $1-1$ correspondence of the primes of K and L. This correspondence restricts to a small equivalence on any finite set of primes of K containing the original set S. Since any global square-class x in K^*/K^{*2} lies in the group of S''-units for some finite set S'' containing S, we can define an isomorphism t from K^*/K^{*2} to L^*/L^{*2} by setting $t(x) = t_{S''}(x)$. This definition does not depend on the choice of S''. Then (t, T) is a reciprocity equivalence extending the given S-equivalence, and tame outside S. This proves Theorem 2.

By Theorem 2, small equivalences can be used to produce reciprocity equivalences with finite wild sets. Conversely, suppose we are given a reciprocity equivalence (t, T) which is known to be tame outside a *finite* set of primes. Then we can always enlarge that set so that both it and its image under T are sufficiently large in the sense of the definition above. Being tame outside this enlarged set,

S, the pair (t, T) maps S-units of K to TS-units of L, by part a) of Lemma 5, producing the ingredients necessary for a small equivalence. Thus we see

COROLLARY 3. *There is a reciprocity equivalence between global fields K and L that is tame outside a finite set if and only if there is a small equivalence between K and L. There is a tame reciprocity equivalence if and only if there is a tame small equivalence.*

REMARK. It remains open whether every reciprocity equivalence with an *infinite* wild set arises by (wildly) extending some small equivalence.

We conclude §6 by showing how Theorem 1 and Theorem 2 combine with Carpenter's results, mentioned in the introduction, to obtain a local-global principle for Witt equivalence of global fields. We claim:

Two global fields are Witt equivalent if and only if their places can be paired so that corresponding completions are Witt equivalent.

Neither implication is trivial. In one direction, if K and L are Witt equivalent global fields, then they are reciprocity equivalent, by Theorem 1, so there are local Hilbert-symbol preserving square-class isomorphisms

$$t_P : K_P^*/K_P^{*2} \longrightarrow L_{TP}^*/L_{TP}^{*2} .$$

By Remark 2 of §2, these maps t_P canonically induce isomorphisms between the Witt rings of the completions. Hence global Witt equivalence implies local Witt equivalence. Conversely, let $P \leftrightarrow TP$ be a bijection of places, and let $\phi_P : W(K_P) \to W(L_{TP})$ be a family of local Witt ring isomorphisms. We do *not* claim that the given ϕ_P's are induced by a global Witt ring isomorphisms. However, the *level* of K is the maximum of the levels of the K_P, which equals one-half the additive order of $\langle 1 \rangle$ in $W(K_P)$. It follows from the local isomorphisms that K and L have the same level. This is Carpenter's condition a). Similarly, the local Witt ring isomorphisms ϕ_P canonically induce local square-class isomorphisms mapping -1 to -1. These local square-class isomorphisms guarantee that the dyadic places P of K correspond to dyadic places TP of L, and real archimedean places of K correspond to real archimedean places of L, by display (5) of §4. By the same display, we conclude that the local degrees at corresponding dyadic places agree. Finally, by Remark 2 at the end of §4, the levels of corresponding dyadic completions agree. Hence all of Carpenter's conditions a), b), and c) are satisfied, allowing us to conclude that there is a small equivalence between K and L. By Theorem 2, the fields K and L are reciprocity equivalent, and then Theorem 1 implies that these fields are Witt equivalent.

7. Examples

We begin this section with an example involving a cubic field. If we forget the multiplication in the Witt ring, then we can still ask for isomorphisms of the additive Witt groups. It is not our purpose here to point out everything that is preserved by an additive group isomorphism, but we wish to recall that the Witt group of a number field K is additively the direct sum of a free abelian group of rank r, where r is the number of real embeddings of K, with the torsion subgroup $W(K)_{tor}$ of $W(K)$ (see [**Lam**, Cor. 3.9, p. 172]). This implies that any *additive isomorphism* of Witt groups of number fields at least preserves the field degree modulo 2. We present the following example showing that the additive structure of the Witt group alone does not preserve the field degree over Q. This situation is typical: Take $K = Q(\sqrt[3]{2})$. Then K has a real embedding so $W(K)$ has a signature.

From this, it follows that every (additive) torsion class in $W(K)$ has even rank, and has additive order either 2 or 4 (see [**C-P**, Lemma I.2.6]). Moreover, there are infinitely many torsion classes of each of these orders, by the Knebusch exact sequence. Thus the additive Witt group $W(K)$ is the direct sum of an infinite cyclic group with countably many cyclic groups of order 2 and countably many cyclic groups of order 4. The same applies to the Witt group $W(Q)$ of the rationals. So $W(K) \cong W(Q)$ as additive groups. By Corollary 2 of §4, we know that $W(K)$ is *not* isomorphic to $W(Q)$ as rings. However, we would not know how to prove this (or even to *suspect* its truth) without the results of this paper.

Quadratic fields have been completely classified up to reciprocity equivalence by Carpenter (see [**C1**], [**C2**]) and by Czogała (see [**Cz1**], [**Cz2**]). This can be done by using the notion of small equivalence directly, or by using Carpenter's reduction of the conditions defining small equivalence given in the Introduction. For the convenience of the reader, we state the classification of quadratic fields up to Witt equivalence here.

THEOREM 3. *There are exactly seven Witt equivalence classes of quadratic number fields, represented by $Q(\sqrt{d})$ for $d = -1, \pm 2, \pm 7$, and ± 17. Given a square-free integer $n \neq 1$, then the quadratic field $Q(\sqrt{n})$ is Witt equivalent to $Q(\sqrt{d})$ with d determined as follows: $d = -1$ if $n = -1$, and if $n \neq -1$, then*

$$d = \begin{cases} \operatorname{sign}(n) \cdot 2 & \text{if } |n| \equiv 2, 3, 5, 6 \pmod 8; \\ \operatorname{sign}(n) \cdot 7 & \text{if } |n| \equiv 7 \pmod 8 \\ \operatorname{sign}(n) \cdot 17 & \text{if } |n| \equiv 1 \pmod 8). \end{cases}$$

We won't prove this theorem here. We do mention that a real quadratic field cannot be Witt equivalent to a complex quadratic field, since Witt equivalence implies the fields have the same number of real embeddings. We also remark that Theorem 3 might lead the observant reader to suspect that when $K(\sqrt{a})$ and $K(\sqrt{b})$ are Witt equivalent, then so are $K(\sqrt{-a})$ and $K(\sqrt{-b})$. This conclusion is not always correct; conditions under which it does hold are discussed in [**Sz3**].

The reader might wish to verify that the quadratic fields $K = Q(\sqrt{-3})$ and $L = Q(\sqrt{-6})$ are Witt equivalent, without appealing to Theorem 3, by explicitly establishing a small equivalence between them. The problem of classifying quadratic fields up to tame equivalence is much more subtle than that of plain Witt equivalence, and has been accomplished by Czogała (see [**Cz2**]). We do not give his classification here, other than to mention that he has shown there are infinitely-many tame classes, in sharp contrast to the seven Witt equivalence classes mentioned above.

The next example is to show that the bijection T in a reciprocity equivalence (t, T) cannot generally be expected to map primes P lying over a given rational prime p to primes TP lying over the same prime p. Let K and L be distinct Witt equivalent quadratic fields, and let (t, T) be a reciprocity equivalence. If T maps primes over p to primes over p for all rational primes p, or even for all p outside a fixed set of rational primes of density zero, then, up to this set of density zero, the same rational primes are totally split in the quadratic fields K and in L, and this forces $K = L$, by well-known results in algebraic number theory (see, *e.g.*, [**L**, Theorem 9, p. 168]).

Next, we recall that two number fields are *arithmetically equivalent* when their Dedekind zeta functions coincide. A reference for what follows is [**P**]. Non-isomorphic quadratic fields cannot be arithmetically equivalent, but can be Witt equivalent. So Witt equivalence does not imply arithmetic equivalence. Conversely, one can ask whether arithmetic equivalence implies Witt equivalence, that is, does the zeta function determine the structure of the Witt ring? In fact, the zeta function *does* determine the number of real embeddings, and the number of dyadic primes. So, by Carpenter's conditions, the only issues are the local dyadic degrees, the global level, and the local dyadic levels. Let

$$K = Q(\sqrt[8]{97}) \quad \text{and} \quad L = Q(\sqrt[8]{16 \cdot 97}) \ .$$

It is shown in [**P**, p. 351], that K and L are nonisomorphic arithmetically equivalent fields of degree 8. Moreover, the rational prime 2 splits into four prime ideals, each of inertia degree 1, in K and in L. However, in K, one of these prime ideal factors of 2 has local dyadic degree 1, while in L, all four factors of 2 have local dyadic degree 2. So, by Carpenter's conditions, these arithmetically equivalent fields are not Witt equivalent.

Simply to illustrate the power of small equivalence, we close by giving the following example:

For any positive prime $p \equiv 3 \pmod 8$ the complex quadratic field $Q(\sqrt{-p})$ is tamely equivalent to $Q(\sqrt{-2})$.

PROOF. The rational prime 2 is inert in $K = Q(\sqrt{-p})$, and thus the set $S = \{(2)\}$ is a sufficiently large set in K in the sense of §6, there being no real infinite primes and the class number already being odd. Then $\{-1, 2\}$ is a set of generators for the group of S-units of K, modulo squares. Let v be a local unit in K_2 whose square root generates the quadratic unramified extension of K_2. Since any unit in K_2 is a norm from an unramified extension, we see that in K_2 the element v makes a Hilbert symbol of 1 with any local unit. We easily check that

$$(-1, \sqrt{-p})_2 = -1,$$

so v does not represent the square-class of either -1 or of $\sqrt{-p}$). Then the 16 square-classes in K_2/K_2^{*2} are generated by

$$-1, \quad \sqrt{-p}, \quad v, \quad 2.$$

Similarly, $S' = \{P\}$ with $P = (\sqrt{-2})$ is a sufficiently large set of primes of $L = Q(\sqrt{-2})$, and $\{-1, \sqrt{-2}\}$ generates the group of S'-units of L, modulo squares. And the local square-class group of L is generated by the four classes

$$-1, \quad 1+\sqrt{2}, \quad v', \quad \sqrt{-2},$$

where v' denotes the square-class generating the unramified quadratic extension of $L_{P'}$. Of course, v' is the square-class of 5, but we won't need to know this. Define a group isomorphism

$$K_2^*/K_2^{*2} \xrightarrow{t_2} L_{P'}^*/L_{P'}^{*2}$$

by mapping each given generator above to the generator directly below it. This then restricts to an isomorphism between the S-units modulo squares of K and the S'-units modulo squares of L. All that remains to be checked is that t_2 preserves symbols, for tameness was built in automatically. One can easily check the following symbols in K_2, where we have omitted the subscript 2:

$$(v, \textit{ local unit}) = (-1,-1) = (-1,2) = 1;$$

$$(v,2) = (-1,\sqrt{-p}) = (\sqrt{-p},2) = -1.$$

The corresponding symbols in $L_{P'}$ read

$$(v', \textit{ local unit}) = (-1,-1) = (-1,\sqrt{-2}) = 1;$$

$$(v',\sqrt{-2}) = (-1, 1+\sqrt{-2}) = (1+\sqrt{-2},\sqrt{-2}) = --1.$$

These symbols generate all the symbols, allowing us to conclude that the map t_2 preserves symbols. This produces a small tame equivalence between K and $Q(\sqrt{-2})$, completing the proof.

References

[A-E-J] J. K. Arason, R. Elman, and B. Jacob, *Rigid elements, valuations and realization of Witt rings*, J. Algebra **110** (1987), 449–467.

[B-M] R. Baeza and R. Moresi, *On the Witt-equivalence of fields of characteristic 2*, J. Algebra **92** (1985), 446–453.

[C1] J. Carpenter, *Finiteness theorems for forms over number fields*, Ph.D. thesis, Louisiana State University, 1989, pp. 1–68.

[C2] ———, *Finiteness theorems for forms over global fields*, Math. Zeit. **209** (1992), 153–166.

[C-P] P. E. Conner and R. Perlis, *A survey of trace forms of algebraic number fields*, Series in Pure Math, vol. 2, World Scientific Publishing Co., 1984.

[Co] C. Cordes, *The Witt group and the equivalence of fields with respect to quadratic forms*, J. Algebra **26** (1973), 400–421.

[Cz1] A. Czogała, *Witt rings of algebraic number fields*, Ph.D. thesis, Silesian University, Katowice, 1987. (Polish)

[Cz2] ———, *On reciprocity equivalence of quadratic number fields*, Acta Arith. **58** (1991), 27–46..

[Cz3] ———, *Witt equivalence of quadratic extensions of global fields*, Math. Slovaca **41** (1991), 251–256.

[H] D. K. Harrison, *Witt rings*, University of Kentucky Notes, Lexington, Kentucky, 1970.

[J-Mar] S. Jakubec and F. Marko, *Witt equivalence classes of quartic number fields*, Math. Comp. **58** (1992), 355–368.

[Lam] T. Y. Lam, *Algebraic theory of quadratic forms*, Benjamin, Reading, Massachusetts, 1973.

[L] S. Lang, *Algebraic number theory*, Addison-Wesley, Massachusetts, 1970.

[Le-Ma] D. Leep and M. Marshall, *Isomorphisms and automorphisms of Witt rings*, Bull. Can. Math. Soc. **31** (1988), 250–256.

[Ma] M. Marshall, *Abstract Witt rings*, Queen's Papers in Pure and Applied Math. No 57., Kingston, Ont., 1980.

[Me] A. S. Merkur'ev, *On the norm residue symbol of degree* 2, Soviet Math. Dokl. **24** (1981), 546–551.

[M] J. Milnor, *Algebraic K-theory and quadratic forms*, Invent. Math. **9** (1970), 318–344.

[M-H] J. Milnor and D. Husemoller, *Symmetric bilinear forms*, Springer-Verlag, Berlin, 1973.

[O'M] O. T. O'Meara, *Introduction to quadratic forms*, Grundlehren der mathematischen Wissenschaften, 117, Springer-Verlag, Berlin, 1963.

[Pa] T. Palfrey, *Density theorems for reciprocity equivalences*, Ph.D. thesis, Louisiana State University, 1989, pp. 1-27.

[P] R. Perlis, *On the equation* $\zeta_K(s) = \zeta_{K'}(s)$,, J. Number Theory **9** (1977), 342–360.

[Sz1] K. Szymiczek, *Witt equivalence of algebraic number fields*, 9th Czech. Colloquium on Number Theory, Račkova Dolina, 1989, pp. 104–113.

[Sz2] ———, *Witt equivalence of global fields*, Comm. in Alg. **19** (1991), 1125–1149.

[Sz3] ———, *Witt equivalence of global fields*, II. Relative quadratic extensions (to appear).

[Sz4] ———, *Matching Witts locally and globally*, Math. Slovaca **41** (1991), 315–330.

[Wa] R. Ware, *Valuation rings and rigid elements in fields*, Can. J. Math. **32** (1981), 1338–1355.

[We] A. Weil, *Basic number theory*, Grundlehren der mathematischen Wissenschaften, 144, Springer-Verlag, Berlin, 1974.

[W] E. Witt, *Theorie der quadratischen Formen in beliebigen Körpern*, J. Reine Angew. Math. **176** (1937), 31–44.

(R. Perlis, P. E. Conner and R. Litherland) DEPARTMENT OF MATHEMATICS, LOUISIANA STATE UNIVERSITY, BATON ROUGE, LA 70803

(K. Szymiczek) INSTITUTE OF MATHEMATICS, SILESIAN UNIVERSITY, BANKOWA 14, 40–007 KATOWICE, POLAND

Contemporary Mathematics
Volume **155**, 1994

On Witt-Kernels of Function Fields of Curves

JONATHAN SHICK

ABSTRACT. We study the Witt-kernel of the algebraic function field of a curve as an extension of its field of constants. We give results relating this kernel to the Witt-group of the curve, as well as to the presence of the weak Hasse-Minkowski principle for the function field. We derive various conditions under which the Witt-kernel vanishes. We show that the kernel is generated by 2-fold Pfister forms in numerous cases when the field of constants is global. We examine the case of elliptic and hyperelliptic curves. For an elliptic curve without rational points, we show that the 2-fold Pfister forms in the Witt-kernel may be described in terms of points on an associated elliptic curve. Various examples are given.

Introduction

Let $W(F)$ be the Witt ring of anisotropic quadratic forms over a field F of characteristic not 2. If $F \subseteq K$ is any extension of fields, let $r^*_{K/F}: W(F) \to W(K)$ denote the scalar extension map and $W(K/F)$ the kernel of $r^*_{K/F}$, the *Witt-kernel* of the extension. Such kernels have been of particular interest in the algebraic theory of quadratic forms. For relevant basic material and notation on quadratic forms we refer the reader to [L] or [Sch].

In this paper we study $W(K/k)$ in the case of K being an algebraic function-field in one variable with field of constants k, i.e., $k \subset K$ is a finitely-generated field extension of transcendence degree one with k algebraically closed in K. Let $\mathcal{C} = C(K,k)$ denote the abstract nonsingular curve consisting of the set of discrete valuation rings of K containing k. Then $K = k(\mathcal{C})$, the function field of $\mathcal{C}$. Now, there is a complex

$$0 \longrightarrow W(k) \xrightarrow{r^*_{k(\mathcal{C})/k}} W(k(\mathcal{C})) \xrightarrow{\partial_{\mathcal{C}}} \coprod_{p \in \mathcal{C}} W(k(p)) \tag{1}$$

1991 *Mathematics Subject Classification.* 11E04 11E81 14H45.

This paper is submitted in final form and no version of it will be submitted for publication elsewhere

where $k(p)$ is the residue field of the discrete valuation ring associated with $p \in \mathcal{C}$, and $\partial_{\mathcal{C}}$ consists of second residue maps with respect to some choice of uniformizing parameters.

We define $H_0(k(\mathcal{C}),k)$ and $H_1(k(\mathcal{C}),k)$ as the homology at $W(k)$ and at $W(k(\mathcal{C}))$, respectively, of the complex (1). So $H_0(k(\mathcal{C}),k) = W(k(\mathcal{C})/k)$. The $H_i(k(\mathcal{C}),k)$ are $W(k)$-modules, independent of the choice of uniformizing parameters, and $H_1(k(\mathcal{C}),k)$ is closely related to the Witt-group of the curve, $W(\mathcal{C})$ (cf. [Kn$_2$]). More precisely,

$$0 \longrightarrow W(\mathcal{C}) \longrightarrow W(k(\mathcal{C})) \xrightarrow{\partial_{\mathcal{C}}} \coprod_{p \in \mathcal{C}} W(k(p))$$

is exact ([Kn$_1$], Satz 13.3.6), so we have an exact sequence:

$$0 \longrightarrow H_0(k(\mathcal{C}),k) \longrightarrow W(k) \longrightarrow W(\mathcal{C}) \longrightarrow H_1(k(\mathcal{C}),k) \longrightarrow 0$$

Let $\gamma = \gamma(x)$ be square-free in $k[x]$. If $\mathcal{C} = \mathcal{C}_\gamma \to \mathbb{P}^1$ is an elliptic or hyperelliptic curve described by the equation $y^2 = \gamma(x)$, let $H_i(\gamma, k)$ denote $H_i(k(\mathcal{C}_\gamma),k)$ in this case.

In section 1 we give some results on H_0 involving H_1 for curves $\mathcal{C}$ in general, and relate this to the validity of the weak Hasse-Minkowski principle for $k(\mathcal{C})$. As a corollary we derive various conditions under which $H_0(k(\mathcal{C}),k)$ vanishes. We also show that $H_0(k(\mathcal{C}),k)$ is generated by 2-fold Pfister forms in numerous cases when k is a global field. In section 2 we obtain some results on $H_0(\gamma,k)$ for the curves $\mathcal{C}_\gamma$, giving many cases in which it is trivial (eg., when $deg(\gamma)$ is odd), or a principal ideal of $W(k)$ generated by a 2-fold Pfister form. In section 3 we examine $H_0(\gamma,k)$ when $deg(\gamma)$ is even. We give a criterion for when a Pfister form is in $H_0(\gamma,k)$, and use this to show that when $deg(\gamma) = 4$, the 2-fold Pfister forms in $H_0(\gamma,k)$ may be described in terms of k-rational points on an associated elliptic curve. This is used to give various examples.

This paper is based on some results from the author's Ph.D. thesis, written under the supervision of Adrian R. Wadsworth. We wish to thank the National Science Foundation for support in part while this research was being carried out, and Professor Wadsworth for helpful discussions.

1. Results for curves in general

Throughout this section $K = k(\mathcal{C})$ will be the function field of a complete nonsingular integral algebraic curve $\mathcal{C}$, having field of constants k. We will identify $\mathcal{C}$ with $C(K,k)$, the set of discrete valuation rings of K containing k. If $p \in \mathcal{C}$ then $\widehat{K_p}$ will denote the completion of K with respect to the discrete valuation of K associated with p, and $k(p)$ will denote the residue field. Let I^nF be the nth power of the fundamental ideal $I(F)$ of the Witt ring of a field F. The ideal I^nF is additively generated by the n-fold Pfister forms $\langle\langle a_1,\ldots,a_n\rangle\rangle := \otimes_{i=1}^n \langle 1, a_i\rangle$, $a_i \in F^*$. Recall that I^2F is equal to the set of Witt-classes of even-dimensional forms q with $disc(q) = 1$ (cf. [**L**, p. 40]).

THEOREM 1.

(*i*) $H_0(K,k) \subseteq \bigcap_{p\in\mathcal{C}} W(k(p)/k)$.

(*ii*) $H_0(K,k) \subseteq I^2k$.

(*iii*) *The following sequences are exact, where* ρ *is reduction modulo* $\mathrm{im}(r^*_{K/k})$.

$$0 \to H_0(K,k) \hookrightarrow \bigcap_{p\in\mathcal{C}} W(k(p)/k) \xrightarrow{r^*_{K/k}} \bigcap_{p\in\mathcal{C}} W(\widehat{K_p}/K) \xrightarrow{\rho} H_1(K,k) \tag{2}$$

$$0 \to H_0(K,k) \hookrightarrow \bigcap_{p\in\mathcal{C}} W(k(p)/k)\cap I^2k \xrightarrow{r^*_{K/k}} \bigcap_{p\in\mathcal{C}} W(\widehat{K_p}/K)\cap I^2K \xrightarrow{\rho} H_1(K,k) \tag{3}$$

PROOF.

(*i*) This follows directly from the well-known fact that, for each $p \in \mathcal{C}$, the following diagram is commutative:

$$\begin{array}{ccc} W(k) & \xrightarrow{r^*_{k(p)/k}} & W\big(k(p)\big) \\ {\scriptstyle r^*_{K/k}}\downarrow & & \uparrow{\scriptstyle \partial^1_p} \\ W(K) & \xrightarrow[r^*_{\widehat{K_p}/K}]{} & W(\widehat{K_p}) \end{array}$$

where ∂^1_p is the first residue map associated with the completion $\widehat{K_p}$.

(*ii*) If $q \in W(k)$ and $r^*_{K/k} \in I^2K$, then $dim(q)$ is even, and $disc(q) \in k^* \cap K^{*2} = k^{*2}$ as k is algebraically closed in K. It follows that $q \in I^2k$. The above holds for all $q \in H_0(K,k)$.

(*iii*) The exactness of (2) at $\bigcap_{p\in\mathcal{C}} W(k(p)/k)$ is immediate from (*i*) and the definition of $H_0(K,k)$. For the exactness at $\bigcap_{p\in\mathcal{C}} W(\widehat{K_p}/K)$, clearly $\rho\circ r^*_{K/k}$ is trivial. Conversely, if $q \in \bigcap_{p\in\mathcal{C}} W(\widehat{K_p}/K)$ and $\rho(q) = 0$, then $q = r^*_{K/k}(z)$, for some $z \in W(k)$. By (4) it follows that $r^*_{k(p)/k}(z) = 0$, for all $p \in \mathcal{C}$, so $z \in \bigcap_{p\in\mathcal{C}} W(k(p)/k)$, as desired. The exactness of (3) now follows from (2), using the argument of (*ii*). □

Note that the term $\bigcap_p W(\widehat{K_p}/K)$, in (2) and (3), is a group of obstructions to the weak Hasse-Minkowski principle for K; by definition, $\bigcap_p W(\widehat{K_p}/K) = 0$ if and only if K satisfies this local-global principle for isometry (i.e., for q_1, q_2 K-forms, $q_1 \simeq q_2$ if and only if $(q_1)_{\widehat{K_p}} \simeq (q_2)_{\widehat{K_p}}$ for all p). Using (2) and (3) we then have

COROLLARY 2. *If K satisfies the weak Hasse-Minkowski principle, then*

$$H_0(K,k) = \bigcap_{p\in\mathcal{C}} W(k(p)/k) \subseteq I^2k.$$

COROLLARY 3.

(*i*) $H_0(K,k) = 0$ *if* C *has an* F*-rational point, where* F *is an odd degree extension of* k. (eg., if C has a k-rational point.)

(*ii*) $H_0(K,k) = 0$ *if* $I^2k = 0$. (This holds, eg., if k is quadratically closed, finite, or a Laurent series field in one indeterminate over a quadratically closed field.)

PROOF.

(*i*) Suppose F/k is of odd degree. If p is an F-rational point then clearly $[k(p):k]$ is odd, so $W(k(p)/k)$ is trivial by Springer's theorem on odd degree extensions or (cf. [**L**, pp. 197–198]). Hence $H_0(K,k) = 0$ by Theorem 1(*i*).

(*ii*) This follows at once from Theorem 1(*ii*). In each example cited we have $|W(k)| \le 4$, so $I^2k = 0$. □

Finally we show that in many arithmetic cases $H_0(K,k)$ is generated by 2-fold Pfister forms.

PROPOSITION 4. *Suppose* K *is the function field of a complete nonsingular absolutely irreducible algebraic curve* C*, having field of constants* k*, with* k *a global field. If* $H_0(\widehat{k_v}\cdot K, \widehat{k_v})$ *is trivial for each real valuation* v *of* k*, then each anisotropic* $q \in H_0(K,k)\setminus\{0\}$ *is similar to a 2-fold Pfister form.*

PROOF. Suppose $q \in H_0(K,k)\setminus\{0\}$ is anisotropic. By Theorem 1(*ii*), $q \in I^2k$. It follows that $dim(q) \ge 4$. Now, for each valuation v of k we have $\widehat{k_v}\cdot K = \widehat{k_v}(C_{\widehat{k_v}})$ is well-defined, since C is absolutely irreducible, and $r^*_{\widehat{k_v}/k}(H_0(K,k)) \subseteq H_0(\widehat{k_v}\cdot K, \widehat{k_v})$. So $r^*_{\widehat{k_v}/k}(q) = 0$ for each real valuation v. If $dim(q) > 4$ then q is isotropic by Meyer's Theorem (cf. [**Sch**, 6.6.6(vii)]). Thus, we must have $dim(q) = 4$. So, since $det(q) = 1$, there are a, b, c with $q \simeq \langle a,b,c,abc\rangle$, which is similar to a 2-fold Pfister form. □

2. Some special cases of $H_0(\gamma, k)$

Henceforth K will be the function field of the curve $C_\gamma\colon y^2 = \gamma(x)$, with $\gamma \in k[x]$ square-free. We show here that in many cases $H_0(\gamma,k)$ is trivial or a principal ideal of the Witt ring $W(k)$.

COROLLARY 5. $H_0(\gamma,k) = 0$ *if* $y^2 = \gamma(x)$ *has a solution with* x *and* y *in some odd degree extension of* k*. This holds, for example, if* γ *has a factor of odd degree.*

PROOF. By hypothesis, the curve C_γ has a rational point in an odd degree extension of k. The conclusion now follows directly from Corollary 3(*i*). □

The next proposition gives many examples of even degree γ with nontrivial principal $H_0(\gamma,k)$. Contained in this is the well-known case of the Witt-kernel for function fields of curves of genus zero since any such curve is birational to one of the form $y^2 = ax^2 + b$.

PROPOSITION 6. *If* $\gamma = af^2 + bg^2$ *for some* $a, b \in k^*$, *and some* $f, g \in k[x] \setminus \{0\}$, *with* $\max\{deg(f), deg(g)\}$ *odd, then*

$$H_0(\gamma, k) = \langle\langle -a, -b\rangle\rangle W(k).$$

PROOF. Let $L := k(u)(\sqrt{au^2+b}) \subset K = k(x)(\sqrt{\gamma})$, where $u = \frac{f(x)}{g(x)}$. Then $k(u)$ is a purely transcendental extension of k as $k(x)$ is algebraic over $k(u)$ of finite degree $\max\{deg(f), deg(g)\}$. So, $[K : L] = [K : L][L : k(u)]/2 = [K : k(x)][k(x) : k(u)]/2 = \max\{deg(f), deg(g)\}$ is odd. Hence $H_0(\gamma, k) = W(K/k) = W(L/k)$ by Springer's theorem on odd degree extensions. But we have $W(L/k) = \langle\langle -a, -b\rangle\rangle W(k)$ since L is isomorphic to the function-field of the form $\langle 1, -a, -b\rangle$ (cf. [**A**], example after Satz 2.1, p. 457). □

3. $H_0(\gamma, k)$ for γ of even degree

By Corollary 5, $H_0(\gamma, k)$ is trivial if γ has odd degree. And although it may be tractable for some even degree γ (cf. Proposition 6), we shall see in Theorem 9 that $H_0(\gamma, k)$ can be more elusive even if γ has degree only four.

Propositions 4 and 6 show it is interesting to determine the (2-fold) Pfister forms in $H_0(\gamma, k)$; this is the concern of the next three results.

If q is a quadratic form over a field F, if $F \subseteq L$ and q_L is the extension of q to L, then $D_L(q)$ denotes $\{a \in L^* \mid a = q_L(v), \text{ for some } v\}$.

LEMMA 7. *Suppose* $\rho \simeq \langle 1\rangle \perp \rho'$ *is an anisotropic Pfister form over* k. *Then* $\rho \in H_0(\gamma, k)$ *if and only if* $-\gamma \in D_{k(x)}(\rho')$.

PROOF. If $\rho \in H_0(\gamma, k)$, then $\rho_{k(x)} \simeq q \otimes \langle 1, -\gamma\rangle$ for some q defined over $k(x)$ by [**Sch**, 2.5.11(i), p. 51]. As $D_{k(x)}(\rho)$ is a group, we may assume $1 \in D_{k(x)}(q)$. By the Witt Cancellation Theorem [**L**, I.4.2], it follows that $-\gamma \in D_{k(x)}(\rho')$.

Conversely, if $-\gamma \in D_{k(x)}(\rho')$, then $\rho_{k(x)} \simeq \langle 1, -\gamma, \ldots\rangle$, which is isotropic over $K = k(x)(\sqrt{\gamma})$. Since ρ is Pfister, ρ is hyperbolic over K, so $\rho \in H_0(\gamma, k)$. □

We now obtain a criterion, in the case of even degree γ, for determining when a Pfister form is in $H_0(\gamma, k)$.

PROPOSITION 8. *Suppose* $\gamma = \sum_{i=0}^{2d} \gamma_i x^i$ *has even degree* $2d$, *and is square-free. Suppose* $\rho \simeq \langle 1\rangle \perp \rho'$ *is an anisotropic* n*-fold Pfister form over* k. *Let* (V, β) *denote the bilinear form of* ρ'. *Then* $\rho \in H_0(\gamma, k)$ *if and only if there are vectors,* $v_i \in V$, *for* $i \in \{0, 1, \ldots, d\}$, *such that for all* $j \in \{0, 1, \ldots, 2d\}$,

$$-\gamma_j = \begin{cases} \rho'(v_{j/2}) + \sum_{i=\max\{0,j-d\}}^{(j-2)/2} 2\beta(v_i, v_{j-i}) & \text{for } j \text{ even} \\ \sum_{i=\max\{0,j-d\}}^{(j-1)/2} 2\beta(v_i, v_{j-i}) & \text{for } j \text{ odd.} \end{cases} \tag{5}$$

PROOF. By Lemma 7, $\rho \in H_0(\gamma, k)$ if and only if $-\gamma \in D_{k(x)}(\rho')$. By the Cassels-Pfister Theorem [**L**, IX.1.3], this occurs if and only if $-\gamma \in D_{k[x]}(\rho')$, if and only if there are $f_1, \ldots, f_{2^n-1} \in k[x]$ such that

$$-\gamma = \sum_{i=1}^{2^n-1} a_i f_i^2\,, \qquad \text{where} \quad \rho' \simeq \langle a_1, \ldots, a_{2^n-1}\rangle. \tag{6}$$

Since ρ' is anisotropic, we must have that each f_i has degree $\leq d$. Write $f_j = \sum_{l=0}^{d} c_{j,l}x^l$. For each $i \in \{0, 1, \ldots, d\}$, let $v_i = (c_{1,i}, \ldots, c_{2^n-1,i})$. We may rewrite (6) as $-\gamma = \rho'(\sum_{i=0}^{d} v_i x^i)$. Then it is easily seen that (6) is equivalent to (5), after equating coefficients. □

We next apply Proposition 8 to determine the set of 2-fold Pfister forms in $H_0(\gamma, k)$ when $\gamma = \sum_{i=0}^{4} \gamma_i x^i$ has degree 4. We may assume that $\gamma_3 = 0$, after making the substitution $x \longmapsto (x - \frac{\gamma_3}{4\gamma_4})$.

THEOREM 9. *Suppose $\gamma = \sum_{i=0}^{4} \gamma_i x^i$ is square-free with $\gamma_3 = 0$. Let $\mathcal{E}$ denote the elliptic curve defined by*

$$\mathcal{E}\colon \tau^2 = \omega^3 - 2\gamma_2\omega^2 + (\gamma_2^2 - 4\gamma_0\gamma_4)\omega + \gamma_4\gamma_1^2$$

and let

$$\Omega = \Big\{ \langle\langle -\gamma_4, -\omega \rangle\rangle \mid (\omega, \tau) \in \mathcal{E}(k),\ \omega \neq 0 \Big\} \cup \{0\}.$$

If $\mathcal{P}_2(k)$ denotes the set of 2-fold Pfister forms over k, then

$$\mathcal{P}_2(k) \cap H_0(\gamma, k) = \begin{cases} \Omega & \text{if } \gamma_1 \neq 0\,; \\ \Omega \cup \{\langle\langle -\gamma_4, -(\gamma_2^2 - 4\gamma_0\gamma_4)\rangle\rangle\} & \text{if } \gamma_1 = 0\,. \end{cases}$$

PROOF. Let $\rho = \langle 1 \rangle \perp \rho'$ be a nonzero, hence anisotropic, 2-fold Pfister form in $W(k)$, and (V, β) be the bilinear form of ρ'. Let $v_0, v_1, v_2 \in V$ and $M := [\beta(v_i, v_j)]$, $i = 2, 1, 0$, $j = 2, 1, 0$. If we set $\omega := -\rho'(v_1)$ then, by Proposition 8, $\rho \in H_0(\gamma, k)$ if and only if

$$M = \begin{bmatrix} -\gamma_4 & 0 & -(\gamma_2 - \omega)/2 \\ 0 & -\omega & -\gamma_1/2 \\ -(\gamma_2 - \omega)/2 & -\gamma_1/2 & -\gamma_0 \end{bmatrix} \tag{7}$$

for some choice of v_0, v_1, and v_2. Letting $g(\omega) := \omega^3 - 2\gamma_2\omega^2 + (\gamma_2^2 - 4\gamma_0\gamma_4)\omega + \gamma_4\gamma_1^2$, we then have $\delta := det(M) = \frac{1}{4}g(\omega)$.

If $\omega \neq 0$ then M is congruent to

$$M_1 := \begin{bmatrix} -\gamma_4 & 0 & 0 \\ 0 & -\omega & 0 \\ 0 & 0 & \delta/\gamma_4\omega \end{bmatrix}$$

by Gram-Schmidt and a determinant comparison. So $\langle -\gamma_4, -\omega \rangle$ is a subform of ρ', showing that $\rho \simeq \langle\langle -\gamma_4, -\omega \rangle\rangle$.

Note that $\mathcal{E}$ is given by the equation $\tau^2 = g(\omega)$, where $g(\omega)$ is as defined above. If $\delta \neq 0$ then M is nonsingular, so $\{v_0, v_1, v_2\}$ is a base of the space of ρ' and $\delta = det(\rho') \in k^2$. If $\delta = 0$, then again $\delta \in k^2$. Choosing $\tau \in k$ with $\tau^2 = 4\delta$, we have $\tau^2 = g(\omega)$, so $(\omega, \tau) \in \mathcal{E}(k)$.

If $\omega = 0$ then $v_1 = 0$ as ρ' is anisotropic, so $\gamma_1 = -2\beta(v_0, v_1) = 0$ by (7). After modifying M by interchanging v_0 and v_1, then applying Gram-Schmidt and a determinant calculation, we have that M is congruent to

$$M_2 := \begin{bmatrix} -\gamma_4 & 0 & 0 \\ 0 & \epsilon/(-\gamma_4) & 0 \\ 0 & 0 & 0 \end{bmatrix}$$

where

$$\epsilon := det \begin{bmatrix} -\gamma_4 & -\gamma_2/2 \\ -\gamma_2/2 & -\gamma_0 \end{bmatrix} = \frac{1}{4}(4\gamma_0\gamma_4 - \gamma_2^2). \tag{8}$$

Since $\gamma(x)$ is square-free, we must have $\epsilon \neq 0$. So $\langle -\gamma_4, -\epsilon/\gamma_4 \rangle$ is a subform of ρ', hence $\rho \simeq \langle\langle -\gamma_4, -\epsilon/\gamma_4 \rangle\rangle \simeq \langle\langle -\gamma_4, 4\epsilon \rangle\rangle \simeq \langle\langle -\gamma_4, -(\gamma_2^2 - 4\gamma_0\gamma_4) \rangle\rangle$.

Conversely, suppose $\rho = \langle\langle -\gamma_4, -\omega \rangle\rangle \in \Omega - \{0\}$ with $(\omega, \tau) \in \mathcal{E}(k)$ and $\omega \neq 0$. Then set $\delta := \frac{1}{4}g(\omega) = \tau^2/4 \in k^2$. Since $\delta \in k^{*2}$ or $\delta = 0$ it is clear that there are vectors w_0, w_1, w_2 in the space of ρ' such that $[\beta(w_i, w_j)] = M_1$ as above. Because M_1 is congruent to M (as $\delta = det(M)$), there are v_0, v_1, v_2 spanning the same space as w_0, w_1, w_2 such that $[\beta(v_i, v_j)] = M$. Hence $\rho \in H_0(\gamma, k)$ by Proposition 8.

Finally, suppose that $\gamma_1 = 0$ and $\rho = \langle\langle -\gamma_4, -(\gamma_2^2 - 4\gamma_0\gamma_4) \rangle\rangle$. Then $\rho \simeq \langle\langle -\gamma_4, -\epsilon/\gamma_4 \rangle\rangle$ with ϵ as in (8). So there are v_0, v_1, v_2 with $[\beta(v_i, v_j)] = M$, as in the preceding paragraph, using M_2 in place of M_1. Thus $\rho \in H_0(\gamma, k)$ by Proposition 8. □

The author is continuing to study the relationship of $\mathcal{E}$ to $\mathcal{C}$. We next use Theorem 9 to give some examples of $H_0(\gamma, k)$ in the following.

EXAMPLES 10.

(*i*) *Let p be a rational prime congruent to* 7 *or* 11 *modulo* 16, *and let* $\gamma(x) = \frac{p}{2}x^4 - \frac{1}{2} \in \mathbb{Q}[x]$. *Then* $H_0(\gamma, \mathbb{Q}) = \{0, \langle\langle -\frac{p}{2}, -p \rangle\rangle\}$.

(*ii*) *Let* $\gamma(x) = \frac{73}{2}x^4 - \frac{1}{2} \in \mathbb{Q}[x]$. *Let* $\mathcal{E}$ *be the elliptic curve defined by* $\tau^2 = \omega^3 + 73\omega$. *Then the rational points* $\mathcal{E}(\mathbb{Q})$ *of* $\mathcal{E}$ *form a finitely-generated abelian group of rank two, and*

$$H_0(\gamma, \mathbb{Q}) = \left\{ \langle\langle -146, -\omega \rangle\rangle \mid (\omega, \tau) \in \mathcal{E}(\mathbb{Q}),\ \omega \neq 0 \right\} \cup \{\langle\langle -146, -73 \rangle\rangle\} \cup \{0\}$$

(*iii*) *Let* $\gamma(x) = 3x^4 + 5 \in \mathbb{Q}[x]$. *Then* $H_0(\gamma, \mathbb{Q}) = \{\langle\langle -3, \pm 1 \rangle\rangle, \langle\langle -3, \pm 5 \rangle\rangle\}$.

PROOF.

(*i*) The rationals $\mathbb{Q}$ has a unique real valuation. The reals $\mathbb{R}$ is the completion. Since the equation $y^2 = \gamma(x)$ has a solution in $\mathbb{R}$, we have $H_0(\gamma, \mathbb{R}) = 0$ by Corollary 5. By Proposition 4, each nonzero anisotropic element of $H_0(\gamma, \mathbb{Q})$ is similar to a 2-fold Pfister form. The 2-fold Pfister forms in $H_0(\gamma, \mathbb{Q})$ are given by Theorem 9. In this case, $\mathcal{E}$ is given by the equation $\tau^2 = \omega^3 + p\omega$. By [**Sil**, Proposition 6.2(c), p. 311], $\mathcal{E}(\mathbb{Q})$ has rank 0. So, by [**Sil**, Proposition 6.1], $\mathcal{E}(\mathbb{Q})$ has only two elements: $(0,0)$ and the identity, the point at infinity. Thus, in Theorem 9, we have $\Omega = \{0\}$, and so the only nonzero 2-fold Pfister form in $H_0(\gamma, \mathbb{Q})$ is $\langle\langle -\frac{p}{2}, -p \rangle\rangle$. By Meyer's Theorem we see that this Pfister form is universal. It follows that the similarity class of this form, $\langle\langle -\frac{p}{2}, -p \rangle\rangle$, is just the form itself.

(*ii*) This follows analogously to (*i*) with $p = 73$. As noted in [**Sil**, Remark 6.4], the rational points $\mathcal{E}(\mathbb{Q})$ of $\mathcal{E}$ here have rank two, two independent elements of infinite order being $(36, 222)$ and $(\frac{9}{16}, \frac{411}{64})$.

(*iii*) By the argument in (*i*), we have that every nonzero anisotropic form in $H_0(\gamma, \mathbb{Q})$ is similar to a 2-fold Pfister form, and the 2-fold Pfister forms in $H_0(\gamma, \mathbb{Q})$ are given by Theorem 9. Here $\mathcal{E}$ is given by $\tau^2 = \omega^3 - 60\omega$, and the nonzero 2-fold Pfister forms in $H_0(\gamma, \mathbb{Q})$ are

$$\left\{\langle\langle -3, -\omega \rangle\rangle \mid (\omega, \tau) \in \mathcal{E}(\mathbb{Q}),\ \omega \neq 0\right\} \cup \{\langle\langle -3, -5 \rangle\rangle\}$$

First note that $\langle\langle -3, -a \rangle\rangle = \langle\langle -3, -b \rangle\rangle$ in $W(\mathbb{Q})$ if and only if a and b are congruent modulo norms from $\mathbb{Q}(\sqrt{3}\,)$. Now suppose $(\omega, \tau) \in \mathcal{E}(\mathbb{Q})$ and $\omega \neq 0$. Then $\tau^2 = \omega^3 - 60\omega$, equivalently $\omega^2 - \left(\frac{\tau}{\omega}\right)^2 \omega - 60 = 0$. This has a rational solution if and only if the discriminant $\left(\frac{\tau}{\omega}\right)^4 + 240 = \mu^2$, for some $\mu \in \mathbb{Q}$. Letting $\frac{\tau}{\omega} = \frac{2a}{b}$ and $\mu = \frac{4c}{b^2}$, for some integers a, b, and c, this equation for the discriminant becomes $a^4 + 15b^4 = c^2$. By the quadratic formula we have $\omega = 2(a^2 \pm c)/b^2$ which is congruent to $-(a^2 \pm c)$ modulo norms from $\mathbb{Q}(\sqrt{3}\,)$. Thus, $\langle\langle -3, -\omega \rangle\rangle = \langle\langle -3, (a^2 \pm c) \rangle\rangle$, where $a^4 + 15b^4 = c^2$, for some integers a, b, c, with $b \neq 0$. Note that this Pfister form depends only on the prime factors of odd multiplicity of $a^2 \pm c$, and then only on the congruence class of these prime factors modulo norms from $\mathbb{Q}(\sqrt{3}\,)$. We will show that if p is such a prime factor, then $p \in \{2, 3, 5\}$, each of which is congruent to ± 1 or 5 modulo norms from $\mathbb{Q}(\sqrt{3}\,)$. For this, we show that any prime factor p of $a^2 \pm c$, not in $\{2, 3, 5\}$, must divide with even multiplicity. We rewrite the equation $a^4 + 15b^4 = c^2$ as follows.

$$(a^2 + c)(a^2 - c) = -15b^4 \tag{9}$$

Suppose firstly that $p \notin \{2, 3, 5\}$, $p | a^2 - c$, but $p \nmid a^2 + c$. By (9), $p | 15b^4$, so p has multiplicity a multiple of 4 in $a^2 - c$. The same argument applies if $p | a^2 + c$, but $p \nmid a^2 - c$.

Suppose instead that $p \notin \{2, 3, 5\}$, $p | a^2 - c$, and $p | a^2 + c$. Then $p | 2a^2, p | c$, and $p | b$, so $p^4 | (a^4 + 15b^4) = c^2$, and hence $p^2 | c$. Writing $a = pa', b = pb', c = p^2c'$ we obtain $((a')^2 + c')((a')^2 - c') = -15(b')^4$, and $p^2((a')^2 \pm c') = a^2 \pm c$. Thus, by descent (i.e., by induction) we find that p must have even multiplicity in $a^2 + c$ and $a^2 - c$.

As above, we now have that any prime factor of odd multiplicity of $a^2 \pm c$ must be in $\{2,3,5\}$, and hence congruent to ± 1 or 5 modulo norms from $\mathbb{Q}(\sqrt{3})$. Thus, $\langle\langle -3, -\omega\rangle\rangle \in \{\langle\langle -3, \pm 1\rangle\rangle, \langle\langle -3, \pm 5\rangle\rangle\}$. One may verify that each of these forms is in fact obtainable in $H_0(\gamma, \mathbb{Q})$. As in (i), an application of Meyer's Theorem shows these Pfister forms to be universal. Thus the similarity class of each form consists of just the form itself. □

In Example 10(ii) it is unknown whether $H_0(\gamma, \mathbb{Q})$ is finite or infinite.

As a corollary to Theorem 9 and Corollary 5 we obtain a curious property of rational points on certain elliptic curves.

COROLLARY 11. *Let $\mathcal{E}$ be the elliptic curve defined by $\tau^2 = \omega^3 + a_2\omega^2 + a_1\omega + a_0$, with a_2, a_1, a_0 in a field k of characteristic not 2.*

(i) *Suppose $a_0 \neq 0$. Define the polynomial $\gamma(x) \in k[x]$ by*

$$\gamma(x) := a_0(x^2 - \frac{a_2}{4a_0})^2 + x - \frac{a_1}{4a_0}.$$

Suppose that $\gamma(x)$ is square-free, and $y^2 = \gamma(x)$ has a solution with x and y in some odd degree extension of k. Then, for each $(\omega, \tau) \in \mathcal{E}(k)$, ω is a norm from $k(\sqrt{a_0})$.

(ii) *Suppose $a_0 = 0$. For each $d \in k^*$, define the polynomial $\gamma_d(x) \in k[x]$ by*

$$\gamma_d(x) := d(x^2 - \frac{a_2}{4d})^2 - \frac{a_1}{4d}.$$

Suppose that, for some $d \in k^*$, $\gamma_d(x)$ is square-free and $y^2 = \gamma_d(x)$ has a solution with x and y in some odd degree extension of k. Then ω is a norm from $k(\sqrt{d})$, for each $(\omega, \tau) \in \mathcal{E}(k)$.

PROOF.

(i) For this γ we have $\gamma_0 = \frac{a_2^2 - 4a_1}{16a_0}$, $\gamma_1 = 1, \gamma_2 = \frac{-a_2}{2}, \gamma_3 = 0, \gamma_4 = a_0$; so the curve $\mathcal{E}$ of Theorem 9 is given by $\tau^2 = \omega^3 + a_2\omega^2 + a_1\omega + a_0$, as in the statement of this corollary. By Corollary 5, $H_0(\gamma, k)$ is trivial. So, by Theorem 9, $\Omega = \left\{\langle\langle -\gamma_4, -\omega\rangle\rangle \mid (\omega, \tau) \in \mathcal{E}(k),\ \omega \neq 0\right\} \cup \{0\}$ is trivial. Suppose $(\omega, \tau) \in \mathcal{E}(k)$. If $\omega = 0$ then of course ω is a norm from $k(\sqrt{a_0})$. If $\omega \neq 0$ then $\langle\langle -a_0, -\omega\rangle\rangle \in \Omega = 0$, so ω is a similarity factor for $\langle 1, -a_0\rangle$, and hence represented by this form since it is a Pfister form. Thus ω is a norm from $k(\sqrt{a_0})$, as desired.

(ii) Suppose d is chosen so that γ_d satisfies the hypotheses of this corollary. The desired conclusion then follows from the same line of reasoning used for (i), with γ_d in place of γ. □

REMARK. Using the Euclidean algorithm one finds that $\gamma_d(x)$ of Corollary 11(ii) is square-free if and only if $a_1 \neq 0$ and $a_2^2 \neq 4a_1$. The same method however does not yield such a succinct general condition for square-freeness on the $\gamma(x)$ of Corollary 11(i).

REFERENCES

[A] J. Kr. Arason, *Cohomologische Invarianten quadratischer Formen*, J. Algebra **36** (1975), 448–491.

[Kn1] M. Knebusch, *Grothendieck- und Wittringe von nichtausgearteten symmetrischen Bilinearformen*, 3.Abh. 1970, Akad. Wiss. math.-naturw. Kl., S.-Ber. Heidelberg, 1969/70.

[Kn2] ———, *Symmetric bilinear forms over algebraic varieties*, Queen's Papers in Pure and Applied Math., vol. 46 (G. Orzech, ed.), 1977, pp. 103–283.

[L] T. Y. Lam, *The algebraic theory of quadratic forms*, Benjamin, Reading, Mass., 1973.

[Sch] W. Scharlau, *Quadratic and Hermitian forms*, Springer, Berlin, 1985.

[Sil] J. Silverman, *The arithmetic of elliptic curves*, Springer, Berlin, 1986.

DEPARTMENT OF MATHEMATICAL SCIENCES, LOYOLA UNIVERSITY, NEW ORLEANS, LA 70118–6195

Contemporary Mathematics
Volume 155, 1994

On the Canonical Class of Hyperelliptic Curves

V. SURESH

In this note, we give a criterion for the canonical class of a hyperelliptic curve, which is a double covering of an anisotropic conic over a field k of characteristic $\neq 2$, to be even. This supplements results of [**PS**] to a complete solution to the problem of when the canonical class of a hyperelliptic curve is even. We also calculate the 2–torsion in the class group of a hyperelliptic curve.

I thank Prof. R. Parimala and Prof. R. Sridharan for their generous help and many valuable discussions.

1. The canonical class of hyperelliptic curves

Throughout, k denotes a field with $\mathrm{char}(k) \neq 2$. Let Y be a smooth projective curve of genus 0 over k. Let $\pi : X \to Y$ be a double covering with genus of X at least 2. We recall the terminology set up in [**PS**]. If $Y = \mathbb{P}^1$, we have $k(Y) = k(T)$, the rational function field in the variable T, $k(X) = k(T)(\sqrt{p(T)})$, p a square free polynomial in T with degree $p = 2d = 2(g+1)$, g = genus of X (assuming without loss of generality that the point at infinity ∞_Y is unramified for π). If Y is an anisotropic conic, we have $k(Y) = k(T)(\sqrt{aT^2 + b})$ where (a, b) is a non-split quaternion algebra over k, $k(X) = k(Y)(\sqrt{p})$ where $p \in k[T, \sqrt{aT^2 + b}]$ is a product of distinct primes. We assume without loss of generality that ∞_Y is unramified for π.

Let ${}_2CL(X)$ denote the 2–torsion subgroup of the divisor class group of X. We define a homomorphism $\eta : {}_2CL(X) \to k^*/k^{*2}$ as follows. Let D be a divisor on X with its class $[D] \in {}_2CL(X)$. Let $2D = (f)_X$, $f \in k(X)^*$, $(f)_X$ denoting the divisor of the function f. Then, if $\mathcal{N} : Div\ X \to Div\ Y$ is the norm map on the group of divisors, $2\mathcal{N}(D) = (\mathcal{N}f)_Y$, so that the degree of $\mathcal{N}(D)$ is zero. Since genus of Y is zero, there exists a function $g \in k(Y)^*$ such that $\mathcal{N}(D) = (g)_Y$.

1991 *Mathematics Subject Classification.* 14C20, 14H05.

This paper is submitted in final form and no version of it will be submitted for publication elsewhere.

Thus $(\mathcal{N}(f)(g^{-2}))_Y = 0$, so that there is a scalar $\varepsilon \in k^*$ with $\mathcal{N}f = \varepsilon g^2$. The class $\overline{\varepsilon}$ of ε in k^*/k^{*2} is uniquely defined by the class $[D] \in {}_2CL(X)$. We define $\eta([D]) = \overline{\varepsilon}$. Since the covering π is unique up to isomorphism for genus of $X \geq 2$ [**Ch**, Ch. IV, §9], the subgroup $\eta({}_2CL(X))$ of k^*/k^{*2} is intrinsically associated to X. We observe that $\eta({}_2CL(X))$ is contained in the subgroup H of k^*/k^{*2} defined by

$$H = \{\overline{\varepsilon} \in k^*/k^{*2} \mid \text{the quaternion algebra } (\varepsilon, p)_{k(Y)} \text{ is split}\}.$$

Let $\pi : X \to Y$ be a hyperelliptic covering with genus $Y = 0$. Let $\Omega_X \in CL(X)$ be the canonical class. We call Ω_X *even*, if there exists $\Theta \in CL(X)$ such that $\Omega_X = 2\Theta$. We say that X is of *type* $(*)$ if π satisfies one of the following conditions:

A) $Y = \mathbb{P}^1$, genus X is even and all ramification points of π are of even degree.

B) Y is an anisotropic conic, genus $X \equiv 3 \pmod 4$ and all ramification points of π have degree $\equiv 0 \pmod 4$.

LEMMA (1.1). *Let $\pi : X \to Y$ be of type $(*)$. Then if $f \in k(X)^*$ is such that $(f)_X = \infty_X + 2D$ for some divisor D on X, then $\mathcal{N}f = \varepsilon g^2$ with $\varepsilon \in k^*$ not a square.*

PROOF. On the same lines as proof of [**PS**, 2.4]. □

THEOREM (1.2). *Let $\pi : X \to Y$ be of type $(*)$. Then Ω_X is even if and only if there exists $\varepsilon \in k^*$, $\overline{\varepsilon} \notin \eta({}_2CL(X))$ such that the quaternion algebra $(\varepsilon, p)_{k(Y)}$ splits.*

PROOF. Since X is of type $(*)$, we have $\Omega_X \equiv \infty_Y \pmod 2$ (cf. [**PS**, §2]). Suppose Ω_X is even. Then there exists a divisor D on X and a function $f \in k(X)^*$ such that $\infty_X = 2D + (f)_X$. Taking norms, we get

$$2\infty_Y = 2\mathcal{N}(D) + (\mathcal{N}f)_Y,$$

so that there exist $\varepsilon \in k^*$ and $g \in k(X)^*$ such that $\mathcal{N}f = \varepsilon g^2$. Since ε is a norm from $k(Y)(\sqrt{p})$, the quaternion algebra $(\varepsilon, p)_{k(Y)}$ splits. We verify that $\overline{\varepsilon} \notin \eta({}_2CL(X))$. Suppose there is a divisor D' on X such that $2D' = (f')_X$ and $\mathcal{N}f = \varepsilon {g'}^2$, $g' \in k(Y)^*$. Then

$$(ff')_X = \infty_X + 2(D' - D) \text{ and } \mathcal{N}(ff') = (\varepsilon g g')^2,$$

contradicting Lemma 1.1.

Suppose conversely that there exists $\varepsilon \in k^*$, $\overline{\varepsilon} \notin \eta({}_2CL(X))$, such that $(\varepsilon, p)_{k(Y)}$ splits. Then there exist r, s, $t \in k[T]$ or $k[T, \sqrt{aT^2 + b}]$ (according as X satisfies A or B) such that $r^2 - \varepsilon s^2 = pt^2$. Let $f = r + t\sqrt{p}$. Arguing as in [**PS**, Proof of Th. 2.4], one can show that $(f)_X = 2D + \delta\infty_X$ for some divisor D on X, with $\delta = 0$ or 1. Suppose $\delta = 0$. Then $(f)_X = 2D$ and $\mathcal{N}f = r^2 - pt^2 = \varepsilon s^2$, which shows that $\bar{\varepsilon}$ belongs to $\eta({}_2CL(X))$, contradicting the assumption on ε. Thus $\delta = 1$ and $(f)_X = \infty_X + 2D \equiv \Omega_X \pmod 2$, so that Ω_X is even. □

COROLLARY (1.3). [**PS**, Theorem 2.4] *If $Y = \mathbb{P}^1$ and $\pi : X \to Y$ is of type $(*)$, then Ω_X is even if and only if there exists $\varepsilon \in k^*$, which is not a square, such that the quaternion algebra $(\varepsilon, p)_{k(Y)}$ splits.*

PROOF. By Theorem 1.2, it suffices to show that $\eta({}_2CL(X))$ is trivial in k^*/k^{*2}. Suppose D is a divisor on X with $2D = (f)_X$ and $\mathcal{N}f = \varepsilon g^2$, $\varepsilon \in k^*$, $g \in K(Y)^*$. If $\varepsilon \notin k^{*2}$, there exists (cf. [**PS**, Proof of Th. 2.4]) $f' \in k(X)^*$, such that $(f')_X = \infty_X + 2D'$ and $\mathcal{N}f' = \varepsilon g'^2$, $g' \in k(Y)^*$. Then, we have $(ff')_X = \infty_X + 2(D + D')$ and $\mathcal{N}(ff') = (\varepsilon g g')^2$, contradicting Lemma 1.1. Thus it follows that ε belongs to k^{*2} and $\eta({}_2CL(X)) = 1$. □

We give an example of a hyperelliptic curve $\pi : X \to Y$ of type $(*)$, with Y an anisotropic conic over k, which satisfies the condition: There exists $\varepsilon \in k^*$, ε not a square, with $(\varepsilon, p)_{k(Y)}$ split but for which Ω_X is not even.

EXAMPLE (1.4). Let Y be the anisotropic conic $U^2 + 3T^2 + 3V^2$ in $\mathbb{P}^2_{\mathbb{Q}_3}$. Let X be the hyperelliptic curve with function field $\mathbb{Q}_3(T, \sqrt{-3(T^2+1)}, \sqrt{p})$, where $p = (T+1)^4 - 3T^4$.

The polynomial p is irreducible over $\mathbb{Q}_3$. In fact, under the change of variable $T \to T + 2$, p maps to $(T+3)^4 - 3(T+2)^4$, which is irreducible over $\mathbb{Z}_3[T]$. Further, p splits into two points p_1, p_2 in Y, each of degree 4, the genus of X is equal to 3 and X satisfies condition B of $(*)$. We have $\mathbb{Q}_3{}^*/\mathbb{Q}_3{}^{*2} = \{1, -1, 3, -3\}$. We claim that neither $(-3, p)$ nor $(-1, p)$ splits over $\mathbb{Q}_3(Y)$. If either of them splits, then there exist $f, g, h \in \mathbb{Q}_3[T, \sqrt{-3(T^2+1)}]$ without common factors such that $f^2 + 3g^2 = h^2 p$ or $f^2 + g^2 = h^2 p$. Thus either -3 or -1 is a square in $\mathbb{Q}_3[T]/(p(T))$. Since 3 is a square in $\mathbb{Q}_3[T]/(p(T))$, we get that both -3 and -1 are squares in $\mathbb{Q}_3[T]/(p(T))$. Since $\mathbb{Q}_3[T]/(p(T))$ is a local field with residue field $\mathbb{F}_3$, it would follow that -1 is a square in $\mathbb{F}_3$, which is not possible. Thus $(-3, p)_{\mathbb{Q}_3(Y)}$ and $(-1, p)_{\mathbb{Q}_3(Y)}$ are non split. On the other hand, $(3, p)_{\mathbb{Q}_3(Y)}$ splits, by the very choice of p, since p is a norm from $\mathbb{Q}_3(T)(\sqrt{3})$. However, 3 belongs to $\eta({}_2CL(X))$. In fact, if $h = (T+1)^2 + \sqrt{p} \in \mathbb{Q}_3(X)$, then it is easily checked that $(h)_X = 2D'$ for some divisor D' on X and $\mathcal{N}(h) = 3T^4$, so that $\eta(D) = 3$. Thus the subgroup $H \subset \mathbb{Q}_3^*/\mathbb{Q}_3^{*2}$ is equal to $\{1\ ,\ 3\}$ and η surjects onto H. Thus by Theorem 1.2, Ω_X is odd.

2. The 2–torsion in the class group

Let $\pi : X \to Y$ be a double covering with genus of Y equal to 0. We follow the same notation as in §1. Let $p = p_1 p_2 \ldots p_n$ be a product of n distinct primes. Let $G = \ker\ (\eta :\ {}_2CL(X) \to H)$.

LEMMA (2.1). *We have*

$$|G| = \begin{cases} 2^{n-2} & \textit{if there is a } p_i \textit{ of odd degree} \\ 2^{n-1} & \textit{if all } p_i \textit{ are of even degree} \end{cases}$$

(*here degree of* p_i *is defined as* $-\vartheta_Y(p_i)$, *where* ϑ_Y *is the valuation at* ∞_Y).

PROOF. Let D be a divisor on X with $\eta([D]) = 1$. Then there exist $f \in k(X)$, $g \in k(Y)$ such that $2D = (f)_X$ and $\mathcal{N}f = g^2$. By Hilbert Theorem 90, there exists $h \in k(Y)$ such that $f = hf_1{}^2$, $f_1 \in k(X)$. Suppose

$$(h)_Y = e_\infty \infty_Y + e_1 p_1 + \cdots + e_n p_n + \sum_{q \neq p_i, \infty} e_q q,$$

then

$$(h)_X = e_\infty \infty_X + 2e_1 P_1 + \cdots + 2e_n P_n + \sum_{q \neq p_i, \infty} e_q \pi^* q.$$

Thus $e_\infty = 2e'_\infty$, $e_q = 2e'_q$. Write $e_i = 2e'_i + \delta_i$, where $\delta_i = 0$ or 1. Let $h' = \prod q^{e'_q} \prod p_i^{e'_i} \in k(Y)$. Then, for some $e''_\infty \in \mathbb{Z}$,

$$(h/h'^2) = 2e''_\infty \infty_X + 2\sum_{i=1}^{n} \delta_i P_i = 2D',$$

where $D' = e''_\infty \infty_X + \sum \delta_i P_i$ and $[D] = [D']$ in $CL(X)$. Thus every element of G is of the form $(1/2)(f)_X$, where $f \in k(Y)$ is a product of distinct p_i with the only condition that the number of p_i of odd degree in the support of $(f)_X$ is even. Thus $\mid G \mid = 2^{n-2}$ if some p_i has odd degree, and $\mid G \mid = 2^{n-1}$ if all p_i are of even degree. □

This group G could be thought as a function field analogue of the *ambiguous narrow classes* defined for quadratic number fields (cf. [**Ha**, Ch. 29, §3]).

LEMMA (2.2). *The index of* $\eta({}_2CL(X))$ *in* H *is at most* 2.

PROOF. Let $\overline{\varepsilon}_i \in H, i = 1, 2$. Then ε_i is a norm from $k(X)$ and one can choose $f_i \in k(X)$ such that $(f_i)_X = \delta_i \infty_X + 2D_i$ for some $D_i \in Div\ X$, $\delta_i = 0$ or 1 (cf. [**PS**, Proof of Th. 2.4]) and $\mathcal{N}f_i = \varepsilon_i g_i{}^2$ for some $g_i \in k(Y)$. If $\delta_i = 0$, $\overline{\varepsilon}_i \in \eta({}_2CL(X))$. Suppose $\delta_i = 1, i = 1, 2$. Then $(f_1 f_2)_X = 2(\infty_X + D_1 + D_2)$ and $\mathcal{N}(f_1 f_2) = \varepsilon_1 \varepsilon_2 (g_1 g_2)^2$. Thus $\eta(\infty_X + D_1 + D_2) = \overline{\varepsilon_1 \varepsilon_2}$, so that $\overline{\varepsilon}_1$ and $\overline{\varepsilon}_2$ define the same class of $H/\eta({}_2CL(X))$. □

Remark (2.3). From the above two lemmas, we get

$$(1/2)\,|G|\,|H| \leq |{}_2CL(X)| \leq |G|\,|H|\,.$$

PROPOSITION (2.4). *Let Y be the projective line and X be a double cover of Y. Then*

$$|{}_2CL(X)| = \begin{cases} 2^{n-2} & \text{if there is a } p_i \text{ of odd degree} \\ 2^{n-1} & \text{if all } p_i \text{ are of even degree} \\ & \quad \text{and genus of } X \text{ is even} \\ 2^{n-1}\,|H| & \text{otherwise.} \end{cases}$$

PROOF. Case 1 is dealt with in [**PS**, Remark 2.5]. In Case 2 the map η is trivial (cf. Proof of 1.3), so that $G = {}_2CL(X)$. In Case 3, for a given $\overline{\varepsilon} \in H$, one can show by arguments similar to [**PS**, Proof of 2.4], that there exists $f \in k(X)$ with $(f)_X = 2D$ and $\mathcal{N} f = \varepsilon s^2$, so that η is surjective. □

PROPOSITION (2.5). *Let Y be an anisotropic conic and let X be a double cover of Y of genus g. Then*

$$|{}_2CL(X)| = \begin{cases} 2^{n-2}\,|H| & \text{if there is a } p_i \text{ of odd degree} \\ 2^{n-1}\,|H| & \text{if all } p_i \text{ are of even degree,} \\ & \quad g \equiv 3 \pmod 4 \text{ and } \Omega_X \text{ odd} \\ 2^{n-2}\,|H| & \text{if all } p_i \text{ are of even degree,} \\ & \quad g \equiv 3 \pmod 4 \text{ and } \Omega_X \text{ even} \\ 2^{n-1}\,|H| & \text{if all } p_i \text{ are of even degree,} \\ & \quad g \equiv 1 \pmod 4 \text{ and } \Omega_X \not\equiv 0 \pmod 4 \end{cases}$$

$$\leq \begin{cases} 2^{n-1}\,|H| & \text{if all } p_i \text{ are of even degree,} \\ & \quad g \equiv 1 \pmod 4 \text{ and } \Omega_X \equiv 0 \pmod 4. \end{cases}$$

PROOF. If there is a p_i of odd degree, then η is surjective. In fact, let $\overline{\varepsilon} \in H, f \in k(X)$ with $(f)_X = \delta\infty_X + 2D$, $\delta = 1$, and $\mathcal{N} f = \varepsilon h^2$, $h \in k(Y)$. Then $(fp_i)_X = 2D'$ and $\mathcal{N}(fp_i) = p^2{}_i h^2 \varepsilon$. Now assume that all p_i are of even degree. If $g \equiv 3 \pmod 4$ and Ω_X is odd, since $\Omega_X \equiv \infty_X \pmod 2$, given $\overline{\varepsilon} \in H$, any choice of $f \in k(X)$ with $(f)_X = \delta\infty_X + 2D$ and $\mathcal{N} f = \varepsilon h^2$, $h \in k(Y)$, $\delta = 0$ or 1 necessarily yields $\delta = 0$. Thus η is surjective. If $g \equiv 3 \pmod 4$ and Ω even, there exists $f \in k(X)$ with $(f)_X = \infty_X + 2D$ and $\mathcal{N} f = \varepsilon h^2$ with $h \in k(Y)$, $\varepsilon \in k^*$ not a square. Arguing as in (2.4), using Lemma 1.1, we have $\overline{\varepsilon} \notin \eta(\,{}_2CL(X))$, so that $\eta(\,{}_2CL(X))$ has index 2 in H. If $g \equiv 1 \pmod 4$ and $\Omega \not\equiv 0 \pmod 4$),then let $\overline{\varepsilon} \in H$ and $f \in k(X)$ be such that $(f)_X = \delta\infty_X + 2D, \delta = 0$ or 1 and $\mathcal{N} f = \varepsilon g^2$. If $\delta = 1$, we have $\Omega_X \equiv 0 \pmod 4$, contradicting our assumption. Thus $\delta = 0$ and η is surjective. Now the proposition follows from Lemmas 2.1, 2.2 and Remark 2.3. □

References

[Ch] C. Chevalley, *Introduction to the theory of algebraic functions of one variable*, Math. Surveys., Amer. Math. Soc., Providence, RI, 1951.

[Ha] H. Hasse, *Number theory*, Grundlehren der mathematischen Wissenschaften 229, Springer -Verlag, Berlin, Heidelberg, and New York, 1980.

[PS] R. Parimala and W. Scharlau, *On the canonical class of a curve and the extension property for quadratic forms* (to appear, this volume).

School of Mathematics, Tata Institute of Fundamental Research, Homi Bhabha Road, Bombay 400 005, India

E-mail address: suresh@tifrvax.bitnet

Epilogue

Putting this volume together was an adventure in electronic communication and modern TEXnology. In the end, all circuits led through Corvallis to Providence. I wish to personally thank both Donna Harmon and Christine Thivierge of the American Mathematical Society for their help, hard work, and necessary prodding, and Bill McMecham of *Professional Editing* in Corvallis for translating some papers into AMS-TEX and adding the necessary AMS-TEX formats to others.

–Robby Robson, Corvallis, June 1993–

E-mail address: robby@math.orst.edu

Recent Titles in This Series

(Continued from the front of this publication)

124 **Darrell Haile and James Osterburg, Editors,** Azumaya algebras, actions, and modules, 1992
123 **Steven L. Kleiman and Anders Thorup, Editors,** Enumerative algebraic geometry, 1991
122 **D. H. Sattinger, C. A. Tracy, and S. Venakides, Editors,** Inverse scattering and applications, 1991
121 **Alex J. Feingold, Igor B. Frenkel, and John F. X. Ries,** Spinor construction of vertex operator algebras, triality, and $E_8^{(1)}$, 1991
120 **Robert S. Doran, Editor,** Selfadjoint and nonselfadjoint operator algebras and operator theory, 1991
119 **Robert A. Melter, Azriel Rosenfeld, and Prabir Bhattacharya, Editors,** Vision geometry, 1991
118 **Yan Shi-Jian, Wang Jiagang, and Yang Chung-chun, Editors,** Probability theory and its applications in China, 1991
117 **Morton Brown, Editor,** Continuum theory and dynamical systems, 1991
116 **Brian Harbourne and Robert Speiser, Editors,** Algebraic geometry: Sundance 1988, 1991
115 **Nancy Flournoy and Robert K. Tsutakawa, Editors,** Statistical multiple integration, 1991
114 **Jeffrey C. Lagarias and Michael J. Todd, Editors,** Mathematical developments arising from linear programming, 1990
113 **Eric Grinberg and Eric Todd Quinto, Editors,** Integral geometry and tomography, 1990
112 **Philip J. Brown and Wayne A. Fuller, Editors,** Statistical analysis of measurement error models and applications, 1990
111 **Earl S. Kramer and Spyros S. Magliveras, Editors,** Finite geometries and combinatorial designs, 1990
110 **Georgia Benkart and J. Marshall Osborn, Editors,** Lie algebras and related topics, 1990
109 **Benjamin Fine, Anthony Gaglione, and Francis C. Y. Tang, Editors,** Combinatorial group theory, 1990
108 **Melvyn S. Berger, Editor,** Mathematics of nonlinear science, 1990
107 **Mario Milman and Tomas Schonbek, Editors,** Harmonic analysis and partial differential equations, 1990
106 **Wilfried Sieg, Editor,** Logic and computation, 1990
105 **Jerome Kaminker, Editor,** Geometric and topological invariants of elliptic operators, 1990
104 **Michael Makkai and Robert Paré,** Accessible categories: The foundations of categorical model theory, 1989
103 **Steve Fisk,** Coloring theories, 1989
102 **Stephen McAdam,** Primes associated to an ideal, 1989
101 **S.-Y. Cheng, H. Choi, and Robert E. Greene, Editors,** Recent developments in geometry, 1989
100 **W. Brent Lindquist, Editor,** Current progress in hyperbolic systems: Riemann problems and computations, 1989
99 **Basil Nicolaenko, Ciprian Foias, and Roger Temam, Editors,** The connection between infinite dimensional and finite dimensional dynamical systems, 1989
98 **Kenneth Appel and Wolfgang Haken,** Every planar map is four colorable, 1989
97 **J. E. Marsden, P. S. Krishnaprasad, and J. C. Simo, Editors,** Dynamics and control of multibody systems, 1989
96 **Mark Mahowald and Stewart Priddy, Editors,** Algebraic topology, 1989

(See the AMS catalog for earlier titles)